Injury and Repair of the Musculoskeletal Soft Tissues

American Academy
of Orthopaedic Surgeons
Symposium

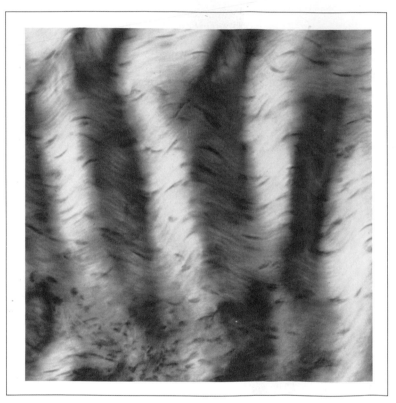

Medial collateral ligament of the dog under polarized light.

Injury and Repair of the Musculoskeletal Soft Tissues

Edited by
Savio L-Y. Woo, PhD
Professor of Surgery and Bioengineering
Division of Orthopaedics and Rehabilitation
University of California, San Diego
La Jolla, California

Joseph A. Buckwalter, MD
Professor
Department of Orthopaedic Surgery
The University of Iowa Hospitals and Clinics
Iowa City, Iowa

with 168 illustrations

Workshop
Savannah, Georgia
June 1987

Supported by the
American Academy of Orthopaedic Surgeons

and the
National Institute of Arthritis and
Musculoskeletal and Skin Diseases

American Academy of Orthopaedic Surgeons
222 South Prospect Avenue
Park Ridge, Illinois 60068

Library of Congress Cataloging in Publication Data:

Symposium on Injury and Repair of the Musculoskeletal Soft Tissues
(1987: Savannah, Ga.)
   Injury and repair of the musculoskeletal soft tissues.

"Symposium on Injury and Repair of the Musculoskeletal Soft Tissues"—Half t.p.
   Includes bibliographies and index.
   1. Musculoskeletal system—Wounds and injuries—Congresses.   2. Musculoskeletal system—Surgery—Congresses. I. Woo, Savio L-Y. II. Buckwalter, Joseph A. III. American Academy of Orthopaedic Surgeons. IV. National Institute of Arthritis and Musculoskeletal and Skin Diseases (U.S.) V. Title
[DNLM: 1. Muscles—injuries—congresses.   2. Wound Healing—congresses.
WE 500 S9884i 1987]
RD680.S96   1987   617'.47   87-72970
ISBN: 089203-023-2

Second Printing, July 1991

Workshop Participants

Mark Adams, MD, FRCPC
Head, Clinical Investigation
CIBA-Geigy Canada, Ltd
Mississauga, Ontario, Canada

Wayne H. Akeson, MD
Professor and Head
Division of Orthopaedics and Rehabilitation
University of California, San Diego
San Diego, California

Kai-Nan An, PhD
Consultant
Biomechanics Research Laboratory
Department of Orthopaedic Surgery
Mayo Clinic
Rochester, Minnesota

Thomas P. Andriacchi, PhD
Professor and Director
Section of Orthopedic Research
Rush Presbyterian/St. Lukes Medical Center
Chicago, Illinois

Steven Arnoczky, DVM
Director, Laboratory of
   Comparative Orthopaedics
The Hospital for Special Surgery
New York, New York

Albert J. Banes, PhD
Assistant Professor
Department of Surgery, School of Medicine
The University of North Carolina
   at Chapel Hill
Chapel Hill, North Carolina

Mark Bolander, MD
Director
Orthopaedic Research Unit
National Institutes of Health
Bethesda, Maryland

Richard Brand, MD
Professor
Department of Orthopaedic Surgery
The University of Iowa Hospitals and Clinics
Iowa City, Iowa

Joseph A. Buckwalter, MD
Professor
Department of Orthopaedic Surgery
The University of Iowa Hospitals and Clinics
Iowa City, Iowa

David L. Butler, PhD
Associate Professor
Aerospace Engineering and
   Engineering Mechanics
University of Cincinnati
Cincinnati, Ohio

Arnold I. Caplan, PhD
Professor
Department of Biology
Case Western Reserve University
Cleveland, Ohio

Bruce M. Carlson, MD, PhD
Professor
Department of Anatomy and Cell Biology
University of Michigan Medical School
Ann Arbor, Michigan

Bruce Caterson, PhD
Professor
Department of Biochemistry
West Virginia University
Medical Center
Morgantown, West Virginia

Reginald R. Cooper, MD
Professor and Chairman
Department of Orthopaedic Surgery
The University of Iowa Hospitals and Clinics
Iowa City, Iowa

Richard D. Coutts, MD
Head, Department of Joint Reconstruction
Malcolm and Dorothy Coutts Institute for
   Joint Reconstruction and Research
San Diego, California

Laurence E. Dahners, MD
Assistant Professor
Division of Orthopaedics
The University of North Carolina
   at Chapel Hill
Chapel Hill, North Carolina

Thomas B. Dameron, Jr., MD
Orthopaedic Surgeon
Raleigh, North Carolina

Kenneth E. DeHaven, MD
Professor
Department of Orthopaedics
University of Rochester Medical Center
Rochester, New York

David R. Eyre, PhD
Ernest M. Burgess Professor
  of Orthopaedic Research
Department of Orthopaedics
University of Washington
Seattle, Washington

Gerald A. Finerman, MD
Dorothy and Leonard Straus Scholar
Professor of Surgery/Division of Orthopaedics
University of California, Los Angeles
Los Angeles, California

Donald A. Fischman, MD
Harvey Klein Professor of
  Biomedical Sciences
Chairman, Department of Cell Biology
  and Anatomy
Cornell University Medical School
New York, New York

David W. Florence, MD
Director of Medical Affairs
Peoples Community Hospital Authority
Wayne, Michigan

Cyril B. Frank, MD
Assistant Professor
Department of Surgery
The University of Calgary
Calgary, Alberta, Canada

Leo T. Furcht, MD
Stone Professor of Pathology
Department of Laboratory Medicine
  and Pathology
University of Minnesota Medical School
Minneapolis, Minnesota

William E. Garrett, Jr., MD, PhD
Assistant Professor
Division of Orthopaedic Surgery
Sports Medicine Section
Duke University Medical Center
Durham, North Carolina

Richard H. Gelberman, MD
Professor
Department of Orthopaedic Surgery
Harvard Medical School
Boston, Massachusetts

Victor M. Goldberg, MD
Professor and Vice-Chairman
Department of Orthopaedics
Case Western Reserve University
School of Medicine
Cleveland, Ohio

Stephen L. Gordon, PhD
Director, Musculoskeletal Diseases Program
National Institute of Arthritis and
  Musculoskeletal and Skin Diseases
Bethesda, Maryland

Edward S. Grood, PhD
Professor
Department of Orthopaedic Surgery
University of Cincinnati
Cincinnati, Ohio

Ernst B. Hunziker, MD
Professor, Institute of Anatomy
University of Bern
Bern, Switzerland

Jack L. Lewis, PhD
Professor of Orthopaedic Surgery
  and Mechanical Engineering
Department of Orthopaedic Surgery
University of Minnesota
Minneapolis, Minnesota

Göran Lundborg, MD, PhD
Professor
Department of Orthopaedics
University Hospital in Lund
Lund, Sweden

Richard Lymn, PhD
Director, Muscle Biology Program
National Institute of Arthritis and
  Musculoskeletal and Skin Diseases
Bethesda, Maryland

Marston Manthorpe, PhD
Associate Research Biologist
Department of Biology
University of California, San Diego
La Jolla, California

Jerry A. Maynard, PhD
Professor and Chairman
Department of Exercise Science
  and Physical Education
The University of Iowa Hospitals
  and Clinics
Iowa City, Iowa

Van C. Mow, PhD
Professor of Mechanical Engineering
  and Orthopaedic Bioengineering
Department of Orthopaedic Surgery
College of Physicians and Surgeons
  of Columbia University
New York, New York

Helen M. Muir, PhD
Director
Mathilda and Terence Kennedy
  Institute of Rheumatology
Hammersmith, London, England

Theodore R. Oegema, PhD
Professor
Departments of Orthopaedic Surgery
  and Biochemistry
University of Minnesota
Minneapolis, Minnesota

Barry W. Oakes, MD
Professor
Department of Anatomy
Monash University
Clayton, Victoria, Australia

Eric L. Radin, MD
Professor and Chairman
Department of Orthopaedic Surgery
West Virginia University Medical Center
Morgantown, West Virginia

A. Hari Reddi, PhD
Chief of Bone Cell Biology
National Institute of Dental Research
Bethesda, Maryland

Lawrence C. Rosenberg, MD
Director of Connective Tissue
  Research Laboratories
Montefiore Hospital and Medical Center
Bronx, New York

Melvin P. Rosenwasser, MD
Assistant Professor
Department of Orthopaedic Surgery
College of Physicians and Surgeons
  of Columbia University
New York, New York

Björn Rydevik, MD, PhD
Associate Professor
Department of Orthopaedics
University of Gothenburg
Sahlgren Hospital
Gothenburg, Sweden

Lawrence E. Shulman, MD
Director
National Institute of Arthritis and
  Musculoskeletal and Skin Diseases
Bethesda, Maryland

Ileen Stewart, MS
Executive Secretary
Orthopaedic Study Section
Division of Research Grants
National Institutes of Health
Bethesda, Maryland

James Tidball, PhD
Assistant Professor
Division of Biomedical Sciences
University of California, Riverside
Riverside, California

Peter Torzilli, PhD
Associate Professor
Department of Biomechanics
The Hospital for Special Surgery
New York, New York

Andrew Weiland, MD
Professor
Department of Orthopaedic Surgery
Johns Hopkins University
Baltimore, Maryland

Savio L-Y. Woo, PhD
Professor of Surgery and Bioengineering
Division of Orthopaedics and Rehabilitation
University of California, San Diego
La Jolla, California

Preface

Motor vehicle and work-related accidents, competitive sports, and vigorous recreational activities frequently cause strains, sprains, contusions, lacerations, and ruptures of the musculoskeletal soft tissues. Many of these injuries are more difficult to treat than fractures and can cause significant pain and impairment. Despite their importance, clinical studies and scientific investigations of musculoskeletal soft-tissue injuries have lagged behind similar work on bone injury and repair. The resulting lack of clinical and basic scientific information has limited progress in the treatment of soft-tissue injuries. Thus, a clear need exists to increase understanding of musculoskeletal soft-tissue injury and repair and thereby facilitate development of methods to accelerate repair and improve the function of the repair tissue.

To help meet this need, the Committee on Research of the American Academy of Orthopaedic Surgeons (AAOS) under the leadership of Dr. Richard N. Stauffer; the National Institute of Arthritis and Musculoskeletal and Skin Diseases (NIAMS) under the leadership of Drs. Lawrence E. Shulman and Stephen L. Gordon; and the Orthopaedic Research Society (ORS) asked us to organize a workshop that would address the problems of ligament and tendon repair. We expanded the scope of the workshop to include skeletal muscle; peripheral nerve; peripheral vessel; ligament, tendon, and joint-capsule insertions; myotendinous junction; articular cartilage; and meniscus. Hence, the title became the "Injury and Repair of the Musculoskeletal Soft Tissues."

Review of recent orthopaedic research convinced us of the need to encourage study of musculoskeletal soft-tissue injury and repair. From 1983 to 1987, 600 of the 1,412 papers presented at the annual ORS meetings related to the musculoskeletal soft tissues. Most of these papers dealt with the structure, composition, and biomechanics of tendon, ligament, cartilage, and meniscus. There were relatively few studies of injury and repair in these tissues and even fewer investigations of the other musculoskeletal soft tissues. In the same period, only 157 out of 683 (23%) NIAMS Orthopaedic Study Section grant applications were for work on the soft tissues.* The investigators working in this field are productive, but a limited number of them seek funding for in-depth investigations. With a few exceptions, the repair of acute musculoskeletal soft-tissue injuries has not been systematically investigated, and study of some of the musculoskeletal soft tissues has been neglected.

The specific aims of the workshop were to (1) review current understanding of the structure and function of the musculoskeletal soft tissues, the injury and repair of these tissues, and methods of facilitating repair; (2) identify the most productive directions for future investigations of soft-tissue injury and repair; (3) promote interchange of ideas and stimulate collaborative investigations between clinicians and basic scientists; and (4) encourage investigators to study musculoskeletal soft-tissue injury and repair. To accomplish these aims, we decided that the workshop participants should include basic scientists and clinicians, the participants should prepare and distribute reviews of current knowledge and suggestions for future work in advance of the workshop, the workshop should emphasize open discussion and the development of new ideas, and the proceedings should be published soon after the workshop.

We divided the musculoskeletal soft tissues into nine groups: (1) tendon; (2) ligament; (3) tendon, ligament, and joint-capsule insertions to bone; (4) myotendinous junction; (5) skeletal muscle; (6) peripheral nerve; (7) peripheral vessel; (8) articular cartilage; (9) meniscus. Drs. Richard Gelberman, Cyril Frank, Wayne Akeson, William Garrett,

*Data supplied by Ileen Stewart of the NIAMS.

Arnold Caplan, Gören Lundborg, Andrew Weiland, Lawrence Rosenberg, and Steven Arnoczky were invited to lead groups responsible for preparing presentations and manuscripts concerning each of these tissues. We asked each group leader to answer the following questions in their manuscripts and presentations: (1) What is the normal structure and function of the tissue? (2) What is the normal injury and repair reaction in the tissue and what are the short-term and long-term results of repair in this tissue? (3) What methods have been used to improve the repair response in this tissue? (4) What investigations should be performed to discover new ways to improve the repair of these tissues? They all agreed to serve in this important and laborious role, helping us to decide the final list of participants and to assign participants to groups.

In January 1987, the group leaders and participants formalized the workshop format. They decided that in the morning there would be plenary presentations by group leaders and group members followed by discussions involving all participants. They agreed to distribute their manuscripts to all participants six weeks before the workshop so that formal presentations could be kept to a minimum and informal exchange of ideas could be encouraged.

The workshop was held in Savannah, Georgia, June 18-20, 1987. The presentations were outstanding and the discussions were sophisticated and intellectually stimulating. We defined *injury* as acute damage or loss of cells and extracellular matrix, *repair* as the replacement of damaged or lost cells and extracellular matrices with new cells and matrices, and *regeneration* as a form of repair that produces new tissue structurally and functionally identical to the normal tissue. Each morning, three tissues were discussed in plenary sessions and specific research questions were identified. In the afternoon sessions, the participants divided into three groups and discussed the questions identified in the morning.

Specific recommendations for future research and treatment of musculoskeletal soft-tissue injuries evolved from the intense discussion sessions. Although each tissue has unique features including specific factors that influence its repair, common principles applicable to all musculoskeletal soft tissues emerged from the discussions. The participants agreed that morphology, biomechanics, biochemistry, and cell biology must be integrated to accelerate progress in understanding normal soft-tissue function, the response of the tissues to injury, and the function of the repair tissue. It became apparent that the extent of the injury, that is, the size of the tissue defect caused by the injury, influences the results of repair in all the tissues. Repair of small defects usually proceeds more rapidly and completely than repair of large defects, and small defects may not affect tissue function even if they are not repaired. For example, animal experiments indicate that skeletal muscle can regenerate normal tissue to repair defects smaller than 3 g, but repair of larger defects is much less successful. Articular cartilage defects less than 1 mm in diameter frequently heal with tissue that morphologically and histochemically appears nearly normal, but defects 3 mm in diameter or larger usually fill with fibrous tissue. Thus, it is not surprising that significant human muscle and cartilage injuries commonly result in permanent impairment despite current treatment. A great deal can be learned from the repair of small defects, but it is the larger defects that cause persistent pain and permanent loss of function in patients. For this reason, the discussions emphasized proposals for future work directed toward improving repair of clinically significant defects, that is, soft-tissue defects that result in loss of function. Pursuing these proposals will be important because only under limited circumstances does repair following clinically significant injuries to human musculoskeletal soft tissue regenerate tissue with normal structure, composition, and long-term function. More often, it either forms tissue grossly resembling uninjured tissue that may provide satisfactory but less than normal function, or tissue consisting of dense scar that can restore near-normal function in meniscus, tendon, and ligament tissues but not in muscle, nerve, or cartilage tissues. Occasionally, repair forms loose fibrovascular tissue or extensive matted scar and fibrous adhesions that do not restore function. In contrast, bone fractures usually heal by regeneration of normal bone tissue. Thus, satisfactory repair of clinically significant soft-tissue defects is generally more difficult than restoration of normal bone function. To improve the repair of musculo-skeletal soft tissues, methods of improving the repair response must be developed. Methods that may be helpful include use of growth factors, modifications of interactions between cells and extracellular matrix, use of extra-cellular matrix molecules to promote repair and regeneration, use of mechanical loading to stimulate and guide repair, and stimulation and control of cell and matrix changes that resemble embryonic tissue formation and development.

Because of enthusiastic participation and hard work by all, the workshop was a tremendous success. Many new ideas were generated, and future directions for research were formulated. The participants learned a great deal from each other, made plans to collaborate in new investigations, and left the meeting with great enthusiasm for future work in musculoskeletal soft-tissue injury and repair. However, to achieve our original specific aims, we needed to publish the proceedings in a timely fashion. Thanks to the efforts of the group leaders, the participants, and the Academy's editorial staff, headed by Dr. Marilyn Fox, we were able to begin editing and revising the manuscripts immediately after the workshop and to assemble them by October 1987. This allowed us to begin the final revisions in November and to publish the text in February 1988.

The workshop and the preparation of the text took a great deal of time from our families and our normal schedules, but we were rewarded by the opportunity to learn from outstanding scientists and clinicians. We hope this text will provide a review of current knowledge for clinicians and investigators working with the problems of soft-tissue injury and repair, stimulate other clinicians and basic scientists to join in the effort to improve the treatment of musculoskeletal soft-tissue injuries, and encourage new clinical and basic research programs. We anticipate that a similar workshop ten years from now will demonstrate exciting progress in the understanding of musculoskeletal soft-tissue injury and repair and in the treatment of injuries to these tissues.

Savio Lau-Yuen Woo, PhD
Joseph Addison Buckwalter, MS, MD

Acknowledgments

The efforts and support of many individuals and groups made possible the workshop and book on the *Injury and Repair of the Musculoskeletal Soft Tissues*. The National Institute of Arthritis and Musculoskeletal and Skin Diseases and the American Academy of Orthopaedic Surgeons provided encouragement, advice, and financial support. We are especially pleased that this publication honors the National Institutes of Health in its centennial year. Our deep appreciation goes to the participating scientists and clinicians for generously contributing their time and expertise. Their efforts and eagerness to share their ideas made the workshop a scientific success. Their enthusiasm and good humor made it a pleasure.

We would like to thank the participants from the National Institutes of Health and the American Academy of Orthopaedic Surgeons, who contributed in numerous ways to the workshop. Our most sincere thanks go to the following AAOS staff members: Dr. Marilyn Fox, assistant director, scientific publications, worked with us as project manager for more than a year in planning, developing, and producing this book. We also greatly appreciate the work of Wendy Schmidt, medical editor, who edited the manuscripts, sized the illustrations, and indexed the book. In addition, we thank Mark Wieting, director of communications and publications, who contributed both advice and guidance in the development of the book, and Catherine Smith and Geraldine Dubberke, who contributed their word processing skills to the project. All of these persons shared our commitment to publish this book on schedule while maintaining the standard of excellence personified by the workshop participants.

We would like to acknowledge Ms. Karen Schneider, the workshop coordinator, who worked with us from the outset in organizing and planning the workshop. Her diligence, enthusiasm, humor, and charm were appreciated by all the participants, and particularly by both of us. Finally, we thank Ms. Lynette Fleck for her help all along and for her extra efforts during the workshop.

Savio Lau-Yuen Woo, PhD
Joseph Addison Buckwalter, MS, MD

Contents

Section One
Tendon

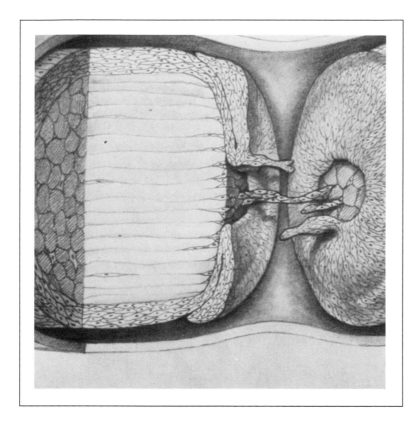

Group Leader

Richard Gelberman, MD

Group Members

Kai-Nan An, PhD
Albert Banes, PhD
Victor Goldberg, MD

Group Participants

Richard Brand, MD
Bruce Carlson, MD, PhD
Laurence Dahners, MD
David Eyre, PhD
Gerald Finerman, MD
William Garrett, Jr., MD,
    PhD
Stephen Gordon, PhD
Göran Lundborg, MD,
    PhD
Richard Lymn, PhD
Marston Manthorpe, PhD
James Tidball, PhD
Andrew Weiland, MD

Synopsis

It has been said that tendon surgery of the hand reflects the history and progress of hand surgery. The repair of injured tendons in the hand is a challenging and clinically important topic for basic research. In this chapter, the discussion of the injury and repair of tendon tissue concentrates on the major flexors of the hand. However, the concepts discussed may be applied to tendons in other anatomic locations as well. We discuss the structure and properties of normal tendons; the mechanical environment and injury of the tendons; response in tendon repair; methods for improving the repair response; and, finally, directions for future research.

The structure, function, and properties of normal tendon in biomechanical and biochemical terms are summarized. This knowledge is important for a discussion of tendon injury and repair. The general function and anatomic structure of the tendon pulley system in the hand are described. The histologic characteristics of the tendon and the collagen structure are reviewed. The blood supply and intrinsic vascularization of the tendon are defined as stemming from four sources. Suggested dual nutrient pathways are perfusion in the vascularized area and diffusion in the avascular region. The mechanical properties of tendon represented by its stress-strain relationships, strength, and viscoelastic characteristics are reviewed and discussed. These properties are related to the biochemical composition of tendon.

The mechanical environments to which tendon is exposed are important for understanding the mechanism of injury and for consideration of a repair and rehabilitation regimen. Therefore, a brief review of the current experimental and analytical methods used for the studies of joint kinematics and kinetics is provided.

These studies lead to the association of the tendon pulley system to joint kinematics as represented by the joint-tendon excursion relationship. The loads experienced by the flexor tendon during the activities of daily living as well as rehabilitation procedures can be characterized. The modes of injury are avulsions directly from the bone and in-substance transections. Each requires different clinical settings and treatment.

We further review the literature on the repair response of flexor tendons. Two distinct theories regarding tendon healing have been developed. Immobilization leads to adhesion ingrowth healing, whereas intrinsic process healing follows earlier controlled passive motion. The tendon healing process can be divided into inflammatory, proliferative, and organizational stages. However, the cellular response in the two healing mechanisms is controversial. The mechanical strength of the repaired tendon at different healing stages relates quite well to its histologic structure and collagen type and organization. Mechanical loading and motion are found to be particularly important to the mechanical characteristics of the healing process.

Current state-of-the-art methods of improving the repair and healing of tendons are also discussed. From a biomechanical point of view, suturing techniques for tendon and sheath and postoperative rehabilitative processes are two major topics of investigation. Various suturing techniques have been developed to increase the strength of the repair site and to reduce possible gap formation. Protected mobilization by either passive or active motion may improve or be detrimental to the healing process depending on the relationship of the cellular response to the mechanical environments encountered. There are several biochemical agents that can prevent scar formation and adhesion. Soluble factors, such as platelet-derived growth factor and transforming growth factor beta, have been considered to regulate fibroblast growth, proliferation, and migration. For certain severely damaged tendons, an artificial tendon implant or augmentation device provides a possible repair solution.

Finally, a discussion of topics that deserve more investigation is presented. Studies of normal tendons and the effects of age, immobilization, recovery, in vivo tendon forces, kinetics, and kinematic measurements of the tendon-sheath apparatus based on rigorous analysis of multiple joint motions will be required. Further study of the vascular anatomy of extrasynovial tendons complementing intrasynovial tendons and examination of nutrient transport among the synovial fluid, tendon, and tendon sheath will help to determine specific future studies. The signal that directs the cellular response to effect repair without scar formation is still unknown. Development of species-specific culture media will aid in gaining this information. In addition, the effects of post-repair rehabilitation modalities, such as the dose-response curve for the use of controlled passive motion or the effect of dynamic splints on the eventual outcome of tendon healing, must be learned. Whether or not the amount of excursions between the repaired tendon and its sheath under these regimens is sufficient along both the proximal and distal ends of the repair sites must be determined. Both short- and long-term studies will help to answer these questions.

Chapter 1
Tendon

Richard Gelberman, MD
Victor Goldberg, MD
Kai-Nan An, PhD
Albert Banes, PhD

Chapter Outline

Structure and Function of Normal Tendons
    Anatomy and Microanatomy
    Nutrition
    Biomechanical Properties of Normal Flexor Tendons
    Biochemical Composition of Normal Flexor Tendons
    Biochemistry and Biomechanics of Tendon Cells
Mechanical Environment of Tendon Injury
    Joint Kinematics
    Effect of Loading on Tissue Composition
    Mechanisms of Tendon Injury
Response in Tendon Repair
    Historical Concepts of Tendon Repair
    Cellular Response in Tendon Repair
    Biomechanical Response in Tendon Repair
    Biochemistry of Tendon Repair Response
Methods of Improving the Repair Response
    Improved Repair Techniques
    Evolving Concepts of Rehabilitation
    Cellular Response to Intermittent Passive
        Mobilization
    Effect of Biochemical Agents on the Cellular Response
    Effect of Soluble Factors on Cells
    Artificial Tendon Implants
Future Directions

Structure and Function of Normal Tendons

This review focuses on the tendons of the flexor digitorum profundus and the flexor digitorum superficialis muscles because they are of major importance to the surgeon.[1,2] Both these tendons are long and narrow structures with excursions of as much as 5 cm between full flexion and full extension. They are generally oval in cross section, have a constant diameter distally, and are arranged in anatomic units that result in a complex dynamic function as they act independently or in units. The flexor digitorum superficialis tendons arise as separate units from a specific bundle, so that they act independently. In contrast, the flexor profundus tendon complex acts as a single unit. The complex flexor tendons of the hand are capable of generating considerable force. It has been estimated that while the resting tension in these tendons may be on the order of 1 N, tension during power grip is considerable and could be as much as 200 N.

Anatomy and Microanatomy

Tendons in the distal palm and digits are enclosed in synovial sheaths lined by a glistening smooth synovial layer continuous with a proximal mesotenon. More proximally, they are surrounded by a thin adventitia called a paratenon. Synovial sheaths enhance the gliding of flexor tendons and are thickened in segments to form pulleys, which are critical components of the flexor sheath. These anatomic units are important in maintaining smooth tendon performance and preventing bow-stringing during function. In addition, there are smaller cruciate pulleys that provide sheath flexibility during function. The pulley system is necessary whenever tendons turn a corner or bend in concert with a neighboring joint. This system is particularly complex in the case of the flexor tendons of the hand. In the area of the pulley, the tendon is avascular and contains tissue that resembles fibrocartilage. Biomechanically, the pulley system maintains an appropriate moment arm for efficient tendon excursion and muscle force. Any aberrations of the pulley system may result in decreased range of joint motion, decreased flexion power, and flexion contractures of the overlying joint, which leads to an increased risk of additional pulley rupture. The flexor digitorum superficialis and the flexor digitorum profundus are examples of the complex interrelationship necessary for an apparently simple function, such as finger flexion. Integration and balance of both flexor and extensor forces are necessary to achieve equilibrium in joint flexion.

Microanatomic studies of tendons reveal the following hierarchical structures: (1) fascicles surrounded by visceral paratenon of collagen fibrils parallel to the long axis of the tendon; (2) one or more fascicles surrounded by parietal paratenon; (3) the endotenon, which surrounds the fascicle, blood vessels, and nerves; (4) the epitenon, continuous with the endotenon and bearing capillaries. The epitenon contains a well-ordered criss-cross pattern of crimped collagen with fibril diameters of 20 to 200 nm $\times$ 87 nm, and is loosely arranged into fibers.[3] The endotenon has the same general arrangement. The paratenon collagen fibrils are small, 20 to 70 nm $\times$ 47 nm, grouped into ill-defined fibers, and widely separated by fibroblasts and a microfibrillar material that may be elastin.[3] The fascicle is composed of collagen fibrils with a wide range of sizes, 20 to 50 nm $\times$ 129 nm; the smaller fibers are nestled among the larger ones.[3] The fascicle surface usually has a layer of uniform collagen fibrils and elastin. Collagen in the parietal paratenon has a small crimp size, whereas that of the fascicle has a larger crimp size. Each fascicle is capable of sliding past its neighbors. There appear to be no direct attachments or cellular communications between neighboring fascicles.[4] The paratenon layers in vivo are probably associated with a fluid environment similar to synovial fluid.[3]

Collagens in different parts of the tendon have different crimp lengths and angles.[4-7] Therefore, at low tensile loads, tendon exhibits great extensibility; at higher tensile loads, its extensibility is reduced. Tendons from the tibial tarsus region of 10- to 14-day embryonic chicks lacked the characteristic crimp pattern that can be discerned in tendons from 15- to 16-day embryonic chicks.[4] For the young tendons, the average crimp length was $8 \pm 2\ \mu$ and the crimp angle, 38 ± 4 degrees, whereas for 9-week-old hens, the values were $48 \pm 7\ \mu$ and 24 ± 5 degrees. Kastelic and associates[4] concluded that fibroblasts that synthesize collagen are programmed to interact in the genesis of crimps. In development, tractional structuring may be responsible for orientation of collagen in tendons.

The histologic anatomy of tendons has been extensively reviewed and reported by Lundborg and Rank,[8,9] Manske and associates,[10-12] and Gelberman and associates.[13-15] Tendons are composed of long spiraling bundles of collagen fibers; in the flexor tendons of the finger, these coalesce into main bundles in the palm. The two bundles of the flexor digitorum superficialis separate in the distal palm, whereas the bundles of the profundus tendon, although not separated, remain as distinct entities throughout their course. The surrounding tendon sheath contains circumferentially aligned fibroblasts and collagen fibers. The lining surface is also a layer of fibroblasts. The tendon itself is composed of an epitenon, endotenon, tendon bundles, and blood vessels. In the hand, the tendons are covered by the paratenon. The epitenon is a fine fibrous and cellular layer adherent to the tendon surface. The cells are synovial-like and are continuous proximally with the mesotenon. The surface of the individual tendon bundles of collagen is covered by the endotenon. The tendon bundles themselves are longitudinally oriented and consist of mature fibroblasts (tenocytes) and type I collagen.

Nutrition

Studies of the intrinsic vascularization of human flexor tendon have stressed the specific segmental nature of tendon's blood supply. Four sources of vascularity have been defined: longitudinal vessels extending down the tendon from the palm, vessels entering the tendon at the level of the proximal synovial reflection, vessels entering through long and short vinculae, and vessels entering the superficialis and profundus tendons at their osseous insertions. Recent vascular injection studies have emphasized zones of avascularity: two in the flexor digitorum profundus tendon and one in the superficialis tendon.[16,17] Lundborg and associates[17] noted one zone of avascularity in the flexor digitorum profundus between the two proximal sources of vascularization and the long vinculum and a second zone between the long and short vinculae. The flexor digitorum superficialis had one zone of avascularity, between the proximal two sources of vascularization and the long vincula. In addition, the palmar fourth of the tendon, particularly in the region of the pulleys, appears devoid of capillary loops. Lundborg and associates[17-19] surmised that there are dual nutrient pathways whereby the vascularized areas of flexor tendon are nourished by perfusion and the avascular areas by diffusion. During recent years there has been increasing evidence that nutrition via diffusional pathways from the synovial fluid is of considerable significance for flexor tendons normally, as well as during tendon healing.[8,18] Manske and associates[20,21] studied the relative roles of synovial fluid diffusion and vascular perfusion in an experiment measuring the uptake of tritiated proline in the tendons of four different animals. They concluded that the process of diffusion functions more quickly and completely than perfusion and is a relatively more important pathway for the nutrition of flexor tendons. Lundborg,[18] in an ingenious study of flexor tendons transferred to the synovial pouches of the knee joint, concluded that flexor tendons have an intrinsic repair capability and are not dependent on extrinsic adhesions for healing. The diffusion pathway from the synovial fluid appeared sufficient for the superficial layers of the tendon graft to proliferate and produce collagen fibers.

Potenza and Herte[22] subsequently placed acellular profundus tendon allografts into the knee joints of dogs. The grafts, harvested at two, three, and six weeks, were covered by a layer of proliferating fibroblast-like cells that healed cuts made in their surfaces. The source of cells for this phenomenon was presumably the synovial lining of the knee joint. These findings stimulated other investigators to explore the healing capability of flexor tendons placed in a cell-free tissue culture environment. Subsequently, in vitro studies of tendon healing have been done to compare the cellular repair activity of various animal species and to determine how their responses compare with previous experimental findings in vivo.[10] A consistent pattern of increased cell proliferation, as well as increased collagen synthesis and incorporation, during flexor tendon repair was noted in rabbit, chicken, dog, and monkey segments in tissue culture media.[11] In each of the cultured tendons, the epitenon demonstrated the initial morphologic change. Epitenon thickening, as a result of fibroblast proliferation, marked the first notable response to injury. After three weeks, fibroblast differentiation into macrophage-type cells in the peripheral cell layer was consistently observed in all animals. New collagen deposition by the deeper layers of epitenon fibroblasts was also responsible for some epitenon thickening. Between six and 12 weeks, there was an increase in cells within the endotenon and in collagen production adjacent to the tendon stumps. The in vitro model demonstrated several similarities in the various species, including an initial epitenon thickening with increases in cells and collagen content, followed later by the appearance of macrophages within the gap between the tendon ends. These findings correlate well with the results of in vivo experiments examining the morphologic changes associated with early passive mobilization of tendon repairs.[13,14] Future studies should emphasize the development of optimal species-specific culture media. In addition, since complete healing of the repair was not seen by 12 weeks in this model, longer intervals should be studied to examine the more advanced stages of repair.

Biomechanical Properties of Normal Flexor Tendons

The primary function of tendons is to transmit muscle force to the skeletal system with limited elongation. Tendon tissue is composed of dense, parallel fibers of connective tissue. Tendons possess one of the highest tensile strengths of all the soft tissues in the body, because collagen is the strongest fibrous protein and also because these fibers are arranged parallel to each other in the direction of the tensile force of the muscle. These properties of tendon depend mainly on the properties and architecture of the collagen and elastin fibers in interaction with each other. Proteoglycans are noncollagenous macromolecules interwoven with collagen fiber and elastin that contribute to the time-dependent and history-dependent viscoelastic properties.[23,24]

Most biomechanical studies of tendon have focused on ultimate tensile strength and strain.[25-28] It has been found that the ultimate tensile strength of tendons ranges from 45 to 125 MPa.[24,28,29] In human tendons, strength has been found to range from 50 to 150 MPa. The ultimate strain value for tendons has been found to range from 9% to 30%.

These large variations result primarily from differences in the species, type, and age of the tendons studied. In addition, the major differences in these findings are a result of the measurement techniques employed. The difficulties have been well recognized in the measurement of cross-sectional area, gauge length, and the elongation of tendons during tensile testing. In strength measurements, clamp designs that produce artificial breakage as a result of stress concentration are still a major problem.[24] For large tendons, the uneven distribution of the stress among the tendon bundles during tensile testing also results in poorer strength measurements.

The stress-strain relationship in the tendon is generally similar in characteristics to those in other collagenous soft tissue such as ligaments and skin. The stress-strain curve always begins with a region called the "toe region," in which the tendon deforms easily without much tensile force. The toe region is rather small because collagen fibers in the tendon are initially parallel, and less realignment is required compared with other soft tissues. The toe region is followed by a fairly linear region, and the slope of the curve in this region has been used to represent the elastic stiffness of tendon. During the course of tensile deformation, small dips are sometimes seen with sharp drops in stress, which may represent sequential microfailure of certain fiber bundles.[28] Finally, with further tensile loading the specimen fails and the stress-strain curve drops sharply to zero.

Tendons are viscoelastic materials. The rate of loading, however, has a minor effect on the stress-strain relationship. With increases in loading rate, the curve shifts slightly toward the left, that is, the tendon becomes stiffer. Also, the area of hysteresis between cyclic loading and unloading is not pronounced.[29] On the other hand, the viscoelastic and viscoplastic characteristics of tendons can be demonstrated by the successive shifting of the stress-strain curve to the right. A majority of this change is recoverable but significant resting time is required. Numerous rheologic analogies and theories of viscoelasticity have been adopted to describe these time- and history-dependent properties.[24,30-33]

Factors that affect the properties of tendon have also been examined by various investigators. The stress-strain behavior of rat tail tendon as a function of age was examined by Kastelic and Baer.[34] The temperature during testing may also be of importance. Rigby and associates[35] found no change in the mechanical properties of rat tail tendon in the temperature range from 0 to 37 C. Others have arrived at a similar qualitative conclusion.[36,37] However, a recent study by Woo and associates[38] clearly showed a quantitative relationship between temperature-dependent properties of the canine medial collateral ligament. These authors believe that experimental errors in previous work may be a leading factor contributing to incorrect results. Further, water content also had an effect on its tensile properties. Tendon tissue becomes stiffer and stronger with the loss of water (Stucke,[39] Wertheim,[40] and K.-N. An and K. Tamai, unpublished data).

Tendon homeostatic responses secondary to mechanical stress have also been determined. The effects of immobilization on tendons are the significant loss of water content, glycosaminoglycan concentration, and strength.[24,41-45] On the other hand, exercise training results in increased collagen fibril size, ultimate strength, and stiffness.[26,28,45-49] In a study of miniature swine under running activity, alteration of the mechanical properties was found to be more pronounced in the extensor than in the flexor tendons.[48,49] The differences in results can be attributed to the biochemical constituents as well as to anatomic differences in these tissues. However, an additional difference in loads experienced by the two tendons may also be a causal factor and verification is required.

Biochemical Composition of Normal Flexor Tendons

Collagen, which is the major component of tendons, is arranged as a triple helix of tropocollagen molecules with a three-fold axis along a single protein molecule. Rat tail tendon collagen has been shown to have a well-developed crystalline structure probably similar to human flexor tendons.[50] The lattice structure of collagen during tension can be extended to give an increase in the spacing of the tropocollagen molecule of about 5%, from a polar reflection of 0.285 nm to approximately 0.3 nm. When the tension is relaxed, this repetitive spacing reverts to its original value of 0.285 nm. This short-term reversibility is critical in the functioning of any tendon structure.

Normal adult flexor tendons are composed largely of genetic type I collagen (>95%).[51] The remaining 5% consists of type III and type V collagen.[52,53]

The rate of collagen synthesis and degradation as a function of age in rats has been studied by Neuberger and Slack.[54] The rate of synthesis in young rats was approximately nine times higher than that for young adults. Using a rabbit model, Cetta and associates[51] demonstrated that the collagen content of Achilles tendon increased linearly from 18% by dry weight in late fetal life to more than 80% by the second post-partum day, and then declined slightly to approximately 75% by the third day. Thereafter, although the collagen content of tendon increased, the level of synthesis was constant. Banes and associates[55] calculated that the collagen content of pristine, adult, avian flexor tendon at the level of the interphalangeal joint was about 75% by dry weight.

Klein and Lewis[56] published a series of experiments dealing with simultaneous collagen destruction and formation in radioactive tail tendon grafts transplanted to the Achilles tendon site in the rat. For isografts, collagen loss was 45% for fresh tendon and 58% for frozen tendon after one month. New collagen synthesis was approximately equal to loss. Allografts demonstrated almost complete replacement of old collagen by newly synthesized collagen, again with little change in total mass. Xenografts had an 88% loss of prelabeled collagen and almost no new synthesis. At three months, both the isografts and the allografts had approximately the same degree of collagen loss, 54% and 63%, respectively. However, collagen replacement for isografts was 64% and that for allografts was only 38%, a net loss of total mass for the allograft. Hence, significant collagen degradation can occur in grafts, presumably caused by the activity of locally invading cells as well as macrophages. Allografts and xenografts clearly demonstrated more collagen degradation and less replacement than did more immunologically compatible grafts. This point is important because it indicated that collagen was metabolized and replaced when grafted in a recipient.

The cross-linking of collagen has also been elucidated during the last two decades. Early experiments by Mechanic[57] indicated that the major reducible (reduced with tritiated sodium borohydride) cross-link in collagen from Achilles tendon was dihydroxylysinonorleucine (DHLNL), a difunctional (united two chains of collagen), Schiff-base cross-link derived from reaction of the epsilon amino group of hydroxylysine and the aldehyde on hydroxyallysine. Similarly, the major reducible cross-link precursor in tendon is the aldehyde of hydroxylysine, dihydroxynorleucine (DHNL). The ratio of the cross-link precursors, DHNL to hydroxynorleucine (HNL), the reduced aldehydic compound derived from allysine, increased with age in bovine tendon.[58] Kuboki and Mechanic[59] showed that, although type I collagens from tendon and dentin had the same amino acid composition with slight variations in hydroxylysine content, the amount and distribution of cross-links throughout the molecules were different. This landmark publication was the first to substantiate the thought that there was a quantitative relationship between the function of a collagenous tissue and its three-dimensional chemical structure. The reduced cross-link hydroxylysinonorleucine (HLNL) formed from lysine and allysine, and its precursor, HNL, are also present in tendon.[58]

Fujimoto and associates[60] reported a new, stable, fluorescent, non-reducible cross-link isolated from bovine Achilles tendon which they termed 3-hydroxypyridinium, or pyridinoline. Pyridinoline is a tri-functional cross-link that probably unites three chains of collagen.[60] Pyridinoline exists in two forms, hydroxylysylpyridinoline (HP) and lysylpyridinoline (LP).[61,62] Human, adult patellar tendon contains both HP and LP in a ratio of at least 10:1 (HP, 0.44 mol/mol of collagen; LP, 0.03 mol/mol of collagen), whereas adult rabbit tendon contains only HP (0.33 mol/mol of collagen).[61]

Elastin, another structural protein, is present in tendons at the fascicle surface.[3] The amount of elastin in most tendons is less than 1% by dry weight and can be monitored by the presence of cross-links unique to the molecule, such as lysine-derived desmosine and isodes-mosine. Glycosaminoglycans also constitute less than 0.5% of tendon by weight. It has been suggested that glycosaminoglycans react with the surface of collagen fibrils but do not penetrate these fibrils. Also, dermatan sulfate has been shown to be the principal glycosaminoglycan in rat tail tendons.

Biochemistry and Biomechanics of Tendon Cells

Ultrastructural studies have indicated the diversity in tendon organization and described the various cell types in the structural units of tendon.[3,6,63–67] Clearly, the morphologic characteristics of the surface cells of tendon are different from those of the fibroblasts that surround the collagen fibrils in fascicles. Riederer-Henderson and associates[68] were the first to isolate two cell populations from 17-day embryonic chick digital flexor tendon. The population isolated from the synovial layer as primary cells was less efficient at attachment to substrate, synthesized less sulfated glycosaminoglycans, and synthesized 10% types I and III collagen. The fibroblasts from the tendon interior attached efficiently to their substrate, produced and secreted abundant glycos-aminoglycans, and synthesized 30% type I collagen. Banes and associates[69] further isolated cell populations from avian tendon with a procedure that permits isolation of cells from the surface of adult tendons (synovial cells), as well as those from the tendon interior. Using a technique employing sequential enzyme digestion of intact tendon and mechanical disruption of the outer architecture of the tendon, they found three cell populations from the tendon surface, based on morphologic properties and size. In the outer tendon surface two hours after plating, one cell population had an average length and width of 71 × 64 μ, a second cell type was 111 × 98 μ, and a third type was 50 × 57 μ.

Moreover, when fibroblasts were subjected to applied tensile deformation in culture (a regimen of ten seconds of 25% tension, ten seconds of relaxation, repeated throughout the growth cycle), cells responded by dividing and by aligning perpendicularly to the direction of stretch. This alignment can occur in as little as four hours in vitro. Banes and associates[70] also showed that internal fibroblasts responded to applied compression (as little as 0.1%, with 25 seconds of compression and five minutes of relaxation) by reducing the amount of tubulin synthesized during a two-hour pulse in vitro. Hence, cells isolated from flexor tendon can respond quickly both morphologically and biochemically to applied force in vitro.

In terms of collagen synthesis, Schwarz and associates[71,72] showed that tendon cells lose their capacity to produce large amounts of collagen when transferred from the organ to culture. The addition of 30 mmol of lactate may also restore the production of higher levels of collagen synthesis. Tendon cells maintained at high density, with added ascorbate, were also shown to retain the capacity to produce collagen at a level equivalent to that in vivo.[73]

Anttinen and associates[74] showed that various metal ions, particularly Zn^{2+} and Co^{2+} were efficient inhibitors of collagen synthesis in tendon cells, probably by competing with Fe^{2+} for the site on prolylhydroxylase. Similarly, Majamaa[75] showed that the addition of cis-hydroxyproline or azetidine–2-carboxylic acid to tendon cells actively synthesizing collagen inhibited procollagen triple-helix formation and increased the amount of hydroxylation in proline. These data indicated that if triple-helix formation is delayed, the extent of hydroxylation of proline is increased.

Kao and associates[76] reported that collagen synthesis in freshly isolated tendon cells followed two, quasi first-order processes, with half-times of 14 and 115 minutes. These authors suggested that the kinetics of collagen secretion were influenced by a two-compartment system in which one metabolic pool of the secretory process was rate-limiting.

Aalto and associates[77] reported that collagen synthesis in chick tendon cells could be inhibited by addition of N-acetylneuraminic acid-containing glycoproteins isolated from granulation tissue. Chick tendon fibroblasts in vitro can synthesize types I, III, and V collagen but only synthesized types I and V in vivo, as reported by Herrmann and associates.[78]

Mechanical Environment of Tendon Injury

Joint Kinematics

Kinematic analysis of joints has been categorized into two main areas. The gross motion of joints can be described by adopting a Eulerian angle system to define generalized three-dimensional rotation.[79,80] With proper selection of axes of rotation between two bone segments, the associated finite rotation is independent of sequence. Second, a detailed analysis of joint articulating surface motion, incorporating generalized three-dimensional unconstrained rotation as well as translation, is performed utilizing the concept of the screw-displacement axis.[79] When the surface geometry and soft-tissue constraints are known, the kinematics of an articulating joint can be analyzed to provide basic information for lubrication and wear studies.

Currently available measurement techniques of human joint kinematics can be classified into three categories:

(1) The electrogoniometer method uses the exoskeletal linkage systems containing rotatory potentiometers fastened to the proximal and distal limb segments of the joint. Depending on the sophistication of the mechanism's design, the electrogoniometer can be used to measure simple hinge-joint motion, three-dimensional joint rotation, and more generalized six-degree-of-freedom rigid-body motion.

(2) Stereometric, photographic, radiographic, and sensing methods have been used extensively in the field of orthopaedics and dentistry. When three noncolinear points fixed to a rigid body are defined within an inertial reference frame, the position and orientation of the rigid body can be specified, and the relative and absolute motion can be calculated. Various types of markers, including light-emitting diodes and reflecting dots, have been used for various sensory systems such as the basic photographic, video, and photodetector systems. More recently, two types of stereometric sensory systems have been developed. One is based on the application of ultrasonic impulse and microphone; the other is based on the electromagnetic pause and sensor. Both systems are able to monitor the generalized three-dimensional motion of the joint. In the same category of measurements, the stereometric radiographic method has been used to measure the precise and detailed joint motion of skeletal systems.

(3) The accelerometric method of three-dimensional joint kinematic analysis has rather limited usage. On the basis of Euler's theorem that the general displacement of a rigid body with one point fixed can be obtained by a rotation about a unique axis, three independent measurements of acceleration can thus provide such data. However, the complex computation involved and the instrument-related problems often make this method difficult to use.

As described, the detailed kinematics, or pattern of joint movement, depends on the geometric shape of the articular surface and/or the surrounding soft-tissue constraints. However, the driving forces of joint movement are either external force applied during passive joint motion or muscle and tendon forces applied during active voluntary movement. The interrelationship of joint rotation and tendon excursion has been carefully studied experimentally as well as analytically. In general, it is well recognized that the further away the tendon is from the joint center or axis of rotation for a given tendon excursion, that is, the larger the moment arm, the less joint rotation will be generated. Conversely, when the tendon is closer to the joint center, the same amount of tendon excursion generates more joint rotation.

The relationship between distance and joint rotation and tendon excursion can be described mathematically.[81] The instantaneous moment arm of the tendon in the plane of motion at a specific joint configuration can be obtained from the slope of the plot of the tendon excursion (vertical axis) versus joint rotational displacement (horizontal axis).

On the basis of this principle, the tendon excursions, joint rotations, and moment arms of the finger flexors, extensors, and intrinsic muscles have been studied. For example, at the metacarpophalangeal joint, extensor excursion is almost linearly related to the joint angle of rotation. This is simply because the extensor tendon passes the joint by wrapping around the dorsal surface of the metacarpal head and, thus, the moment arm is relatively constant throughout the arc of motion. The relationship for both flexors is relatively linear but slightly curved when the joint is close to full flexion. In other words, the moment arms of these two flexors increase as the joint flexion angle increases. The relationship between tendon excursion and joint rotation of the intrinsic muscles at the metacarpophalangeal joint is highly nonlinear, as reflected by the variation in the moment arms as a function of the joint angle.

For flexor tendons, the moment arm and excursion, as related to joint rotation, are governed by the constraint of the pulley system.[80] Therefore, any alteration in the flexor pulley system will result in a change in the normal relationship.

Understanding the relationship of tendon to joint excursion will aid in the design of rehabilitation programs for patients who have undergone tendon repair. It was found that the moment arms of the flexor digitorum profundus tendon at the distal interphalangeal, proximal interphalangeal, and metacarpophalangeal joints were 0.41, 0.79, and 1.11, respectively.[82,83] The moment arms of the flexor digitorum superficialis tendon at the proximal interphalangeal and metacarpophalangeal joints were 0.62 and 1.19 cm, respectively. Since the profundus tendon has a larger moment arm than the superficialis tendon at the distal interphalangeal and proximal interphalangeal joints, any movement of these two joints will generate gliding motion between these two flexor tendons. When the distal interphalangeal joint is flexed to 60 degrees, there will be a 4-mm gliding motion between the two tendons in the zone 2 region. If the proximal interphalangeal joint is flexed to 60 degrees, an additional 1.8 mm of relative gliding motion will be added. Large motion of the distal interphalangeal and proximal interphalangeal joints, such as in the hook-and-fist position, will generate the most gliding motion between the two flexors.[84]

Effect of Loading on Tissue Composition

When joints are performing static or dynamic functions, three types of force are involved: muscle-tendon force, joint-articulating force, and capsuloligamentous force. These forces are of interest but difficult to measure. Most studies have determined these forces by analytical methods based on kinematic measurements, defining the line of action and moment arms of muscles and tendons passing the joint.[82] In addition, the physiologic constraints of muscles (length-tension and force-velocity relationships) must also be considered. The problem is an indeterminate one, that is, the number of unknown variables exceeds the number of available equations, primarily because of the redundant nature of the anatomic system. Mathematically, this problem has no unique solution, and systematic reduction of the redundant known variables and elimination of the unknown variables by electromyography in order to make the system determinate and solvable are the most used methods. The optimization method, based on either linear or nonlinear objective functions, has also been widely used to obtain the solution. The finger-tendon forces involved in various hand functions have been determined using this method.[85] The flexor forces range from one to three times the force applied during pinch-and-grasp functions. In other words, 100 N to 300 N of flexor force is required to generate 100 N of pinch force.

Experimentally, these muscle and tendon forces have been measured directly with the tendon-tension transducer and indirectly with electromyographic signals. For isometric contraction, the rectified, integrated electromyographic signals correlate well with the muscle forces. These relationships can be established and calibrated with the mathematical model described above in conjunction with the data obtained from a set of isolated tests. The muscle and tendon forces involved in other isometric activities can then be determined.[86,87] Strain-gauge transducers have also been used to measure forces in the major flexors of the index finger and thumb.[85,88] During active extension and flexion without resistance, flexor tendons have approximately 2 to 4 N of force, respectively. Active flexion with mild resistance produced up to 10 N of force; in firm pinch, these forces can range up to 120 N.

Tendons, in general, encounter tensile force. However, at the region of the pulley or where they wrap around the articular surface, some compressive force may be experienced. Measurements obtained with a pressure transducer demonstrated that the pressure between the pulley and flexor tendon can reach 700 mm Hg in active flexion.[86] This type of compressive force can also alter the histologic structure of a tendon.[89] Under a pulley, a cartilage-like material appears in the tendon.

Mechanisms of Tendon Injury

There are two major mechanisms of flexor tendon injury: avulsion directly from bone and in-substance transection. Avulsion of the flexor tendon insertion is not as common as laceration. It occurs most frequently during participation in athletic events by young adults, and its occurrence is often missed initially by the examining physician.

Although the ring finger is affected 75% of the time, avulsion injuries have been reported in all the fingers and the thumb. Manske and Lesker[90] demonstrated that the strength of the profundus insertion of the ring finger is usually significantly less than that of the long finger.

Three types of profundus avulsions have been described.[91] In type 1 injuries, the tendon retracts to the palm where it is held by the lumbrical muscle. In type 2, the tendon retracts to the level of the proximal interphalangeal joint but the vinculum remains intact, preserving more blood supply to the tendon. In type 3, there is a large bony fragment beyond the distal portion of the middle phalanx. Both vinculae are intact in this injury. Treatment of the avulsion injury in the finger consists of reinsertion of the tendon or osseous fragment into the base of the distal phalanx either by bone fixation or tendon suture through a raised osteoperiosteal flap with a pull-out wire and button.

Most tendon injuries occur by way of transection with or without associated crush injury and injuries to other structures. Verdan[92] favored primary repair of tendon injuries in the digital sheath region of the hand. In clean lacerations repaired within three weeks of injury, the results indicated that this approach is superior to primary flexor tendon grafting. Advantages of primary repair include returning the tendon to its normal length, reducing joint stiffness, maintaining a satisfactory and improved tendon bed, preserving a blood supply where possible, and reducing the period of disability necessary for wound healing and secondary grafting. Associated injuries, however, do affect the ultimate result. Tendon repairs are carried out at the time of associated severe injuries, including crush injuries to soft tissue, phalangeal fractures, and digital reimplantations. The results are not comparable to those obtained after repair of simple, clean lacerations.

For the most part, experimental studies in tendon repair have not addressed the most difficult complex types of tendon injury. Specifically, most experimental models of tendon injury include incomplete or complete tendon transection, sharp transection of the digital sheath, and immediate meticulous repair. A more complex injury model is important, as 75% of tendon injuries have other associated significant injuries.

Response in Tendon Repair

Historical Concepts of Tendon Repair

In 1941, the experiments of Mason and Allen[93] introduced the concept that the gliding surface between the flexor tendon and the fibrous digital sheath is obliterated after tendon injury and repair. Their experimental findings influenced the clinical and research communities for the next four decades. They divided the healing process into two stages, the exudative and the formative or remodeling stages, correlating well with the three stages of soft-tissue repair described by Ross[94]: the inflammatory, the proliferative, and the organizational or remodeling stages. Mason and Allen's observations on the early stages of tendon repair encouraged clinicians to immobilize tendons, using the remodeling stage for stimulating increased tendon strength and gliding.

Potenza,[95] in a study using dogs, showed that there was no evidence of an intrinsic fibroblastic response following tendon injury and concluded that tendon healing depended on extrinsic cellular ingrowth. Potenza's observations, based on a total immobilization model after repair of cut tendons within the digital sheath, dictated a healing mode dominated by adhesions. Granulation tissue, including capillary buds and fibroblasts, proliferated from synovial sheath cells in the first several days following repair, and occupied the injury site. Perpendicularly oriented fibroblasts and collagen fibers documented the peripheral origin of the ingrowing tissue. From the third to fourth week, fibroblasts became oriented in a longitudinal direction as remodeling took place. By the fifth week, the newly formed collagen appeared more mature, and by 128 to 135 days the collagen appeared indistinguishable from normal tendon collagen. Attempts at mechanical separation of the injury site from the surrounding sheath only delayed the invasion of peripheral fibroblasts into the wound, as they most frequently circumvented any mechanical barriers.

As Lindsay and associates[96-98] had observed earlier, Potenza[95] noted that fibrous adhesions attached themselves to the tendon surface at each point where the surface had been disrupted. Lindsay and associates considered the fibrous attachment of the peripheral tissue to be an integral part of the healing process: a process essential for vascular and nutritional support of the healing tendon. Both Lindsay and associates and Potenza believed that collagen fibers within the adhesions were capable of remodeling as the synovial layer of the tendon sheath was restored. Peacock[99] subsequently proposed the one-wound concept, indicating that successful healing depended entirely on cells migrating into the repair area from outside the tendon itself.

The increasing evidence gathered by Matthews and Richards,[100,101] Ketchum and associates,[102,103] Lindsay and associates,[97,98] and Lundborg and Rank[8] that tendon healing may progress without flexor tendon sheath ingrowth opened up new avenues of investigation. Revascularization of tendons within the sheath, based on a delicate vinculum system, was believed essential to the preservation of blood flow to tendon. Peacock[99] stressed the importance of preserving the tendon's segmental blood supply, stating that tendons that are permanently denied their segmental vascularity undergo necrosis and replacement with undifferentiated scar tissue.

Over the past two decades, many of the details of the tendon healing process, including information on the relative contributions to healing made by both adhesion ingrowth and intrinsic processes, have been provided. Experiments have concentrated on three areas: tendon suturing techniques, tendon sheath repair and reconstruction, and the post-repair rehabilitation process.

Cellular Response in Tendon Repair

There is no area in tendon physiology that has resulted in so much controversy as the cellular sequence of events that occurs after flexor tendon injury. Potenza[1,2] argued that tendon healing occurs from an extrinsic source and that it requires the development of an inflammatory response and scarring for ultimate reconstitution. By contrast, recent studies[8,9,11,12,14,104] have suggested that the inflammatory response is nonessential and that tendons do possess an intrinsic capacity to heal. It is now believed that tendons possess both extrinsic and intrinsic capabilities to heal and the contribution of each may depend on the type and site of the tendon injury. A general overview is helpful in understanding the repair process after tendon laceration; although this review emphasizes flexor tendon healing, the cellular events can be applied generally.

Tendon healing proceeds in three phases: an inflammatory stage, a reparative or collagen-production stage, and, finally, a remodeling stage influenced by normal loads. Inflammation occurs during the first three days after injury and is highlighted by cells that originate from extrinsic peritendinous tissue and intrinsic tissue from the epitenon and endotenon. Collagen synthesis is seen within the first week, and collagen fibers are formed initially in a random, disorganized pattern. Fibroblasts are soon the predominant cell type and collagen content increases through the first four weeks. The cells that migrate into the wound gap and the collagen fibrils are usually oriented at this time in a plane perpendicular to the long axis of the tendon. Collagen maturation and functional linear realignment is usually seen by two months after injury. The ultimate tensile load of repaired tendons occurs only after appropriate physiologic loading has occurred. In order to understand further the healing process of wounded tendons, it is essential to increase our knowledge of the cellular basis of this process. Although one can easily make an argument for exclusive participation of either the intrinsic or extrinsic tissue in repair, it appears that the clinical circumstances may determine which sequence is predominant.

The specific cellular events that occur after a laceration of a flexor tendon within its tendon sheath have been well documented by Lindsay and associates[97,98] and by Potenza.[1,2] All the structures surrounding the tendon appear to provide a fibroblastic and vascular component for healing. These include the synovial sheath, the periosteum of the underlying bony structures, subcutaneous tissue, and deep adventitia and fascia. It appears that the intrinsic tenocytes are overwhelmed and play little role in the repair during immobilization. Granulation tissue from the synovial sheath migrates into the wound gap along with the fibroblasts derived from extrinsic tissues and intrinsic cells from the epitenon and endotenon during the first few days after injury. Many of the cells that proliferate and migrate into the wound at this time have a phagocytic function. The wound gap is filled rapidly by these cells. The fibroblasts in the wound lie perpendicularly to the long axis of the tendon and by the fifth day actively secrete collagen. These fibrils are disorganized and also oriented in a plane perpendicular to the long axis of the tendon. Although tenocytes from the epitenon and endotenon proliferate during this stage, they appear to be overwhelmed rapidly by the ingrowth of granulation tissues and fibroblasts migrating from the digital sheath and surrounding connective tissues. There is a marked fibroblastic proliferation from the endotenon by 21 days into the repair process. By 42 days, fibroblasts originating from the endotenon play the most active part in the healing process. These fibroblasts participate actively in collagen resorption, although they are actively synthesizing collagen at the same time. It appears that the dense collagen tissue formation seen in vivo is stimulated by trauma to the tendon sheath and immobilization. The synovial layer of the digital sheath is reconstituted by the third week after wounding; as the wound matures, a smooth gliding surface develops. The fibroblasts in the wound scar begin to orient themselves in line with the long axis of the tendon at about the same time. Collagen reorientation is complete by 28 days after injury, when the active fibroblasts and collagen orient themselves in line with the long axis of the tendon. By 112 days after the injury, there appears to be complete maturation of the repair site and the fibroblasts have reverted to quiescent tenocytes.

The intrinsic healing of flexor tendons has been studied by Matthews and Richards[100] and Lundborg and Rank[8,9] in vivo and by Manske and associates[10-12] and Gelberman and associates[14] in vitro. Although there are some differences between species, the cellular proliferation and sequence are similar but the rates of healing differ. In vitro, tendons demonstrate a graded response after 12 weeks; the rabbit tendon is the closest to complete repair and the chicken, dog, and monkey respond in that order. Initial morphologic changes with thickening and fibroblastic proliferation of the epitenon occur by three weeks. The fibroblasts differentiate into macrophage-type cells and migrate into the wound gap. Although there seems to be cellular activity in the endotenon, few of the migrating cells appear to originate from this layer. These cells appear to have a well-developed endoplasmic reticulum and mitochondria and to contain a considerable amount of extracellular material. The macrophage migration found in the rabbit and chicken is absent in the monkey, although macrophage function in the latter case is probably provided by endotenon fibroblasts. This balance between macrophages and fibroblasts in vivo may be under the influence of local soluble factors. By six weeks after injury, the epitenon cells form a continuous bridge across the repair site and the gap appears to be filled incompletely with collagen. There is an increase in collagen synthesis by the endotenon, which is associated with an increase in fibroblasts between six and 12 weeks in vitro. Between nine and 12 weeks in the rabbit, the wound gap in vitro is almost closed by collagen fiber deposition, probably from endotenon fibroblasts. Lesser tendon responses are seen in the other species.

These in vitro data parallel those of in vivo studies. Lundborg and Rank[8,9] were able to demonstrate that lacerated rabbit flexor tendons devoid of the tendon sheath, when repaired and reimplanted in the knee joint, healed by an intrinsic tenocyte response. A continuous layer of tenocytes was present two weeks after repair and after six weeks was highly developed. This proliferation of fibroblasts occurs peripherally but to a minimum degree centrally.

Matthews and Richards[101] were able to show that rabbit flexor tendons exhibited intrinsic healing capability when the synovial sheath was not violated. There was a marked proliferation of fibroblasts of the epitenon and endotenon eight days after wounding. Subsequent increased synthesis of collagen and maturation of the cells resulted in a healed defect by 14 weeks. There was never any indication of a synovial-sheath response during any period.

This pattern of cellular response after tendon laceration may be summarized as follows: There is an early increase in epitenon fibroblastic response with the migration of these cells into the gap and the synthesis of collagen and protein. This response is rapidly followed by migration of macrophage-like cells from the epitenon. Endotenon fibroblasts are also capable of exhibiting macrophage activity. In the circumstances of immobilization, the parietal sheath appears to respond actively. Immobilized tendons appear to heal by ingrowth of tenoblasts from the digital sheath and cellular proliferation of the epitenon, which results in a dense scar. When the synovial sheath is not injured or is excluded, the same general sequence of cellular response occurs; however, there is a balance between the contributions of epitenon and endotenon tenocytes and adhesions are not prominent. These two controversial concepts of tendon repair are in themselves not exclusive. The determination of which process is paramount appears to be influenced by local phenomena, most importantly by the presence or absence of the synovial sheath. Healing under the latter circumstance appears to be entirely by tendon cell proliferation. This is reflected by an increase in cell density in the gap and a change from the normally elongated shape of quiescent tendon cells to a predominance of active tenoblasts. These proliferating tenoblasts are contributed by the epitenon and the endotenon. If, however, the synovial sheath is injured, then the extrinsic mechanism of repair overwhelms the intrinsic tendon abilities. The control of these mechanisms in vivo is surely under the direction of both biologic and nonbiologic factors.

Biomechanical Response in Tendon Repair

The mechanical strength of the healing tendon is closely related to the three histologic phases of the healing process. The inflammatory phase takes place soon after injury. The defect is filled with clotted blood, tissue debris, and fluid. The strength of the healing tendon, therefore, is derived only from fibrin in the clot and is not sufficient to keep the lacerated tendon together.

The strength of an injured tendon with proper suture repair increases rapidly in the fibroplasia phase during which granulation tissue is formed to repair the defect. Tendon healing is achieved either by extrinsic healing (mediated by fibroblasts and dependent on the formation of adhesions between the tendon and surrounding tissues) or by intrinsic healing (mediated by tenocytes with nutrition provided by synovial fluid). Collagen and ground substance are secreted and synthesized. The collagen fibrils are laid down in a random pattern and thus possess little strength. Gradually, a more systematic pattern of collagen fibrils begins to emerge. The strength of healing tendon thus increases when the collagen in the microfibrils is stabilized by cross-linking and the fibrils are assembled into fibers.

In the maturation phase, the mechanical strength of healing tendon continues to increase because of remodeling and organization of the fiber architecture along the direction of muscle force. In addition, a gradual shift of collagen production from type III to type I may also contribute to the increased mechanical strength.

Scar tissue, which usually forms in the early stage of the healing process, possesses some random patterns of waves and does not achieve strength comparable to that of intact tendon even after prolonged periods of healing.[105,106] The remodeling process takes place in response to mechanical stimuli. In behavior similar to that described by Wolff's law for bone, tension on the maturing granulation tissue may cause the development of a preferential direction of the collagen molecules in the fibers.[107]

Experimentally, the development of strength in a totally immobilized wound is somewhat retarded.[108] In a study of healing of partial tenotomy, Hutton and Ferris[106] observed that the tissue elements become successively more organized and the fibroblastic cells begin to orient in the direction of the force line. Early weightbearing during healing after repaired tenotomy, however, results in rupture in most cases, but also results in increased strength if the continuity is not broken. The tensile load at failure increases during tendon healing and is more enhanced by controlled passive motion than by immobilization.[14,93,109]

Another effect that early controlled passive motion and stress have on healing of the repaired tendon is related to nutrient provision. This effect is especially important for those tendons healing within the fibro-osseous sheath. Two possible pathways have been defined by which nutrients may reach the flexor tendon.[21,110] Besides direct delivery by the vasculature, nutrients can also be provided by diffusion of synovial fluid into the interstices of the tendon. As Weber[110] observed, there are ridges in the tendon and pulleys oriented at 90 degrees to each other that may provide a pumping action to propel the fluid into the tendon.

In summary, from the biomechanical viewpoint, the optimal tendon healing procedure is to minimize scar formation by using strong suture repair to reduce gap formation. After the initial healing phase, earlier controlled passive motion and tensile stress on the repaired tendon seem to be desirable to promote earlier organization and remodeling of the collagen fiber tissue so that higher strength can be achieved. However, numerous basic and practical questions must yet be answered. Are there better suture materials and suturing techniques that will reduce or minimize foreign-body reaction, diminished circulation, and gap formation? If a permanent suture is used, the collagen juncture is protected from mechanical stimulation. Does stretching of the tendon juncture induce polarization of collagen along the axis of the tendon? Will these effects be blocked by permanent sutures that provide stress shielding? Controlled passive motion augments the quality of the healing mode. How do magnitude and duration affect the results? How do these factors relate to the detailed pumping mechanism of nutrient transport? The mechanism of finger-joint motion and the diffusion of nutrients have been studied experimentally by the isotope uptake technique.[19] More experimental and analytic investigations are required.

Biochemistry of Tendon Repair Response

The biochemistry of the repair response in tendons has centered around the kinetics of collagen synthesis. In 1970 Munro and associates[111] performed the first thorough analysis of both mucopolysaccharide and collagen synthesis in the digital tendon of the chicken, under circumstances including simple immobilization only, suture only, excision of the superficialis tendon, profundus excision and suture, and tendon grafted to the digital tendon site. They reported that mucopolysaccharide synthesis varied with collagen synthesis and demonstrated that (1) immobilization alone reduced collagen synthesis, (2) suturing the tendon, without an excision, increased collagen synthesis at the region of the suture, but not adjacent to it, (3) excising the superficialis or profundus induced exuberant collagen synthesis, and (4) grafting a tendon produced a maximal stimulatory effect on synthesis. These investigators concluded that synthesis of collagen and mucopolysaccharides was an interdependent process and that trauma to the tendon induced new collagen synthesis.

The effects of devascularization on total protein synthesis, and on collagen synthesis in particular, in the flexor tendon of the chicken was investigated by Banes and associates.[69] In the divided and sutured tendon, collagen synthesis was maximal on day 10 but was detected as early as day 4. Protein synthesis in the tendons that were incised and had a diminished blood supply was reduced considerably, and on day 21 after trauma it was 40% of the value for the group with an intact blood supply. At day 21, collagen synthesis per unit of DNA in the incised tendon group was stimulated 37-fold beyond the level of the resting tendon group. Therefore, it appears that devascularization reduces collagen synthesis, and that tendon cells possess the ability to metabolize, both on a maintenance schedule and in response to trauma.

In a separate study of avian tendon healing, Banes and associates[55] hypothesized that either a chemical or physical property of the wound region could stimulate the cells in adjacent, normal matrix to migrate over their collagenous substrate and reorganize the collagen, as occurs within hours in cell-populated collagen gels in vitro. When the cells in vivo reorganize their collagen, they reveal cross-link precursor sites (aldehydes and amino groups of lysine and hydroxylysine), previously buried within the matrix. These groups can then react with precursors on newly synthesized collagen to unite, covalently, collagen in established matrix with that in the new matrix. Perhaps it is this reaction, the cell-driven integration of new and old collagen, that contributes to the final stability of matrix at the union.

Methods of Improving the Repair Response

Improved Repair Techniques

More important than the strength of the suture material itself is the type of tendon repair that is carried out. In 1980, Nicoladoni[112] recognized that tendon fascicles are oriented longitudinally and that sutures inserted in line with tendon fascicles tended to pull out of the tendon. He placed the proximal and distal portions of the suture at right angles to the tendon substance. In an effort to avoid the tendency of longitudinally oriented sutures to pull through tendon, Bunnell[113] developed a criss-cross tendon suture configuration specifically designed to avoid the shearing of suture through tendon. Most modern techniques, however, are variations of the one described by Kessler[114] in 1973. This method utilizes two strands of suture material, one inserted into each tendon stump. The sutures are initially passed perpendicularly to the longitudinal axis of the tendon and then within the tendon, parallel to the longitudinal fibers in the area of tendon injury. The knots are tied on opposite ends of the repair site approximately 1 cm from the site of injury. Other techniques have been described for extrasynovial flexor tendon repair. In 1965, Pulvertaft[115] described end-to-side and fish-mouth repair techniques designed to provide increased repair strength in those regions.

Three recent evaluations of tendon repair techniques have been carried out. Urbaniak and associates[116] determined the relative strength of flexor tendon repairs in adult mongrel dogs at five-day intervals for 20 days. The tendon specimens were held in pneumatic clamps and attached to displacement transducers. A constant speed gear was used to drive the clamps apart and the load was continuously recorded. Four types of sutures and six different suture techniques were tested. The investigators found that, of the synthetic materials tested, nylon had the lowest load at failure and braided polyester fiber had the highest (20 N unknotted vs 40 N unknotted). Tying a knot in the suture altered its load at failure (a 22% reduction for nylon, a 36% reduction for braided Dacron, and a 46% reduction for Tevdek). None of the synthetic materials was as strong as a 4–0 stainless steel suture, which was least affected by knotting with only a 10% to 13% reduction in load at failure.

The investigators divided the different methods of tendon repair into three basic groups. Group 1 included those sutures in which the pull of the suture was parallel to the collagen bundles of the tendon and the stress of the repair was transmitted to the opposing tendon ends. Group 2 included repairs in which the stress was applied directly across the repair by the suture material itself and thus depended on the strength of the suture. Group 3 included tendon repairs in which the suture was placed perpendicularly to both the collagen bundles of the tendon and to the applied stress (fish-mouth side-weave). The tendons in group 1 were the weakest. In all cases the suture pulled out of the tendon without breaking the suture itself. Interrupted sutures around the circumference of the tendon had the lowest value (17 N) and the Nicoladoni repair was only slightly stronger (27 N). In group 2, because of improved purchase by the suture, the suture itself broke most often before it pulled out of the tendon substance. Bunnell's criss-cross stitch withstood a load of 39 N before the suture material failed. The Kessler and Mason-Allen sutures withstood a slightly greater load (40 N). In 80% of specimens, failure was a result of the suture breaking; in only 20% did the suture pull out of the tendon substance. The strength of repair approached the tensile strength of the suture material itself.

In group 3, there was no direct pull on the suture because the suture had been placed perpendicularly to both the collagen bundles and to the stress applied to the tendon. The suture applied a compressive force to the tendon at right angles to the longitudinal shearing force when loaded. A very strong repair (64 N) was achieved with the end-weave technique; less strength was achieved with the fish-mouth (40 N). The suture did not fail; thus, the investigators concluded that the strength of the repair was related to both the redistribution of shearing forces across the repair and the inherent properties of this type of repair.

Wray and Weeks[117] performed a similar experiment comparing tendon gliding and strength in the Bunnell-type suture and three more modern sutures: the Kessler, the Kleinert, and the Tsuge. Repairs performed in chickens were evaluated by measurements of angular motion and longitudinal tendon movement. They found equal strength, gliding, and rupture rates with the Bunnell, the Kessler, and the Kleinert repairs. The investigators reported a high rupture rate, probably as a result of postoperative management. The animals were allowed to ambulate in an unrestricted fashion with the toes bandaged in acute flexion. In addition, all the animals were tested at one interval, 28 days postoperatively. No special rehabilitation, such as controlled passive motion, was used. In 1977, Ketchum and associates[103] reported the results of a similar study in canine flexor tendons comparing the Bunnell suture with the lateral trap technique. Immobilization was used. The greatest resistance to rupture was found with the lateral trap repair at 28 days.

As Wade and associates[118] pointed out, many investigators have attempted to improve suturing techniques, often with the hope of producing a repaired tendon strong enough to allow active movement.[113,114,119–123] More important, however, may be the technique needed to resist gap formation. Seradge[124] illustrated the problem of gaps clinically by marking and measuring the gap between the repaired tendon ends. He found that poor results can be expected if even a small gap forms between the repaired ends of tendons. Poorer results are obtained if the gap is larger. Running or peripheral stitches have been suggested for invaginating free tendon ends. The contribution of these sutures to gap formation resistance was measured by Wade and associates.[118] In vitro mechanical tests revealed that with a running suture along the repair, the load at failure can be as high as some of the core suture techniques. The dynamic-compression suturing technique is another method of reducing gap formation and is currently under active investigation in the laboratory.

The properties of suture materials may be of significance. For example, friction between the suture and tendon manipulates the tension distribution among stitches in certain types of suture techniques. Absorbable sutures have been developed for tendon repair; the advantages are the low long-term foreign-body tissue reaction as well as the reduction of stress-shielding effects on the host tissue. However, the optimal rates of material absorption and strength reduction are yet to be determined. The following questions must be answered: Are there better suture materials and suturing techniques that will reduce or minimize foreign-body reaction, diminished circulation, and gap formation? Does stretching of the tendon junction induce polarization of collagen along the axis of the tendon? Will these effects be blocked by permanent sutures that provide stress-shielding?

Evolving Concepts of Rehabilitation

Over the past two decades, many of the details of the tendon healing process, including information on the relative contributions to healing by both adhesion ingrowth and intrinsic processes, have been provided. Experimental studies have concentrated on three areas: tendon suturing techniques, tendon sheath repair and reconstruction, and early motion following repair. Results indicating that contact tendon healing, stimulated by protective passive motion, occurs in the absence of tendon sheath adhesions have offered an exciting avenue for renewed research in this field.

Attempts at optimization of the postoperative mobilization regimen after flexor tendon repair have been essentially empirical. The duration and level of exercise have lacked clear, conceptual guidelines. It was not known to what extent early motion affected the roles of vascular and synovial fluid nutrition and influenced the gliding surface of tendon repair sites and sheaths. It was believed that improperly applied early motion could be harmful, leading to gap formation and rupture, particularly if the magnitude of stress was too great.

In an experimental study in chickens, Lindsay and associates[97,98] explored the causes and results of gap formation between tendon ends. One of the major theoretical factors associated with gap formation, inadequate immobilization, was soon to be associated with increased callus size, increased adhesion formation, delayed collagen maturation, and disoriented fibroblastic proliferation. As a method of controlling the gap, they suggested prolonged immobilization, along with dividing the flexor tendon at the wrist.

Ketchum and associates[102,103] also found that gap formation increased adhesions, and decreased tendon gliding function. If this were the case under all conditions, then early motion would appear deleterious and likely to result in repair site failure. Seradge[124] showed that the incidence of failure after flexor tendon repair was directly related to elongation of the repair site.

In a canine experimental study, Gelberman and associates[125] demonstrated that carefully applied early protected motion can produce a consistent range of gliding in the early postoperative period that is not associated with significant repair site deformation. Morphologically, the healing of mobilized tendons differed considerably from that of immobilized tendons.[13,14] In all of the mobilization groups, the thickened epitenon cellular layer migrated into the depths of the repair site. The predominant origin of the repair site cell was consistently the surface layer rather than the endotenon. This early epitenon response was much more dramatic than the delayed proliferative response of the endotenon seen in the immobilized tendons. Significant adhesions did not extend from the flexor sheath to the repaired surface in the mobilized groups.

The mechanism by which this significant alteration of healing mode was stimulated has not been conclusively determined. Duran and Houser[126] showed that limited relative motion (3 to 5 mm) was sufficient to prevent adhesion formation between the repaired tendons and the digital sheath. An explanation for this effect is the rapid re-institution of tendon function, alternating stress and relaxation with gliding of the tendon, which induces a cellular response from the tendon itself and the epitenon layer. If this is correct, small loads and small increments of motion may be sufficient to provide the stimulus for cellular activation of the epitenon. Another possibility is that the ingrowth of the cells from the tendon sheath in the proliferative stage overwhelms the repair process of the epitenon. In this case, early motion may block an inhibitory response by denying inflammatory tissue acess to the repair site. Greater magnitudes and durations of motion may have an acceleratory and/or a magnifying effect.

In experimental studies[14] comparing immobilization, delayed mobilization, and early mobilization in animal tendons, the load at failure of immediately mobilized tendons tested at three weeks was twice that of immobilized tendons while the linear slope was almost three times greater. The differences continued at each interval through 12 weeks. There were also significant differences in the angle of rotation of the distal interphalangeal joints after the application of a small load. At 12 weeks, the angular rotation of tendons in the immobilized group averaged only 19% $\pm$ 2% of their intact contralateral controls. The tendons undergoing delayed mobilization produced 95% $\pm$ 10% of the controls' joint motions. These data support the concept that a limited, immediate mobilization program improves the quality of the biologic repair response after flexor tendon repair within the digital sheath by effectively eliminating the associated adhesion formation.

Hitchcock and associates[127] recently examined the effect of immediate controlled mobilization on the strength of flexor repairs. In the chicken flexor tendon, they found, an initial loss of tendon repair strength is not inevitable. Immediate controlled mobilization of tendon repairs allowed progressive tendon healing while eliminating an intervening phase of tendon softening. Controlled mobilization after tendon repair induced a change in the healing mode by stimulating an intrinsic tendon healing response. Salter and associates[128-132] demonstrated the effect of continuous compression on living articular cartilage. The augmentation of extrasynovial flexor tendon healing by continuous passive motion techniques in rabbits has been demonstrated by the same investigators. The effects of continuous passive motion on the healing of intrasynovial flexor tendons has not yet been determined.

There are many factors that influence the cellular response to tendon injury. Although mechanical and environmental factors may overwhelm the entire process, there are biologic influences, such as soluble factors, that are important. These factors include mitogens such as platelet-derived growth factors, chemotactic factors, and differentiating factors, such as transforming growth factor beta (TGF-β).[133-139] Motion has been shown to influence the cellular response of injured tendons.[133,136] Additionally, agents that affect cellular responses have been used to modify the formation of adhesions, but to date no drug has consistently improved the clinical results without delaying wound or tendon healing or causing systemic complications.[140-143] A brief overview of these agents will be presented.

Cellular Response to Intermittent Passive Mobilization

The cellular response to repair in mobilized tendons within the first ten days appears to be the result of proliferation and migration of fibroblasts from the epitenon. These cells penetrate the repair site and actively synthesize collagen. Twenty-one days after repair, mobilized tendons demonstrate a mild proliferation of endotenon fibroblasts; however, their cellular activity is significantly reduced compared to the immobilized tendon. Additionally, these fibroblasts have a well-developed endoplasmic reticulum and Golgi apparatus, indicating protein synthesis; however, they do not appear to be resorbing collagen actively. Thus, there is some cellular activity in the endotenon, but it is significantly reduced compared to immobilized tendons. At 42 days after tendon repair and mobilization, fibroblasts within the epitenon continue to proliferate and synthesize collagen bundles oriented parallel to the underlying collagen. Some fibroblastic response is seen in the endotenon, but again this is markedly inhibited compared to immobilized tendons. The ultrastructure of the fibroblasts within the area of the repair demonstrates cytoplasmic processes consistent with active protein synthesis. New vasculature is seen throughout the repair site and there are no adhesions. It appears, therefore, that mobilization of tendon results in an effective repair without overwhelming scar formation. The signal that directs the cell response to this activity is still unknown and offers an interesting area of investigation.

Effect of Biochemical Agents on the Cellular Response

Various biochemical agents have been utilized in an attempt to modify the excessive scar formation and adhesions that develop after tendon repair with immobilization. These agents include antihistamines, anabolic agents, and corticosteroids. Douglas and associates[144] tried to produce changes in collagen production that would correlate with improved healing in tendons. The anabolic, anti-inflammatory steroid dexamethasone and the steroid norethandrolone were used, as was promethazine, an antihistamine. Radioactive proline, incorporated in regions of incised, avian flexor tendons was quantified. Maximum collagen synthesis was achieved in the norethandrolone-treated group by ten to 14 days after trauma, whereas the dexamethasone- and promethazine-treated groups were equivalent and maximum synthesis occurred between days 14 and 28 in both.

Tension relieved by severing the tendon resulted in a significant reduction in collagen synthesis. In the dexamethasone-treated group, the tendon healing sites ruptured easily by day 14 and adhesions were considerable by day 84. Therefore, norethandrolone seems to be promising in the treatment of tendon injuries.

Hyaluronate has also been used after tendon surgery to limit the development of tendon adhesions.[145,146] Sodium hyaluronate prepared from rooster combs or umbilical cord (average molecular weight, 1.2 $\times$ 10^6 daltons), was applied as a paste (2% dilution in physiologic buffer) to the flexor tendons of owl monkeys after surgery. The data indicated that hyaluronate-treated tendons had 27% less flexion deformity than did saline-treated controls. The hyaluronate paste not only reduced adhesions to the repair site but also failed to inhibit the diffusion of nutrients.

A lathyrogenic agent, β-aminopropionitrile, demonstrated interesting early results with inhibition of adhesions after digital flexor repair in chickens.[140,143] This drug blocks the conversion of lysine in peptide linkage to an aldehyde. It appears to affect newly synthesized collagen without affecting mature collagen. The drug has been used clinically with some encouraging results; however, because of its toxic side effects, it has been abandoned.[143] Newer analogues of this drug may prove useful since the side effects are reported to have been eliminated. Other drugs that affect the inflammatory cycle, such as ibuprofen, have been used to inhibit adhesion formation in primates.[141,142] This nonsteroidal anti-inflammatory agent acts at the cell level by blocking prostaglandin synthesis through the inhibition of the enzyme cyclooxygenase. Another possible mechanism is the decrease of fibroblastic proliferation and reduction in chemotaxis of unnecessary cells. Early results indicate a significant reduction in adhesions after repair of flexor tendon lacerations in cynomolgus monkeys but an attendant reduction of repair strength at six weeks after surgery. Additional investigations are necessary to understand the mechanism of action and to develop means of inhibiting scar formation without compromising strength.

Effect of Soluble Factors on Cells

Any injury and repair sequence must be affected by soluble biologic factors in vivo. These polypeptides include mitogens, chemotactic factors, and differentiating factors. A brief overview of the role of each of these factors in influencing the possible cellular response is important because these factors are implicated in a wide variety of physiologic and pathologic processes.[133–139] A better understanding of the mechanism of action of these factors may help us to develop techniques for improving the repair response of tendons. Mitogens are polypeptide growth factors that act alone or in synergy to induce mitosis in cultured fibroblasts. These mitogens may include purified polypeptide growth factors, mitogenic hormones, and other pharmacologic agents. One of these factors, known as platelet-derived growth factor, has considerable significance in the regulation of fibroblast growth. Platelet-derived growth factor stimulates DNA synthesis and cell division in the absence of any other factor. The first step in this interaction with the fibroblast is a binding of platelet-derived growth factor to specific high-affinity cell membrane receptors that undergo rapid phosphorylation and redistribution by endocytosis. This signal is propagated into the nucleus, where these events are followed by molecular responses leading to DNA replication and cell division. The initial interaction of this mitogenic factor with the cell may provide important clues to the primary regulatory mechanisms that control fibroblastic growth during tendon repair. The cascade of early events appears to require the stimulation of sodium, potassium, and hydrogen ion flux across the plasma membrane. Calcium ions are quickly mobilized, which is necessary for the continued activation of this ion flux. In addition, phosphorylation of the epidermal growth-factor receptor occurs, which is important for the enhanced expression of messenger RNA. This leads to DNA synthesis and cell proliferation. In parallel, there is an elevation of the intracellular level of cyclic adenosine monophosphate. This model of DNA synthesis and mitogenesis provides the framework for a better understanding of the complex processes by which extracellular factors regulate cell proliferation. A broader understanding of these mechanisms may have wide application in modulating the repair of tendons on a cellular level.

Shortly after proliferation of cells, there must be migration (chemotaxis) of these cells into the repair site. This migration is also under the influence of soluble polypeptides. One factor, TGF-β, is a 2-disulfide-linked polypeptide chain, each of approximately 12.5 kD. It is present in a variety of cells, including T-lymphocytes, monocytes, and platelets. It has been shown previously to stimulate the production of fibronectin and collagen by dermal fibroblasts in vitro and has been noted to induce fibroblast chemotaxis. This latter function may be critical to an understanding of the cellular sequence of tendon repair. Previous studies indicate[134,136,137,139,147] that a dissociation of the chains of TGF-β causes a significant loss of chemotactic potency and that polyvalent antibodies effectively block this activity. The release of TGF-β from platelets, lymphocytes, monocytes, and macrophages, which are present in tendon repair, may play a critical role in effecting the migration of epitenon fibroblasts. In addition, it can also act as a stimulus for these fibroblasts to produce increased quantities of collagen. Also, types I, II, and III human collagens, as well as collagen-derived peptides, have been shown to induce a chemotactic response of human dermal fibroblasts. Because collagen is degraded in the process of tendon repair, it is possible that collagen and collagen degradation peptides function as additional chemotactic stimuli for epitenon and endotenon cells.

TGF-β has also been shown to induce further differentiation of fibroblastic cells for enhanced synthesis of collagen. When TGF-β is injected into wire-mesh wound-healing chambers implanted in rats, it stimulates collagen formation and the wound healing response. This action is effected through direct action on fibroblasts and is specific. Lymphocytes present in the granulation tissue in the repair sequence are known to secrete increased amounts of TGF-β when they are activated. The conditioned media from such activated lymphocytes stimulate proline incorporation into collagen in rat fibroblasts. This stimulatory action can be blocked by specific antibodies to TGF-β. TGF-β can also cause a striking angiogenic response, which is also important in tendon repair. Although the action of TGF-β may be the result of an increased number of fibroblasts, other data suggest a direct effect of TGF-β on collagen synthesis by fibroblasts. Although TGF-β may act by itself, it is clear that in vivo it acts in concert with other polypeptide growth factors, such as epidermal growth factor, transforming growth factor alpha, platelet-derived growth factor, and basic fibroblastic growth factor. These recent findings suggest a significant role for these polypeptide growth factors in wound healing responses. These factors can be obtained on a large enough scale (nanograms) to warrant further investigation of their possible role on the cellular level in the healing response of tendons.

Artificial Tendon Implants

The reconstruction of severely damaged tendon has taken three directions: the use of devices to prevent scar formation, the implantation of implants to enhance pseudosheath formation before tendon grafting, and the use of artificial tendon substitutes to bridge tendon grafts. Early work used either celluloid in tubes or stainless steel implants in scarred tendon beds.[148-151] These methods were usually unsuccessful because they ultimately caused joint stiffness during the period of implantation. Additionally, failure of fixation to the muscle was common. Teflon has also been used as a tendon substitute and as a blocking sheath to prevent adhesions. The response of the host bed in each instance was detrimental, with a massive inflammatory reaction in many of the experiments.

Nylon has also been used as a tendon substitute; however, there are problems with its use, including failure at the proximal and distal attachments, and long-term biocompatibility. Further, tubes of this material provide a dead space for fluid collection and may fail because they are subjected to excessive frictional wear. Bassett and Carroll[152] showed that silicone-rubber rods could be implanted safely and that they could stimulate the development of a pseudosheath. Hunter and Salisbury[150,153] used silicone Dacron-reinforced tendon prostheses for the reconstruction of flexor and extensor tendons. They showed that mesothelial cells are aligned on the surface of the implant as early as five days; by three weeks after implantation, a well-developed bursa is present. A more advanced sheath with fluid resembling synovial fluid was formed when these Dacron-reinforced silicone tendons were allowed to move. These implants were used as an active tendon for many years. However, in many instances the repair failed. The sheath was found to be well-formed, with no evidence of synovitis during the second-stage reconstruction with a tendon graft. This observation suggested the use of an artificial tendon substitute to enhance pseudo-sheath formation before tendon grafting. These tendon substitutes are presently used as a passive gliding prosthesis to ensure the development of a tendon sheath to be used in a subsequent procedure for tendon grafting. However, there is still a place for an active tendon substitute because many of the problems previously seen at the distal bone or proximal muscle attachments may have solutions.

Future Directions

In order to gain further knowledge about tendon healing and repair processes, fundamental knowledge about normal tendon is required. Suggested studies include the effects of aging, immobilization, and recovery from immobilization, determinations of in vivo tensile forces, and evaluations of the relative motions between the flexor tendon and its sheath at areas around the three interphalangeal joint levels (excursions) based on rigorous kinematic and experimental methods.

Additional basic science studies should involve factors and mechanisms that may affect tendon healing. The vascular anatomy of intrasynovial flexor tendons has been described, but little work has centered on extrasynovial tendon with demonstrated clinical evidence of failure attributable to ischemic changes. Nutrient transport within the tendon sheath, produced by a pumping mechanism induced by the movement of joints and tendon excursion, may be an important factor in improving intrinsic healing. The mechanism of finger-joint motion and the diffusion of nutrients have been studied experimentally with an isotope uptake technique. More data are required on the characterization of the synovial sheath and synovial fluid. Specifically, the function of synovial fluid as it relates to lubrication (for example, the presence of factors responsible for enhancing gliding function) should be studied.

Tendon can encounter large tensile forces. In the pulley region and where tendon wraps around articular surfaces, increased compressive stresses may be experienced. It has been demonstrated that, in these areas, the histologic structure of tendons is modulated into cartilage-like cells and matrix. Does this cartilage-like tendon, therefore, respond to injury as does the rest of the tendon? More experimental and analytical studies on the interaction of forces between the tendons and constraining pulleys, as well as the relationship of pulleys to the healing of repaired tendon, must be encouraged to gain further understanding of the tendon apparatus.

It appears that mobilization of tendon results in an effective repair without overwhelming scar formation. The signal that directs the cell response to this activity is still unknown and offers an interesting area of investigation. There is a need to understand the responses of different cellular elements within the tendon to injury and repair and specific cell markers must first be developed in an in vitro model for identifying factors responsible for cellular activities. Emphasis should also be given to the development of optimal, species-specific culture media. In addition, the quantification of the natural levels of growth factors at the tendon wound site must be performed to assess the general level of mitogenic, chemotactic, and differentiating activities. Once this has been done, the capacity of endogenous cells to respond to the pharmacologic addition of growth factors can be evaluated. A protein data base must be established for cells specific to tendon. These cells include those resident in the epitenon, endotenon, paratenon, and fasicles as well as in the sheath. Proteins specific to these cells can be isolated and monoclonal antibodies prepared. These antibodies may be used to identify the target cells involved in the repair response.

Although controlled passive motion has been shown to augment the quality of the healing mode, the optimal magnitude and duration of motion for the best healing process has not been determined. Are the strengths of current suturing techniques sufficient for immediate controlled mobilization? Also, there are numerous methods of tendon mobilization after repair, such as continuous passive motion devices and dynamic splints. There are, however, many factors that affect the results of these devices. Understanding the relationship of the effects of these factors on the excursion and tension in the repaired tendon is of paramount importance to the success of the tendon repair. Most tendons, particularly the digital flexor tendons, pass through multiple joints. Tendon excursion, therefore, is highly dependent on the rotation of the joints spanned by the tendon. Analytical studies and the development of devices that promote interphalangeal joint motion (for example, by moving the wrist joint) should be encouraged.

For the most part, experimental studies in tendon repair have not addressed the more difficult types of complex tendon injury. Slightly more complex but clinically relevant injury models must be developed. For example, 75% of tendon injuries involve associated injuries of surrounding soft and hard tissues. It should also be noted that current data on tendon repair and healing have been limited to short-term results. Since complete healing has not been seen, longer-interval studies should be carried out to examine the more advanced stages of repair.

References

1. Potenza AD: Concepts of tendon healing and repair, in American Academy of Orthopaedic Surgeons *Symposium on Tendon Surgery in the Hand.* St. Louis, CV Mosby Co, 1975, pp 18–47.
2. Potenza AD: Tendon and ligament healing, in Owen R, Goodfellow J, Bullough P (eds): *Scientific Foundations of Orthopaedics and Traumatology.* London, William Heinemann Medical Books, 1980, pp 300–305.

3. Rowe RWD: The structure of rat tail tendon. *Connect Tissue Res* 1985;14:9–20.

4. Kastelic J, Galeski A, Baer E: The multicomposite structure of tendon. *Connect Tissue Res* 1978;6:11–23.

5. Nicholls Sp, Gathercole LJ, Keller A, et al: Crimping in rat tail tendon collagen: Morphology and transverse mechanical anisotropy. *Int J Biol Macromol* 1983;5:283–288.

6. Rowe RWD: The structure of rat tail tendon fascicles. *Connect Tissue Res* 1985;14:21–30.

7. Shah JS, Palacios E, Palacios L: Development of crimp morphology and cellular changes in chick tendons. *Dev Biol* 1982;94:499–504.

8. Lundborg G, Rank F: Experimental intrinsic healing of flexor tendons based upon synovial fluid nutrition. *J Hand Surg* 1978;3:21–31.

9. Lundborg G, Rank F: Experimental studies on cellular mechanisms involved in healing of animal and human flexor tendon in synovial environment. *Hand* 1980;12:3–11.

10. Manske PR, Gelberman RH, Vande Berg JS, et al: Intrinsic flexor tendon repair: Morphological study *in vitro*. *J Bone Joint Surg* 1984;66A:385–396.

11. Manske PR, Lesker PA: Biochemical evidence of flexor tendon participation in the repair process: An in vitro study. *J Hand Surg* 1984;9B:117–120.

12. Manske PR, Lesker PA: Histologic evidence of intrinsic flexor tendon repair in various experimental animals: An *in vitro* study. *Clin Orthop* 1984;182:297–304.

13. Gelberman RH, Vande Berg JS, Lundborg GN, et al: Flexor tendon healing and restoration of the gliding surface: An ultrastructural study in dogs. *J Bone Joint Surg* 1983;65A:70–80.

14. Gelberman RH, Woo SL-Y, Lothringer K, et al: Effects of early intermittent passive mobilization on healing canine flexor tendons. *J Hand Surg* 1982;7:170–175.

15. Gelberman RH, Manske PR, Vande Berg JS, et al: Flexor tendon repair in vitro: A comparative histologic study of the rabbit, chicken, dog and monkey. *J Orthop Res* 1984;2:39–48.

16. Gelberman RH, Posch JL, Jurist JM: High-pressure injection injuries of the hand. *J Bone Joint Surg* 1975;57A:935–937.

17. Lundborg G, Myrhage R, Rydevik B: The vascularization of human flexor tendons within the digital synovial sheath region—Structural and functional aspects. *J Hand Surg* 1977;2:417–427.

18. Lundborg G: Experimental flexor tendon healing without adhesion formation—A new concept of tendon nutrition and intrinsic healing mechanisms: A preliminary report. *Hand* 1976;8:235–238.

19. Lundborg G, Holm S, Myrhage R: The role of the synovial fluid and tendon sheath for flexor tendon nutrition: An experimental tracer study on diffusional pathways in dogs. *Scand J Plast Reconstr Surg* 1980;14:99–107.

20. Manske PR, Bridwell K, Lesker PA: Nutrient pathways to flexor tendons of chickens using tritiated proline. *J Hand Surg* 1978;3:352–357.

21. Manske PR, Lesker PA: Nutrient pathways of flexor tendons in primates. *J Hand Surg* 1982;7:436–444.

22. Potenza AD, Herte MC: The synovial cavity as a "tissue culture in situ"—Science or nonsense? *J Hand Surg* 1982;7:196–199.

23. Woo SL-Y: Biomechanics of tendons and ligaments, in Schmid-Schönbein GW, Woo SL-Y, Zweifach BW (eds): *Frontiers in Biomechanics*. New York, Springer-Verlag, 1986, pp 180–195.

24. Woo SL-Y: Mechanical properties of tendons and ligaments: I. Quasi-static and nonlinear viscoelastic properties. *Biorheology* 1982;19:385–396.

25. Abrahams M: Mechanical behaviour of tendon in *vitro*: A preliminary report. *Med Biol Eng* 1967;5:433–443.

26. Barfred T: Experimental rupture of the Achilles tendon: Comparison of experimental sutures in rats of different ages and living under different conditions. *Acta Orthop Scand* 1971;42:406–428.

27. Cronkite AE: The tendon strength of human tendons. *Anat Rec* 1936;64:173–186.

28. Viidik A: Tensile strength properties of Achilles tendon systems in trained and untrained rabbits. *Acta Orthop Scand* 1969;40:261–272.

29. Viidik A: Biomechanical behavior of soft connective tissues, in Akkas N (ed): *Progress in Biomechanics*. Amsterdam, Sijthoff and Noordhoff, 1979, pp 75–113.

30. Fung Y-CB: Stress-strain-history relations of soft tissues in simple elongation, in Fung YC, Perrone N, Anliker M (eds): *Biomechanics: Its Foundations and Objectives*. Engelwood Cliffs, Prentice-Hall, 1972, pp 181–208.

31. Haut RC, Little RW: A constitutive equation for collagen fibers. *J Biomech* 1972;5:423–430.

32. Jenkins RB, Little RW: A constitutive equation for parallel-fibered elastic tissue. *J Biomech* 1974;7:397–402.

33. Viidik AV, Gottrup F: Mechanics of healing soft tissue wounds, in Schmid-Schönbein GW, Woo SL-Y, Zweifach BW (eds): *Frontiers in Biomechanics*. New York, Springer-Verlag, 1986, pp 263–279.

34. Kastelic J, Baer E: Reformation in tendon collagen: The mechanical properties of biological materials. Presented at the XXXIV Symposium of the Society for Experimental Biology, 1980.

35. Rigby CBJ, Hirai N, Spikes JD, ct al: The mechanical properties of rat tail tendon. *J Gen Physiol* 1959;43:265–283.

36. Apter JT, Rabinowitz M, Cummings DH: Correlation of visco-elastic properties of large arteries with microscopic structure: I. Methods used and their justification; II. Collagen, elastin and muscle determined chemically, histologically, and physiologically; III. Circumferential viscous and elastic constants measured in vitro. *Circ Res* 1966;19:104–121.

37. Riedl H, Nemetschek TH: *Molekular Struktur und Mechanisches Verhalten von Kollagen*. Berlin, Springer-Verlag, 1977.

38. Woo SL-Y, Lee TQ, Gomez MA, et al: Temperature dependent behavior of the canine medial collateral ligament. *J Biomech Eng* 1987;109:68–71.

39. Stucke K: Über das elastische Verhalten der Achilolessehne im Belastungsversuch. *Arch Klin Chir* 1950;265:579–599.

40. Wertheim HG: Mémoire sur l'elasticité et al cohésion des principeaux tissus du corps humain. *Chim Phys* 1847;21:385–414.

41. Akeson WH: An experimental study of joint stiffness. *J Bone Joint Surg* 1961;43A:1022–1034.

42. Akeson WH, Amiel D, LaViolette D: The connective-tissue response to immobility: A study of the chondroitin-4 and 6 sulfate and dermatan sulfate changes in periarticular connective tissue of control and immobilized knees of dogs. *Clin Orthop* 1967;51:183–197.

43. Akeson WH, Amiel D, LaViolette D, et al: The connective tissue response to immobility: An accelerated ageing response? *Exp Gerontol* 1968;3:289–301.

44. Akeson WH, Amiel D, Mechanic GL, et al: Collagen cross-linking alterations in joint contractures: Changes in the reducible cross-links in periarticular connective tissue collagen after nine weeks of immobilization. *Connect Tisue Res* 1977;5:15–19.

45. Woo SL-Y, Gomez MA, Woo Y-K, et al: Mechanical properties of tendons and ligaments: II. The relationships of immobilization and exercise on tissue remodeling. *Biorheology* 1982;19:397–408.

46. Ingelmark BE: The structure of tendons at various ages and under different functional conditions: II. An electron-microscopic investigation of Achilles tendons from white rats. *Acta Anat* 1948;6:193–225.

47. Kiiskinen A: Physical training and connective tissues in young mice—Physical properties of Achilles tendons and long bones. *Growth* 1977;41:123–137.

48. Woo SL-Y, Ritter MA, Amiel D, et al: The biomechanical and biochemical properties of swine tendons: Long term effects of exercise on the digital extensors. *Connect Tissue Res* 1980;7:177–183.

49. Woo SL-Y, Gomez MA, Amiel D, et al: The effects of exercise on the biomechanical and biochemical properties of swine digital flexor tendons. *J Biomech Eng* 1981;103:51–56.

50. Idler RS: Anatomy and biomechanics of the digital flexor tendons. *Hand Clin* 1985;1:3–11.

51. Cetta G, Tenni R, Zanaboni G, et al: Biochemical and morphological modifications in rabbit Achilles tendon during maturation and ageing. *Biochem J* 1982;204:61–67.

52. Jimenez SA, Yankowski R, Bashey RI: Identification of two new collagen α-chains in extracts of lathyritic chick embryo tendons. *Biochem Biophys Res Commun* 1978;81:1298–1306.

53. Piez KA, Miller EJ, Lane JM, et al: The order of the CNBr peptides from the $\alpha1$ chain of collagen. *Biochem Biophys Res Commun* 1969;37:801–805.

54. Neuberger A, Slack HGB: Metabolism of collagen from liver, bone, skin and tendon in normal rat. *Biochem J* 1953;53:47–52.

55. Banes AJ, Link GW, Peterson HD, et al: Temporal changes in collagen crosslink formation at the focus of trauma and at sites distant to a wound, in *The Pathophysiology of Combined Injury and Trauma*. New York, Academic Press, 1987, pp 257–273.

56. Klein L, Lewis JA: Simultaneous quantification of ^{3}H-collagen loss and ^{1}H-collagen replacement during healing of rat tendon grafts. *J Bone Joint Surg* 1972;54A:137–146.

57. Mechanic GL: The intermolecular cross-link precursors in collagens as related to function: Identification of γ-hydroxy α-amino adipic semialdehyde in tendon collagen. *Isr J Med Sci* 1971;7:458–462.

58. Mechanic GL: The existence of chemically distinct soluble collagens in pre-natal and post-natal bovine skin and tendon, in Slavkin HC, Greulich RC (eds): *Extracellular Matrix Influences on Gene Expression*. New York, Academic Press, 1975, pp 347–354.

59. Kuboki Y, Mechanic GL: The distribution of δ, δ'-dihydroxylysinonorleucine in bovine tendon and dentin. *Connect Tissue Res* 1974;2:223–230.

60. Fujimoto D, Akiba K-Y, Nakamura N: Isolation and characterization of a fluorescent material in bovine Achilles tendon collagen. *Biochem Biophys Res Commun* 1977;76:1124–1129.

61. Eyre DR, Koob TJ, Van Ness KP: Quantitation of hydroxypyridinium crosslinks in collagen by high-performance liquid chromatography. *Anal Biochem* 1984;137:380–388.

62. Ogawa T, Ono T, Tsuda M, et al: A novel fluor in insoluble collagen: A crosslinking moiety in collagen molecule. *Biochem Biophys Res Commun* 1982;107:1252–1257.

63. Chaplin DM, Greenlee TK Jr: The development of human digital tendons. *J Anat* 1975;120:253–274.

64. Elliott DH: Structure and function of mammalian tendon. *Biol Rev* 1965;40:392–421.

65. Greenlee TK Jr, Beckham C, Pike D: A fine structural study of the development of the chick flexor digital tendon: A model for synovial sheathed tendon healing. *Am J Anat* 1975;143:303–314.

66. Greenlee TK, Ross R: The development of the rat flexor digital tendon: A fine structure study. *J Ultrastruct Res* 1967;18:354–376.

67. Ippolito E, Natali PG, Postacchini F, et al: Morphological, immunological, and biochemical study of rabbit Achilles tendon at various ages. *J Bone Joint Surg* 1980;62A:583–598.

68. Riederer-Henderson MA, Gauger A, Olson L, et al: Attachment and extracellular matrix differences between tendon and synovial fibroblastic cells. *In Vitro* 1983;19:127–133.

69. Banes AJ, Enterline D, Bevin AG, et al: Effects of trauma and partial devascularization on protein synthesis in the avian flexor profundus tendon. *J Trauma* 1981;21:505–512.

70. Banes AJ, Gilbert J, Taylor D, et al: A new vacuum-operated stress-providing instrument that applies static or variable duration cyclic tension or compression to cells *in vitro*. *J Cell Sci* 1985;75:35–42.

71. Schwarz R, Colarusso L, Doty P: Maintenance of differentiation in primary cultures of avian tendon cells. *Exp Cell Res* 1976;102:63–71.

72. Schwarz RI, Farson DA, Soo WJ, et al: Primary avian tendon cells in culture: An improved system for understanding malignant transformation. *J Cell Biol* 1978;79:672–679.

73. Schwarz RI, Bissell MJ: Dependence of the differentiated state on the cellular environment: Modulation of collagen synthesis in tendon cells. *Proc Natl Acad Sci USA* 1977;74:4453–4457.

74. Anttinen H, Ryhänen L, Oikarinen A: Effects of divalent cations on collagen biosynthesis in isolated chick embryo tendon cells. *Biochim Biophys Acta* 1980;609:321–328.

75. Majamaa K: Effect of prevention of procollagen triple-helix formation on proline 3-hydroxylation in freshly isolated chick-embryo tendon cells. *Biochem J* 1981;196:203–206.

76. Kao WW-Y, Berg RA, Prockop DJ: Kinetics for the secretion of procollagen by freshly isolated tendon cells. *J Biol Chem* 1977;252:8391–8397.

77. Aalto M, Potila M, Kulonen E: Glycoproteins from experimental granulation tissue and their effects on collagen synthesis in embryonic chick tendon cells. *Biochim Biophys Acta* 1979;587:606–617.

78. Herrmann H, Dessau W, Fessler LI, et al: Synthesis of types I, III and AB_2 collagen by chick tendon fibroblasts *in vitro*. *Eur J Biochem* 1980;105:63–74.

79. An KN, Chao EY: Kinematic analysis of human movement. *Ann Biomed Eng* 1984;12:585–597.

80. An KN, Chao EY, Cooney WP III, et al: Normative model of human hand for biomechanical analysis. *J Biomech* 1979;12:775–788.

81. An KN, Cooney WP, Chao EY, et al: Determination of forces in extensor pollicis longus and flexor pollicis longus of the thumb. *J Appl Physiol* 1983;54:714–719.

82. An KN, Takahashi K, Harrigan TP, et al: Determination of muscle orientations and moment arms. *J Biomech Eng* 1984;106:280–282.

83. An KN, Ueba Y, Chao EY, et al: Tendon excursion and moment arm of index finger muscles. *J Biomech* 1983;16:419–425.

84. Wehbé MA, Hunter JM: Flexor tendon gliding in the hand: Part II. Differential gliding. *J Hand Surg* 1985;10A:575–579.

85. An KN, Cooney WP, Chao EYS, et al: In vivo measurement of tendon forces in the hand. Presented at the Third Congress of the International Federation of Societies of the Hand, Tokyo, Nov 3–8, 1986.

86. Azar CA, Fleegler EJ, Culver JE: Dynamic anatomy of the flexor pulley system of the fingers and thumb, abstract. *J Hand Surg* 1984;9A:595.

87. Cooney WP III, An K-N, Daube JR, et al: Electromyographic analysis of the thumb: A study of isometric forces in pinch and grasp. *J Hand Surg* 1985;10A:202–210.

88. Cooney WP, An KN, Chao EYS: Direct measurement of tendon forces in the hand. *Trans Orthop Res Soc* 1986;11:53.

89. Okuda Y, Gorski JP, An K-N, et al: Biochemical, histological, and biomechanical analyses of canine tendon. *J Orthop Res* 1987;5:60–68.

90. Manske PR, Lesker PA: Avulsion of the ring finger flexor digitorum profundus tendon: An experimental study. *Hand* 1978;10:52–55.

91. Leddy JP, Packer JW: Avulsion of the profundus insertion in athletes. *J Hand Surg* 1977;2:66–69.

92. Verdan CE: Half a century of flexor-tendon surgery: Current status and changing philosophies. *J Bone Joint Surg* 1972;54A:472–491.

93. Mason ML, Allen HS: The rate of healing of tendons: An experimental study of tensile strength. *Ann Surg* 1941;113:424–459.

94. Ross R: The fibroblast and wound repair. *Biol Rev* 1968;43:51–96.

95. Potenza AD: Tendon healing within the flexor digital sheath in the dog. *J Bone Joint Surg* 1962;44A:49–64.

96. Lindsay WK, Birch JR: The fibroblast in flexor tendon healing. *Plast Reconstr Surg* 1964;34:223–232.

97. Lindsay WK, Thomson HG: Digital flexor tendons: An experimental study: Part I. The significance of each component of the flexor mechanism in tendon healing. *Br J Plast Surg* 1959–1960;12:289–316.

98. Lindsay WK, Thomson HG, Walker FG: Digital flexor tendons: An experimental study: Part II. The significance of a gap occurring at the line of suture. *Br J Plast Surg* 1960;13:1–9.

99. Peacock EE Jr: Fundamental aspects of wound healing relating to the restoration of gliding function after tendon repair. *Surg Gynecol Obstet* 1964;119:241–250.

100. Matthews P, Richards H: Factors in the adherence of flexor tendon after repair: An experimental study in the rabbit. *J Bone Joint Surg* 1976;58B:230–236.

101. Matthews P, Richards H: The repair potential of digital flexor tendons: An experimental study. *J Bone Joint Surg* 1974;56B:618–625.

102. Ketchum LD: Primary tendon healing: A review. *J Hand Surg* 1977;2:428–435.

103. Ketchum LD, Martin NL, Kappel DA: Experimental evaluation of factors affecting the strength of tendon repairs. *Plast Reconstr Surg* 1977;59:708–719.

104. Graham MF, Becker H, Cohen IK, et al: Intrinsic tendon fibroplasia: Documentation by in vitro studies. *J Orthop Res* 1984;1:251–256.

105. Frank C, Woo SL-Y, Amiel D, et al: Medial collateral ligament healing: A multidisciplinary assessment in rabbits. *Am J Sports Med* 1983;11:379–389.

106. Hutton P, Ferris B: Tendons, in Bucknall TE, Ellis H (eds): *Wound Healing for Surgeons.* London, Balliere Tindall, 1984, pp 286–296.

107. Forrester JC, Zederfeldt BH, Hayes TL, et al: Wolff's law in relation to the healing skin wound. *J Trauma* 1970;10:770–779.

108. Hunt TK, Van Winkle W Jr: Normal repair, in Hunt TK, Dunphy JE (eds): *Fundamentals of Wound Management.* New York, Appleton-Century-Crofts, 1979, pp 2–67.

109. Woo SL-Y, Gelberman RH, Cobb NG, et al: The importance of controlled passive mobilization on flexor tendon healing: A biomechanical study. *Acta Orthop Scand* 1981;52:615–622.

110. Weber ER: Nutritional pathways for flexor tendons in the digital theca, in Hunter JM, Schneider LH, Mackin EJ (eds): *Tendon Surgery in the Hand.* St. Louis, CV Mosby Co, 1987, pp 91–99.

111. Munro IR, Lindsay WK, Jackson SH: A synchronous study of collagen and mucopolysaccharide in healing flexor tendons of chickens. *Plast Reconstr Surg* 1970;45:493–501.

112. Nicolandoni C: Ein orchlag zur Sehnennaht. *Wien Klin Wochenschr* 1980;52:1413–1417.

113. Bunnell S: Repair of tendons in the fingers and description of two new instruments. *Surg Gynecol Obstet* 1918;26:103–110.

114. Kessler I: The "grasping" technique for tendon repair. *Hand* 1973;5:253–255.

115. Pulvertaft RG: Suture materials and tendon junctures. *Am J Surg* 1965;109:346–352.

116. Urbaniak JR, Cahill JD Jr, Mortenson RA: Tendon suturing methods: Analysis of tensile strengths, in American Academy of Orthopaedic Surgeons *Symposium on Tendon Surgery in the Hand.* St. Louis, CV Mosby Co, 1975, pp 70–80.

117. Wray RC, Weeks PM: Experimental comparison of technics of tendon repair. *J Hand Surg* 1980;5:144–148.

118. Wade PJF, Muir IFK, Hutcheon LL: Primary flexor tendon repair: The mechanical limitations of the modified Kessler technique. *J Hand Surg* 1986;11B:71–76.

119. Becker H: Primary repair of flexor tendons in the hand without immobilisation: Preliminary report. *Hand* 1978;10:37–47.

120. McKenzie AR: An experimental multiple barbed suture for the long flexor tendons of the palm and fingers: Preliminary report. *J Bone Joint Surg* 1967;49B:440–447.

121. Murray G: A method of tendon repair. *Am J Surg* 1960;99:334–335.

122. Shaw PC: A method of flexor tendon suture. *J Bone Joint Surg* 1968;50B:578–587.

123. Tsuge K, Ikuta Y, Matsuishi Y: Intra-tendinous tendon suture in the hand: A new technique. *Hand* 1975;7:250–255.

124. Seradge H: Elongation of the repair configuration following flexor tendon repair. *J Hand Surg* 1983;8:182–185.

125. Gelberman RH, Botte MJ, Spiegelman JJ, et al: The excursion and deformation of repaired flexor tendons treated with protected early motion. *J Hand Surg* 1986;11A:106–110.

126. Duran RJ, Houser RG: Controlled passive motion following flexor tendon repair in zones 2 and 3, in American Academy of Orthopaedic Surgeons *Symposium on Tendon Surgery in the Hand.* St. Louis, CV Mosby Co, 1975, pp 105–114.

127. Hitchcock TF, Light TR, Bunch WH, et al: The effect of immediate controlled mobilization on the strength of flexor tendon repairs. *Trans Orthop Res Soc* 1986;11:216.

128. Salter RB, Bell RS, Keeley FW: The protective effect of continuous passive motion on living articular cartilage in acute septic arthritis: An experimental investigation in the rabbit. *Clin Orthop* 1981;159:223–247.

129. Salter RB, Minister RR, Bell RS, et al: Continuous passive motion and the repair of full-thickness articular cartilage defects: A one-year follow-up. *Trans Orthop Res Soc* 1982;7:167.

130. Salter RB, O'Driscoll SW: The effects of continuous passive motion on the repair of major full-thickness defects in a joint surface with autogenous osteoperiosteal grafts, abstract. *Ann R Coll Phys Surg Can* 1983;16:360.

131. Salter RB, Ogilvie-Harris DJ: Healing of intra-articular fractures with continuous passive motion, in American Academy of Orthopaedic Surgeons *Instructional Course Lectures, XXVIII.* St. Louis, CV Mosby Co, 1979, pp 102–117.

132. Salter RB, Simmonds DF, Malcolm BW, et al: The biological effect of continuous passive motion on the healing of full-thickness defects in articular cartilage: An experimental investigation in the rabbit. *J Bone Joint Surg* 1980;62A:1232–1251.

133. Davidson JM, Klagsbrun M, Hill KE, et al: Accelerated wound repair, cell proliferation, and collagen accumulation are produced by a cartilage-derived growth factor. *J Cell Biol* 1985;100:1219–1227.

134. Postlethwaite AE, Keski-Oja J, Moses HL, et al: Stimulation of the chemotactic migration of human fibroblasts by transforming growth factor β. *J Exp Med* 1987;165:251–256.

135. Postlethwaite AE, Seyer JM, Kang AH: Chemotactic attraction of human fibroblasts to type I, II, and III collagens and collagen-derived peptides. *Proc Natl Acad Sci USA* 1978;75:871–875.

136. Roberts AB, Sporn MB, Assoian RK, et al: Transforming growth factor type β: Rapid induction of fibrosis and angiogenesis in vivo and stimulation of collagen formation in vitro. *Proc Natl Acad Sci USA* 1986;83:4167–4171.

137. Rozengurt E: Early signals in the mitogenic response. *Science* 1986;234:161–166.

138. Sporn MB, Roberts AB, Wakefield LM, et al: Transforming growth factor-β: Biological function and chemical structure. *Science* 1985;233:532–534.

139. Sporn M, Roberts A, et al: Polypeptide transforming growth factor isolated from bovine sources and used for wound healing in vivo. *Science* 1986;234:161–166.

140. Herzog M, Lindsay WK, McCain WG: Effect of beta-aminopropionitrile on adhesions following digital flexor tendon repair in chickens. *Surg Forum* 1970;21:509–511.

141. Kulick MI, Brazlow R, Smith S, et al: Injectable ibuprofen: Preliminary evaluation of its ability to decrease peritendinous adhesions. *Ann Plast Surg* 1984;13:459–467.

142. Kulick MI, Smith HS, Hadler K: Oral ibuprofen: Evaluation of its effect on peritendinous adhesions and the breaking strength of a tenorrhaphy. *J Hand Surg* 1986;11A:110–120.

143. Peacock EE Jr, Madden JW: Some studies on the effects of β-aminopropionitrile in patients with injured flexor tendons. *Surgery* 1969;66:215–223.

144. Douglas LG, Jackson SH, Lindsay WK: The effects of dexamethasone, norethandrolone, promethazine and a tension-relieving procedure on collagen synthesis in healing flexor tendons as estimated by tritiated proline uptake studies. *Can J Surg* 1967;10:36–46.

145. St. Onge R, Weiss C, Denlinger JL, et al: A preliminary assessment of Na-hyaluronate injection into "no man's land" for primary flexor tendon repair. *Clin Orthop* 1980;146:269–275.

146. Weiss C, Balazs EA, St. Onge R, et al: Clinical studies of the intra-articular injection of Healon® (sodium hyaluronate) in the treatment of osteoarthritis of human knees. *Semin Arthritis Rheum* 1981;11(suppl 1):143–144.

147. Davison PF, Brennan M: The organization of cross-linking in collagen fibrils. *Connect Tissue Res* 1983;11:135–151.

148. Gonzalez RI: Experimental use of Teflon in tendon surgery. *Plast Reconstr Surg* 1959;23:535–539.

149. Grau HR: The artificial tendon: An experimental study. *Plast Reconstr Surg* 1958;22:562–566.

150. Hunter JM: Artificial tendons: Early development and application. *Am J Surg* 1965;109:325–338.

151. Sarkin TL: The plastic replacement of severed flexor tendons of the fingers. *Br J Surg* 1956;44:232–240.

152. Bassett AL, Carroll RE: Formation of tendon sheath by silicone-rod implants, abstract. *J Bone Joint Surg* 1963;45A:884–885.

153. Hunter JM, Salisbury RE: Flexor-tendon reconstruction in severely damaged hands: A two-stage procedure using a silicone-Dacron reinforced gliding prosthesis prior to tendon grafting. *J Bone Joint Surg* 1971;53A:829.858.

Section Two
Ligament

Group Leader

Cyril Frank, MD

Group Members

Thomas Andriacchi, PhD
Richard Brand, MD
Laurence Dahners, MD
Kenneth DeHaven, MD
Barry Oakes, MD
Savio Woo, PhD

Group Participants

Steven Arnoczky, DVM
Mark Bolander, MD
David Butler, PhD
Reginald Cooper, MD
Thomas Dameron, Jr., MD
David Eyre, PhD
Gerald Finerman, MD
Victor Goldberg, MD
Stephen Gordon, PhD
Edward Grood, PhD
Peter Torzilli, PhD

Synopsis

Skeletal ligaments are highly specialized, dynamic, dense connective tissues that connect bones. They serve a passive mechanical function in stabilizing joints and in guiding joint motion. An increasing body of evidence also suggests that ligaments have an important neurosensory role, supplying important proprioceptive information, or serving as important transducers of dynamic information to muscles.

Changing joint position alters the passive stability of joints because of complex interactions between ligaments and the joints they protect. The knee joint, in particular, is known to have a complicated kinematic pattern controlled, in part, by its ligaments. Recent information suggests a connection between knee kinematics and muscle efficiency through the changing of lever arms during flexion. Thus, an alteration in kinematics may be able to influence the dynamics of the joints as well as their passive characteristics.

The mechanical behavior of ligaments is similar to that of other nonlinear viscoelastic soft tissues, but with adaptations that allow joints to be flexible, yet stable. Specialized testing systems and protocols have been developed to examine the properties of certain model ligaments, but many other ligaments have yet to be analyzed. Factors known to have a major influence on the structural and mechanical properties of ligaments and their insertions are age and stress. Ligaments are somewhat strain-rate sensitive, becoming slightly stronger and stiffer at higher loading rates. Temperature is important to their mechanical behavior, but careful freezing (without drying) before mechanical testing does not seem to affect their properties significantly.

The structure of ligaments is complex. Grossly, ligaments are dense, white, fibrillar, usually easily distinguished, avascular structures attached to very specific sites on bones at either end. Functional subdivisions of ligaments have been identified and are known to act in different joint positions. The microscopic anatomy of ligaments shows them to be relatively hypocellular, containing interconnected fibroblastic cells in their midsubstance, with more chondroid cells near their insertions. Histochemical stains reveal that most ligaments are composed of collagen, with small proportions of glycosaminoglycans and elastin. Antibody stains have also revealed the presence of some actin and fibronectin. Not all ligaments are the same histologically and they differ from tendons as well. Polarized light microscopy demonstrates an important structural feature of ligaments involving a periodic wave of collagen known as its "crimp." Crimp varies anatomically within any ligament or tendon. Although its functional significance remains somewhat debatable, crimp is thought to contribute to the nonlinear mechanical properties of ligament substance.

Ultrastructural studies of ligaments have confirmed that the majority of ligament substance consists of fibrillar collagen, with smaller quantities of the substances noted above. Cells appear to be metabolically active and to have long interconnected processes. Small proteoglycans are seen in specific areas around cells and their processes, forming a latticework throughout the ligament substance. Developmental and exercise studies suggest that new collagen fibrils have small cross-sectional diameters, possibly contributing to the relative weakness and compliance of newly formed ligament tissues. Similarly, small collagen fibrils have also been seen in human tendon grafts, years after they have replaced damaged ligaments.

Individual ligaments grow and develop in specialized ways. Some growth is certainly interstitial and probably reflects the pericellular addition of collagen and other matrix components. An active, vascular outer layer or "periligament" has also been identified in the rabbit medial collateral ligament and possibly contributes to its growth. Ligament growth may not be symmetric and may be a function of many factors, such as position relative to growth plates and ligament strain.

Biochemical analysis of ligaments confirms the above observations that ligament consists of water, types I and III collagen, several proteoglycans, elastin, and a variety of other substances. Ligament collagen is partly stabilized by covalent intramolecular and inter-molecular cross-links, and is also probably stabilized by the stable, covalent trihydrox-ypyridinium. Very little is known about the functional significance of these cross-links, or their interactions with other matrix components. A great deal of work is needed to define better the molecular nature of ligaments and to correlate this information with what is known about their mechanical properties.

Ligament healing is a long, complex process involving the same basic phases as wound healing and is apparently subject to the same external influences. Both local and systemic factors can influence the process, particularly motion or immobilization. A host of other, potentially alterable variables (mediators of inflammation, growth factors) offer significant promise for optimizing the ligament healing process. Biologic and artificial replacements of ligaments have had only limited study with both favorable and unfavorable results. Ligament replacement remains a controversial topic.

Clinically, ligament injuries have proven equally difficult to evaluate. Some ligaments do not seem to heal as well as others, with combined injuries involving the anterior cruciate ligament in "high-risk" patients doing particularly poorly. A variety of ligament reconstructive and replacement procedures are being used clinically but long-term quantitative data are very limited.

A great deal of study is still required in all areas of ligament research, including method development, model development and validation, and a variety of biologic studies. A more detailed analysis of ligament structure and function and more rigorous clinical studies are necessary to provide a rational approach to the clinical treatment of all ligament injuries.

Chapter 2
Normal Ligament: Structure, Function, and Composition

Cyril Frank, MD
Savio Woo, PhD
Thomas Andriacchi, PhD
Richard Brand, MD
Barry Oakes, MD
Laurence Dahners, MD
Kenneth DeHaven, MD
Jack Lewis, PhD
Paul Sabiston, MD

Chapter Outline

Overview of Ligament Structure and Function

Ligaments are short bands of tough but flexible fibrous connective tissue that bind the bones of the body together and support the organs in place.[1] We will discuss only what is known about the ligaments that connect bone to bone, stabilizing and guiding joints; these are the so-called skeletal ligaments.

The gross anatomy and function of skeletal ligaments have been studied for many centuries, but only within the past 20 years have they received detailed scientific attention. At first they were thought to be simple and biologically inert tissues, but it has become increasingly obvious that this is not the case. Clinical recognition of the frequency of ligament injuries and growing awareness of their variable healing potentials have intensified investigations of their nature and have indicated clearly the need for a better understanding of normal ligament structure and function.[2]

The first description of ligament structure appeared in the Smith Papyrus (3000–1700 BC)[3] and the first description of ligament abnormality resulting from joint dislocations can be attributed to Hippocrates (400 BC). The first skeletal ligament to be defined anatomically was the ligamentum teres of the hip joint. Hegator of Alexandria (100 BC) was probably the first, therefore, to distinguish ligaments (*ligare* = Latin, to bind) that connect bone to bone from tendons (*tendere* = Latin, to stretch; *tenon* = Greek, a sinew) that connect muscles to bones. All ligaments and tendons were thus classified on the basis of these differences. Despite this clear anatomic separation, however, subsequent gross and histologic classifications of ligaments, tendons, and fascia as "dense connective tissues" perpetuated the long-standing concept that they were identical. This concept also probably fostered the thought that these tissues might be surgically interchangeable and probably contributed to many of the popular "biologic substitution" operations (for example, tendons or fascia being substituted for ligaments). Although it is clear that these dense connective tissues have certain similarities, there is also evidence to suggest the existence of important differences.[4]

The several hundred skeletal ligaments in the human body have been named on the basis of gross, structural, and functional features.[5] Most are identified by their points of bone attachment (for example, coracoacromial), their shape (deltoid), their gross functions ("capsular"), their relationships to a joint (collateral), or their relationships to each other (cruciates). Functional subdivisions of some of these ligaments have also been recognized,[6] probably at least doubling the number of functionally discrete ligaments. Only a few of these ligaments have been studied scientifically to date. We must, therefore, use those structures that have been quantified as models, with the clear understanding that generalizations may not be valid and that a great deal of work still needs to be done in defining normal ligament structure and biology. More thorough overviews of these perspectives are available elsewhere.[7-11] In this chapter, we will review what is known about the function, structure, and biochemistry of the normal ligament.

Normal Ligament Function

Mechanical Functions of Ligaments and Joint Kinematics

In order to function, the musculoskeletal system must have articulating joints that are both stable and free to move. These two mutually exclusive conditions are reconciled mechanically through the interaction between the passive soft tissues surrounding the joints, primarily the ligaments, and the articulating surfaces. Usually, changing joint position changes the joint's passive stability.[12-14]

Knee Kinematics   The knee can move in six independent directions (three translational and three rotational). Thus, kinematically, it is a joint with six degrees of freedom. Knee motion must be analyzed carefully because of its multiphasic movement.[13,15,16] The predominant kinematic characteristics of the knee are determined by the curvature of the femoral and tibial articulating surfaces, as well as by the orientation of the four major ligaments of the knee.[17] The distal portion of the lateral femoral condyle has a larger radius of curvature than does the distal portion of the medial femoral condyle, causing the femur to roll posteriorly a greater distance on the lateral tibial plateau than on the medial tibial plateau during the first 10 to 15 degrees of flexion (Fig. 2-1). Thus, during flexion, the region of contact between the tibia and femur moves posteriorly a greater distance on the lateral tibial plateau than on the medial tibial plateau.[18-20] During flexion, this difference in posterior movement is associated with a coupled passive internal axial rotation of the tibia with respect to the femur. The reverse movement to full extension produces a coupled external tibial rotation sometimes described as the screw-home movement.[21] With flexion, the posterior movement of the tibiofemoral contact also changes the lever arms of the muscles crossing the knee joint and can have a substantial influence on function. Thus, knee kinematics involves not only passive ligament stability, but dynamic muscle efficiency as well.[22]

The cruciate ligaments significantly influence the movements of the knee joint. The length and degree of tension developed in the cruciate ligaments change during flexion[12,23-25] and extension. The passive action of the cruciate ligaments is responsible, in part, for the sliding movement that occurs when the knee flexes beyond 20 degrees. As the knee continues to roll back, a portion of the anterior cruciate ligament (ACL) becomes taut and prevents the pure rolling of the femur from continuing, thus initiating sliding between the femur and the tibia at approximately 20 degrees. During extension, the posterior cruciate ligament (PCL) is responsible for the sliding movement of the condyle posteriorly.[17]

The collateral ligaments, while playing a primary role in stabilizing the knee, also influence the kinematics of the joint. Thus, the ligaments can provide stability while permitting the normal instantaneous movement of the knee. This observation is also consistent with three-dimensional studies[26-30] of the instantaneous axis of motion that runs

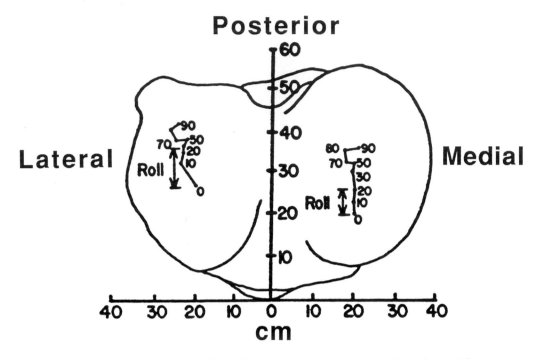

Fig. 2–1 *The tibial-femoral contact moves posteriorly with knee flexion. The contact on the lateral side moves posteriorly more during flexion (0 to 20 degrees) than does the medial side because the lateral femoral condyle is rolling on a larger radius than the medial condyle. Beyond 20 degrees, sliding begins on both condyles.*

obliquely with respect to the frontal plane and moves posteriorly with flexion.

Passive Knee Joint Stability   The passive stability of the knee joint has been described in terms of its resistance to anteroposterior drawer movements, varus-valgus bending movements, and internal-external rotation movements. The knee has nearly negligible resistance to movement in the flexion-extension direction within its normal range of movement. However, normal functional demands on the joint require varying degrees of passive stability in directions other than flexion, depending on the angle of knee flexion. The stability of the joint can be described in terms of stiffness (unit load divided by unit displacement). The stability (or laxity) of the knee has been described by a number of investigators.[14,31–37] Some common observations from these studies include the following:

(1) The stiffness of the joint depends on the magnitude of the force or moment applied. At low forces (moments), the joint has a relatively low stiffness (higher laxity); as the force (moment) is increased, the stiffness increases. The initial lower stiffness has been described as initial stiffness and the final stiffness as "terminal" stiffness (Fig. 2–2).[34]

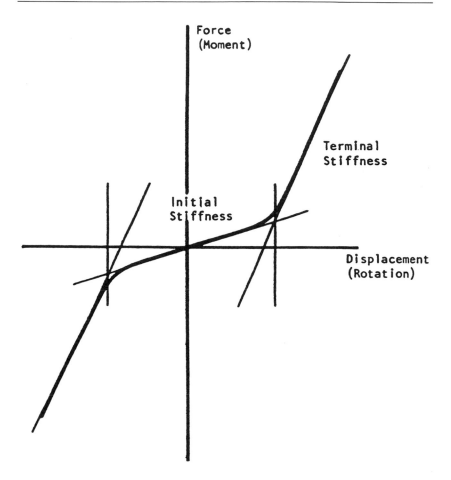

Fig. 2–2 *Typical characteristics of the load-displacement relationship in the knee. The slope of the curves represents the stiffness. Under low loads or displacement, the slope (stiffness) is considerably less than under high loads.*

(2) Stiffness (stability) is greatest at full extension and reduces rapidly at approximately 20 degrees of flexion (Fig. 2–3). At this position of knee flexion, the joint kinematics changes from pure rolling to sliding, and the cruciate ligaments must have some tension in order to maintain the sliding movement.

(3) Varus-valgus laxity is typically ten times greater than rotational stiffness. Varus stability is important because of the large adduction moment applied to the knee during normal walking. A reduction in varus stiffness results in a varus angular movement during function, causing lift-off of the lateral femoral condyle and transmitting all of the load across the knee through the medial femoral condyle.

These general observations are common to most reports describing the passive characteristics of the knee. However, studies differ in the

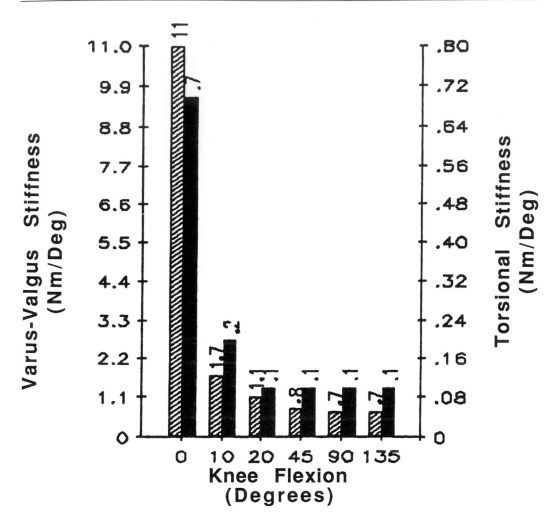

Fig. 2–3 *The varus-valgus and torsional stiffness declines rapidly as the knee flexes from 0 to 20 degrees. The varus-valgus stiffness (diagonally lined columns) is substantially higher than the torsional stiffness (solid columns). (Adapted from Markolf and associates.[34])*

quantitative descriptions of these characteristics. Some differences can be attributed to the testing method and to the degree of constraint applied during the testing procedure[12,32]; others are based on normal variations among specimens. The precise quantitative description of the passive stability of the knee is nevertheless important in explaining the intricate relationships between stability and freedom of movement as the knee moves through its range of motion. An important additional step is the extrapolation of these data to the functional behavior of the knee.

Dynamic Loads There is little information on the dynamic loads sustained by the ACL during function. Data are derived indirectly from human subjects by using mathematical models to relate external forces acting on the limb to internal forces sustained by muscle and soft tissue. Maximum forces in the ligaments range from 25% to 60% of body weight, depending on the activity.[38]

Muscular contraction produces substantial ligamentous forces that, in most cases, are not accounted for in mathematical models. For example, the contraction of the quadriceps muscle applies an anterior force to the tibia between 0 and 60 degrees of knee flexion.[18,19] Thus, an isolated contraction of the quadriceps produces an additional force in the ACL.

Similarly, the contraction of other muscle groups substantially alters tension in other passive, soft-tissue structures around the joint and alters the overall stability of the joint.[19] Muscle contractions can increase knee stiffness an average of two to four times over the passive stiffness of the joint.[39] This dynamic stabilization[40] plays an important role in protecting the joint during functional activities. However, in most cases muscular contraction cannot occur rapidly enough to prevent a traumatic load from damaging the ligaments.

In summary, the mechanics of the normal knee must provide for the delicate interchange between freedom of movement and stability during a range of normal functional activities. When the knee joint is placed in positions of greater passive stability, the ligaments are more vulnerable to stretching and rupture than in positions where the joint has greater flexibility. Muscular contraction can substantially increase the dynamic stability of the joint. At present, there are still few quantitative data on the role of the muscles in producing the overall stability of the joint. Recent information has suggested a connection between knee kinematics and muscle efficiency through the changing of lever arms during flexion. Thus, altered kinematics may be able to influence the dynamic and the passive characteristics of the joints.

A Neurosensory Hypothesis of Ligament Function

As noted, ligaments are generally viewed as passive structural elements, the primary purpose of which is to hold the bones together and to guide (restrict) the relative motions of opposed articular surfaces.[31,34,41-56] Substantial evidence exists, however, to suggest that ligaments and other periarticular structures (capsule, menisci, retinacula) serve another major role: as signal sources for reflex systems of the locomotor apparatus, or neuromuscular system. If this role is important, if it is disrupted during injury, producing symptoms and dysfunction, then our current treatment modalities have done little or nothing to restore that lost function.

What is the specific evidence for a neurosensory role of ligaments? Demonstrations of mechanoreceptors in pericapsular tissues began shortly after the development in the middle 1800s of the compound microscope and appropriate stains by Krause,[51,52] Rauber,[53] and others. Many subsequent studies unequivocally documented mechanorecep-

tomuscular reflexes," but they used relaxed supine subjects, well-known inhibitory and facilitatory characteristics,[101] which may explain their results. Thus, most investigators conclude that (1) there are mechanoreceptors in ligament and capsule and they are different, (2) pulling on ligaments and probing capsule, in some circumstances, can produce muscle activity, and (3) loss of joint sensory function can impair motor performance in animals. The mediating pathways, mechanisms of function, and importance to human function, however, are still unknown.

Other evidence has more recently surfaced to support a neurosensory role in ligament function. Two groups recently reported altered electromyographic patterns after ACL disruption. Shiavi and associates[102] noted "moderate to major changes in the activity in a majority of muscles" (gastrocnemius, vastus lateralis, biceps femoris, semitendinosus), particularly during cutting activity. Tibone and associates[103] found few differences in electromyographic activity in the vastus medialis obliquus, medial hamstring, biceps femoris longus, and gastrocnemius medialis during level free walking, but did find differences during running; cutting was not studied. Carlsöö and Nordstrand,[104] on the other hand, demonstrated no qualitative differences between the electromyographic patterns of five normal subjects and five patients with cruciate ligament injury; they did not use the opposite limb for an internal control and studied only one dynamic activity, level walking.

Additionally, two groups recently reported minimal loading of knee ligaments in animals during certain activities, and lower levels than during simulated clinical stress testing.[105,106] Lewis and Shybut,[105] in a simulated clinical stress test, noted that the recorded loads in the ligament were less than 18 N (compared with presumed tendon loads in major canine muscles of 196 N, a load in the range of one animal's body weight) during functional activities (walking, trotting, being dropped from a height). Brand and associates[106] reported similar results, that is, functional loads in the ligaments less than one half of those recorded during a clinical stress test. They estimated functional loads of as much as 56 N, but in somewhat larger dogs the loads approached body weight, or approximately 292 N. These observations do not directly support the existence of a neurosensory role of ligaments, but they do suggest that during normal activities the structural role of the ligament may be minimal, and that a joint's motion is controlled largely through its geometry and muscular activity.

Finally, Brand[2] reviewed some indirect evidence consistent with a neurosensory role of ligaments. These included (1) variability of reported structural roles for given ligaments; (2) general (albeit variable) functional recovery following knee-ligament injuries without surgery; (3) the infrequency of degenerative changes sufficient to require total knee replacement following ligament injuries; (4) the lack of clear documentation from controlled, randomized studies of the long-term efficacy of most knee-ligament surgery. To this list might be added the generally similar results of surgical and nonsurgical treatment of certain knee-ligament injuries.[107–110] These results are consistent with a neurosensory role for ligaments because neither surgical nor nonsurgical methods restore neural structures or functions, and both treatment methods might be expected to result in roughly similar "natural" restoration of neural function. Brand[2] emphasized that he was not implying that surgical restoration of ligaments structure is never indicated; abundant evidence suggests that selected patients benefit substantially from knee-ligament reconstructions. It is clear that ligaments have mechanoreceptors that take part in motor control. The function and importance of this role are, however, not clear.

These ideas, although well documented, have not been incorporated into the broad concepts of treatment for the following reasons: (1) a neurosensory role of ligaments is less obvious and less intuitively evident than a structural role; (2) it is not clear which sorts of receptors contribute to which specific functions (proprioception is not a single function)[93]; (3) it is even less obvious how nonsurgical and surgical interventions may facilitate restoration of those elements; (4) there has been no substantial discourse among investigators of the potential neurosensory role of ligaments and clinicians.

It is not difficult to envision a variety of experimental and clinical studies that could test the neurosensory role of ligaments, potentially altering the way we treat ligament injuries. Such methods might include rehabilitation programs designed to improve proprioception,[111,112] although we do not currently have an appropriate conceptual framework to develop such techniques. Because cutaneous receptors facilitate signals from muscular and possibly joint receptors,[95] we might design more effective and less encumbering braces to take advantage of this physiologic feat. Even more speculative at this time is the possible introduction of methods to facilitate nerve ingrowth, a concept with some experimental support.[113] It is even conceivable, although again purely speculative, that future implants of new (possibly biodegradable) materials may facilitate nerve ingrowth into new scar tissue more than other materials.

Structural and Mechanical Properties of Normal Ligaments

Viidik[114] asserted that the geometric configuration of the constituent collagen fibers and the interaction of collagen fibers with the noncollagenous tissue component are the basis of the mechanical behavior of soft tissues. Readers are encouraged to read this classic work, as well as the comprehensive list of references included. In this section, we will discuss the current refinements and improvements of techniques for studying the biomechanical behavior of normal ligaments in uniaxial tension.

Quasistatic Tensile Properties   A typical nonlinear load-deformation curve (called structural properties) from uniaxial tensile testing of a bone MCL-bone complex is shown in Figure 2–4. The factors contributing to these structural properties are (1) the mechanical properties of ligament substance (ligament being evaluated as a material), which depend on the organization and/or orientation of their collagen

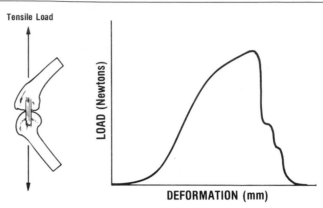

Fig. 2–4 *A typical load-deformation curve for bone-ligament complex representing its structural properties.*

fibers as well as the percentages of various constituent materials; (2) the geometry of the ligament tissues, that is, cross-sectional area, length, and shape; and (3) the complexity of the bone and ligament junctions. Despite the fact that ligaments have well-organized collagen fibers, there exists a small amount of longitudinal recruitment of collagen fibers that contribute to nonlinear stress-strain behaviors.[115] Thus, the slope of the load-deformation curve increases with stretch.

The reported mechanical properties of ligaments have been widely divergent. For example, ultimate strains can range from 12% to more than 50%.[116-121] These differences can be attributed in part to factors such as age, species, and type of specimens (obtained from different anatomic locations). However, experimental techniques such as (1) cross-sectional area measurements (for stress calculations), (2) measurement of original length for strain calculations, (3) clamping of the test specimens, (4) length-width (aspect) ratio for uniform stress, (5) measurements of tissue deformation (for strain determinations), and (6) the test environment also contributed to this discrepancy.

To illustrate, ligaments are usually too short (poor aspect ratio) and testing these tissues in their isolated state probably yields questionable results.[119] As a result, the bone-ligament-bone preparation is preferred for tensile tests.[118-120] The nonuniform properties of the ligament substance and its insertions to bone, as well as the difficulties in defining the original length of the ligament, are, however, real technical problems. Details of newly developed methods of overcoming some of these problems are available elsewhere.[11,120-122]

Special clamps have been designed for specific bone-ligament-bone composites. The bones should be affixed to clamps so that the tensile load can be applied directly along the longitudinal axis of the ligament with no bending (Fig. 2–5). An experimental apparatus, such as that shown in Figure 2–6, is one of the most convenient and accurately designed to measure the structural and mechanical properties of lig-

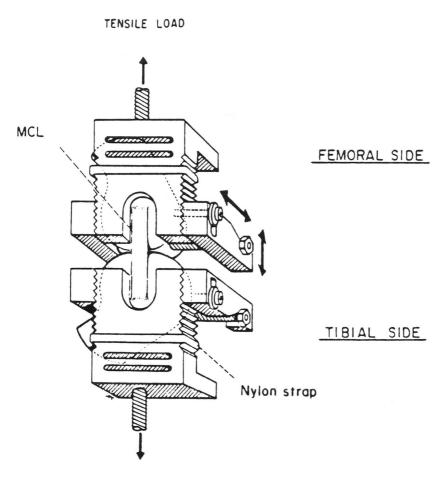

TENSILE LOAD

MCL

FEMORAL SIDE

TIBIAL SIDE

Nylon strap

Fig. 2–5 *The L-shaped clamps designed for tensile testing of a bone-ligament complex. Note that the clamp should be positioned so that the tensile load can be applied along the longitudinal direction of the ligament.*

aments. Other investigators have used special devices such as the ω-shaped strain gauge[123] and magnetic field displacement transducers,[124] designed to measure the ligament force and strain during tensile loading. However, these systems must make direct contact with the ligament and this may produce undesired artifacts. Using noncontact methods, together with the video dimensional analyzer (VDA) system to measure the surface strains of the ligament substance, has proved to be accurate and convenient (Fig. 2–6). The frequency response of the VDA system can be as high as 120 Hz and errors in linearity and accuracy are less than 0.5%.[120,122]

With improved experimental techniques, a few fundamental studies of the properties of the ligament have been performed and the discrepancies in previously published results have been clarified.

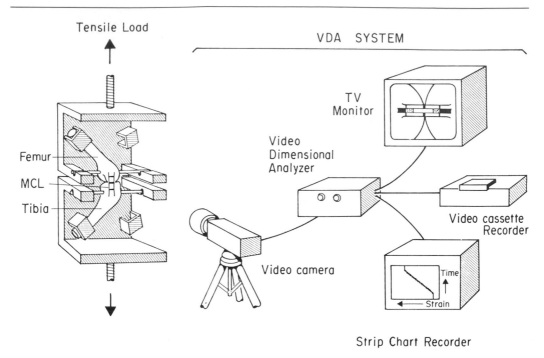

Fig. 2–6 *The experimental apparatus designed to measure the tensile characteristics of the ligament-bone complex as well as the ligament substance. The VDA system is used to determine ligament strains.*

The Effects on Ligaments of Storage by Freezing Several studies comparing the properties of ligaments after storage have been conducted, but conflicting results have been reported.[35,125,126] Recently, the mechanical properties of the rabbit MCL substance and the structural properties of the bone-ligament unit were compared in fresh preparations and preparations carefully stored for three month at −20 C (without possible drying).[127] In this study, Woo and associates found that freezing significantly decreased the area of hysteresis during the first few cycles of loading-unloading but did not appear to affect other mechanical and structural properties of the ligament and its insertion sites. Thus, the area of hysteresis may be a sensitive indicator of minor changes in the bone-ligament complex secondary to storage by freezing.

Temperature-Dependent Behavior An important consideration when testing ligaments is the environmental temperature. A canine femur-MCL-tibia preparation was clamped and submerged in a 0.9% saline bath equipped with a heating and cooling system.[128] The solution temperature was monitored to be within 0.5 C. In one study, temperatures ranging from 22 C to 37 C were used. Each specimen was sequentially tested at 22, 22, 27, 32, and 22 C. The specimen was cycled to 2-mm extension (approximately 3% ligament strain) for ten cycles and then unloaded for one hour at 22 C to allow recovery from its viscoelastic effects, after which the cycle test was repeated at a different temperature. In a second study, the temperatures ranged from 2 C to 22 C. Each specimen was sequentially tested at 22, 22, 2, 6, 14, and 22 C under the same protocol.

The ligament was found to have significant temperature-dependent properties.[128] The measured area of hysteresis decreased with increasing temperature; the cyclic load relaxation behavior leveled off to a higher value at lower temperatures; and the tensile load at a predetermined strain level had an inversely proportional relationship with respect to temperature.

Effects of Age on Tensile Properties   The effects of age (maturation) on ligaments and their insertions have been previously demonstrated.[35,129] Age-related changes in the mechanical properties of the rabbit MCL substance and the bone-ligament complex were recently evaluated by Woo and associates.[130] The animals were 1½ and 4 to 5 months old (open epiphyses by radiologic and gross examination) and 6 to 7 and 12 to 15 months old (closed epiphyses). The structural properties of the bone-ligament-bone complex increased dramatically between 1½ and 6 months of age, at which time the rate of increase diminished. The mechanical properties of the ligament substance matured relatively early, that is, by 4 to 5 months of age the stress-strain curves were similar to those of the adults.

In the rabbits with open epiphyses, all preparations failed by tibial avulsion, whereas in those with closed epiphyses, most failed by ligament substance tear. Thus, it seems that the maturation of the animals has a significant effect on the relative strength of the various components along the bone-ligament complex (Fig. 2–7).

Quasilinear Viscoelastic Properties   Advanced theories of viscoelasticity such as the quasilinear viscoelastic (QLV) theory of Fung[131] can be used to describe time- and history-dependent properties of ligaments. The theory assumes that the stress relaxation function can be written as:

$$\sigma[\epsilon(t);t] = G(t) * \sigma^e(\epsilon)$$

(1)

Where $\sigma^e(\epsilon)$ is the nonlinear "elastic response" (function of ϵ only), and $G(t) = \sigma(t)/\sigma(0)$ is the reduced relaxation function (function of t only). The stress at time t, $\sigma(t)$, is thus the convolution integral of the reduced relaxation function and the rate of elastic stress, or:

$$\sigma(t) = \int_0^t G(t - \tau) \frac{\partial \sigma^e(\epsilon)}{\partial \epsilon} \frac{\partial \epsilon}{\partial \tau} d\tau.$$

(2)

Therefore, when G(t), $\sigma^e(\epsilon)$, and the strain history, $\epsilon(t)$, are known, the

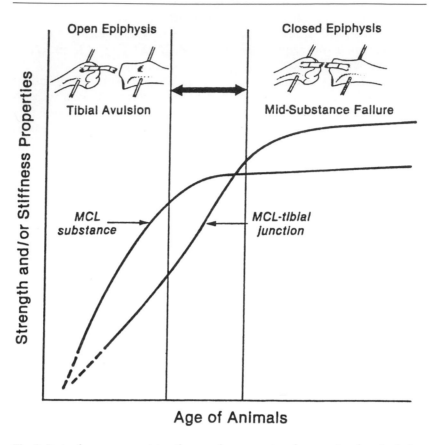

Fig. 2–7 *A schema summarizing the asynchronous rates of maturation from both the structural characteristics of the bone-ligament complex and the mechanical properties of the ligament substance.*

time- and history-dependent stress, $\sigma(t)$, is completely described by equation 2.

Stress and relaxation tests were performed on canine MCL by rapidly stretching it to 2.5% strain and then allowing it to stress relax for as long as 16 hours. These data, together with tensile tests of canine bone-ligament complex at a constant strain rate, the reduced relaxation function $G(t)$, and the elastic response $\sigma^e(\epsilon)$ of the MCL, were determined by means of a nonlinear curve-fitting procedure.[132]

These results were further confirmed by an independent cyclic testing experiment. If $G(t)$, $\sigma^e(\epsilon)$, and $\delta\epsilon/\delta\tau$ are known, the peak and valley stresses, $\epsilon(t)$, can be calculated by using equation 2. As can be seen from Figure 2–8, the stresses calculated by the QLV theory match well with those obtained by the experiments. Thus, the time- and history-dependent viscoelastic properties of canine MCL have been completely described.[132]

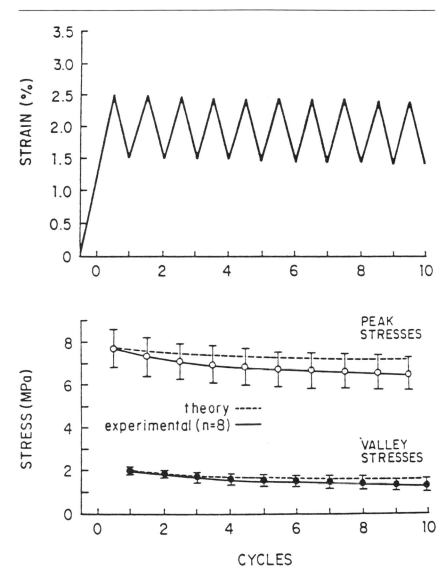

Fig. 2–8 *Curves show the experimental peak-and-valley stresses and those predicted by the quasilinear viscoelastic theory for a canine MCL subjected to cyclic loading and unloading.*

Ligament Failure Modes   The functional support provided by a ligament is as a bone-ligament-bone unit. Clinically, failure in adults by substance tear is more common than failure by avulsion. However, in the laboratory, it is difficult to produce ligament substance injuries. This discrepancy has stimulated interest in investigating factors, such as species and ligament type, axis of loading, animal activity levels, strain-rate, and age, that may influence failure modes.

(1) The load can be applied along the longitudinal axis of the ligament or along the longitudinal axis of the tibia.[133] In the rabbit and porcine ACL, a tensile load directed along the ligament axis with knee flexion angle between 0 and 90 degrees had no significant effect on the ultimate load values and most failures were by bony avulsion. However, directing the tensile forces along the tibial axis produced progressively inferior structural properties as knee flexion increased, and failures occurred primarily by progressive fiber failure of the ligament substance.[133]

(2) Laboratory findings have demonstrated that immobilization of a joint for clinically relevant periods leads to a precipitous decline in ligament-bone junction strength, particularly in those collateral ligaments that insert via the periosteum.[120,134] Most failures occur by bony avulsion resulting from subperiosteal resorption.

(3) Considerable attention has been given to the effects of rate of stretch on the failure mode of a bone-ligament complex. Some investigators reasoned that avulsion failures result primarily from slow extension rates.[135-137] Recent investigations, however, showed that the effects of age on the failure modes can be much more significant than strain rate. In two groups of rabbits—one with open epiphyses (3½ months old) and the other with closed epiphyses (8½ months old)—the bone-ligament complex was subjected to tensile tests at five different extension rates ranging from 0.01 to 100 mm/sec (corresponding strain rates for ligament substance, 0.1%/sec to 200%/sec). For high strain rates, a high-speed video recording system was used (2,000 frames/sec compared with the normal speed of 60 frames/sec).[138]

The ultimate load and the energy absorbed at failure of the bone-ligament complex both increased with increasing strain rate. Similar changes were also observed for the stress-strain curves; however, the differences between the slowest strain rate and the fastest were not large (Fig. 2–9). Conversely, the differences in these properties between the two age groups were very large. Failure modes were also found to be independent of strain rate but dependent on age. In animals with open epiphyses, all failure was by tibial avulsion, whereas in animals with closed epiphyses all failure was by ligament disruption.[139]

Normal Ligament Composition

Ligament Morphology

Gross Anatomy Ligaments have a number of gross forms. In the hip joint, ligaments are primarily capsular (broad and flat); they extend from the edge of the acetabulum to the base of the femoral neck in a circumferential fashion so that they can successfully restrain the joint by becoming taut at the limits of flexion, extension, abduction, and adduction. The fibers of the ligament are spirally wound so that they can successfully restrain the joint from excessive internal and external rotation. In other joints the situation becomes somewhat more com-

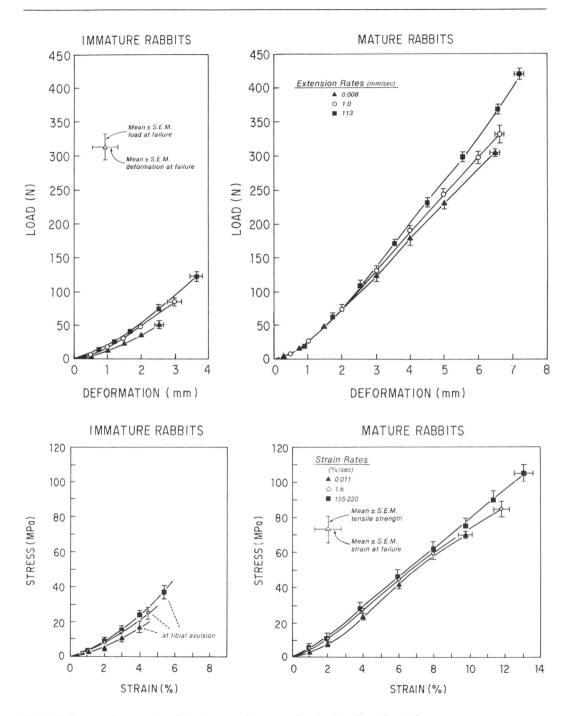

Fig. 2–9 *Structural properties of the ligament-bone complex **(top)** and mechanical properties of ligament substance **(bottom)** in skeletally immature and mature rabbits tested at different extension rates.*

birth. They probably represent a cellular phagocytic mechanism of rapid collagen remodeling similar to that seen in the periodontal ligament and Achilles tendon scar tissue.[161,162] Vacuoles adjacent to the Golgi apparatus contained thin filaments about 300 nm long, suggesting that these filaments are procollagen molecules aggregated in secretory vacuoles similar to those described in the odontoblast.[163,164] The cells may "excavate" a space for themselves in the rapidly deposited collagen matrix by this method of direct phagocytosis. In the adult-ligament "chondroid" cells, synthesizing organelles were decreased but there was a marked increase in intermediate filaments 8 to 10 nm in diameter similar to that described for aging articular chondrocytes.[165]

The 14-day fetal rat ACL contained two distinct populations of fibrils, one group with diameters of 10 to 15 nm and a second group of fibrils 10 to 12 nm in diameter, collected into discrete longitudinal bundles and close to the fibroblast cell membrane. These fibrils exhibited an irregular beaded periodicity and are probably the anionic structural glycoprotein precursor framework for future elastic fibers.[166] At 18 days in the fetal rat, the collagen crimp was evident across the whole width of the developing ACL. The appearance of this crimp before the deposition of elastin indicates that crimp may be intrinsic to the collagen matrix or may be the result of stress created by a fibroblast-matrix interaction.[167,168]

Collagen fibrils were measured at various ages of the rat. When the mean diameter and the range of fibrils from the largest to the smallest for each interval were plotted against age, it was shown that the fibrils begin to plateau at about seven weeks after birth (Fig. 2–13). When the force required to rupture the ACL in the rat as a function of age was plotted,[169] it was apparent that a strong correlation probably exists between the size of the collagen fibrils and the ultimate load of the ACL, as was suggested by Parry and associates.[170]

Elastic fiber growth and maturation in the rat ACL are disappointing compared with developing tendons.[171] A "true" cross-linked elastin as an amorphous electron-lucent material is never deposited in relation to the fibrils (12 nm in diameter) (Fig. 2–14).

Using ruthenium red, Luft[172] visualized aggregates of proteoglycans and glycoproteins in the matrix of the developing rat ACL and compared them with those in similar developing tendons. He observed a close regular association of proteoglycan granules with ACL collagen, as well as fine filaments linking the granules with themselves and collagen fibrils. Similar observations have been made in other tissues. Hascall[173] determined that the granules observed ultrastructurally represent proteoglycan monomer and that the fine filaments linking them correspond to the linear hyaluronate thread onto which the monomers are polymerized (Figs. 2–15 and 2–16).[174] Larger branching filaments found in the tendons were not seen in the ACL. Scott and Orford[175] have shown that dermatan sulfate, which is the most abundant glycosaminoglycan of tendon, is associated closely with the D period on

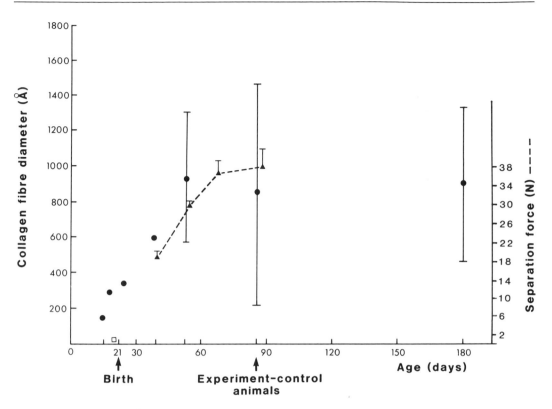

Fig. 2-13 *Collagen fiber diameter in the rat ACL plotted against age. Also shown is the separation force for the rat ACL. (Reproduced with permission from Larson N, Parker AW: Physical activity and its influence on the strength and elastic stiffness of knee ligaments, in Howell ML, Parker AW (eds):* Sports Medicine: Medical and Scientific Aspects of Elitism in Sport. *Brisbane, Australian Sports Medicine Federation, 1982, vol 8, pp 63–73.)*

the outside of the collagen fibril. How these glycosaminoglycans influence collagen fibril growth is still not well understood.

Collagen Fibril Populations in Intensively Exercised Young Rat Cruciate Ligaments On the basis of previous observations that exercise can influence ligament-bone junction strength[176-178] and collagen ultrastructure[179] the following studies were conducted.[180,181] Young rats were subjected to an intensive four-week exercise program with alternate daily periods of running and swimming. Caged rats of the same age were used as controls. The exercise schedule was designed to produce a graded increase in exercise load, finishing with a treadmill speed of 26 m/min on a gradient incline of 10%. After the rats were killed, the ACL, PCL, the collateral ligaments, and the patellar tendon were removed for biochemical or microscopic study. Thin sections were cut at a 90-degree angle to the longitudinal axis of the structure and three random micrographs of collagen fibrils were manually measured with

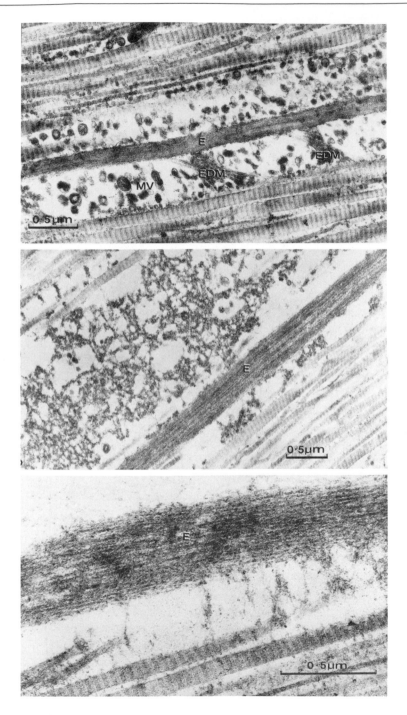

Fig. 2–14 *Variations in the ultrastructure of elastic fibers in the adult rat ACL.* **Top:** *A moderately electron-dense elastic fiber (E) surrounded by matrix vesicles (MV) and linked to the collagen fibers by electron-dense material (EDM) (× 40,200).* **Center:** *An elastic fiber (E) 0.34 μ in diameter surrounded by aggregates of amorphous material. Note the microfibril substructure in the fiber (× 34,840).* **Bottom:** *High-power view of an elastic fiber. Note that the beaded longitudinal microfibril substructure can be seen clearly (× 80,400).*

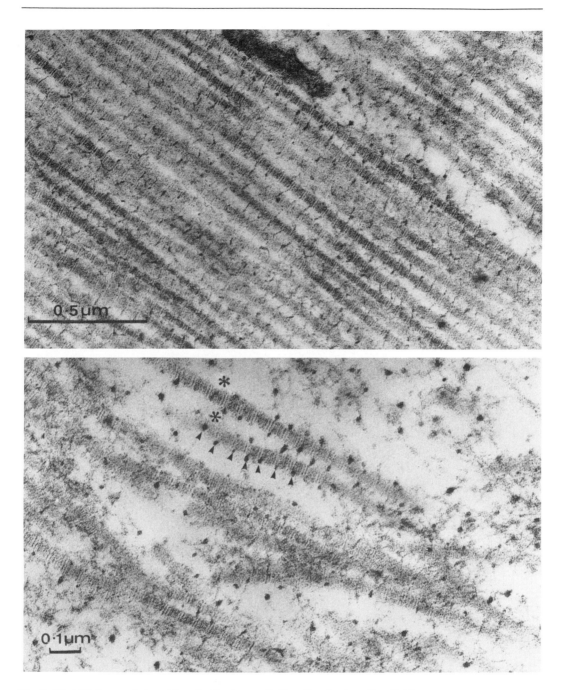

Fig. 2–15 *ACL from a 6-day-old rat (ruthenium red).* **Top:** *Regular interfiber linking filaments (× 69,000).* **Bottom:** *Relationship of proteoglycan granules to collagen fibers. A regular association can be seen in the two marked fibers (**). Granules are associated with the C and D bands of the collagen fibers (× 103,500).*

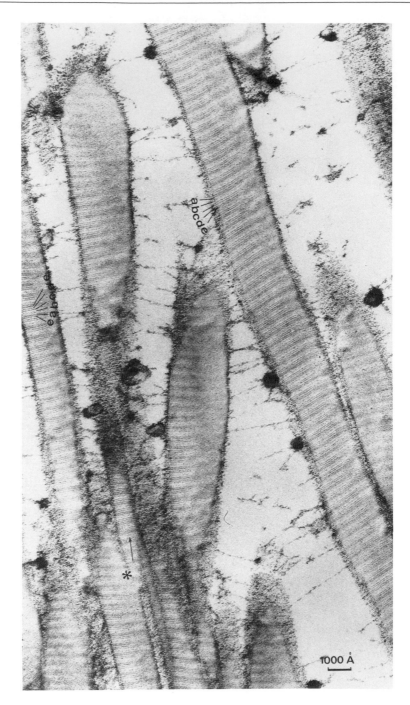

Fig. 2–16 *Adult rat ACL collagen fixed with ruthenium red after treatment with elastase at pH 8.8 for 12 hours. Note the typical regular banding and linking filaments, probably hyaluronate, that link the fibers between the C and D bands. Asterisk, Fiber appears to be divided into a smaller fiber. Proteoglycan granules are attached to the fibers in the region of the C and D bands (× 98,000).*

a calibrated image analysis system. The percentage area occupied for the number of fibrils in each diameter grouping was calculated[179] to quantify the "mass" of collagen represented for each diameter group; this correlates better with the mechanical properties of the tissue than does diameter alone.

Figure 2–17 shows profiles for control MCL, ACL, PCL, lateral collateral ligament (LCL), and patellar tendon. The profiles are similar except for the ACL and PCL, which had fewer fibrils and a small number of large fibrils. This is reflected more clearly in the percentage area covered.

Typical transverse sections through exercised and nonexercised control ACL are shown in Figure 2–18. There are more smaller-diameter (80 to 100 nm) fibrils (with the greatest number in the 125-nm range) in the exercised animals. The control ACL has more fibrils with diameters of 125 to 162 nm. Quantitative histograms are shown in Figure 2–19. Statistical analysis showed a significant increase ($P < .05$) in the number of small-diameter fibrils in exercised ACL but there was a simultaneous significant decrease ($P < .05$) in the number of larger fibrils.

Figure 2–19 also shows the population of mean fibril diameters for the PCL. The controls had a bimodal fibril distribution with the greatest number of fibrils having diameters of 37 or 125 nm. Clearly, this bimodal distribution was lost in the exercised PCL and there was a shift to the smaller-diameter fibrils. Statistical analysis of the three fibril groupings described above showed that there was a difference ($P < .05$) between exercised PCL and the controls in the middle group of fibrils (Fig. 2–20). Figures 2–20 and 2–21 show the mean percentage area occupied by each diameter group of fibrils. There was also a significant change in the maximum area occupied by the fibril groupings. The larger-diameter fibrils predominated in the control group, whereas in the exercise group there was a significant increase in the medium- and smaller-diameter fibrils.

Biochemical analysis indicated that the exercised PCL had almost twice as much normalized content of collagen as the control (Table 2–1). The other ligaments did not change their collagen content per unit of DNA. The marked increase in the PCL resulted from an increase in collagen as the DNA content of both the control and exercise PCL was approximately the same.

It is clear from the data presented that there was a training response in both the ACL and the PCL of these young rats. With four-week training, there was an accumulation of small-diameter fibrils in the exercised animal and a marked increase in the number of fibrils. There was an increase of 29% for the exercised ACL compared to 16% for the nonexercised controls. Similar changes occurred in the PCL (Fig. 2–21). However, this increase in small-diameter fibrils may be responsible for the decreased elastic stiffness in the ligaments of exercised rats.[182] There was also a reduction in the number of large-diameter fibrils in both ligaments (Fig. 2–19); this was reflected in the decrease in percentage area occupied by the large-diameter group (125 to 175

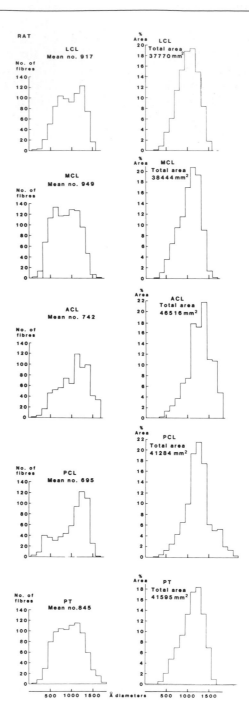

Fig. 2–17 *Histograms of 9-week-old rat control ligaments and patellar tendon (PT) for comparison.* **Left:** *Number of fibers plotted against fiber diameters.* **Right:** *Percentage area covered for each diameter group.*

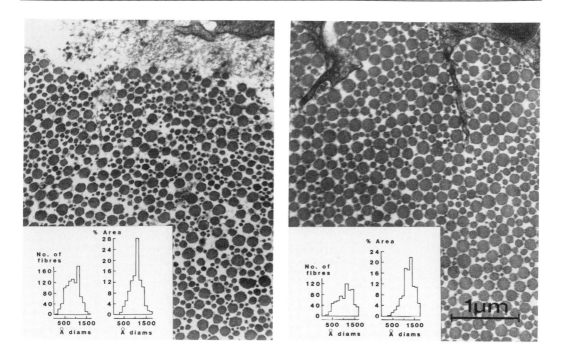

Fig. 2-18 *Transverse sections through exercised ACL* ***(left)*** *and nonexercised control ACL* ***(right)*** *(× 21,600).* **Insets:** *Histogram profiles of mean data for number of fibers and percentage area occupied for each diameter group.*

nm). The areas occupied for the middle group of fibrils in both ligaments was, however, increased (Figs. 2–20 and 2–21) and may be responsible for the increased tensile strength with training for longer periods.[183,184]

Although the number of fibrils in the exercised ACL, compared to the control ACL, was increased, the total cross-sectional area in both the control PCL and ACL was actually greater than that of the exercised group. These results indicate that little change in the tensile strength of exercised ligaments should be detected.[182-184] The increase in small-diameter fibrils seen in this study would support decreased stiffness in the ACL and PCL similar to that observed by Tipton and associates.[185] Recent work by Binkley and Peat[186] has shown the converse of that reported here: after six weeks of immobilization of the rat MCL the small fibrils diminished in number.

Collagen Fibril Populations in Human Knee Ligaments and Grafts
To gain insight into collagen repair mechanisms within human cruciate ligament grafts, biopsy specimens were obtained from autogenous ACL grafts from patients hospitalized for various reasons (for example, staples, arthroscopy). Most of the ACL grafts were from the central third of the patellar tendon as a free graft (33 patients), patellar tendons left

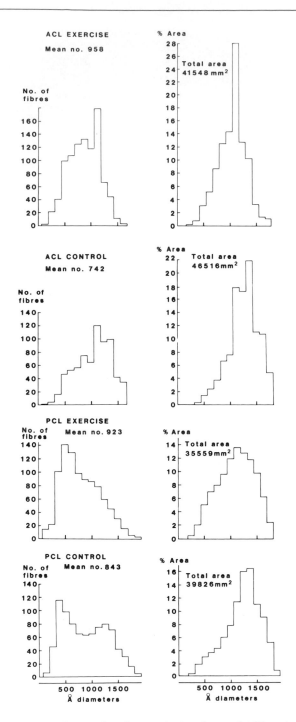

Fig. 2–19 *Histograms of mean data for exercised and control ACL and PCL.* **Left:** *Number of fibers in each diameter group.* **Right:** *Percentage area occupied by each diameter group.*

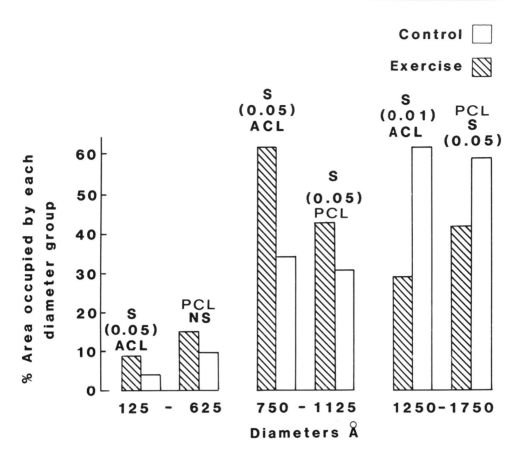

Fig. 2–20 *Comparison of percentage area occupied by three diameter groupings used for statistical analysis in exercised and control ACL and PCL.*

attached distally (eight patients), or from hamstrings or the iliotibial tract (seven patients). All had grades II or III pivot shifts and gross anterior drawer preoperatively; these were eliminated postoperatively in 87% of patients. Clinical review after three years showed an increase in drawer with a return in 20% of grade I pivot shift.

Collagen fibril diameters were quantitatively analyzed in 48 specimens from patients aged 19 to 42 years and the data compared with collagen fibril populations obtained from five cadaver ACL specimens, ten ACL specimens from young (<30 years) patients, and six specimens from older patients (>30 years) who had sustained recent tears. Specimens were also obtained from seven normal patellar tendons at operation and also from three cadavers.

The results (Fig. 2–22) of morphometric analysis in all grafts indicated a predominance of small-diameter collagen fibrils. Absence of a regular crimping of collagen fibrils was observed by both light and

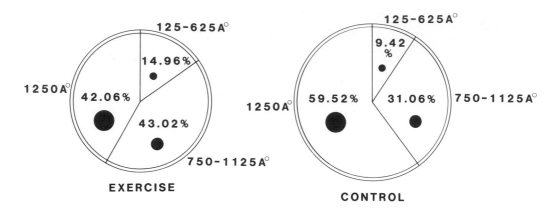

Fig. 2–21 *Comparison of percentage area occupied by the three diameter groupings used for statistical analysis in the exercised and control PCL.*

Table 2–1 Mean collagen content of pooled ligaments in control and exercised rats

Ligament	Collagen (μ/μ of DNA)
Anterior cruciate	
Control	111.0 ± 4.9
Exercised	119.0 ± 8.0
Posterior cruciate	
Control	130.8 ± 5.6
Exercised	232.1 ± 33.2
Lateral collateral	
Control	149.9 ± 7.9
Exercised	133.8 ± 18.4
Medial collateral	
Control	147.4 ± 16.4
Exercised	134.1 ± 0.4
Patellar tendon	
Control	229.0 ± 29.4
Exercised	250.7 ± 18.5

electron microscopy, as was a less ordered parallel arrangement of fibrils.

Quantitative findings (Figs. 2–22 and 2–23) were as follows:

(1) Large-diameter collagen fibrils (>100 nm) form a large proportion (approximately 45%) of the percentage cross-sectional area in normal human patellar tendon.

(2) Collagen fibrils less than 100 nm in diameter form a large proportion (approximately 85%) of the percentage cross-sectional area in the normal human ACL.

(3) In all ACL grafts, collagen fibrils less than 100 nm in diameter are the major contributors to the collagen fibril cross-sectional area, regardless of length of time in vivo (nine months to six years).

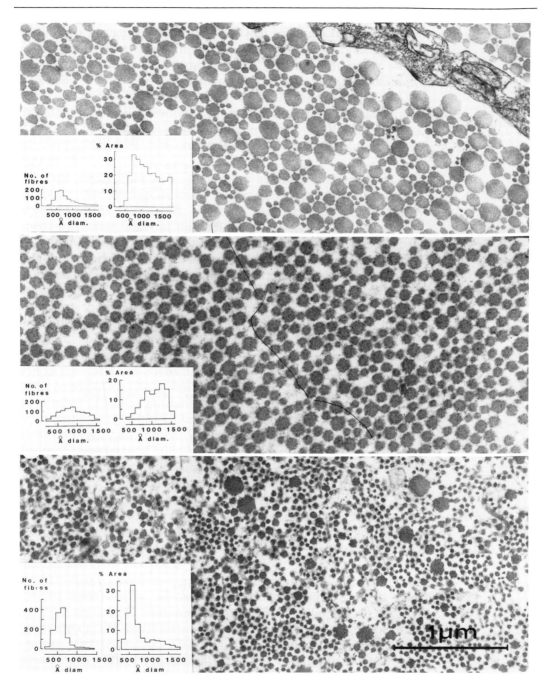

Fig. 2–22 *Transferase sections through collagen fibers of the normal young adult patellar tendon (top), normal young adult ACL (center), and Jones free graft (bottom) (× 34,100).* **Insets:** *Number of fibers plotted against and percentage area occupied for each diameter group. Note preponderance of small-diameter fibers in the graft and large fibers in the patellar tendon but not in the normal ACL.*

"more mature" with fibrils ranging from 30 to 60 nm in diameter. These observations indicate a faster maturation and remodeling on the mobile side and demonstrated the probable importance of loading on remodeling fibroblasts in determining collagen fibril profiles.[198]

Is gentle mechanical loading in the ACL grafts an important stimulus to fibroblast proliferation and collagen disposition? Inadequate mechanical stimulus may occur, especially if grafts are nonisometric and are "stretched out" by the patient before they have adequate tensile strength. A lax ACL graft may not induce sufficient mechanical loading on graft fibroblasts to alter the glycosaminoglycan-collagen biosynthesis ratios to favor large-diameter fibril formation. Certainly in this study ACL graft laxity increased postoperatively, lending credence to the above idea. However, use of continuous passive motion in grafted primates did not increase the strength of grafts.[199] Another more likely possibility is that the "replacement fibroblasts" in the ACL grafts are derived from stem cells from the synovium (and synovial perivascular cells), which are known to synthesize hyaluronate, which in turn favors small-diameter fibril formation.[200]

The conclusion from the above studies is that the predominance of the small-diameter collagen fibrils and their poor packing and alignment in all ACL grafts irrespective of the type of graft, their age, and the surgeon, may explain the clinical and experimental evidence of decreased tensile strength in such grafts compared to the normal ACL. It appears that in the adult, the "replacement fibroblasts" in the remodeled ACL graft cannot re-form the large-diameter, regularly crimped, and tightly packed fibrils seen in the normal ACL.

Normal Ligament Growth

Although ligament growth may be important in pathologic musculoskeletal conditions, little knowledge is currently available. It is easy to postulate that excessive ligament growth about a joint may result in joint laxity and that insufficient joint ligament growth may result in joint contracture. There are currently a number of childhood conditions involving joint contracture or laxity whose origins remain unexplained. To investigate these pathologic conditions, we must first define normal growth.

It is important to recognize that some ligaments cross physeal bone growth plates, and so must not only grow enough to accommodate enlargement of the joint but also rapid bone growth.

Anatomic Location of Growth   Current knowledge of the anatomic location of growth is derived from rabbit models.[201,202] In one model, wire markers implanted in ligaments during growth demonstrated elongation of all parts of the ligament. There was no "growth plate" responsible for the majority of ligament enlargement such as has been demonstrated in other musculoskeletal structures (bone, muscle, tendon).[203,204] In the MCL, which originates on the femoral epiphysis and inserts on the tibial metaphysis, growth was more rapid near the tibial insertion than near the femoral origin (Fig. 2–24). Growth near the

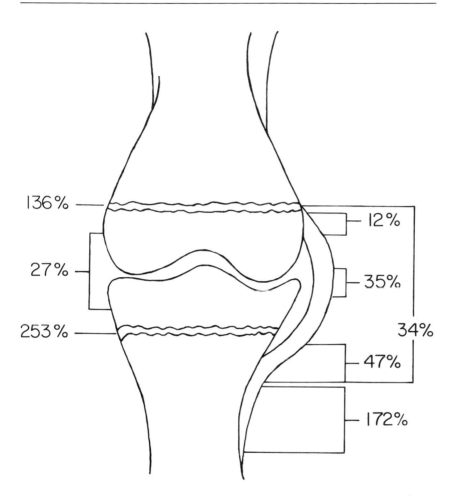

Fig. 2–24 *Anatomic location of growth in the rabbit MCL over a six-week period.*

origin was 12%. Growth near the insertion was significantly greater at 47%. Growth in the center of the ligament (over the joint) was 35%. Metal markers implanted in the bone of the femoral and tibial epiphyses showed that growth secondary to enlargement of the bone and cartilage of the joint itself was approximately 27%. The growth of the entire ligament from the origin down to the area of insertion was 34%, quite similar to growth in the central portion of the ligament (35%) and the bony structure of the joint (27%).

  In this model, the most distal suture (implanted in the ligament directly over its insertion) became incorporated into bone and was displaced away from the next suture at a rapid rate (172% growth) approximating physeal growth (Fig. 2–24). Frank and associates[202] noted rapid cell division at the metaphyseal insertion of the rabbit MCL, indicating growth in this area. This phenomenon is similar to the phenomenon documented by Videman,[204] who used radioactive collagen markers to examine tendon insertions and showed that the collagen at the insertion became incorporated into bone. This collagen was subsequently found far from the current tendon insertion. This incorporation of ligament and tendon insertions into bone, with a rapid growth area just above the insertion, is the process by which the metaphyseal insertions of ligaments migrate along the metaphysis to maintain their appropriate location near the joint during growth. If this migration did not occur, the ligament would eventually have a diaphyseal insertion.

A study of growth of the rabbit deltoid ligament at the ankle revealed that this ligament originates on the tibial epiphysis and inserts for the most part into the calcaneus.[205] It does not cross a physeal growth plate and its origin does not have to migrate. The ligament was divided into four segments with marking sutures, and growth was very similar at the origin, at the insertion, and in the central two segments as well. There is no rapid growth area as there is no need to displace an origin or insertion in this ligament.

Frank and associates,[202] using autoradiographic techniques, documented that growth in width and thickness of the rabbit MCL occurs interstitially and perhaps on the surface by the "periligament."

Factors Influencing Growth   More recently, work by Muller and Dahners[206] has been directed toward determining whether or not mechanical distraction influences growth. It seems necessary that ligament growth be coordinated in some sense with the enlargement of the bones beneath the ligament. Therefore, it seemed plausible that the ligament might respond in some way to the mechanical stretching produced by enlargement of the underlying bone.

An initial experiment was carried out with growing rabbits. The lateral collateral ligament was marked at several locations with sutures and the distance between the sutures measured. The fibular head was cut free from the tibia and the fibular shaft resected. A rubber band was then attached between the fibular head and the distal fibula, and either placed under tension (experimental group) or left in a relaxed condition (sham group) to evaluate the effect of mechanical tension on the growth of the fibular collateral ligament. In each rabbit the contralateral ligament was marked as above but the fibular head was left attached to the proximal tibia (control group).

Results showed that the tension applied by the rubber band produced greater growth in the experimental ligament than in the control contralateral ligament in eight of ten rabbits. This indicates that mechanical stretching does have an effect on ligament growth. However, in the rabbits in which the ligament was not placed under tension (sham group), growth still occurred and was essentially the same as growth in the contralateral control ligament. The contralateral control ligament in these rabbits was still attached to a fibular head, which was itself attached to the proximal tibia, so that mechanical stretching was applied to the ligament by the growth of the joint. It appears from this experiment that, although mechanical stretching can increase the rate of growth of the ligament, tension is not necessary (in the growing animal) for the ligament to enlarge itself at an appropriate rate.

When the same experiment was carried out in adult rabbits, it was found that the experimental group loaded under tension was essentially unchanged in length, as was the control group in which the fibular head was not cut loose. The sham group in which the fibular head was cut loose and not loaded under tension showed contraction of the ligament. This contraction was consistent with previous experiments in which adult ligaments, relieved of tension, routinely contacted.[207]

Thus, it appears that ligaments from growing animals can be induced to lengthen more rapidly by the application of mechanical distraction, and that distraction does not result in lengthening ligaments in adult animals. However, it also seems that distraction is not necessary for the growth of ligaments to proceed at an apparently normal rate.

These findings raise many unanswered questions. It seems, from the fact that ligaments continue to grow without mechanical stretch, that a ligament spanning a physis (such as the MCL) should become excessively lax after a physeal arrest. It may be that the portions of the MCL and the capsular ligaments that attach to the epiphysis are sufficient to keep the joint itself tight; however, more investigation is needed in this area. The fact that mechanical stress results in increased growth leads to the question of why extremely athletic children do not have excessively lax joints. It may be that it is necessary to have a constant stretching force, such as was applied by the rubber band in the above experiment, rather than intermittent stretching forces, such as are produced by sports. This question also requires further investigation. It would be very interesting to determine whether adult ligaments, placed under stretch and under the influence of growth hormone, could be induced to begin lengthening themselves again.

Normal Ligament Matrix Biochemistry

Ligaments are composed of a complex macromolecular network similar to other dense regular connective tissues.[7] Water makes up about two thirds of the weight of a normal ligament and the fibrillar protein collagen makes up the majority of the remaining dry weight. Collagen is thought to be the principal component that resists tensile stress in a ligament.[208]

Collagen in Ligaments   The most abundant substance in normal ligaments is collagen, which constitutes between 70% and 80% of the ligament's dry weight.[4,209,210] Ligament collagen is biochemically similar to collagen found in tendon and skin. It has been shown that a normal ligament is more than 90% fibrillar type I collagen with less than 10% being type III collagen (Fig. 2–25). Other collagen types are probably also present in smaller quantities.

As noted elsewhere,[211] collagen is synthesized in connective tissues by fibroblasts in a series of specific intracellular and extracellular events.

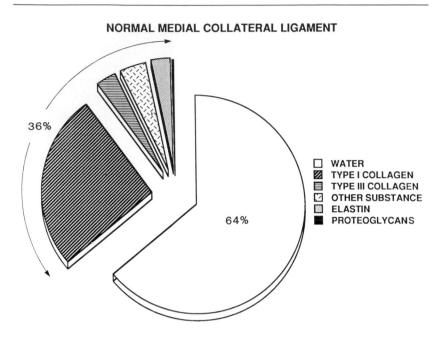

Fig. 2–25 *Approximate biochemical composition of a rabbit ligament.*

Several events in the synthesizing sequence have particular relevance to ligament collagen structure. Specific hydroxylases, which require iron, ascorbate, and α-ketoglutarate for activity, alter the amino acids proline and lysine to their hydroxylated forms in the cells. After hydroxylation, some residues are further glycosylated in the rough endoplasmic reticulum. Glycosylation of these residues is thought to be important in the subsequent regulation of collagen synthesis, the proteoglycan-collagen interaction of collagen synthesis, and the proteoglycan-collagen interaction with fiber size; all these factors are probably relevant to subsequent ligament function.[7,212]

Another important process in the formation of ligament microfibrils is the formation of collagen cross-links. After exocytosis of procollagen molecules from the cells and during the formation of ligament microfibrils, the enzyme lysyl oxidase acts on peptide-bound lysines and hydroxylysines to form aldehydes (Fig. 2–26). The formation of aldehydes is an important preliminary step in the formation of covalent intramolecular (aldol) and intermolecular (Schiff base) cross-links. It is these Schiff-base cross-links that are thought to have the greatest functional significance in soft tissues. In ligaments, these reducible intermolecular cross-links are dihydroxylysinonorleucine, hydroxylysinonorleucine (Fig. 2–27), and histidinohydroxymerodesmosine (Fig. 2–28).[7] With maturation of connective tissue, the concentration of these reducible cross-links falls.[213] A stable trivalent amino acid cross-

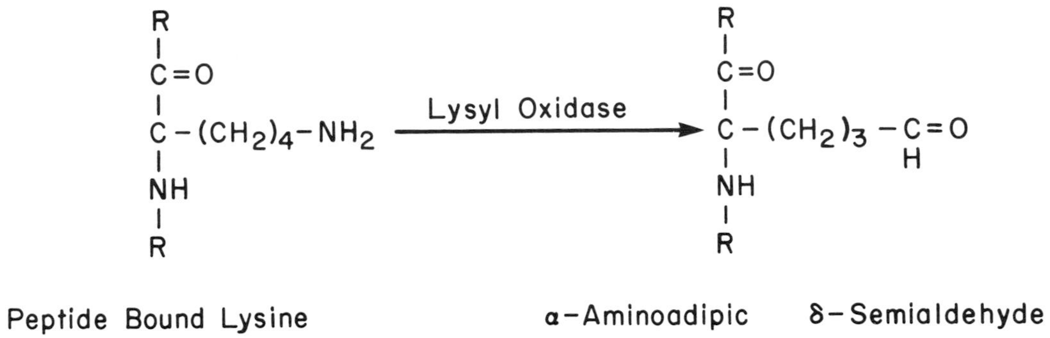

$$R \quad\quad\quad\quad\quad\quad\quad\quad\quad\quad\quad\quad R$$

R
|
C=O
|
C – (CH$_2$)$_4$ – NH$_2$ **Lysyl Oxidase** ⟶
|
NH
|
R

R
|
C=O
|
C – (CH$_2$)$_3$ – C=O
| H
NH
|
R

Peptide Bound Lysine   α – Aminoadipic   δ – Semialdehyde

Fig. 2–26 *As one step in collagen cross-linking, lysyl oxidase oxidatively deaminates peptide-bound lysine to generate aldehydes.*

link, 3-hydroxypridinium, has also been found in many collagenous tissues, including ligaments. It has two chemical forms, hydroxylysylpyridinoline, which is thought to be formed from three residues of hydroxylysine, and lysylpyridinoline, which is thought to be formed from two residues of hydroxylysine and one of lysine (Fig. 2–28).[214] At the present time, the specific functional significance of these stable cross-links in ligament tissue remains unknown.

After secretion, modification, and self-assembly of the procollagen molecules into microfibrils, there are several levels of collagen organization. A number of microfibrils aggregate to form subfibrils and a number of subfibrils form fibrils. Collagen fibrils are then associated into collagen fibers that are visible on light microscopy and that under polarized light have a longitudinal waveform or crimp pattern (Fig. 2–29).[215]

In mature ligament tissues there is generally a balance between collagen synthesis and degradation, with the half-life of collagen estimated to be between 300 and 500 days.[216] Collagen degradation in ligaments occurs when specific collagenases are activated. Neutral collagenases are synthesized by fibroblasts and secreted in an inactive form bound to inhibitors. Collagenases can be activated by a variety of substances and, when activated, cleave the collagen triple helix into two pieces, rendering the remaining collagen molecule susceptible to other proteases. Collagen turnover is probably mediated by a number of factors that require further elucidation.

Ligament Proteoglycans Although proteoglycans constitute less than 1% of the ligament's total dry weight, they probably play a key role in ligament function. Periarticular and ligament proteoglycans have a similar general arrangement with a protein core and glycosaminoglycan side chains consisting of chondroitin–4 sulfate, chondroitin–6 sulfate, and dermatan sulfate.[217] While some proteoglycans are aggregated with

72. Boyd IA: The histological structure of the receptors in the knee-joint of the cat correlated with their physiological response. *J Physiol* 1954;124:476–488.

73. Boyd IA, Roberts TDM: Proprioceptive discharges from stretch-receptors in the knee-joint of the cat. *J Physiol* 1953;122:38–58.

74. Andersson S, Stener B: Experimental evaluation of the hypothesis of ligamento-muscular protective reflexes: II. A study in cat using the medial collateral ligament of the knee joint. *Acta Physiol Scand* 1959;48(suppl 166):27–49.

75. Stener B: Experimental evaluation of the hypothesis of ligamento-muscular protective reflexes: I. A method for adequate stimulation of tension receptors in the medial collateral ligament of the knee joint of the cat, and studies of the innervation of the ligament. *Acta Physiol Scand* 1959;48(suppl 166):5–26.

76. Millar J: Joint afferent discharge following muscle contraction in the absence of joint movement, abstract. *J Physiol* 1972;266:72P.

77. Inoue H: An electrophysiological study on the mechanoreceptor of canine knee joints. *Nippon Seikeigeka Gakkai Zasshi* 1985;59:641–649.

78. Clark FJ: Information signaled by sensory fibers in medial articular nerve. *J Neurophysiol* 1975;38:1464–1472.

79. Grigg P: Mechanical factors influencing response of joint afferent neurons from cat knee. *J Neurophysiol* 1975;38:1473–1484.

80. Grigg, P, Harrigan EP, Fogarty KE: Segmental reflexes mediated by joint afferent neurons in cat knee. *J Neurophysiol* 1978;41:9–14.

81. Grigg P, Hoffman AH: Properties of Ruffini afferents revealed by stress analysis of isolated sections of cat knee capsule. *J Neurophysiol* 1982;47:41–54.

82. Rossi A, Grigg P: Characteristics of hip joint mechanoreceptors in the cat. *J Neurophysiol* 1982;47:1029–1042.

83. de Andrade JR, Grant C, Dixon ASJ: Joint distension and reflex muscle inhibition in the knee. *J Bone Joint Surg* 1965;47A:313–322.

84. Adrian ED: Afferent areas in the cerebellum connected with the limbs. *Brain* 1943;66:289–315.

85. Mountcastle VB: Modality and topographic properties of single neurons of cat's somatic sensory cortex. *J Neurophysiol* 1957;20:408–434.

86. Powell TPS, Mountcastle VB: Some aspects of the functional organization of the cortex of the postcentral gyrus of the monkey: A correlation of findings obtained in a single unit analysis with cytoarchitecture. *Bull Johns Hopkins Hosp* 1959;105:133–162.

87. Freeman MAR, Wyke BD: Articular contributions to limb muscle reflexes: The effects of partial neurectomy of the knee-joint on postural reflexes. *Br J Surg* 1966;53:61–69.

88. Samuel EP: The autonomic and somatic innervation of the articular capsule. *Anat Rec* 1952;113:53–70.

89. Dee R: The innervation of joints, in Sokoloff L (ed): *The Joints and Synovial Fluid.* New York, Academic Press, 1978, vol 1, pp 177–204.

90. Eklund G, Skoglund S: On the specificity of the Ruffini like joint receptors. *Acta Physiol Scand* 1960;49:184–191.

91. Grigg P, Hoffman AH, Fogarty KE: Properties of Golgi-Mazzoni afferents in cat knee joint capsule, as revealed by mechanical studies of isolated joint capsule. *J Neurophysiol* 1982;47:31–40.

92. Dykes RW: Central consequences of peripheral nerve injuries. *Ann Plast Surg* 1984;13:412–422.

93. Gandevia SC, McCloskey DI: Joint sense, muscle sense, and their combination as position sense, measured at the distal interphalangeal joint of the middle finger. *J Physiol* 1976;260:387–407.

94. Cafarelli E, Bigland-Ritchie B: Sensation of static force in muscles of different length. *Exp Neurol* 1979;65:511–525.

95. McCloskey DI: Kinesthetic sensibility. *Physiol Rev* 1978;58:763–820.

96. Forsberg H, Wallberg H: Infant locomotion: A preliminary movement and electromyographic study, in Berg K, Ericson BO (eds): *Children and Exercise*. Baltimore, University Park Press, 1980, vol 4.

97. Grillner S: Control of locomotion in bipeds, tetrapods, and fish, in Brookhartt JM, Mountcastle VB, Brooks VB (ed): *Handbook of Physiology: Section 1. The Nervous System*. Bethesda, American Physiological Society, 1981, vol 2, pp 1179–1236.

98. Halbertsma JM, Miller S, van der Meché FGA: Basic programs for the phasing of flexion and extension movements of the limbs during locomotion, in Herman RM, Grillner S, Stein PSG, et al (eds): *Neural Control of Locomotion*. New York, Plenum Press, 1976, pp 489–517.

99. Petersén I, Stener B: Experimental evaluation of the hypothesis of ligamento-muscular protective reflexes: III. A study in man using the medial collateral ligament of the knee joint. *Acta Physiol Scand* 1959;48(suppl 166):51–61.

100. Stener B, Petersén I: Electromyographic investigation of reflex effects upon stretching the partially ruptured medial collateral ligament of the knee joint. *Acta Chir Scand* 1962;124:396–415.

101. Jensen D: *The Principles of Physiology*. New York, Appleton-Century-Crofts, 1976, p 366.

102. Shiavi R, Borra H, Frazer M, et al: EMG envelopes from normal and anterior cruciate ligament deficient individuals. Presented at the North American Congress on Biomechanics, Montreal, Canada, 1986.

103. Tibone JE, Antich TJ, Fanton GS, et al: Functional analysis of anterior cruciate ligament instability. *Am J Sports Med* 1986;14:276–284.

104. Carlsöö S, Nordstrand A: The coordination of the knee-muscles in some voluntary movements and in the gait in cases with and without knee joint injuries. *Acta Chir Scand* 1968;134:423–426.

105. Lewis JL, Shybut GT: In vivo forces in the collateral ligaments of canine knees. *Trans Orthop Res Soc* 1981;6:4.

106. Brand RA, Rubin CT, Seeherman HJ, et al: In vivo measurement of strains in the medial collateral ligament of the sheep and horses. Presented at the Fifth Meeting of the European Society of Biomechanics, Berlin, Sept 8–10, 1986.

107. Jokl P, Kaplan N, Stovell P, et al: Non-operative treatment of severe injuries to the medial and anterior cruciate ligaments of the knee. *J Bone Joint Surg* 1984;66A:741–744.

108. Indelicato PA: Non-operative treatment of complete tears of the medial collateral ligament of the knee. *J Bone Joint Surg* 1983;65A:323–329.

109. Jones RE, Henley, MB, Francis P: Nonoperative management of isolated grade III collateral ligament injury in high school football players. *Clin Orthop* 1986;213:137–140.

110. Odensten M, Hamberg P, Nordin M, et al: Surgical or conservative treatment of the acutely torn anterior cruciate ligament: A randomized study with short-term follow-up observations. *Clin Orthop* 1985;198:87–93.

111. Curl WW, Markey KL, Mitchell WA: Agility training following anterior cruciate ligament reconstruction. *Clin Orthop* 1983;172:133–136.

112. Eldridge VL: Development of pivot shift control. *Phys Ther* 1984;64:751.

113. Pollack ED, Muhlach WL, Liebig V: Neurotropic influence of mesenchymal limb target tissue on spinal cord neurite growth in vitro. *J Comp Neurol* 1981;200:393–405.

114. Viidik A: Biomechanical behavior of soft connective tissues, in Akkas N (ed): *Progress in Biomechanics*. Amsterdam, Sijthoff & Noordhoff, 1979, pp 75–113.

Chapter 3
Ligament: Injury and Repair

Thomas Andriacchi, PhD
Paul Sabiston, MD
Kenneth DeHaven, MD
Laurence Dahners, MD
Savio Woo, PhD
Cyril Frank, MD
Barry Oakes, MD
Richard Brand, MD
Jack Lewis, PhD

Chapter Outline

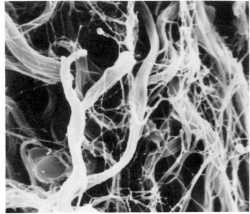

Fig. 3–3 *Normal ligament **(left)** and healing scar at two weeks **(right)** (SEM, × 7,000).*

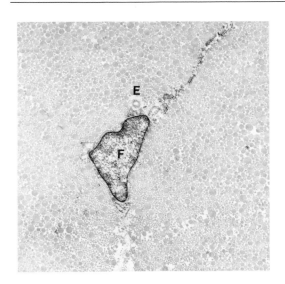
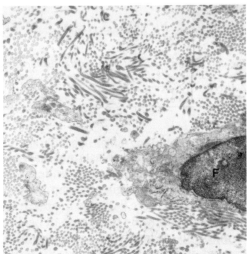

Fig. 3–4 *Normal ligament **(left)** and healing scar at three weeks **(right)** (TEM, × 15,000).*

This maturation phase of healing is probably quite variable in duration and its endpoint and is, no doubt, model- and ligament-specific. In canine MCL, remodeling is quite advanced one year postoperatively, and the tensile strength (a mechanical property representing the quality

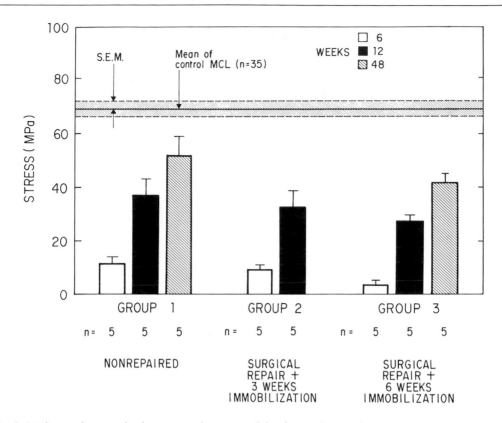

Fig. 3–5 *The tensile strength of nonrepaired, nonimmobilized MCL (group 1) was substantially higher than that of either of the two surgically repaired and immobilized groups (group 2 and group 3). After 48 weeks, its value was two thirds that of the contralateral controls. (Reproduced with permission from Woo SL-Y, Inoue M, McCurk-Burleson E, et al: Treatment of the medial collateral ligament injury: II. Structure and function of canine knees in response to differing treatment regimens. Am J Sports Med 1987;15:22–29.)*

of this tissue) of the unrepaired and untreated MCL may be as much as two thirds that of the contralateral, intact control (Fig. 3–5). Other environmental influences (such as immobilization and activity level) are also particularly important.[15,16,35]

Primary surgical repair appears to decrease the mass of the scar tissue being formed, possibly by induction of a more "primary" healing response[23,55] or possibly by relative immobilization of the torn ligament ends. Immobilization of the scar may, however, decrease both scar quantity and quality. Further, Woo and associates[35] demonstrated that suturing and immobilization of isolated MCL injuries was detrimental to the valgus laxity of the knee. The true functional benefits of suturing torn ligament ends are, therefore, somewhat doubtful, with the possible exception of situations in which torn ligament ends are known to remain separated (for example, the ACL).

Ligament healing is a morphologically complex process marked by a series of overlapping events. Evidence to date suggests that extra-articular ligaments heal by the same processes occurring in normal wound healing and that, therefore, they should be subject to similar control. A better understanding of the subtleties both of these processes and of their controls will enhance the likelihood of optimal ligament healing.

Biochemistry of Ligament Healing

Local biochemical changes after a ligament injury are similar to those observed after injuries in other soft tissues.[56] The following is a brief description of the substances and events involved in the healing process with the rabbit MCL used as a model.[17] For the purpose of consistency, the biochemical events are divided into the four phases already described.

Phase I: Inflammation In response to injury and exposure to fibrin, platelets and most cells release a potent vasodilator—histamine. This is followed by a more prolonged vasodilatory effect mediated by serotonin, bradykinins, and prostaglandins.[45] Bradykinins also increase capillary permeability, allowing transudation of fluid and attraction of inflammatory cells into the injured area. Capillary endothelial buds proliferate into the ligament wound in this early stage, probably in response to an angiogenesis factor[57] secreted by macrophages. These inflammatory mediators, in combination with the injured tissue and coagulum forming in the gap between the injured ligament ends, help to initiate the healing process.

During the inflammatory phase important biochemical changes occur in the area of the forming scar.[17] As expected, the edema associated with the initial phase of healing results in an increased water content. Collagen changes are probably extremely important to scar function. Although collagen concentration is normal or slightly decreased in this early stage,[17,58] the total mass of ligament scar tissue is increased, resulting in increased total collagen content compared with a normal ligament.[17,58,59] Collagen turnover studies show that active collagen synthesis and degradation occur simultaneously during this phase, with synthesis being slightly greater than degradation. Most newly synthesized collagen in the ligament scar is initially type III[60]; there is a smaller proportion of type I collagen (Fig. 3–6). Type III collagen is thought to be responsible for early stabilization of the extracellular collagen meshwork, while type I collagen is probably more important to long-term matrix properties.[61]

Other, less concentrated constituents of newly formed extracellular matrix change even more dramatically in this early phase of healing. Glycosaminoglycan content, as measured by hexosamine, is markedly increased.[17,45,56] Fibronectin content[62] and DNA content (Fig. 3–6) are also increased.

Phase II: Proliferation Water content remains increased. Although the total collagen content continues to increase because of increased

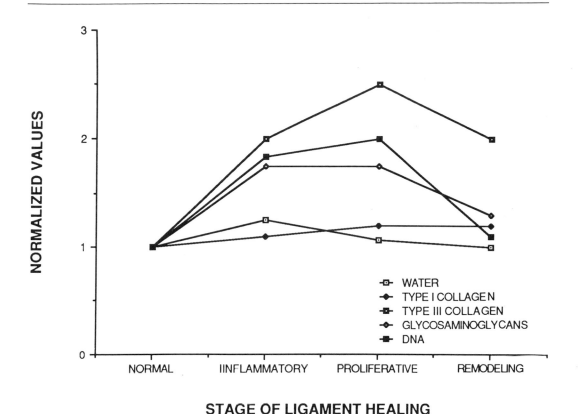

Fig. 3–6 *Changes in certain components of rabbit MCL at various stages of healing. Values are normalized to uninjured ligament (normal = 1).*

scar volume, the collagen concentration remains low because of less dense packing of the collagen framework. Spaces in the matrix are apparently filled with excess water and other extracellular components. Collagen turnover studies indicate increased synthesis in the proliferating scar as well as in the adjacent normal-appearing ligament tissue, reflecting more widespread injury and repair than is apparent morphologically. Collagen turnover, synthesis plus degradation, peaks early in the proliferative phase, but remains increased throughout.[58] Type I collagen is now the predominant matrix component and the reducible cross-link profile of the scar is significantly altered (more embryonic) in comparison with normal ligament.[63]

Glycosaminoglycan concentration, as measured by hexosamines, remains increased. The increased DNA concentration continues to reflect the cellularity of the proliferative stage. Figure 3–7 depicts the relative proportions of matrix components during this stage of ligament healing.

These collective changes indicate a proliferative cellular tissue, ac-

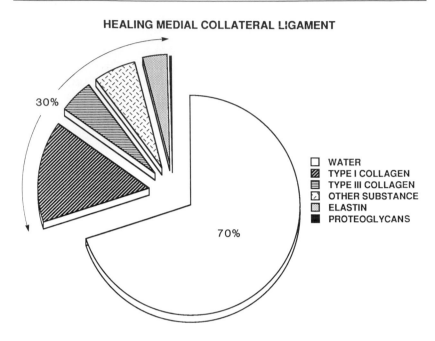

Fig. 3–7 *Approximate proportions of important biochemical components of rabbit MCL 21 days after surgery.*

tively synthesizing extracellular matrix and undergoing changes in collagen cross-linking profiles. The biochemical changes during the proliferative stage of wound healing have been correlated with increasing tensile strength of the matrix.[7,16,17,35,64]

Phases III and IV: Remodeling and Maturation   During these phases there is decreased cellularity, with an accompanying decrease in synthetic activity. The extracellular matrix synthesized during the proliferative phase becomes better organized and the biochemical profile appears more like that of normal ligament tissue. Water content declines to normal ligament values. Collagen concentration stabilizes slightly below that of a normal ligament, while total collagen content remains slightly increased. The relative proportion of type III to type I collagen declines toward normal. The reducible cross-linking profile reverts to that of normal ligaments. Collagen turnover approaches normal levels. DNA content reverts to normal values and reflects the decrease in cellularity. Proteoglycan content remains slightly increased (Fig. 3–6). The possible functional implications of these changes have been discussed elsewhere.[65–68]

To summarize, the biochemical changes in healing ligaments represent part of a dynamic process that correlates with changing morphologic appearance. During the inflammatory stage there is edema, hypercellularity, and early active synthesis of matrix. This is followed by active proliferation of cells and an increase in the synthesis of extracellular matrix. Collagen turnover peaks during this phase and then gradually returns toward normal. With remodeling, there is a trend for the components of the scar tissue to revert toward normal, although even at long-term follow-up some abnormalities remain. The matrix in a ligament continues to mature slowly over the ensuing months and years.

Finally, it should be emphasized that many factors, both local and systemic, may influence healing of ligaments and thus the quality and quantity of the scar matrix components. Further work is needed.

Knee Biomechanics in Ligament Injuries

Injury by stretching or complete tear of any of the major ligaments of the knee alters both the passive stability and the kinematics of the joint. In addition, ligament loss can influence dynamic muscle activity through kinematic changes influencing muscle lever arms, muscular substitution for loss of passive stability, and dynamic changes in movement patterns resulting from changes in proprioception.

Change in passive stability caused by ligament loss is the best documented of these mechanical changes.[27,69-71] Recent studies, however, have demonstrated that care must be taken in the interpretation of changes in joint stability after ligament sectioning. Stability can be substantially influenced by constraints imposed during the testing procedure. For example, increases in varus and valgus stiffness of more than 50% can result solely from constraining the tibia from axial rotation during the testing procedure.[27,72] Similar results have been demonstrated for differences in anteroposterior laxity resulting from constraints.[27,72] These results demonstrate the importance of evaluating the knee as a multidimensional joint in which motion in one degree of freedom is influenced by constraint to other degrees of freedom.

The importance of motion coupling in the evaluation of joint stability testing was shown in recent studies of treatments for canine MCL injury.[26,27] Sectioning the MCL substantially increased (171%) valgus laxity when axial rotation and anteroposterior displacement were constrained. However, MCL sectioning had much less effect on valgus laxity (an increase of only 21%) when constraints were removed. The coupled axial (tibial) rotation with the varus and valgus rotation allowed the ACL to function as an important restraint to varus/valgus rotation. This observation, applied in the interpretation of a subsequent study,[35] indicated that the MCL-deficient knee is kept in check by the ACL and other periarticular structures. Thus, surgical intervention with immobilization is unnecessary, and animals can be treated conservatively with early mobilization to produce better results (Fig. 3–8).

Joint kinematics is clearly influenced by the cruciate ligaments and the collateral ligaments. Ligament sectioning has been shown to influence changes in the instantaneous axis of motion as well as in the

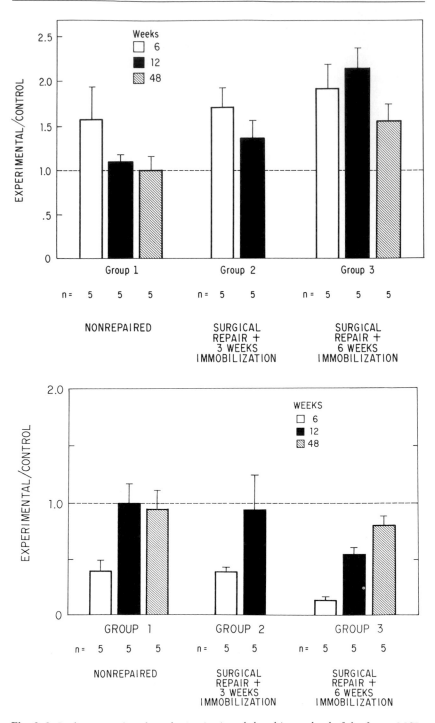

Fig. 3–8 *Both varus-valgus knee laxity **(top)** and the ultimate load of the femur-MCL-tibial complex **(bottom)** were closer to those of the contralateral controls in the nonrepaired, nonimmobilized group (group 1) than in repaired and immobilized groups (groups 2 and 3). These data were consistent with biomechanical properties of the MCL substance, indicating that the MCL-deficient knee is stable and that knee laxity is kept in check sufficiently by the intact ACL and, therefore, prolonged immobilization is not necessary.*

movement of the tibiofemoral contact.[73–75] Sectioning of the ACL and PCL substantially influences the anteroposterior movement of the tibiofemoral contact with flexion. These kinematic changes, while at times quite subtle, can have a substantial influence on muscular lever arms[76,77] and, ultimately, on functional changes through compensatory adaptations to adjust for changes in the mechanical efficiency of the muscles.[78]

Dynamic changes in the mechanics of the joint can also occur through muscular compensation for loss of passive stability. ACL-deficient subjects tested during pivoting maneuvers[79] demonstrated a consistent compensatory mechanism associated with sustaining higher loads at the knee joint by placing the joint in a higher degree of flexion than did normal subjects. Mechanical testing of the muscles in cadaver studies indicated that the hamstring muscle group provides increased stability to axial rotation[77] and anterior drawer by increasing the angle of knee flexion. Thus, it is possible to compensate functionally for loss of passive stability through dynamic compensation. However, these adaptations are not necessarily predictable and require testing for evaluation. Further studies in this area are needed, as are studies related to the loss of proprioception associated with ligament disruption.

Understanding the biomechanics of the joint after repair and the mechanics of the joint during healing is important in improving treatment. The results of initial repair in human cadaver knees vary widely for any given repair and among repairs.[80] Sensitivity to the position of insertion[81] and the initial tension during the repair appear to be critical. There is no well-established connection between the mechanics of initial repair and the final mechanics after healing, although it has been shown that mechanical properties recover more slowly during healing than do structural properties.[17,35,82]

In summary, at present there is a substantial need for further information about changes in function after ligament disruption. Passive changes have been well documented in cadaver studies and this information must be correlated at some point with functional changes. In addition, the mechanics of healing and repair require further documentation.

Ligament Contraction

Because the ligament is a relatively short band of connective tissue, its structures must return to their appropriate length as well as their appropriate strength if an acceptable result is to be achieved after injury. Even nonligamentous injuries, such as periarticular fractures, can produce ligament-length problems if the joint is extensively immobilized during treatment of the fracture. After immobilization the joint is frequently stiff. Although some of the stiffness results from intra-articular adhesions to surfaces that normally glide past one another,[83] there is also good evidence that ligament structures can shorten (contract) and limit the motion of the joint.[84,85]

Conversely, some joint injuries result in direct ligament injury. This can led to complete rupture, partial rupture, or a stretch injury probably resulting from multiple microruptures. After these injuries, some ligaments may remain lax indefinitely whereas others seem to be able to "tighten themselves" back to a relatively normal length. Understanding the mechanisms of this phenomenon is of great importance to the clinical practice of orthopaedics.

There are two mechanisms to be elucidated: (1) the mechanism by which the structure of the ligament is actually shortened and (2) the signal that stimulates the shortening process. Mechanisms proposed for the shortening include biochemical phenomena by which a ligament resting at a relatively shortened length has its collagen restructured or cross-linked at that length, so that it cannot be elongated later. This can be considered a passive phenomenon. Another possibility is an active phenomenon in which the cells in the ligament actually apply tension during the restructuring process.

Candidates for the signal that initiates ligament contraction are mechanical (for example, a lack of mechanical tension or perhaps a low level of cyclic tension such as that applied by joint motion), chemical (for example, stimulation of contraction by local agents produced by the injury or the healing process), or electrical (for example, response to the presence or absence of stress-generated electrical potentials, that is, streaming potentials produced by mechanical loading of collagenous structures).

In evaluations of the role of cross-linking in ligament contraction, Peacock[86,87] induced additional cross-linking by chemical methods in ligaments that were allowed to rest in a shortened state and was able to show that this resulted in minimal ligament shortening. Akeson,[88] Woo and associates,[89] and Akeson and associates[90] carried out extensive studies on the changes associated with joint immobilization and subsequent stiffness. They found alterations in collagen cross-linking, synthesis, and degradation as well as in the water content and proteoglycan content of tissue surrounding the joints. Many of these alterations are compatible with a passive phenomenon in which the soft tissues about the immobilized joint undergo changes during the period of disuse. These changes decrease stiffness and strength in these tissues as well as in the bone-ligament junction.[91] The paradox of increased joint stiffness and decreased tissue stiffness secondary to joint immobility can be explained by adhesion, pannus formation, decreased lubricity, and binding of the loosely interwoven collagen fibers, possibly caused by fixed contact at strategic sites. The random formation of newly produced collagen, it has been hypothesized, forms these interfibrillar contacts and restricts the sliding and motion normal in extending structures such as a joint capsule (for example, in the posterior aspect of the knee).[14]

Dahners and Wilson[84,85] showed that ligaments that no longer had tension applied to them actually shortened, as measured by a diminished distance between marking sutures implanted in the ligament. This shortening process is associated with increased staining for the contractile protein actin (the same protein responsible for muscle contraction). Contraction of such a ligament tissue has been inhibited by artificially simulated, stress-generated electrical potentials (electrical potentials normally produced by mechanical loading of the ligament).[92] The fact that ligament contraction is associated with an increase in a contractile protein and can be inhibited by simulated electrical signals suggests that there is some active component to this contraction involving intracellular contractile proteins. Only an active process would be inhibited by extrinsically applied simulation of the normally occurring stress-generated potentials.

It seems probable that joint stiffness results from a combination of several phenomena. These include (1) the formation of adhesions between normally gliding surfaces, especially those of periarticular connective tissue with nonparallel fibers, secondary to modifications of the collagen content (increased collagen in normally areolar tissues) and ground substance (altered water and proteoglycan contents) and collagen cross-linking and (2) active changes in the length of dense collagenous structures (ligament) caused by fibroblasts (a process in which they actively degrade the current collagen, contract to shorten the collagen in their local area, and then synthesize new collagen at the shortened length).

It may be that immobilizing the joint has several results, including loss of the normal pumping phenomena produced by intermittent stretching of the tissue. This could significantly alter the nutritional state of the tissues, as happens in cartilage. This effect appears to be more prominent in loose connective tissue, resulting in the observed phenomenon of fibrosis through diminished quantities of relatively slippery proteoglycans and production of new collagen at a shortened length. At the same time, stress-generated electrical potentials in the denser connective tissues of ligament would decrease, which could be interpreted by the fibroblasts as a signal to undertake an active contraction process involving degradation of old "longer" collagen molecules and synthesis of new "shorter" collagen molecules after the contractile actin cytoskeleton of the fibroblasts shortens the structure of the ligament.

The current results may have significance in the treatment of patients in the sense that surgeons could (even while immobilizing the joint) perhaps simulate the pumping action of joint motion, administer hyaluronic acid,[93] and mimic the electrical effect of stress-generated potentials, thus diminishing contracture in joints that must be immobilized. Hyaluronic acid also has water-binding and lubricating capacities, both helpful in limited cross-linking or fusion of randomly deposited new collagen fibrils. In patients with excessively lax ligaments, it may be possible to prevent application of stress to the ligament over a period of time, thus eliminating the production of stress-generated potentials and possibly resulting in a desirable contraction of that ligament.

Future Directions

A number of multidisciplinary studies are needed if we are to add to our knowledge of ligaments and optimize clinical ligament repair

41. Jackson RW: Anterior cruciate ligament injuries, in Casscells SW (ed): *Arthroscopy: Diagnostic and Surgical Practice*. Philadelphia, Lee and Febiger, 1984, pp 52–63.

42. Noyes FR, Bassett RW, Grood ES, et al: Arthroscopy in acute traumatic hemarthrosis of the knee: Incidence of anterior cruciate tears and other injuries. *J Bone Joint Surg* 1980;62A:687–695.

43. Clancy WG Jr, Shelbourne KD, Zoellner GB, et al: Treatment of knee joint instability secondary to rupture of the posterior cruciate ligament: Report of a new procedure. *J Bone Joint Surg* 1983;65A:310–322.

44. McDaniel WJ Jr, Dameron TB Jr: Untreated anterior ruptures of the cruciate ligament: A follow-up study. *J Bone Joint Surg* 1980;62A:696–705.

45. Irvin TT: The healing wound, in Bucknall TE, Ellis H (eds): *Wound Healing for Surgeons*. London, Bailliere Tindall, 1982, pp 3–28.

46. Ross R: The fibroblast and wound repair. *Biol Rev* 1968;43:51–96.

47. Akeson WH, Frank CB, Amiel D, et al: Ligament biology and biomechanics, in Finerman G (ed): American Academy of Orthopaedic Surgeons *Symposium on Sports Medicine: The Knee*. St. Louis, CV Mosby Co, 1985, pp 111–151.

48. Frank C, Amiel D, Akeson WH: Healing of the medial collateral ligament of the knee: A morphological and biochemical assessment in rabbits. *Acta Orthop Scand* 1983;54:917–923.

49. Frank C, Amiel D, Woo SL-Y, et al: Normal ligament properties and ligament healing. *Clin Orthop* 1985;196:15–25.

50. Frank C, Schachar N, Dittrich D: Natural history of healing of the repaired medial collateral ligament. *J Orthop Res* 1983;1:179–188.

51. Jack EA: Experimental rupture of the medial collateral ligament of the knee. *J Bone Joint Surg* 1950;32B:396–402.

52. Mason ML, Shearon CG: The process of tendon repair: An experimental study of tendon suture and tendon graft. *Arch Surg* 1932;25:615–692.

53. Flynn JE, Graham JH: Healing following tendon suture and tendon transplants. *Surg Gynecol Obstet* 1962;115:467–472.

54. Manske PR, Gelberman RH, Vande Berg JS, et al: Intrinsic flexor-tendon repair: A morphological study *in vitro*. *J Bone Joint Surg* 1984;66A:385–396.

55. O'Donoghue DH, Rockwood CA, Zaricznyj B, et al: Repair of knee ligaments in dogs: I. The lateral collateral ligament. *J Bone Joint Surg* 1961;43A:1167–1178.

56. Dunphy JE, Udupa KN: Chemical and histochemical sequences in the normal healing of wounds. *N Engl J Med* 1955;253:847–851.

57. Polverini PJ, Cotran RS, Gimbrone MA Jr, et al: Activated macrophages induce vascular proliferation. *Nature* 1977;269:804–806.

58. Amiel D, Frank C, Harwood FL, et al: Collagen alterations in medial collateral healing in a rabbit model. *Connect Tissue Res*, in press.

59. Delaunay A, Bazin S: Mucopolysaccharides, collagen, and nonfibrillar proteins in inflammation. *Int Rev Connect Tissue Res* 1964;2:301–325.

60. Williams IF, McCullagh KG, Silver IA: The distribution of types I and III collagen and fibronectin in the healing equine tendon. *Connect Tissue Res* 1984;12:211–227.

61. Nimni ME: Collagen: Structure, function, and metabolism in normal and fibrotic tissues. *Semin Arthritis Rheum* 1983;13:1–86.

62. Grinnell F: Fibronectin and wound healing. *J Cell Biochem* 1984;26:107–116.

63. Bailey AJ, Bazin S, Delaunay A: Changes in the nature of the collagen during development and resorption of granulation tissue. *Biochim Biophys Acta* 1973;328:383–390.

64. Tipton CM, James SL, Mergner W, et al: Influence of exercise on strength of medial collateral knee ligaments of dogs. *Am J Physiol* 1970;218:894–902.

65. Adamsons RJ, Musco F, Enquist IF: The relationship of collagen content to wound strength in normal and scorbutic animals. *Surg Gynecol Obstet* 1964;119:323–329.

66. Vogel HG: Correlation between tensile strength and collagen content in rat skin: Effect of age and cortisol treatment. *Connect Tissue Res* 1974;2:177–182.

67. Peacock EE Jr: Some aspects of fibrogenesis during the healing of primary and secondary wounds. *Surg Gynecol Obstet* 1962;115:408–414.

68. Bryant WM, Weeks PM: Secondary wound tensile strength gain: A function of collagen and mucopolysaccharide interaction. *Plast Reconstr Surg* 1967;39:84–91.

69. Fukubayashi T, Torzilli PA, Sherman MF, et al: An *in vitro* biomechanical evaluation of anterior-posterior motion of the knee. *J Bone Joint Surg* 1982;64A:258–264.

70. Grood ES, Noyes FR, Butler DL, et al: Ligamentous and capsular restraints preventing straight medial and lateral laxity in intact human cadaver knees. *J Bone Joint Surg* 1981;63A:1257–1269.

71. Markolf KL, Mensch JS, Amstutz HC: Stiffness and laxity of the knee—The contributions of the supporting structures. *J Bone Joint Surg* 1976;58A:583–594.

72. Andriacchi TP, Mikosz RP, Hampton SJ, et al: Model studies of the stiffness characteristics of the human knee joint. *J Biomech* 1983;16:23–29.

73. Blacharski PA, Somerset JH, Murray DG: A three-dimensional study of the kinematics of the human knee. *J Biomech* 1975;8:375–384.

74. Blankevoort L, Huiskes R: The effects of ACL substitute location on knee joint motion and cruciate ligament strains. *Trans Orthop Res Soc* 1987;12:268.

75. Lewis L, Lew WD: A method for locating an optimal "fixed" axis of rotation for the human knee joint. *J Biomech Eng* 1978;100:187–193.

76. Andriacchi TP, Stanwyck TS, Galante JO: Knee biomechanics and total knee replacement. *J Arthrop* 1986;1:211–219.

77. Draganich LF: *The Influence of the Cruciate Ligaments, Knee Musculature and Anatomy on Knee Joint Loading,* thesis. University of Illinois at Chicago, 1985.

78. Andriacchi TP, Galante JO, Fermier RW: The influence of total knee-replacement design on walking and stair-climbing. *J Bone Joint Surg* 1982;64A:1328–1335.

79. Andriacchi TP, Kramer GM, Landon GC: The biomechanics of running and knee injuries, in Finerman G (ed): American Academy of Orthopaedic Surgeons *Symposium on Sports Medicine: The Knee.* St. Louis, CV Mosby Co, 1985, pp 23–32.

80. Lewis JL, Hill JA, Ohland K, et al: Knee joint laxity and ligament force changes with ACL repairs. *Trans Orthop Res Soc* 1987;12:270.

81. Van Dijk R: *The Behavior of the Cruciate Ligaments in the Human Knee,* thesis. Nijmegen, The Netherlands, 1983.

82. Hart DP, Dahners LE: Medial collateral ligament healing in rats: The effects of repair, motion and secondary stabilizing ligaments. *Trans Orthop Res Soc* 1987;12:68.

83. Evans EB, Eggers GWN, Butler JK, et al: Experimental immobilization and remobilization of rat knee joints. *J Bone Joint Surg* 1960;42A:737–758.

84. Dahners LE, Wilson CJ: An examination of the mechanism of ligament and joint contracture. *Trans Orthop Res Soc,* in press.

85. Dahners LE: Ligament contraction—A correlation with cellularity and actin staining. *Trans Orthop Res Soc* 1986;11:56.

86. Peacock EE Jr: Some biochemical and biophysical aspects of joint stiffness: Role of collagen synthesis as opposed to altered molecular bonding. *Ann Surg* 1966;164:1–12.

87. Peacock EE Jr: Comparison of collagenous tissue surrounding normal and immobilized joints. *Surg Forum* 1963;14:440–441.

88. Akeson WH: An experimental study of joint stiffness. *J Bone Joint Surg* 1961;43A:1022–1034.

89. Woo SL-Y, Matthews JV, Akeson WH, et al: Connective tissue response to immobility: Correlative study of biomechanical and biochemical measurements of normal and immobilized rabbit knees. *Arthritis Rheum* 1975;18:257–264.

90. Akeson WH, Woo SL-Y, Amiel D, et al: Rapid recovery from contracture in rabbit hindlimb: A correlative biomechanical and biochemical study. *Clin Orthop* 1977;122:359–365.

91. Woo SL-Y, Gomez MA, Sites TJ, et al: The biomechanical and morphological changes of the MCL following immobilization and remobilization. *J Bone Joint Surg* 1987;69A:1200–1211.

92. Campion ER, Dahners LE: Electrical stimulation and ligament shortening: An *in vivo* study. *Trans Orthop Res Soc*, in press.

93. Amiel D, Frey C, Woo SL-Y, et al: Value of hyaluronic acid in the prevention of contracture formation. *Clin Orthop* 1985;196:306–311.

Section Three
Insertions

Group Leaders	**Group Members**	**Group Participants**
Savio Woo, PhD	David Butler, PhD	Thomas Andriacchi, PhD
Wayne Akeson, MD	Reginald Cooper, MD	Steven Arnoczky, DVM
	Jerry Maynard, PhD	Thomas Dameron, Jr., MD
	Barry Oakes, MD	Cyril Frank, MD
	Peter Torzilli, MD	Leo Furcht, MD
		Edward Grood, PhD
		Jack Lewis, PhD
		A. Hari Reddi, PhD
		Björn Rydevik, MD, PhD
		Ileen Stewart, MS

Synopsis

Structurally, the junction between soft tissues and bones is among the most complex of all biologic tissues. Within a distance of 1 mm, soft tissue is "transformed" into hard tissue by a progression of tissue types. These tissues include collagen fibers that blend into fibrocartilage, mineralized fibrocartilage, and finally bone. Further, the geometric attachments of tendons, ligaments, and joint capsules to bone can vary a great deal. In many instances, the microstructure of these insertion sites differs from ligament to ligament as well as between the two ends of the same ligament. Functionally, these junctions are frequent sites of injury, especially during skeletal immaturity when rapid remodeling of the insertion sites takes place. The serial (and gradual) changes of material from soft to hard tissue, which are designed to reduce stress concentration, cannot overcome the fact that these regions remain areas of weakness during loading.

We will review the structural and functional properties of the insertion sites in histologic, biochemical, and biomechanical terms. The topics include the gross morphology and anatomy of the different types of attachments to bone by tendon, ligaments, and joint capsules and more detailed microscopic descriptions of these insertion sites. We will discuss various microstructures, that is, the direct transformation from soft tissues to bone via the four zones and the indirect insertion consisting of superficial and deep components. This review will also include the current, albeit limited, information available about the vascular and nerve supply to the insertion sites. Unfortunately, there is little or no information on the biochemical properties of the insertion sites; this is an area where much work must be done.

We will examine the embryology, growth, and maturation as well as the aging of soft tissues and bone. Much has been published on these subjects, with the exception of quantitative data on the longitudinal growth of soft tissues around the insertion sites. Any discussion of the functional (or biomechanical) properties of the insertion sites should include the way in which the junction of hard and soft tissue is designed to dissipate tensile forces and to minimize stress concentrations. It should be recognized, however, that the complex geometry of the insertion sites limits the possibility of obtaining direct measurements of tensile deformation and other biomechanical properties. For example, measurements of cross-sectional area and deformation for detailed stress and strain analysis are extremely difficult to obtain. The only methods currently available use indirect approaches to measure these biomechanical properties.

With the recent discovery that the direction of applied loads has a significant effect on the tensile behavior as well as the failure modes of the insertion sites, evaluation and comparison of existing data become almost impossible. As a result, the available values for functional properties are subjects of controversy.

It has long been recognized that joint immobility has profound deleterious effects on insertion sites. The advance of experimental methodology has enabled us to study the effects of immobilization, remobilization after immobilization, and exercise on the morphologic and biomechanical changes at the insertion sites. There appears to be a correlation between the change in strength of the insertion site secondary to immobilization and its osteoclastic activities, but the detailed mechanism causing these homeostatic responses secondary to stress in motion remains undiscovered.

A significant amount of research remains to be performed if we are to gain a better understanding of the insertion sites of soft tissues into bone. The data available at present are limited to those obtained from clinical studies. Basic science studies of an interdisciplinary nature regarding the injury and repair of tendons, ligaments, and capsule insertions to bone and their long-term prognosis are excellent subjects for future investigations.

Four specific and crucial areas of research have been outlined. Important information to be gained from such studies includes an improved understanding of the biomechanical and biochemical properties and the morphologic and histologic appearances of the repair sites after different types of surgical intervention and postoperative treatment. Only when such a database is available can the use of synthetic materials and their fixation to bone be considered.

Chapter 4

Ligament, Tendon, and Joint Capsule Insertions to Bone

Savio Woo, PhD
Jerry Maynard, PhD
David Butler, PhD
Roger Lyon, MD
Peter Torzilli, PhD
Wayne Akeson, MD
Reginald Cooper, MD
Barry Oakes, MD

Chapter Outline

Types of Tissue Inserting to Bone

Gross Morphology

Tendon, ligament, and joint capsule attachments to bone are complex. Their macrostructures vary considerably. Muscle attachments to bone vary between their origin (usually a fleshy diaphyseal attachment) and their insertion (usually a well-circumscribed attachment). Tendons generally have large, parallel fibers that insert uniformly into bone. Ligaments have smaller-diameter fibers that can be either parallel, as in the medial collateral ligament (MCL) of the knee, or branching and interwoven, as in the cruciate ligaments. Joint-capsule fibers are also small and have more of a cross-hatched or random pattern. These differences, which are also present near their attachments to bone, result in different insertion site properties.

Most insertion sites appear grossly to be abrupt, well-defined areas with a relatively sharp boundary between the bone and the attaching soft tissue. This type of insertion can be termed a direct insertion and is the most common type of insertion as well as the most thoroughly studied. Examples include the supraspinatus muscle insertion to bone (Fig. 4–1, *top*) and the proximal insertion of the MCL to the distal femur (Fig. 4–1, *bottom*). There are also broader indirect insertion sites at which the junction of soft tissue with bone is more gradual and diffuse. These indirect insertions usually cover more bone surface than the direct type. Direct and indirect insertion sites differ greatly in their microscopic appearance.

Microscopic Anatomy

Direct Insertions Microscopically, both tendon and ligament fibers forming the insertion site can be divided into superficial and deep groups. Most of the fibers in direct insertion sites are deep fibers. They usually meet the bone at a right angle. The superficial fibers at both the proximal end and the distal end of the structures continue with the periosteum of the bone surface. The deep fibers inserting into the bone at direct insertion sites have four distinct zones.[1,2] These zones constitute the transition from soft tissue to bone at the insertion site (Fig. 4–1).

Zone 1 consists of the tendon or ligament proper. Type I collagen in an extracellular matrix, which includes dermatan sulfate proteoglycans, constitutes most of this zone.[1,3] The collagen fibrils vary in diameter from 25 to 300 nm and are the major tension-bearing elements. Interspersed among the collagen fibrils are a few elastin fibers. The only cell type present in this zone is the fibroblast. Small vessels characteristic of capillaries, small arterioles, or venules run parallel to the collagen fibrils. No nerves or nerve endings have been identified in this area.

Zone 2 consists of fibrocartilage. The collagen fibrils extend without any notable change in orientation or size into this zone. The major difference between zone 1 and zone 2 is that the cells become larger,

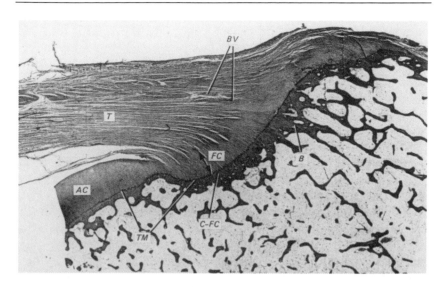

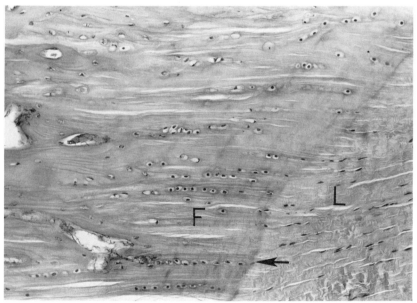

Fig. 4-1 *Direct insertions.* **Top:** *The insertion of the supraspinatus shares certain features with other tendons that have prominent fibrocartilage plugs at their attachment sites. There are four zones: tendon (T), uncalcified fibrocartilage (FC), calcified fibrocartilage (C-FC), and bone (B). Blood vessels (BV) are not present in the fibrocartilaginous zones. The tidemark (TM) between the calcified and uncalcified articular cartilage (AC) is continuous with that at the insertion of the tendon (hematoxylin and eosin, × 9). (Reproduced with permission from Benjamin M, Evans EJ, Copp L: The histology of tendon attachments to bone in man. J Anat 1986;149:89–100, Cambridge University Press.)* **Bottom:** *Femoral insertion of a rabbit MCL is a typical direct insertion. The deep fibers of the ligament (L) pass into bone through a well-defined zone of fibrocartilage (F). Arrow, the line of calcification or tidemark. (Reproduced with permission from Woo SL-Y, Gomez MA, Sites TJ, et al: The biomechanical and morphological changes in the medial collateral ligament of the rabbit after immobilization and remobilization. J Bone Joint Surg 1987;69A:1200–1211.)*

changing from thin, flat fibroblasts to more rounded or oval chondro-cyte-like cells lying in pairs and arranged in rows. The Golgi organelles become prominent and the cell processes are short, with most appearing to remain in a lacunar region that extends 1 to 2 μ around the cells. Chondrocytes are more numerous on this side of the line of mineralization. Small 10- to 20-nm filaments lie in the lacunar region as do dense, 10- to 30-nm granules that appear to attach to the filaments. Because of the narrow extent of this zone, little is known about possible differences in the composition of the matrix ground substance from that in zone 1. Microscopically, this zone appears somewhat similar to other fibrocartilage tissues; however, whether it also corresponds biochemically remains to be determined. In dogs, zone 2 is approximately 150 to 400 μ wide.

Zone 3 is characterized by mineralized fibrocartilage. Between the unmineralized and mineralized fibrocartilage, light microscopy shows the "blue line," cement line, or tidemark. This prominent basophilic line marks the outer limit of mineralization (calcification). The line is usually smooth in contour but occasionally it is irregular. The transition from nonmineralized to mineralized tissue is sudden. The first mineral crystals appear between the collagen fibril and their surfaces. Approximately 12 μ into the zone, the mineral crystals increase dramatically in number and aggregate into masses so that individual crystals are less obvious. This deeper part of zone 3, therefore, is characterized by dense mineral deposits both within and between the collagen fibrils. The mineral infiltration seems to start at the fibril edge and spread into the interior of the fibrils along the C2, D, E, and A intraperiod bands, eventually covering these bands so that they are no longer distinguishable. Mineral also spreads into the B1, B2, and C1 intraperiod bands, but less densely than at the mineralization front. Rarely does mineral appear to permeate the area of the cell lacunae. Although cartilage cells often appear to degenerate in other tissues when surrounded by mineral, many of the fibrocartilaginous cells in this zone apparently remain active despite being enclosed in lacunae surrounded by a mineralized matrix. The total width of this zone is approximately 100 to 300 μ.

Zone 4 is represented by bone. Here, the collagen of the inserting tissues blends imperceptibly with the collagen of the bone-matrix fibrils. There is no separating layer between the two. Hence, the collagen, although embedded in a bone matrix, remains type I. However, it should be noted that bone-matrix proteoglycans are represented by chondroitin sulfate proteoglycans similar to the small proteoglycans of cartilage, whereas the proteoglycans of ligaments and tendons are represented by dermatan sulfate proteoglycans.

Hence, within the short distance of less than 1 mm, the inserting tissues change to form four morphologically distinguishable zones and the ground substance constituting these zones changes from dermatan sulfate proteoglycans to chondroitin sulfate proteoglycans. The nature of these proteoglycans in this transition area remains to be defined.

Immunocytochemical techniques are needed to gain a better understanding.

Indirect Insertions   The fibers from tendons and ligaments at indirect insertion sites, like those at direct insertion sites, have both a superficial component and a deep component.[4,5] At indirect sites, however, the superficial fibers predominate and insertion to bone is mainly via fibers blending with the periosteum. The periosteum is composed of a superficial fibrous layer and a deeper osteogenic layer that maintains its continuity with bone.[6] The perforating fibers of Sharpey may also play an important role in periosteal attachment to bone. Although controversy continues over whether or not Sharpey's fibers (perforating fibers) are anatomic and functional structures,[1] most texts refer to collagen bundles extending from the periosteum and other soft tissues (ligaments and tendons) into the bone as Sharpey's fibers. Sharpey described fibers originating in the periosteum and perforating the underlying bone to extend across the multiple superficial lamellae of bone anchoring the periosteum to the lamellar bone. Sharpey's fibers originate when collagen bundles from the periosteum become buried in bone as the result of new subperiosteal bone formation. These fibers are surrounded by a narrow zone of uncalcified or partially calcified matrix.

The deep fibers of indirect insertions attach to bone with little or none of the transitional zone of fibrocartilage seen in direct insertions.[4,5] The collagen fibers at indirect insertion sites approach the bone at acute angles. Despite the lack of fibrocartilage, a tidemark still exists between the mineralized and nonmineralized tissue. Specifically, Woo and associates[5] reported that the tibial insertion of the rabbit MCL has both superficial fibers that run parallel to the bone surface and deep fibers that insert more obliquely into bone (Fig. 4–2). The superficial fibers of the MCL insert to bone in conjunction with the periosteum. In addition to the tibial insertions of the MCL, the insertions of the pronator teres and deltoid tendons are indirect.

Thus, the insertions of ligaments, tendons, and joint capsules to bone may vary markedly because of their anatomy and mechanical requirements. Despite these differences, all these structures insert to bone by similar mechanisms. Each soft tissue inserting to bone is composed of superficial and deep fibers near its insertion site. The superficial fibers become continuous with the adjacent periosteum and the deep fibers insert to bone either directly or through fibrocartilage. The junction between mineralized and nonmineralized tissues is identified by a tidemark. Direct insertion sites have a tidemark between the zones of mineralized and nonmineralized fibrocartilage, whereas indirect insertion sites have a tidemark between the soft-tissue fibrils and the bone without a fibrocartilage transition zone. The proximal attachment of collateral ligaments is commonly almost perpendicular to the epiphyseal surface of bone, whereas the distal attachment is more oblique with the superficial fibers blending with periosteal fibers. Tendons may insert almost perpendicularly or quite obliquely, depending on the particular muscle involved.

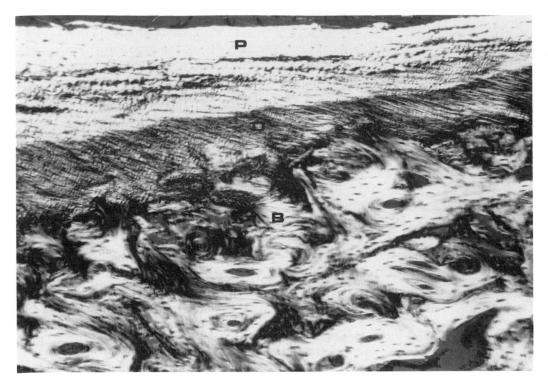

Fig. 4–2 *Tibial insertion of a rabbit MCL is a typical indirect insertion. The superficial fibers (P) predominate and run parallel to the bone (B), inserting into the periosteum. The deeper fibers (D) run obliquely, inserting into the underlying bone without a prominent plug of fibrocartilage (hematoxylin and eosin, polarized light, × 50). (Reproduced with permission from Woo SL-Y, Gomez MA, Sites TJ, et al: The biomechanical and morphological changes in the medial collateral ligament of the rabbit after immobilization and remobilization. J Bone Joint Surg 1987;69A:1200–1211.)*

Vascular and Nerve Supply of Insertion Sites

Although ligaments, tendons, and joint capsules are well vascularized, studies have shown that their insertion sites to bone are relatively avascular.[4,7–9] No vascular channels traverse the zone of fibrocartilage in insertion sites. Underlying bone, surrounding soft tissue (periosteum, synovium, and muscles), and adjacent tendons, ligaments, or joint capsules contribute to the blood supply of the areas immediately surrounding the insertion site (Fig. 4–3).

In a detailed study of vascularity in tendon insertions to bone, Dörfl[10,11] found that vascular patterns differed depending on the type of tendinous insertion. In chondroapophyseal insertions (inserting to an apophysis that develops from cartilaginous anlagen), the intratendinous vessels are separated from those of the bone in all developmental stages. These vessels end blindly in capillary loops near the

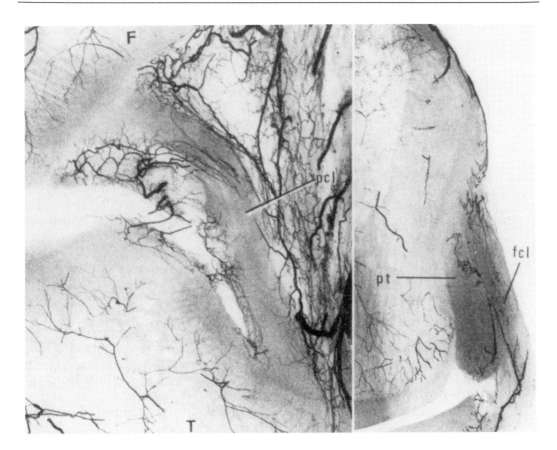

Fig. 4–3 *Ligament and tendon attachments to bone.* **Left:** *Side view of the posterior cruciate ligament (pcl) in a 33-year-old man. A number of vessels penetrate at various levels into the ligament, where they split upward and downward, but not into the osseous attachments. F, femur; T, tibia.* **Right:** *A section 3 mm thick from a 46-year-old man shows the same avascular zone between bone and fibrous bundles. The lateral condyle of the femur is attached to the fibular collateral ligament (fcl) and the popliteus (pt) beneath it. (Reproduced with permission from S. Karger AG, Basel (publisher): Scapinelli R: Studies on the vasculature of the human knee joint.* Acta Anat *1968;70:305–331.)*

insertion site. The vessels of the external peritendineum pass into the vascular network of the periosteum covering the apophysis, thus forming an anastomosis between the vessels of the tendon and those of the periosteum.

However, in diaphysoperiosteal insertions (inserting to the diaphysis of the bone), Dörfl described anastomoses between bone and intratendinous vessels. There are two types of diaphysoperiosteal insertions, planar and circumscribed. In a planar insertion, the tendon joins the periosteum directly. In the young individual, the periosteum has three layers: the cambial, the precursor, and the adventitial. In the adult, when periosteal opposition stops, only the outer two thirds persist. The vessels of the adventitial layer of the periosteum anastomose with vessels of the tendon insertion. Hence, the muscle insertion is connected with the bone vessels by means of the periosteal vessels in planar diaphysoperiosteal insertions. In circumscribed insertions, tendons attach directly to the bone, mostly to the spinae, trabeculae, or tuberosities. The typical periosteum associated with planar insertions is missing at these sites. Dörfl found that, in contrast to chondroapophyseal insertions, the intratendinous vascular bed in circumscribed insertions is connected directly with the vascular system of the bone. The connection between the vessels of the tendon and those of the bone varies with different sites (for example, only the peripheral parts of the biceps brachii insertion show the connection, whereas the vessels of the bone and tendons inserting on the linea aspera are richly connected). In summary, the one common feature of tendon insertions is that the vessels in the marginal parts of the insertion area of the tendon connect with those of the periosteum. Hence, the vessels of the external peritendineum and of the periosteum are the major vessels disrupted in avulsion injuries at the bone surface.

Superficial blood vessels in the peritendineum anastomose with vessels of the periosteum in most tendon-bone junctions, but intratendinous vessels remain separated from the vasculature of bone except in circumscribed diaphysoperiosteal insertions characterized by insertions to spinae, trabeculae, or tuberosities of bone. Dörfl's classification of tendon insertions does not address the presence or absence of a fibrocartilage transition zone. Presumably, vessels cross the insertion site (circumscribed tendon insertions) only at indirect insertion sites where there is little or no fibrocartilage.

The nerve supply of insertion sites presumably has a pattern similar to that of their blood supply. Apparently, no nerves cross the zone of fibrocartilage to innervate both sides of the insertion site.[1] However, bone, tendons, ligaments, and joint capsules have extensive neural elements. These elements, including nerves and free nerve endings, are in close proximity to insertion areas and thus may transmit information important in the analysis of joint motion, position, and acceleration.[12]

Development of Insertion Sites

Embryology

The development of soft-tissue insertions to bone is an important consideration when examining either structural or functional properties of these junctions. At 7 to 8 weeks of embryonic life, the knee joint in humans has cellular condensations forming knee ligaments, tendons, and a joint capsule in contact with chondrifying skeletal elements (Fig. 4–4).[13] Thus, ligaments and skeletal elements form as a unit, in situ, without migration. The development of the insertion site progresses along with that of the bone in general.

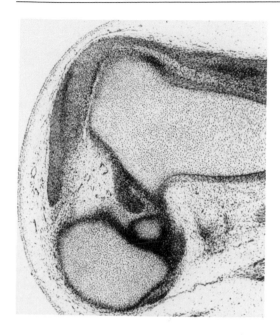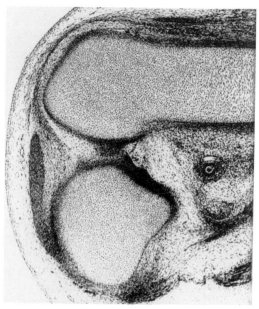

Fig. 4–4 *The embryonic human knee joint.* **Left:** *At the end of the embryonic period, the joint cavity, patellar tendon, joint capsule, and cruciate ligaments are in contact with skeletal elements.* **Right:** *The cruciate ligaments are more apparent between the femur (above) and the tibia (below). Skeletal elements and soft tissues simultaneously develop in situ. (Reproduced with permission from Gardner E, O'Rahilly R: The early development of the knee joint in staged human embryos. J Anat 1968;102:289–299, Cambridge University Press.)*

Growth

During longitudinal bone growth, the insertion sites of tendons and ligaments maintain a constant position relative to the growth plate and adjacent joint despite growth between the joint and the insertion site. Therefore, insertion sites must migrate toward the ends of long bones relative to the surrounding bone. Insertion sites would otherwise lie in the mid-diaphyseal region in mature animals.

Dörfl[14] sought to explain the cause and method of migration of tendon insertions in immature rabbits. Periosteal and bone markers were used to evaluate the migration of insertion sites and periosteum on the bony diaphysis during its growth. Tetracycline was used as an indicator of osteogenesis at the insertion sites. He concluded that insertion site migration is caused by dragging of the insertion by the periosteum, which is itself pulled by the epiphyses as they grow away from the diaphyses of the bone. Dörfl also identified three characteristics of the bone's surface at the insertion sites: osteogenesis, resorption, and combined osteogenesis and resorption. He suggested that these characteristics play some role in local mechanisms governing insertion-site migration.

Table 4–1 Tensile failure modes of the FMT complex*

			Tibial Avulsion		Midsubstance		Femoral Fracture	
Groups	Age (mos)	No.	No.	%	No.	%	No.	%
Open epiphyses								
Group 1	1.5	11	11	100	0	0	0	0
Group 2	4 to 5	20	20	100	0	0	0	0
Closed epiphyses								
Group 3	6 to 7	18	4	22	12	67	2†	11
Group 4	12 to 15	16	0	0	9	56	7†	44

*Reproduced with permission from Woo SL-Y, Orlando CA, Gomez MA, et al: Tensile properties of the medial collateral ligament as a function of age. *J Orthop Res* 1986;4:133–141.
†Failure at femoral transfixing pin sites.

Hurov[15] also studied a variety of soft-tissue insertions to bone during growth in the rabbit knee. He reported that these soft-tissue structures attached principally to fibrous periosteum or perichondrium, which then linked with subjacent bone or cartilage matrices. This precludes periosteal sliding directly along the bone surface as suggested by Dörfl.[14] Hurov concluded that differentiation and remodeling of cartilage and bone at insertion sites vary with age, attachment type, and attachment site. He suggested that more information concerning periosteal growth sites and timing of cartilage and bone differentiation at soft tissue-bone junctions would help explain the interactions of long bone growth and periosteal expansion that maintain the relative positions of attached soft-tissue structures.

Maturation

The effects of maturation on tendons, ligaments, and their insertions have been demonstrated.[16–30] Tipton and associates[30] and Woo and associates[17] demonstrated changes in the structural properties of the femur-MCL-tibia (FMT) complex during maturation of the rat and rabbit. In the case of rabbits, the strength of the FMT complex did not plateau until closure of the epiphyses, at about 7 months of age. Histologic examination demonstrated that the insertion site of the MCL on the tibial side is affected by its proximity to the growth plate where (1) osteoclastic activity in this region weakens the subperiosteal attachment and (2) part of the ligament insertion is at the area of the metaphysis. FMT complexes from rabbits with open epiphyses were tension-tested to failure. They all failed by tibial avulsion. During and after closure of the physis, only 12% (four of 34) failed by tibial avulsion (Table 4–1). Thus, it appears that the insertion sites are directly affected by growth plate activity and remain structurally inferior until epiphyseal closure (Fig. 4–5).

Aging

The changes in insertion sites with age after maturation have yet to be studied in detail. Studies of soft tissue-bone junctions in cadaver

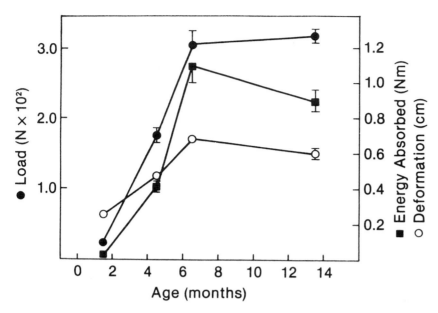

Fig. 4–5 *The structural properties of the FMT complex at failure. In groups 3 and 4 (Table 4–1), specimens that failed through the femoral pin site are excluded. (Reproduced with permission from Woo SL-Y, Orlando CA, Gomez MA, et al: Tensile properties of the medial collateral ligament as a function of age. J Orthop Res 1986;4:133–141.)*

specimens from older donors (more than 50 years of age) have demonstrated an increased incidence of bone avulsion failure compared with younger mature donors[31] (Fig. 4–6). Because many of these junctions involve a blending of the superficial fibers of the tendon or ligament with the periosteum, one must consider the periosteum as part of the insertion. There are considerable changes in the periosteum with aging.[6] The periosteum has a superficial fibrous layer and a deeper osteogenic layer. Fibroblasts and fibrocytes interposed between multidirectional collagen layers permeated by elastic tissue constitute the fibrous layer. The deeper osteogenic layer contains precursor osteogenic cells and mature osteoblasts adjacent to the bone surface with a decreased quantity of elastic fibers. Aging is characterized by a loss of cells and a loss of organelles associated with normal functional activities. Lipofuscin appears in all of the aged cells, including osteocytes. Despite severe ultrastructural alterations with age, it appears that sufficient organelles remain in a few cells of each layer to maintain viability. These cells may account for the reactivation of cell proliferation in response to trauma, even in the oldest animals.[6] The effects of aging on the insertion sites of tendons or ligaments as they mesh with the periosteum have yet to be described, despite the importance of the changes in soft tissue with senescence. To resolve these issues of aging of insertion sites, additional studies must be performed.

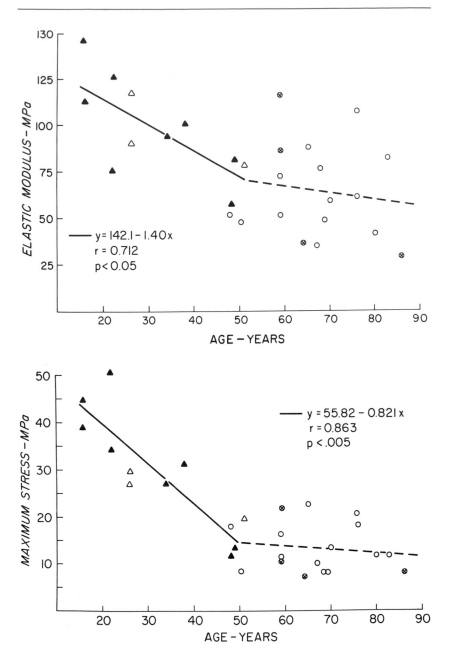

Fig. 4–6 *Human femur-ACL-tibia (FAT) complex calculations.* **Top:** *Elastic modulus correlated against donor age. The elastic modulus was determined in the linear prefailure region of the force elongation curve.* **Bottom:** *Maximum stress correlated against donor age. Solid lines, statistically significant correlations for the specimens that failed by ligament tear. Broken lines, no statistically significant correlations for the specimens that failed by bone avulsion. (Reproduced with permission from Noyes FR, Grood ES: The strength of the anterior cruciate ligament in humans and rhesus monkeys: Age-related and species-related changes.* J Bone Joint Surg *1976;58A:1074–1082.)*

Function of Insertion Sites

Soft-tissue insertions to bone provide a means of attaching a flexible yet strong, tension-bearing structure into a rigid, noncompliant bone. Through these attachments, forces are transmitted between bone and soft tissue. This function of force transmittal is important in joint kinetics and kinematics.

The junctions of ligaments, tendons, and joint capsules to bone have different structural organizations and are subject to different levels of force. For example, in vivo forces on tendon insertions are large because of muscular contraction.[32] By comparison, physiologic forces on ligament and joint capsule insertions may be smaller because of the relatively constant distances between their proximal and distal attachments during normal joint motion. Differences in fiber macrostructure near attachments can create different force distributions across the attachments and, thus, different insertion-site mechanics.

Tendon and ligament insertions to bone are adapted to force dissipation by passing through fibrocartilage to bone and are less susceptible to disruption in the transition area than the extremes on either side (bone or peri-insertional ligament substance). The composite material of insertion sites provides a gradual increase in stiffness from soft tissue to bone, thus diminishing stress concentrations and tearing or shearing at the interface.[31,33,34]

Although failures in the ligament, tendon, and joint capsule are common clinical injuries, there are also frequent failures through and near the insertion site. These failures can involve the actual bone-soft tissue junction or periosteal region, the underlying bone (producing bone avulsion), or the soft tissue adjacent to the insertion. Although these failure mechanisms are classified separately, it is likely that significant injuries involve several regions to some extent, including the actual insertion sites.

Tensile Deformation

The irregular shapes and geometry as well as the complex anatomic construction of insertion sites contribute to the difficulties of testing their biomechanical properties. One particular problem is separating the properties of the insertion site from those of the ligament. Also, accurate measurements of cross-sectional area at the insertion site are difficult. These problems make meaningful measurements of insertion-site deformation and cross-sectional area extremely difficult. Currently, the deformation of these insertion sites can only be estimated. One way to measure end-region (insertional area) deformation of a bone-ligament complex is to use bone and soft-tissue surface markers on either side of the insertion site.

It has recently been recognized that the deformation of an insertion site as a result of applied tensile load can be much greater than that of the ligament substance. Woo and associates[35] demonstrated that the strains at failure for the MCL substance were 14%, 12%, and 7% for dogs, swine, and rabbits, respectively, but that those for the FMT

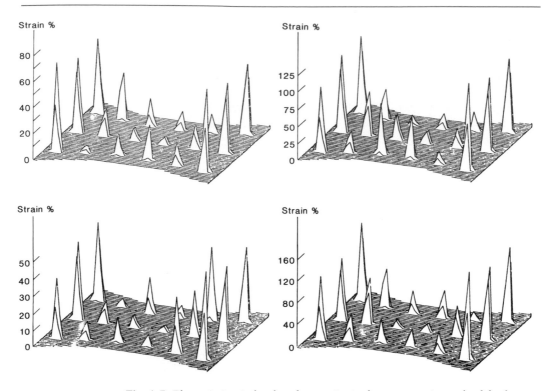

Fig. 4–7 *The variation in local surface strains is shown at maximum load for four patellar tendon-bone units. Each peak is proportional to the local strain (change in length per unit of length) in that region. Note the very large local strains at the patellar (left) and tibial (right) ends of the insertions. The double peaks seen at each location on the surface are simply a result of the computer code that was used for strain determination. (Reproduced with permission from Noyes FR, Butler DL, Grood ES, et al: Biomechanical analysis of human ligament grafts used in knee ligament repairs and reconstruction. J Bone Joint Surg 1984;66A:344–352.)*

complex were 21%, 30%, and 16%. These data suggest that the strains at or around the insertion site (or end region) may be at least twofold higher than in the midsubstance. These properties of ligaments and patellar tendon have also been investigated in human knees.[36–40] Strains at the insertion sites of both ends of the patellar tendon are contrasted with midsubstance strains in Figure 4–7.[33] The height of each peak is proportional to the local strain along and across the tissue. Note that at maximum load, the end-region strains for all four tendon-bone units are three to four times larger than those in the midsubstance. Because the force in individual fascicular bundles is constant, the end region (including zone 1 collagen fibers) cannot be as stiff as the mid-region collagen fibers.[36,37,39]

This pattern of large end-region strains was also seen when multi-bundle-bone units or subunits from human patellar tendon and anterior and posterior cruciate ligaments (ACL and PCL) were tested.[38,41–43]

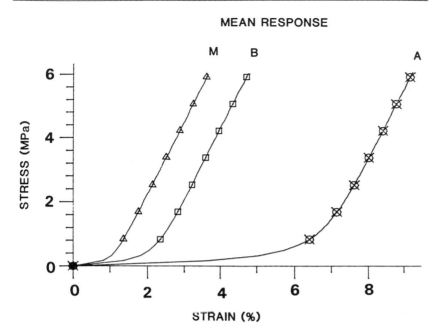

MEAN RESPONSE

Fig. 4-8 *The mean response of patellar tendon subunits. Average stress-strain curves are shown for ten subunits up to a constant peak stress of 6 MPa. The calculated attachment site response (A) is compared with the mean of seven local responses along the length (M) and the measured bone-to-bone response (B). Note the much larger strains near the attachment site. (Reproduced with permission from Stouffer DC, Butler DL, Hosny D: The relationship between crimp pattern and mechanical response of human patellar tendon-bone units.* J Biomech Eng *1985;108:158–165.)*

Figure 4–8 shows average stress-strain responses for ten patellar tendon subunits from three donors.[34] The mean of seven local measurements all along the bundle lengths (M) are plotted against the average bone end-to-bone end readings (B) and the calculated end-region or attachment responses (A). Note that the end regions undergo a much larger initial strain before receiving significant stresses, indicating again that strains in this region are much larger than anywhere else in the tissue.

What is the cause of this strain differential? At present, there are a few suggestions. Differences in collagen-fiber crimp period and angle exist between the midsubstance and the end regions.[38,40-43] In the barely taut subunit, crimp periods are smaller and crimp angles somewhat larger near the attachment site. As load is applied, larger strains are required to remove the end-region crimp than the midsubstance crimp.[38,41,42] Another possible explanation of this phenomenon may be differences in the types of constituents present in zone 1, but this has not been examined to date.

It is important to realize that this method and the results just presented provide only approximations of actual insertion-site behavior. That is, the data give an average response of all four zones plus a few millimeters of zone 1 collagen fiber bundles. The markers simply cannot be placed close enough to the beginning of zone 2 to isolate the fibrocartilaginous response to loading. This probably results in a domination of zone 1 collagen behavior. Another limitation is that these results represent only tissue surface behavior, which may differ significantly from properties in the deeper region.

Direction of Loading

The direction of applied force through the soft tissue influences the behavior of the insertion site. The tensile forces in the ligament, tendon, and joint capsule can produce tensile, compressive, and shear stresses at the insertion. Tissue fibers generally insert into bone at acute angles; that is, the angle between the fibers and the bone surface is small. A simple free-body analysis of the insertion suggests that the principal force acting is shear along the zones with a small component of tension normal to the tidemark. If this is true, the collagen fibers, which can bear only tension, create shear forces through zones 2 and 3. This shear force can be assumed to transmit both shear and compressive stresses to the unmineralized and mineralized matrices. Both proteoglycan and mineral matrix support compression rather well. The response of zones 2 and 3 to shear forces, however, is not known. Given the typical reported failure modes, these zones must resist shear forces rather well.

A study by Woo and associates[44] found a significant change in the structural properties of a bone-ligament-bone complex when the load was applied along the longitudinal axis of the ligament instead of along the longitudinal axis of the tibia. In the rabbit femur-ACL-tibia (FAT) complex, when the load is applied along the ACL axis, the angle of knee flexion has no significant effect on the ultimate load values and the failure mode is primarily bone avulsion. However, when loading is along the axis of the tibia, the FAT complex has progressively less strength as the knee flexion angle increases and the primary failure mode changes to ligament rupture. Similar results have also been seen in porcine and human ACL specimens.[45] The kinds of forces on the insertion site were not addressed in this study. However, these data demonstrated that changes in direction of force application on the insertion site alter its ultimate load and mode of failure.

Failure Modes

Ligaments and tendons can fail through their midsubstance, by bone avulsion (within zone 4), and, sometimes, at the insertion site (through zones 2 and 3). Probably the most common is the midsubstance or interstitial failure. These soft-tissue disruptions can be localized but can also involve collagen all along the tissue. The next most frequent failure mechanism, avulsion, typically occurs with large fragments of bone, that is, well away from the insertion of fibers into lamellar bone in zone 4. True junction failures are not very common. In such a junction separation, no appreciable amount of bone can be present on the failed ends and zones 2 and/or 3 must be disrupted (Fig. 4–9).

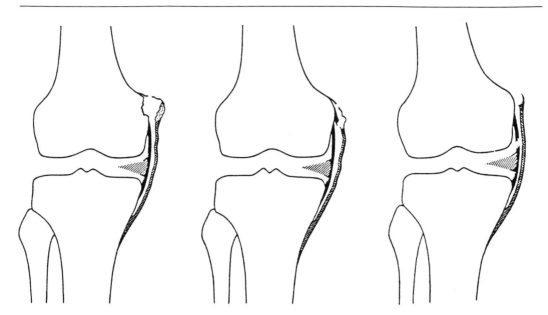

Fig. 4–9 *Schema of injury types near insertion sites.* **Left:** *Failure through lamellar bone well below the insertion site. The ligament end has a large bone fragment attached without disruption of the ligament substance.* **Center:** *Midsubstance failure of the MCL. Torn ligament ends remain attached to bone.* **Right:** *A true avulsion of the MCL. The MCL substance and bone remain intact with the ligament separating from the bone through the insertion junction (between zones 2 and 3), leaving a denuded area of bone. (Reproduced with permission from Muller W:* The Knee. *Berlin, Springer-Verlag, 1983, p 193.)*

Historically, the extension rate used to evaluate the strength of the bone-ligament complex has been slow, primarily because of the limited apparatus available for these experiments. Tipton and associates[46] evaluated the means by which rat MCL fails in vivo. They estimated that more than 70% of all ligament-bone junction failures occur in mineralized fibrocartilage and in bone, with the remaining 30% beginning in the nonmineralized fibrocartilage or in the ligament near the insertion. It should be noted that they conducted their tests at a slow extension rate.

Noyes and associates[34] conducted a study to examine failure mechanisms in ACL-bone complex taken from adult rhesus monkeys captured in the wild. When the specimens were tested at a fast extension rate (8.5 mm/sec), these authors observed three principal failure modes. The first type was a ligament failure through the tissue midsubstance, resulting in a mop-end appearance because of serial failure of fibers (Fig. 4–10). This mode occurred in 66% of the ACL-bone complexes tested. In the second mode, the bone avulsed, primarily at the tibial end, through cancellous bone beneath the insertion site. This mode occurred in 28% of the preparations. The remaining 6% consisted of combined ligament and bone failures. In these specimens, failure oc-

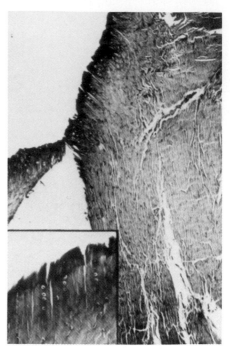

Fig. 4–10 *ACL failure in rhesus monkeys.* **Left:** *Failure by rupture of collagen fiber bundles. Failure at different points along the ligament produces a mop-end appearance (trichrome, × 45).* **Right:** *Failure through zone 3 (mineralized fibrocartilage).* **Inset,** *Columnar arrangement of chondrocytes (PAS, × 8). (Reproduced with permission from Noyes FR, DeLucas JL, Torvik PJ: Biomechanics of anterior cruciate ligament failure: An analysis of strain-rate sensitivity and mechanisms of failure in primates.* J Bone Joint Surg *1974:56A:236–253.)*

curred at several levels through the junction, but infrequently through the mineralized fibrocartilage (zone 3). Crowninshield and Pope[47] tested the rat FMT complex through a wide range of extension rates (from 8.25×10^{-6} m/sec to 5.1 m/sec) and found that the mode of failure depended on the extension rate. During fast rate tests, most failures occurred as midsubstance tears, whereas during slow extension rates, most occurred by tibial avulsion. However, in a recent study of rabbits (skeletally immature, $3\frac{1}{2}$ months old with open epiphyses, and skeletally mature, $8\frac{1}{2}$ months old with closed epiphyses), Peterson and Woo[48] found that the extension rates had no effect on the mode of failure. At five different extension rates (0.008 to 0.113 m/sec), all the samples from the skeletally immature group failed by tibial avulsion, whereas all those from skeletally mature animals failed by MCL substance tear. Strain rate was found to affect the biomechanical properties of the MCL and its insertions to bone, especially in skeletally immature animals. Therefore, it appears that the age of the animal has a more significant effect on the mode of failure than does the strain rate. These findings were supported by histologic evidence demonstrating that the tibial insertion for the MCL changes significantly during the closure of epiphyses. In addition, these authors cautioned that their conclusions were based on MCL-bone complex only and that these findings may not apply to other ligaments because the microstructure of the insertion site for other tissues, such as the ACL, is quite different. In the case of rat FMT complex,[47] one should note that the physes often do not close until senescence and that avulsion failure is, therefore, more likely.

Immobilization and Remobilization

Stress and joint motion are important in maintaining the functional integrity of insertion sites. The effects of joint immobility on soft connective tissue are profound, as exemplified by studies showing deterioration in cartilage, tendon, and ligament morphology, biochemistry, and biomechanics.[49–59] Mechanical and biochemical changes occur in the soft tissues, producing significant increases in joint stiffness and contracture.[53] Decreases in water and total hexosamine content parallel these mechanical changes.[50–53] Underlying bone becomes osteoporotic, resulting in increased compliance and susceptibility to premature failure, and the joint space is replaced by fibrofatty connective tissue.[56]

The effects of immobility on the insertion site are not well understood and have received limited attention. Biomechanical studies using animal models have demonstrated a rapid decline of soft tissue-bone junctional strength with joint immobilization. Specifically, Barfred[54] noted differences in the locations of failure and in the ultimate forces in Achilles tendon-bone units taken from wild and domesticated rats. Failures more often occurred through the bone than through the soft tissue when the animals were caged. Laros and associates[38] found the same effect in dogs. In this case, FMT complexes failed at the ligament-bone attachment site. Additionally, Noyes and associates[34] performed failure tests on immobilized and wild rhesus monkey ACL-bone unit preparations. After eight weeks of immobilization, ACL-bone unit stiffness, maximum force, and energy to failure decreased to 61% to 69% of control values (Fig. 4–11). Avulsion failure increased with immobilization to 60% from a control value of 40%. They concluded that the loss of cortex immediately beneath the ligament insertion reduced the strength of the entire ligament-bone unit. Woo and associates[5] also noted increased bone avulsion failure in FMT complexes after nine weeks of knee immobilization (Fig. 4–12). There was also a softening effect on ligament substance with immobilization. This and other studies have shown that stress deprivation more dramatically affects the indirect (or periosteal) type of insertion site. These authors observed subperiosteal resorption of bone, causing disruption of the fibers that attach to bone at indirect insertion sites (Fig. 4–13). This accounts for the increased avulsion failure in ligament-bone complexes.

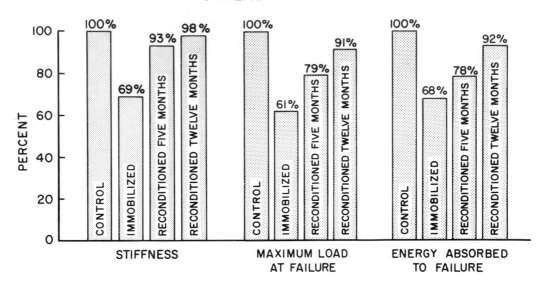

STRENGTH PROPERTIES

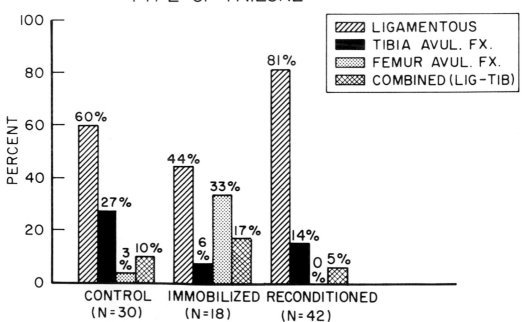

TYPE OF FAILURE

Fig. 4–11 *Immobilization and reconditioning.* **Top:** *Stiffness, maximum load at failure, and energy to failure. The control values are 100%. Decreases in all three factors are statistically significant after eight weeks of immobilization. Recovery is only partial after five months of reconditioning.* **Bottom:** *The most common failure mechanism in femur-ACL-tibia complex units in the three groups is ligamentous (five- to 12-month values combined). Femoral avulsion fractures increased in the immobilized group but not in the reconditioned group. (Reproduced with permission from Noyes FR: Functional properties of knee ligaments and alterations induced by immobilization: A correlative biomechanical and histological study in primates.* Clin Orthop *1977;123:210–242.)*

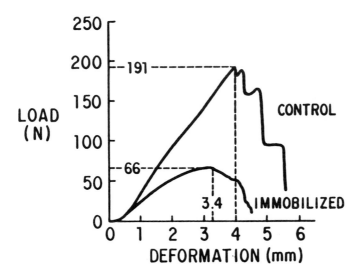

Fig. 4–12 *Typical load-deformation curves of rabbit femur-medial collateral ligament-tibia (FMT) complexes in both control and nine-week immobilization groups. The ultimate load and energy-absorbing capabilities in the immobilized FMT complexes were approximately one third those of the contralateral nonimmobilized control. (Reproduced with permission from Woo SL-Y, Gomez MA, Sites TJ, et al: The biomechanical and morphological changes in the medial collateral ligament of the rabbit after immobilization and remobilization.* J Bone Joint Surg *1987;69A:1200–1211.)*

These studies also examined the effects of reconditioning on ligament-bone complexes. In all there was a reversal of immobilization effects with increased activity levels (remobilization). Woo and associates[5] found that the mechanical properties of the MCL returned to the control values after nine weeks of remobilization but that the structural properties of the FMT complex remained inferior to those of controls. Failures continued to occur at bone insertion sites, and the histologic results supported these findings (Fig. 4–14). Noyes and associates[34] reported a return to normal stiffness and to a control distribution of failure modes in ACL-bone units after five months of remobilization. However, maximum force and energy to failure returned to only 80% of control values (Fig. 4–11). Histologically, the insertion site showed evidence of bone formation at sites of previous resorption. Laros and associates[58] reported that ligament-bone separation forces were equal to controls after three to four months of remobilization and that new bone formation at insertion sites was visible. Thus, the detrimental changes at insertion sites caused by immobilization are generally reversible with an increased activity level after a range of four months to one year.

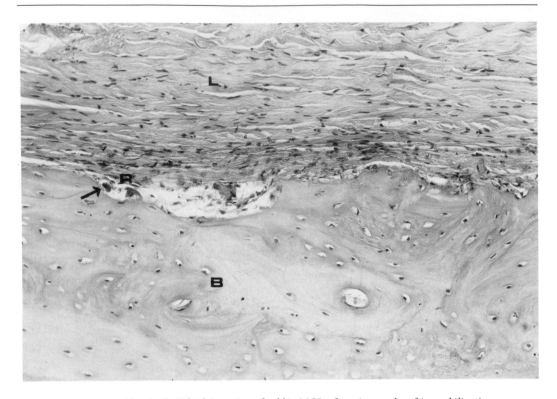

Fig. 4–13 *Tibial insertion of rabbit MCL after nine weeks of immobilization. Osteoclasts (arrow) have resorbed subperiosteal bone (B) and disrupted the ligament's attachment to bone (R). The superficial fibrils (L) maintain continuity with the periosteum (hematoxylin and eosin, × 50). (Reproduced with permission from Woo SL-Y, Gomez MA, Sites TJ, et al: The biomechanical and morphological changes in the medial collateral ligament of the rabbit after immobilization and remobilization. J Bone Joint Surg 1987;69A:1200–1211.)*

Effects of Exercise

The relationship of activity to soft-tissue and insertion-site properties has aroused interest in studying the effects of exercise or stress enhancement on ligaments and tendons at their insertion sites.[58,60–71] Many of these studies have shown that increasing activity level has a beneficial effect on the properties of bone-ligament complexes. Some of these studies have produced confusing or contradictory results. These discrepancies may be the result of different animal models, inadequate intergroup matching of important variables, different experimental procedures, or inconsistently defined controls and exercised animals. Despite these problems, it appears that ligament-bone units and bone-tendon complexes become stronger with exercise. The strength of the unit appears to be related to the type and duration of the activity. An activity must stress the specific insertion site in order to have a positive effect on its strength.

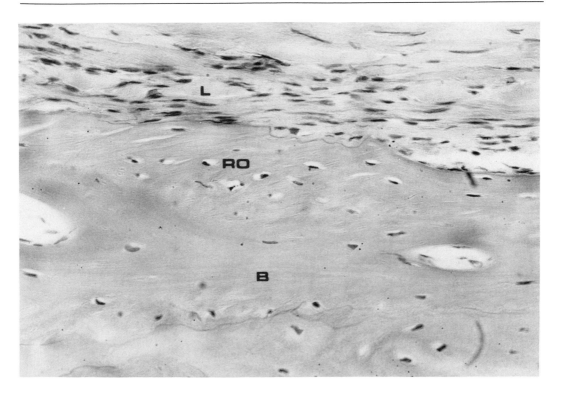

Fig. 4–14 *Rabbit MCL tibial insertion after 12 weeks of immobilization and nine weeks of remobilization. Much of the previously resorbed bone has reossified (RO) to establish continuity with the deeper ligament fibrils (L). The more mature bone lies below (B) (hematoxylin and eosin, × 50). (Reproduced with permission from Woo SL-Y, Gomez MA, Sites TJ, et al: The biomechanical and morphological changes in the medial collateral ligament of the rabbit after immobilization and remobilization.* J Bone Joint Surg *1987;69A:1200–1211.)*

Specifically, Laros and associates[58] demonstrated an increased bone-ligament-bone complex in dogs by keeping the animals in large, open pens. However, histologic findings at ligament insertion sites of exercised dogs were inconclusive, with some sites showing bone resorption and others not. They stated that exercise may only represent protection against the weakening effects of inactivity. Noyes and associates[34] found that exercising one limb in an immobilized primate model did not prevent the detrimental effects of immobilization.

Tipton and associates[66] demonstrated changes in ligaments and their insertion sites after endurance exercise training. Trained animals had ligaments with larger-diameter collagen fiber bundles and higher collagen content. They also demonstrated a higher separation force per body weight for the ligament-bone complex. Animals undergoing non-endurance exercise training did not exhibit similar increases in separation forces. Woo and associates[60,61] found an increase in the ultimate load of swine digital flexor tendons secondary to changes at the bone insertion sites after 12 months of exercise training.

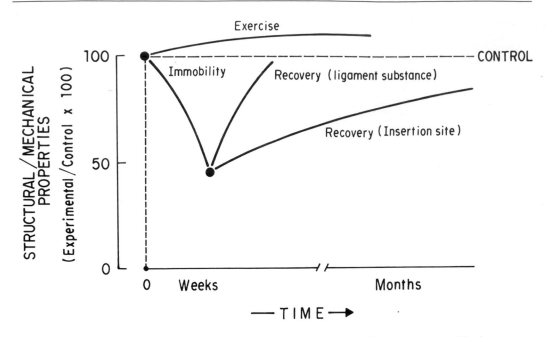

Fig. 4–15 *Summary of the homeostatic responses of the components of the bone-ligament-bone complex when subjected to different levels of physical activity.*

The detailed mechanisms affecting insertion sites are still unresolved. Changes in collagen fibrils (number and size), ground substance, and bone near the junction site may change junctional strength.

It is apparent that ligament insertions are particularly sensitive to stress and motion, especially those that insert in concert with the periosteum. Many of these changes are probably influenced by specific ligaments or types, but there may be a spectrum of tissue response to activity levels, that is, each tendon or ligament may require a unique threshold level of activity to maintain normal homeostatic conditions (Fig. 4–15).

Injury and Repair

Of the few pathologic conditions of soft-tissue insertions to bone, avulsion injuries are the most common.[72–74] These injuries result from acute overloading of the insertion interface. When the applied stress exceeds the tensile strength of the soft tissue-bone junction, separation of the soft tissue from bone results. Failure occurs most commonly in the soft tissue or bone adjacent to the junction. True junction failures occur, but are rare (Fig. 4–9). Junction and bone avulsion failures have significantly better outcomes than disruptions in the soft-tissue substance.

Repair of soft tissue-bone junction and avulsion injuries depends

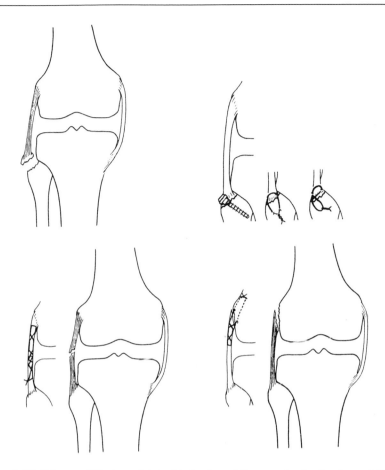

Fig. 4–16 *Schema of the knee joint with the three failure modes for the femur-LCL-tibia complex and methods of repair. (Reproduced with permission from Muller W: The Knee. Berlin, Springer-Verlag, 1983, p 229.)*

on the location and severity of the injury. Some investigators advocate nonsurgical therapy for nondisplaced or minimally displaced avulsion injuries.[75,76] Others recommend immediate surgical treatment.[77,78] Surgical treatment of junction separations consists of reattachment of the soft tissue end (tendon, ligament, or joint capsule) to the denuded bone with transosseous nonabsorbable sutures, screws with toothed or spiked washers, and staples (Fig. 4–16). Good functional results have been achieved with all of these methods. A study of soft-tissue fixation to bone by Robertson and associates[79] showed that screw fixation with a spiked plastic washer or soft-tissue plate has initial loading capabilities superior to those of staple and suture fixation. O'Donoghue and associates[80] showed that primary repair after transection of the ACL at its tibial insertion resulted in reestablishment of near-normal insertion-site structure (Fig. 4–17). However, the strength of the repair

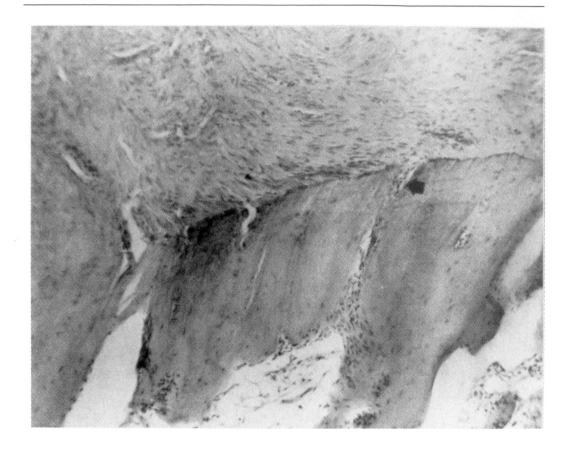

Fig. 4–17 *Repair of an ACL-bone unit (avulsion injury) in a dog after four weeks of immobilization. Fibroblastic activity continues, with fibroblastic tissue apparently invading bone adjacent to the repair (arrow). There is a definite line of demarcation between the healed ligament (above) and the bone (below), resembling that seen in a normal junction (× 40). (Reproduced with permission from O'Donoghue DH, Rockwood CA Jr, Frank GR, et al: Repairs of the anterior cruciate ligament in dogs. J Bone Joint Surg 1966;48A:503–519.)*

at one year was shown to be at most 50% of that at uninjured insertion sites.

If the initial treatment of insertion-site avulsion injuries fails, reconstructive procedures are frequently used to restore normal joint function. Both biologic and synthetic materials are used to replace native ligaments and tendons. These graft tissues require secure fixation to bone to achieve a good result. Frequently, bone tunnels are used to help achieve strong graft-to-bone bonding. The success of this fixation depends on the graft composition. Studies have suggested that using both autograft and allograft tissues leads to reestablishment of normal insertion-site architecture with fibrocartilage present at the interface.[81–86] The remodeling of the bone surrounding these grafts re-

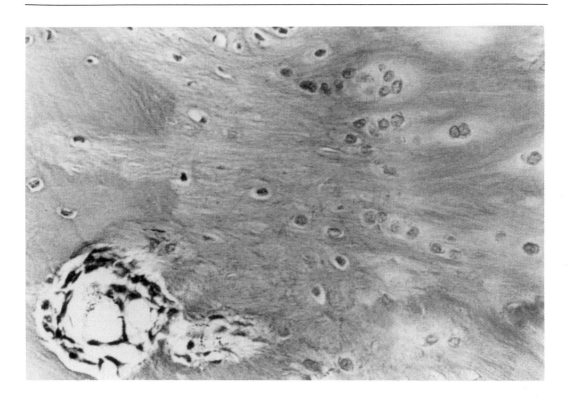

Fig. 4–18 *The intraosseus portion of a tendon allograft used for reconstruction of a canine ACL 30 weeks postoperatively. Note the fibrocartilage between the graft and the bone. This resembles a normal type of insertion. This situation is similar to that seen in patellar tendon autograft specimens. (Reproduced with permission from Shino K, Kawasaki T, Hirose H, et al: Replacement of the anterior cruciate ligament by an allogenic tendon graft: An experimental study in the dog.* J Bone Joint Surg *1984;66B:672–681.)*

sembles normal fracture healing. Within a few weeks, the grafts undergo an ingrowth of chondroblasts with subsequent new bone formation. Eventually, at about 30 weeks, a fibrocartilage junction typical of many normal insertion sites is established between the bone and the graft (Fig. 4–18).[84,85] However, soft-tissue reconstruction using xenografts has instead resulted in fibrous attachments to bone.[87]

Synthetic grafts are also used in soft-tissue reconstructive procedures. A study of carbon and Dacron fiber grafts demonstrated fibrous attachment to bone with bone ingrowth into only the most peripheral fibers of the grafts.[87] In other attempts to improve bone fixation of the synthetic grafts to bone, porous-coated plugs on the ends of the grafts and fenestration of graft ends have produced better results. Further evaluation of these techniques is needed to determine their feasibility.

Trauma at tendon insertion sites may also cause bone growth or ossification, commonly referred to as myositis ossificans. Usually, traumatic osseous growths are present at those insertions exposed to either a single, sudden, excessive force (as in sports injuries or accidents) or to repetitive loading (as in occupational injuries). Several case studies of these injuries have been reported by Hirsch and Morgan.[88] They found that in the early stages of the injury, the insertion site has an ossification center containing large amounts of fibrocartilage or hyaline cartilage continuous into bone but with varying degrees of differentiation. There appeared to be some regions with ossifying cartilage and others with lamellar trabecular bone with residual cartilage. In the later stages of the lesion, there was more lamellar bone and only small patches of cartilage. They postulated that this arrangement implied that the bone trabeculae formed through a process of endochondral ossification, originating in the fibrocartilage that normally composes the insertion site. Trauma, in this case, appears to disrupt the cartilaginous architecture of the insertion site, probably causing vascular ingrowth and ossification. With repeated trauma, osteophytes probably form in response to the force pattern or continual disruption of the fibrocartilage-bone junction.

One chronic abnormal state involving insertion sites is diffuse idiopathic skeletal hyperostosis of the spine (Forestier's disease).[89] It is characterized by bone osteophytes at the vertebral insertion sites of the anterior longitudinal ligament. Involvement is at multiple levels of the anterior thoracic spine, with osteophytes extending from the anterior margins of adjacent vertebrae coinciding with the anterior longitudinal ligament fibers. At the insertion sites, the fibrocartilage zones are thin and poorly organized and the fibrochondrocytes are more randomly oriented. However, the collagen fibers still appear roughly parallel, but are poorly organized. The boundary line is also thin and often indistinguishable from the fibrocartilage zone. In general, the osteophytes are representative of immature bone. The ligaments appear to grow into the osteophytes, and probably represent a functional change to redirect the ligament forces to stabilize the vertebral bodies. Whether this condition actually originates at insertion sites or only involves them secondarily is unresolved.

Future Directions

A more comprehensive understanding of insertion sites will require a number of important studies: the structural and mechanical properties, the composition, and the functions of normal insertion sites must be determined. Variations in these characteristics within and among different ligaments, tendons, and capsules must be identified in humans as well as in experimental animals. The microstructure of the insertions—including the dimensions of the zones, the orientation of the collagen in the zones, the locations and numbers of blood vessels and nerves, and the presence of mechanoreceptors—must be described. Measuring the mechanical properties of each zone will require the development of miniaturized force transducers. It may be possible to determine the functions of insertion sites by using mechanoreceptors as force transducers to measure tissue forces in vivo. In addition, the significance of multiple tidemarks must be explored. Specific biochemical methods—such as assays designed to quantify small amounts of substances—must be developed to determine the composition of different regions so that structure, biomechanical properties, and composition can be correlated.

We have emphasized the similarities among the different soft-tissue insertions to bone. In large part, we have assumed that these similarities exist on the basis of similar morphologic findings. However, anatomic, biomechanical, and biochemical evaluations of ligament, tendon, and joint capsule attachments to bone may allow us to discern important differences among the junctions.

A second area of focus is the identification of the clinical problems involving ligament, tendon, and joint capsule insertions into bone. There are apparently many conditions (for example, infrapatellar tendonitis, tennis elbow, and diffuse idiopathic skeletal hyperostosis) in which disruptions occur near or at insertion sites. Whether these injuries are caused by overuse or result from a single traumatic event should be clarified. Epidemiologic studies are needed to determine the precise locations and frequency of these problems.

The third major area in which research is needed is quantification of the effects of various factors on normal insertion sites. We need additional studies of the effects of growth, maturation, and aging on the mechanical properties, structure, and chemistry of the different insertion sites. How the rate of loading affects specific areas of the insertion site is still unknown. The effects of immobilization, remobilization, and exercise on insertion sites (other than those of the MCL) also require further examination. Specifically, at what rate and by what mechanism does reversal of the deleterious effects secondary to stress deprivation occur? The level of loading required to maintain homeostasis also deserves study.

The final important research area is insertion-site behavior after injury or graft replacement. Little is known about how the structure and function of insertion sites respond to overuse injuries or blunt trauma. Our understanding of short- and long-term cellular and vascular responses in an injured insertion site is limited. Information is needed about the interaction between underlying bone and autografts, allografts, and xenografts. In particular, do these grafts modify with time so that they come to resemble normal insertion sites?

One of the main problems in tendon and ligament reconstruction continues to be graft attachment to bone. This problem is most apparent when xenografts or synthetic grafts are used.

Ultimately, we must determine the biomechanical and biochemical environment that weakens and injures these insertion sites. Quantitative information on forces and force distributions within the joint structure and tissues during both normal and traumatic situations must be established. Problems related to repetitive injuries must also be addressed, particularly those relating to sports and recreational activities. Factors that can enhance healing under such conditions must be defined. The investigation of healing and repair of injuries ranging from minor damage to catastrophic failure of the insertion sites and their relationship to the stages of healing from inflammation to remodeling are subjects of major importance.

Acknowledgments

Supported by the RR&D of the Veteran's Administration Medical Center; NIH grants AM1498, AM00965, AM21172, AM27517, HD02788, TIAM5401, AM-08890-10; NSF grants MEA8118140, ECE8518836; the Malcolm & Dorothy Coutts Institute for Joint Reconstruction and Research; and SUPHS grant AM-08893-05.

References

1. Cooper RR, Misol S: Tendon and ligament insertion: A light and electron microscopic study. *J Bone Joint Surg* 1970;52A:1–21.
2. Dolgo-Saburoff B: Über Ursprung und Insertion der Skelettmuskeln: Mikroskopische Untersuchung. *Anat Anz* 1929;68:30–87.
3. Heinegaard D, Paulsson M: Structure and metabolism of proteoglycans, in Piez KA, Reddi AH (eds): *Extracellular Matrix Biochemistry*. New York, Elsevier, 1984, pp 278–328.
4. Benjamin M, Evans EJ, Copp L: The histology of tendon attachments to bone in man. *J Anat* 1986;149:89–100.
5. Woo SL-Y, Gomez MA, Sites TJ, et al: The biomechanical and morphological changes in the medial collateral ligament of the rabbit after immobilization and remobilization. *J Bone Joint Surg* 1987;69A:1200–1211.
6. Tonna EA: Electron microscopy of aging skeletal cells: III. The periosteum. *Lab Invest* 1974:31:609–632.
7. Scapinelli R: Studies on the vasculature of the human knee joint. *Acta Anat* 1968;70:305–331.
8. Arnoczky SP: Anatomy of the anterior cruciate ligament. *Clin Orthop* 1983;172:19–25.
9. Arnoczky SP, Rubin RM, Marshall JL: Microvasculature of the cruciate ligaments and its response to injury: An experimental study in dogs. *J Bone Joint Surg* 1979;61A:1221–1229.
10. Dörfl J: Vessels in the region of tendinous insertions: I. Chondroapophyseal insertion. *Folia Morphol* 1969;17:74–78.
11. Dörfl J: Vessels in the region of tendinous insertions: II. Diaphysoperiosteal insertion. *Folia Morphol* 1969;17:79–82.
12. Schutte MJ, Dabezies EJ, Zimny ML, et al: Neural anatomy of the human anterior cruciate ligament. *J Bone Joint Surg* 1987;69A:243–247.
13. Gardner E, O'Rahilly R: The early development of the knee joint in staged human embryos. *J Anat* 1968;102:289–299.
14. Dörfl J: Migration of tendinous insertions: I. Cause and mechanism. *J Anat* 1980;131:179–195.
15. Hurov JR: Soft-tissue bone interface: How do attachments of muscles, tendons, and ligaments change during growth?: A light microscopic study. *J Morphol* 1986;189:313–325.
16. Yamada H: Mechanical properties of locomotor organs and tissues, in Evans G (ed): *Strength of Biological Materials*. Baltimore, Williams & Wilkins, 1970, pp 80–81.

17. Woo SL-Y, Orlando CA, Gomez MA, et al: Tensile properties of the medial collateral ligament as a function of age. *J Orthop Res* 1986;4:133–141.

18. Blanton PL, Biggs NL: Ultimate tensile strength of fetal and adult human tendons. *J Biomech* 1970;3:181–189.

19. Benedict JV, Walker LB, Harris EH: Stress-strain characteristics and tensile strength of unembalmed human tendon. *J Biomech* 1968;1:53–63.

20. Galeski A, Kastelic J, Baer E, et al: Mechanical and structural changes in rat tail tendon induced by alloxan diabetes and aging. *J Biomech* 1977;10:775–782.

21. Hall DA: *The Ageing of Connective Tissue.* London, Academic Press, 1976, pp 13–67.

22. Rigby BJ, Hirai N, Spikes JD, et al: The mechanical properties of rat tail tendon. *J Gen Physiol* 1959;43:265–283.

23. Sinex FM: The role of collagen in aging, in Gould BS (ed): *Treatise on Collagen: Vol 2. Biology of Collagen, Part B.* London, Academic Press, 1968, pp 409–448.

24. Vogel HG: Correlation between tensile strength and collagen content in rat skin: Effect of age and cortisol treatment. *Connect Tissue Res* 1974;2:177–182.

25. Booth FW, Tipton CM: Ligamentous strength measurements in prepubescent and pubescent rats. *Growth* 1970;34:177–185.

26. Morein G, Goldgefter L, Kobyliansky E, et al: Changes in the mechanical properties of rat tail tendon during postnatal ontogenesis. *Anat Embryol* 1978;154:121–124.

27. Nathan H, Goldgefter L, Kobyliansky E, et al: Energy absorbing capacity of rat tail tendon at various ages. *J Anat* 1978;127:589–593.

28. Vogel HG: Tensile strength, relaxation and mechanical recovery in rat skin as influenced by maturation and age. *J Med* 1976;7:177–188.

29. Vogel HG: Influence of maturation and age on mechanical and biochemical parameters of connective tissue of various organs in the rat. *Connect Tissue Res* 1978;6:161–166.

30. Tipton CM, Matthes RD, Martin RK: Influence of age and sex on the strength of bone-ligament junctions in knee joints of rats. *J Bone Joint Surg* 1978;60A:230–234.

31. Noyes FR, Grood ES: The strength of the anterior cruciate ligament in humans and rhesus monkeys: Age-related and species-related changes. *J Bone Joint Surg* 1976;58A:1074–1082.

32. Grood ES, Butler DL, Noyes FR: Models of ligament repairs and grafts, in American Academy of Orthopaedic Surgeons *Symposium on Sports Medicine: The Knee.* St. Louis, CV Mosby Co, 1985, pp 169–181.

33. Noyes FR, Grood ES, Butler DL, et al: Clinical biomechanics of the knee—Ligament restraints and functional stability, in Funk FJ Jr (ed): American Academy of Orthopaedic Surgeons *Symposium on the Athlete's Knee: Surgical Repair and Reconstruction.* St. Louis, CV Mosby Co, 1980, pp 1–55.

34. Noyes FR, DeLucas JL, Torvik PJ: Biomechanics of anterior cruciate ligament failure: An analysis of strain-rate sensitivity and mechanisms of failure in primates. *J Bone Joint Surg* 1974;56A:236–253.

35. Woo SL-Y, Gomez MA, Seguchi Y, et al: Measurement of mechanical properties of ligament substance from a bone-ligament-bone preparation. *J Orthop Res* 1983;1:22–29.

36. Noyes FR, Butler DL, Grood ES, et al: Biomechanical analysis of human ligament grafts used in knee-ligament repairs and reconstructions. *J Bone Joint Surg* 1984;66A:344–352.

37. Zernicke RF, Butler DL, Grood ES, et al: Strain topography of human tendon and fascia. *J Biomech Eng* 1984;106:177–180.

38. Stouffer DC, Butler DL, Hosny D: The relationship between crimp pattern and mechanical response of human patellar tendon-bone units. *J Biomech Eng* 1985;107:158–165.

39. Butler DL, Grood ES, Noyes FR, et al: Effects of structure and strain measurement technique on the material properties of young human tendons and fascia. *J Biomech* 1984;17:579–596.

40. Butler DL, Stouffer DC, Wukusick PM, et al: Analysis on nonhomogeneous strain response of human patellar tendon. Presented at the American Society of Mechanical Engineers Joint Biomechanics Symposium, Houston, Texas, 1983.

41. Sheh M, Butler DL, Stouffer DC: Mechanical and structural properties of the human cruciate ligaments and patellar tendon. *Trans Orthop Res Soc* 1986;11:236.

42. Sheh M, Butler DL, Stoffer DC: Correlation between structure and material properties in human knee tendons and ligaments. *Am Soc Mech Eng AMD* 1985;68:17–20.

43. Sheh MY: *Correlation of Structure and Material Properties in Human Patellar Tendon and Knee Ligaments*, thesis. University of Cincinnati, 1986.

44. Woo SL-Y, Hollis JM, Roux RD, et al: The effects of knee flexion on the structural properties of the rabbit femur-anterior cruciate ligament-tibia complex (FATC). *J Biomech*, in press.

45. Hollis JM, Lee EB, Ballock RT, et al: Variation of structural properties of the anterior cruciate ligament as a function of loading direction. *Trans Orthop Res Soc* 1987;12:196.

46. Tipton CM, Matthes RD, Sandage DS: In situ measurement of junction strength and ligament elongation in rats. *J Appl Physiol* 1974;37:758–761.

47. Crowninshield RD, Pope MH: The strength and failure characteristics of rat medial collateral ligaments. *J Trauma* 1976;16:99–105.

48. Peterson RH, Woo SL-Y: A new methodology to determine the mechanical properties of ligaments at high strain rates. *J Biomech Eng* 1986;108:365–367.

49. Akeson WH, Amiel D, Woo SL-Y: Immobility effects on synovial joints: The pathomechanics of joint contracture. *Biorheology* 1980;17:95–110.

50. Akeson WH: An experimental study of joint stiffness. *J Bone Joint Surg* 1961;43A:1022–1034.

51. Akeson WH, Amiel D, LaViolette D: The connective-tissue response to immobility: A study of the chondroitin–4 and 6-sulfate and dermatan sulfate changes in periarticular connective tissue of control and immobilized knees of dogs. *Clin Orthop* 1967;51:183–197.

52. Akeson WH, Woo SLY, Amiel D, et al: The connective tissue response to immobility: Biochemical changes in periarticular connective tissue of the immobilized rabbit knee. *Clin Orthop* 1973;93:356–362.

53. Akeson WH, Woo SL-Y, Amiel D, et al: Biomechanical and biochemical changes in the periarticular connective tissue during contracture development in the immobilized rabbit knee. *Connect Tissue Res* 1974;2:315–323.

54. Barfred T: Experimental rupture of the Achilles tendon: Comparison of various types of experimental rupture in rats. *Acta Orthop Scand* 1971;42:528–543.

55. Cooper RR: Alterations during immobilization and regeneration of skeletal muscle in cats. *J Bone Joint Surg* 1972;54A:919–953.

56. Enneking WF, Horowitz M: The intra-articular effects of immobilization on the human knee. *J Bone Joint Surg* 1972;54A:973–985.

57. Geiser M, Trueta J: Muscle action, bone rarefaction and bone formation: An experimental study. *J Bone Joint Surg* 1958;40B:282–311.

58. Laros GS, Tipton CM, Cooper RR: Influence of physical activity on ligament insertions in the knees of dogs. *J Bone Joint Surg* 1971;53A:275–286.

59. Woo SL-Y, Matthews JV, Akeson WH, et al: Connective tissue response to immobility: Correlative study of biomechanical and biochemical measurements of normal and immobilized rabbit knees. *Arthritis Rheum* 1975;18:257–264.

60. Woo SL-Y, Gomez MA, Amiel D, et al: The effects of exercise on the biomechanical and biochemical properties of swine digital flexor tendons. *J Biomech Eng* 1981;103:51–61.

61. Woo SL-Y, Ritter MA, Amiel D, et al: The biomechanical and biochemical properties of swine tendons—Long term effects of exercise on the digital extensors. *Connect Tissue Res* 1980;7:177–183.

62. Tipton CM, Matthes RD, Maynard JA, et al: The influence of physical activity on ligaments and tendons. *Med Sci Sports* 1975;7:165–175.

63. Viidik A: The effect of training on the tensile strength of isolated rabbit tendons. *Scand J Plast Reconstr Surg* 1967;1:141–147.

64. Viidik A: Tensile strength properties of Achilles tendon systems in trained and untrained rabbits. *Acta Orthop Scand* 1969;40:261–272.

65. Viidik A: Functional properties of collagenous tissues. *Int Rev Connect Tissue Res* 1973;6:127–215.

66. Tipton CM, James SL, Mergner W, et al: Influence of exercise on strength of medial collateral knee ligaments of dogs. *Am J Physiol* 1970;218:894–902.

67. Tipton CM, Schild RJ, Tomanek RJ: Influence of physical activity on the strength of knee ligaments in rats. *Am J Physiol* 1967;212:783–787.

68. Adams A: Effect of exercise upon ligament strength. *Res Q* 1966;37:163–167.

69. Zuckerman J, Stull GA: Effects of exercise on knee ligament separation force in rats. *J Appl Physiol* 1969;26:716–719.

70. Zuckerman J, Stull GA: Ligamentous separation force in rats as influenced by training, detraining, and cage restriction. *Med Sci Sports* 1973;5:44–49.

71. Cabaud HE, Chatty A, Gildengorin V, et al: Exercise effects on the strength of the rat anterior cruciate ligament. *Am J Sports Med* 1980;8:79–86.

72. Norman WH: Repair of avulsion of insertion of biceps brachii tendon. *Clin Orthop* 1985;193:189–194.

73. Torisu T: Avulsion fracture of the tibial attachment of the posterior cruciate ligament: Indications and results of delayed repair. *Clin Orthop* 1979;143:107–114.

74. Leddy JP, Packer JW: Avulsion of the profundus tendon insertion in athletes. *J Hand Surg* 1977;2:66–69.

75. Torisu T: Isolated avulsion fracture of the tibial attachment of the posterior cruciate ligament. *J Bone Joint Surg* 1977;59A:68–72.

76. Heckman JD: Fractures and dislocations of the foot, in Rockwood CA Jr, Green DP (eds): *Fractures in Adults*, ed 2. Philadelphia, JB Lippincott, 1984, vol 2, pp 1703–1832.

77. Larson RL, Jones DC: Fractures and dislocations of the knee: Part 2. Dislocation and ligamentous injuries of the knee, in Rockwood CA Jr, Green DP (eds): *Fractures in Adults*, ed 2. Philadelphia, JB Lippincott, 1984, vol 2, pp 1480–1591.

78. Muller W: *The Knee.* Berlin, Springer-Verlag, 1983, pp 148–222.

79. Robertson DB, Daniel DM, Biden E: Soft tissue fixation to bone. *Am J Sports Med* 1986;14:398–403.

80. O'Donoghue DH, Rockwood CA Jr, Frank GR, et al: Repairs of the anterior cruciate ligament in dogs. *J Bone Joint Surg* 1966;48A:503–519.

81. Forward AD, Cowan RJ: Tendon suture to bone: An experimental investigation in rabbits. *J Bone Joint Surg* 1963;45A:807–823.

82. Arnoczky SP, Warren RF, Ashlock MA: Replacement of the anterior cruciate ligament using a patellar tendon allograft: An experimental study. *J Bone Joint Surg* 1986;68A:376–385.

83. Arnoczky SP, Tarvin GB, Marshall JL: Anterior cruciate ligament replacement using patellar tendon: An evaluation of graft revascularization in the dog. *J Bone Joint Surg* 1982;64A:217–224.

84. Clancy WG Jr, Narechania RG, Rosenberg TD, et al: Anterior and posterior cruciate ligament reconstruction in rhesus monkeys: A histological, microangiographic, and biomechanical analysis. *J Bone Joint Surg* 1981;63A:1270–1284.

85. Shino K, Kawasaki T, Hirose H, et al: Replacement of the anterior cruciate ligament by an allogenic tendon graft: An experimental study in the dog. *J Bone Joint Surg* 1984;66B:672–681.

86. Kernwein GA: A study of tendon implantations into bone. *Surg Gynecol Obstet* 1942;75:794–796.

87. Thomas NP, Turner IG, Jones CB: Prosthetic anterior cruciate ligaments in the rabbit: A comparison of four types of replacement. *J Bone Joint Surg* 1987;69B:312–316.

88. Hirsch EF, Morgan RH: Causal significance to traumatic ossification of the fibrocartilage in tendon insertions. *Arch Surg* 1939;39:824–837.

89. Matyas JR: *The Structure and Function of Tendon and Ligament Insertions to Bone*, thesis. Cornell University Medical College, New York, New York, 1985.

Section Four
Myotendinous Junction

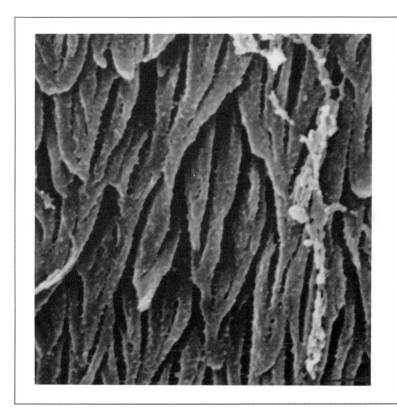

Group Leader	**Group Members**	**Group Participants**
William Garrett, Jr., MD, PhD	Laurence Dahners, MD	Albert Banes, PhD
	Jerry Maynard, PhD	Richard Brand, MD
	James Tidball, PhD	David Butler, PhD
		Arnold Caplan, PhD
		Bruce Caterson, PhD
		Reginald Cooper, MD
		Gerald Finerman, MD
		Donald Fischman, MD
		Richard Gelberman, MD
		Stephen Gordon, PhD
		Richard Lymn, PhD
		Lawrence Shulman, MD
		Savio Woo, PhD

Synopsis

The tension produced in skeletal muscle is transmitted to tendon through the highly specialized region of the myotendinous junction. Although this region has been studied far less than muscle or tendon, information is now available regarding the structure and molecular composition of this region. Data also suggest that indirect muscle injuries or strains occur in the region of the myotendinous junction. The importance of muscle strains makes study of this region exciting.

Structural studies show that the muscle cell surface is highly folded at its termination in the tendon. The surface area for tension transfer from the cell to the extracellular matrix is greatly amplified by these interdigitations. The membrane folding also means that tension is transferred primarily as a shear stress because the angle between the membrane and the direction of the tensile force is small. The foldings effectively increase the area of contact between the cell and extracellular matrix, thus increasing the adhesive strength as well as decreasing the stress—both of which help in the transmission of large forces.

We are gradually coming to understand the ultrastructure and molecular composition of the myotendinous junction. Myofibrils end in thin filaments that pass from the terminal Z-disk to a dense region beneath the sarcolemma. Immunocytochemical studies have localized at least two cytoskeletal proteins, vinculin and talin, to this region. Force must be transmitted from the subsarcolemmal density across the membrane. A large transmembrane glycoprotein, integrin, is probably involved. On electron microscopy, filaments can be seen to connect the cell membrane to the basement membrane surrounding the cell. The basement membrane is an uninterrupted sheath around the cell that links the muscle to the extracellular space. Among the components of the basement membrane are type IV collagen, fibronectin, and laminin, all of which are high-molecular-weight glycoproteins. The basement membrane, through these and other molecules, transmits force to the type I collagen fibers that ultimately connect into the tendon.

Specializations of the myotendinous junction may well involve the terminal few millimeters of the muscle fibers themselves. The fibers in the terminal regions are stiffer than the majority of the muscle fibers, as is shown by shorter sarcomere lengths near the myotendinous junction as muscle is stretched.

Biomechanical studies of whole-muscle preparations have been done to define what happens in muscle with a strain injury. Interestingly, the myotendinous junction has been consistently shown to be the weakest link. These studies have been repeated at several strain rates, with or without muscle activation, and in muscles of different fiber architecture. The site of failure appears to be within the terminal fibers near the myotendinous junction, although not necessarily at the ultrastructural myotendinous junction.

Incomplete muscle injuries produced by muscle strain in the plastic region of the length-tension relationship appear as disruptions of a small number of fibers near the myotendinous junction. The muscle recovers its active ability to generate tension after an initial decline. Healing occurs by an inflammatory reaction followed by a proliferation of cells resulting in increased fibrosis at the region of injury.

Clinical studies demonstrate that muscle is frequently injured during eccentric contraction, that is, activation of the muscle while it is being lengthened by an opposing force greater than the force in the muscle. Animal studies employing a model of muscle injury by eccentric contraction reveal that active muscle activity did not change the length at which muscle disruption occurred compared to passively stretched muscle. There was a small increase in the maximum force when the activated muscle was disrupted. There was, however, a large increase in the energy absorbed by an activated muscle before disruption. Failure occurred in the region of the myotendinous junction. These studies emphasize the importance of energy absorption by muscle in the prevention of injury to muscle or joint structures.

Few studies have been done pertaining to healing or prevention of muscle injuries. A warm-up routine has been simulated in an animal model by a single maximum contraction before biomechanical testing. This preconditioning had a protective effect by allowing a greater increase in length and force before disruption. The use of nonsteroidal anti-inflammatory medications did not significantly alter the load at failure of healing muscle, although it did change the histologic response at the myotendinous junction. The tensile load at failure of immobilized muscle was quite dependent on the position of immobilization. Muscles immobilized under tension in a stretched position developed more force and stretched more than control non-immobilized muscle before disruption. Muscle immobilized without tension developed less force and required less change in length for disruption compared with control muscle.

Animal models of muscle injury have been introduced only recently and there are many basic questions to be asked and answered concerning muscle injury. The field is only in its infancy in terms of understanding the pathophysiology of indirect muscle injury. The problems of treatment, prevention, and recovery from these injuries must be approached in basic and clinical studies. At present, the region of muscle near the myotendinous junction appears to be important in the development of this field of study.

Chapter 5

Myotendinous Junction: Structure, Function, and Failure

William Garrett, Jr.,
MD, PhD
James Tidball, PhD

Chapter Outline

Introduction

The major components of the musculoskeletal system—muscle, tendon, and bone—have been subjected to extensive studies that revealed many fundamental aspects of their structure and physiologic performance in healthy and diseased tissue. However, the junctions among these components have been studied relatively little and the possibility that these junctions are involved in specific soft-tissue injuries has begun to be explored only recently.

We intend to bring together and evaluate information on one of these essential, yet poorly understood junctions, the attachment sites between skeletal muscle and tendon. These sites, called myotendinous junctions, will be considered as structures involved in the transmission of force from the contracting cell to the extracellular matrix. From this premise, myotendinous junctions will be analyzed as mechanical structures. Questions concerning the mechanical function of myotendinous junctions include: How does the structure of the junction relate to its mechanical function? What is the molecular composition of the junction and which of these molecules are involved in force transmission? How do components of the myotendinous junction transmit force and how effective are they at this? How can perturbations of the structure, composition, or mechanical behavior of junctional components determine whether forces are successfully transmitted from cell to extracellular molecules or whether failure occurs at the junction?

We will discuss available data concerning the performance of myotendinous junctions during loading of whole muscle-tendon preparations to failure. What is the potential for involvement of components of the myotendinous junction in soft-tissue injuries? What is the mechanism of failure at myotendinous junctions? What are appropriate approaches for repair of injuries involving myotendinous junctions? In asking these questions, we hope not only to show what progress has been made in understanding myotendinous junction structure and function, but also to clarify the many areas requiring further study if the normal function of these structures and their involvement in soft-tissue injuries is to be understood.

Morphologic Specialization of Load-Bearing Membranes of Myotendinous Junctions

Sites at cell surfaces involved in transmitting forces between intracellular and extracellular proteins are frequently characterized by extensive membrane folding and by dense, subplasmalemmal material that appears to be involved in linking cytoskeletal elements to the membrane.[1-24] The presumed increased metabolic cost to the cell of producing these large specialized junctions and the broad, phylogenetic occurrence of these structures at load-bearing membranes indicate that these specializations are functionally important. Although there is no direct, physiologic evidence to show that myotendinous junctions are sites of force transmission, their location at the end of muscle cells where myofibrils are attached to the cell membrane supports this proposed function for these sites. Furthermore, myotendinous junctions display extensive membrane folding and dense subsarcolemmal material, and thereby resemble load-bearing membranes in both their structure and location.

Morphologic Definition

Myotendinous junctions occur at the ends of long, cylindrical muscle fibers of skeletal muscle (Figs. 5–1 and 5–2). Vertebrate skeletal muscle contains no demonstrated end-to-end connections between any two skeletal muscle cells such as those seen at intercalated disks of cardiac cells. Each skeletal muscle cell, therefore, appears to terminate with connections directly to the tendons on which it acts at origin and insertion ends of the cell.

The morphologic features that most obviously characterize myotendinous junctions at both origin and insertion ends of the cell are most clearly seen in longitudinal sections of the tissue (Fig. 5–3). These sections reveal that the cell membrane is a continuous interface between intracellular and extracellular compartments.[25] The membrane at these sites is extensively folded so that the cell and extracellular connective tissue appear to interdigitate.[1,4,6–10,13–15,17–24]

Thin filaments from the terminal sarcomere extend toward the junctional membrane and lie at low angles to the membrane (Fig. 5–4). Thin filaments that approach the sarcolemma may be longer than those that lie in complete I bands.[6] These filaments enter a dense, subsarcolemmal material where they appear to be cross-linked to one another and attached to the membrane. In this region, Z-disks from the final few sarcomeres extend toward and are structurally continuous with subsarcolemmal densities. It is also noteworthy that the length of sarcomeres near the myotendinous junctions is less than that of sarcomeres throughout the cell.[26,27]

Other quantitative differences in cell structure near myotendinous junctions relative to the nonjunctional region include greater density of mitochondria, nuclei, sarcoplasmic reticulum, and Golgi complexes.[18] Eisenberg and Milton[18] suggested that the increased number of these organelles may reflect increased synthetic activity of the cell at its ends. These investigators also noted that the densities of T-tubules near junctional and within nonjunctional regions are not significantly different.

The external surface of the myotendinous junction is covered by a basement membrane that is continuous with and morphologically identical to basement membrane elsewhere on the cell[6,28] (Fig. 5–4). The basement membrane here, like other basement membranes, can be morphologically subdivided into a dense region close to the cell, called the basal lamina, and an external region, the reticular lamina. The reticular lamina provides structural connections to tendon collagen fibers. The basement membrane is separated from the surface of the sarcolemma by a region of low electron density, the lamina

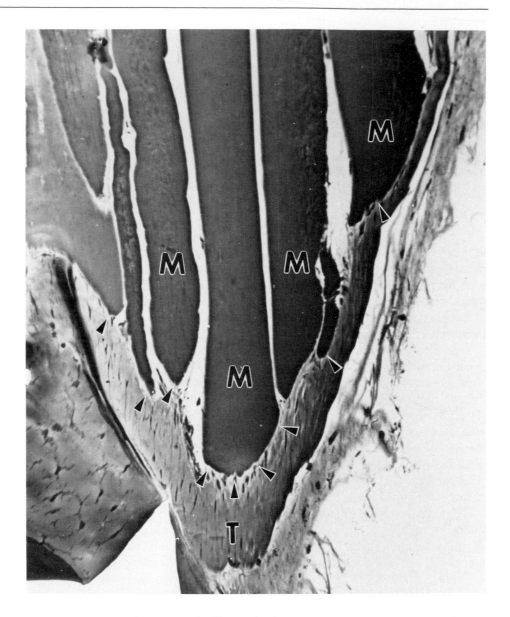

Fig. 5–1 *Light micrograph of longitudinal section through several frog semitendinosus muscle cells (M) attached to their tendon of insertion (T) at myotendinous junctions (arrowheads) (× 250).*

lucida, which is traversed by fine filaments passing from the cell membrane to the basement membrane (Fig. 5–4).

The following discussion is directed toward an analysis of these morphologic features of myotendinous junctions as force-transmitting structures. For successful force transmission across this interface: (1)

Fig. 5–2 *Two frog semitendinosus muscle cells terminating on tendon. Note bundles of collagen fibers passing from tendon to ends of the cylindrical cells (SEM, × 600). (Reproduced with permission from Tidball JG: Myotendinous junction: Morphological changes and mechanical failure associated with muscle cell atrophy.* Exp Mol Pathol *1984;40:1–12.)*

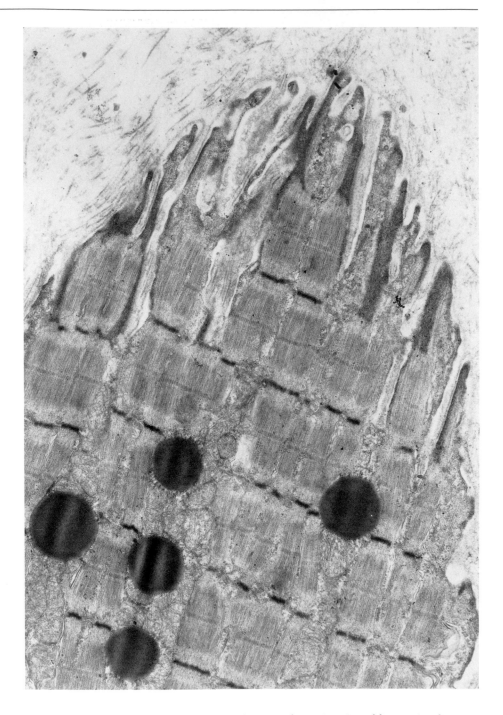

Fig. 5–3 *Longitudinal section through myotendinous junction of frog semitendinosus muscle cell. Note that the cell and surrounding connective tissue appear to interdigitate with one another at the end of the cell. Also, note that dense, fibrillar material extends from the terminal Z-disk to the junctional plasma membrane (TEM, × 12,5000). (Reproduced with permission from Tidball JG: Myotendinous junction: Morphological changes and mechanical failure associated with muscle cell atrophy.* Exp Mol Pathol *1984;40:1–12.)*

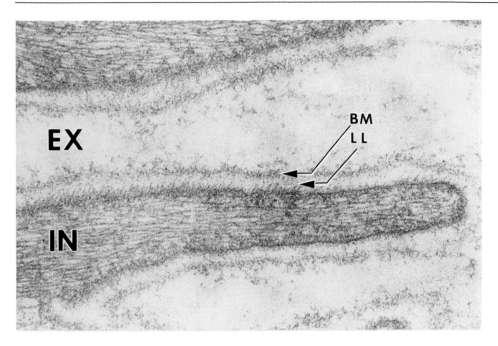

Fig. 5–4 *Single, digit-like process of skeletal muscle cell at myotendinous junction in longitudinal section. Note that thin filaments run along the length of the process (IN). External to the process (EX) the basement membrane (BM) is separated from the junctional membrane by the lamina lucida (LL) (TEM, × 60,000). (Reproduced with permission from Tidball JG: Myotendinous junction: Morphological changes and mechanical failure associated with muscle cell atrophy. Exp Mol Pathol 1984;40:1–12.)*

the cell and tendon must form an adhesive junction at these sites; (2) the load placed on the interface must not exceed the strength of the interface under those loading conditions; and (3) the mechanical properties of the elastic and viscous elements must permit some mechanical energy to be transmitted at the junction. Our attempt to evaluate the structure and physiologic behavior of myotendinous junctions from this mechanical point of view does not consider in detail other possible functions of the junctional region. One of these possibly important functions is that longitudinal growth of the muscle cell occurs at the ends of the muscle cells.[29-33] However, since analyses of adult myotendinous junctions show that the structural features discussed here persist in adult tissues in which little or no longitudinal growth occurs, these features are not exclusively involved in stages of cell growth.

Topography

The three-dimensional structure of myotendinous junctions has recently been analyzed by scanning microscopy[20,22,34] (Figs. 5–5 and 5–6). The cell surface displays at least three different morphologies at

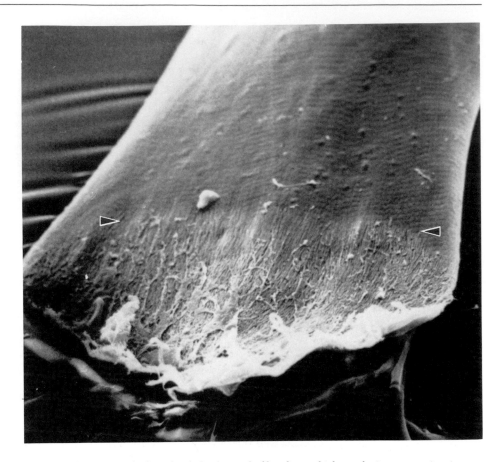

Fig. 5–5 *End of single, skeletal muscle fiber from which enveloping connective tissue has been removed. Note the nonjunctional region of the cell where the regular, transverse banding of underlying sarcomeres is apparent. Transition from nonjunctional to junctional regions of the fiber is abrupt (between arrowheads). The junctional region displays extensive membrane folding (SEM, × 1,000).*

the junctions of skeletal muscle cells: (1) it is invaginated by extracellular matrix extending into clefts at the cell surface[34]; (2) it displays digit-like (circular paraboloid) extensions of the cell into the extracellular matrix[23,24]; and (3) it has cylindrical extensions of the cell into the extracellular matrix.[20] Although these studies used muscles from a variety of vertebrates, differences in cell surface topography do not obviously correlate with phylogenetic differences in the muscle source and may instead correlate with physiologic differences that have not yet been identified.

Precise descriptions of the relationships between the structure and function of myotendinous junctions require quantitative descriptions of the surface area and topography of these junctions. Four geometric models have been used in attempts to quantify junction architecture.

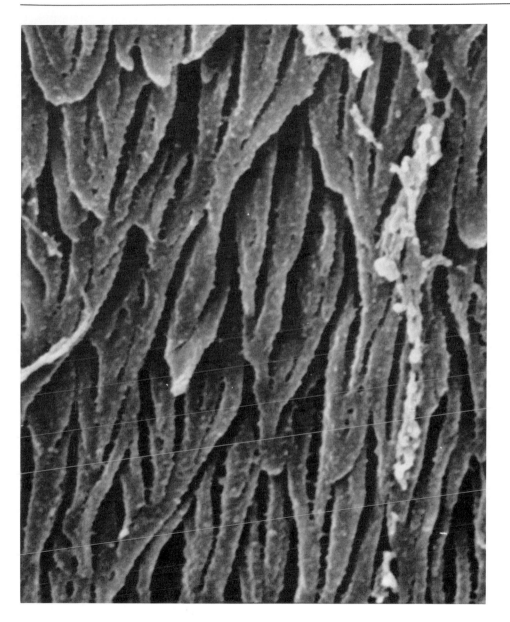

Fig. 5–6 *Junctional region of cell seen in Figure 5–5. Note that the cell extends digit-like processes at the junction. In vivo, the tendon would lie toward the bottom and the nonjunctional portion of the cell would lie toward the top (SEM, × 2,600). (Reproduced with permission from Tidball JG, Daniel TL: Myotendinous junctions of tonic muscle cells: Structure and loading. Cell Tissue Res 1986;245:315–322.)*

The earliest of these modeled the extensions of the cell at the junction as cones.[14] This model produced an estimated increase in junctional membrane surface area resulting from membrane folding by a factor of 13.2. However, the conic model was flawed in that subsequent scanning electron microscopic studies showed it to be an oversimplified geometric estimation. Furthermore, an analytic error led to an underestimate of actual membrane surface area. Subsequent analyses used two different analytic approaches based on the more accurate generalizations of the junctional processes as cylinders[18,19] or, in a later analysis, as circular paraboloids.[22] Although each of these models relies on different mathematical and geometric assumptions, they yield similar estimates of the amount of membrane folding at myotendinous junctions. Twitch cells from a variety of sources and presumably containing a variety of physiologic subgroups (for example, fast-twitch and slow-twitch) increase their junctional surface area through folding by a factor of about ten to 20.[18,19,22] Tonic cells, however, increase junctional surface area by a factor of about 50.[22] These differences in folding may have an important bearing on the physiologic behavior of the junction during cell contraction.

Mechanical Implications of Membrane Folding

Folding of junctional membrane and associated surface coats may have at least three general functional implications. First, folding increases membrane surface area and thereby reduces the apparent stresses on the membrane. Second, folding places the membrane at very low angles relative to the force vector, thereby placing the membrane primarily under shear forces.[6,14,15] For at least some cells, such as erythrocytes, membranes are highly resistant to shear stresses that tend to increase their surface area.[35] Finally, folding places the adhesive interface between cell and extracellular matrix at low angles to the force vector; this can increase the adhesive strength of cell to tendon compared with the same junction when loaded under tension.[36,37]

Increased Surface Area

The mechanical implications of increased junctional area may be vitally important for a contracting cell. For example, a typical, vertebrate fast-twitch cell can generate about 0.33 MPa of stress across the cell.[27,38] The stress placed on the cell's junctional membrane is, however, reduced by folding so that the membrane experiences a maximum stress of 1.5×10^4 Pa.[22] Conceivably, the difference between 1.5×10^4 Pa and 33×10^4 Pa may determine whether mechanical failure at the junction occurs. This is speculative because no data exist to indicate the stress level at which myotendinous junctions would fail. However, these values approach those at which erythrocytes membranes fail under shear loading (<1 MPa),[39,40] suggesting that the amount of membrane folding may be an important structural correlate to the magnitude of the stress placed on the junction.

Simple correlation between stress magnitude and junction structure is inadequate to explain the functional significance of myotendinous junction structure because it, like all biologic materials, is viscoelastic.

As a viscoelastic structure, its behavior under loading varies with frequency, duration, and rate of loading as well as magnitude. This suggests that junction failure occurs at a different stress during relatively long-term loading than during rapidly applied, short-duration loads. This viscoelastic behavior is relevant to muscle performance because the junctions of physiologically distinct muscle fibers (for example, fast-twitch and slow-twitch) experience distinct loading patterns.

The possible spectrum of loading patterns placed on myotendinous junctions is suggested in many studies. For example,[41] feline gastrocnemius contains three physiologically distinct muscle fiber types. One type contracts in 34 msec to produce 0.6 N of tetanic tension but fatigues rapidly under repeated stimulation. A second group contracts in 40 msec and produces 0.2 N of tension and is fatigue-resistant. A third group contracts in 73 msec and produces only 0.05 N of tetanic force.

Myotendinous junction architecture may reflect these differences in loading patterns. This view is supported by the observations of Trotter and associates[19] that muscles containing almost exclusively fast-twitch fibers have slightly more junctional membrane folding than those muscles containing a mixture of fast- and slow-twitch fibers. This observation is consistent with the hypothesis that increased stress at the junction corresponds to increased folding at the junctional membrane because fast-twitch fibers typically generate higher stresses than slow-twitch fibers. This correlation is not, however, always true. For example, single tonic cells generate about one half the maximum stress of twitch cells (0.17 MPa vs 0.33 MPa), although membrane folding at tonic myotendinous junctions is 250% of that of twitch cells.[22] Calculations and measurements indicate that the stress placed on tonic cell junctions (3×10^3 Pa) is about 20% of that placed on twitch junctions.[22] Assuming that junction structure reflects its mechanical role, these observations may seem paradoxical in that there is a negative correlation between stress magnitude and membrane folding.

However, since cell membranes have both viscous and elastic components, membrane architecture may reflect both stress magnitude and loading duration. The significance of this point of view can be placed in the context of membrane fluidity proposed by Eyring's theory.[42,43] This theory postulates that each molecule in a membrane resides in an energetic trough and that a certain amount of energy, called the activation energy, is required to move that molecule from its trough. Once moved, that molecule can promote the displacement of adjacent molecules, leading to membrane flow and nonreversible structural rearrangements. For living cells, these rearrangements may result in cell lysis. In interpreting membrane behavior it is important to note that activation energy can be reached either by applying a relatively large load for a short duration or by applying a small load for a relatively long duration.

Few investigations have been directed toward exploring the relationships among stress magnitude, loading duration, and cell membrane lysis. One investigation that approached these questions used

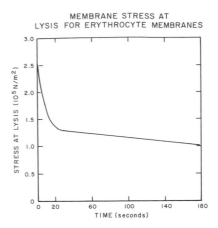

Fig. 5–7 *Rand[39,40] measured the stress required for erythrocyte membrane lysis over a range of loading times. His data have been adapted in the figure shown here. Qualitatively, Rand found that the longer the duration of loading, the smaller the load required for membrane failure. Interestingly enough, the maximum stress these membranes can bear as loading time approaches zero is equivalent to the stress that myotendinous junction membranes would bear if no membrane folding occurred. If myotendinous junction membranes and erythrocyte membranes are of comparable strength and time-dependent behavior, then tonic cell membranes would require greater folding than twitch cells to correspond to greater load durations. Otherwise, membrane failure would occur during long-term loading in tonus.*

erythrocyte membranes for a model and showed that an erythrocyte membrane 10 nm thick could endure 0.2 MPa of shear for eight seconds before lysis but stresses one half that magnitude could cause lysis if applied for 70 seconds.[39,40] These values for cell lysis are similar to the maximum stress experienced by twitch cell junctions during contraction[22] and may indicate a general upper limit of cell membrane stress that varies with loading duration. If this is generally true, the functional significance of increasing junctional membrane area in relatively low-stress tonic fibers may relate to the loading time-dependent function of membrane failure. The importance of this relationship among stress magnitude, loading duration, and membrane failure is emphasized graphically in Figure 5–7, which is derived and extrapolated from Rand's[39,40] data on erythrocyte membranes and assumes a linear viscoelastic material. Note that the "safe zone" for membranes lies below the curve, and that failure can occur by increasing either stress magnitude or loading duration.

Other factors relating to membrane loading or nonmechanical features of myotendinous junction may be involved in determining junction structure. For example, in a single muscle (mouse soleus) there are small but significant differences in the degree of junctional membrane folding at cell origins and insertions.[19] Do these differences reflect differences in membrane loading at these sites? Do they indicate differences in rates of cell growth at origin and insertion of muscle? Accounting for the effects of longitudinal growth of muscle cells on junction structure may be especially important in understanding junction structure and function because these cells undergo sarcomere addition and/or deletion near these sites.

Low Angle of Junction Loading

Membrane folding at myotendinous junctions not only decreases the stress on cell membranes, it also modifies the type of stress placed on the cell membrane. If the cell membrane were not folded and the cell ended as a right circular cylinder, the junctional membrane would experience loading at right angles to the joint interface. This would be tensile loading of the junction. However, membrane folding places the junction at low angles to the force vector so that the junction experiences primarily shear loading.

If the junction is viewed as an adhesive joint between two adherends, the cell and the extracellular matrix, the principles of adhesive science can be applied to myotendinous junctions.[36,37] These principles suggest that the strength of adhesion at an interface varies with the angle of loading of the interface. Joints that are loaded under neither simple tension (force trajectory across joint = θ = 90 degrees) nor shear (θ = 0 degrees) are called scarf joints ($0 < \theta < 90$ degrees). According to this terminology, myotendinous junctions are scarf joints. The strength of scarf joints varies greatly with changes in θ.[14,36] The sensitivity of joint strength to changes in θ is greatest when θ is small. This may be relevant to myotendinous junction strength in which θ is normally near zero. It has been noted that disuse atrophy of muscle results in an increase in θ and an increase of failure at myotendinous junctions at physiologic loads.[14] However, no data have been gathered to show that there is any causal relationship in this correlation and other explanations for myotendinous junction failure during atrophy, such as changes in the adhering surface, may be the actual basis for junction failure.

Further application of adhesive science principles to myotendinous junction mechanics is limited by the morphologic complexity of the myotendinous junction. The myotendinous junction comprises several distinct layers, each of which is characterized by its own behavior during loading. To interpret myotendinous junction mechanics as a complex adhesive joint will require mechanical descriptions of each component of the adhesive junction during loading.

Mechanical Implications of Variable Sarcomere Length

Sarcomeres near myotendinous junctions are shorter than those throughout most skeletal muscle under static loading.[26,27] This indicates that the terminal few millimeters of the muscle fibers have biomechanical properties different from those of portions of the fibers not adjacent to the myotendinous junction. As a result, a modification of the pattern of loading at myotendinous junctions during contraction

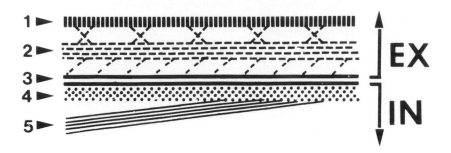

Fig. 5–8 *Schema of the structures involved in force transmission between tendon and contractile proteins of muscle cell. Extracellular components (EX) include tendon collagen fibers (1) and basement membrane (2). The junctional plasma membrane (3) separates extracellular (EX) and intracellular (IN) force-transmitting structures. Within the cell, thin actin filaments (5) are attached to the cell membrane by dense, subsarcolemmal material (4).*

when the cells are stimulated simultaneously along their lengths can be expected.

Force magnitude, change in cell length, and contraction rate all vary as functions of sarcomere length. Decreasing sarcomere length at this region of the cell decreases the force generated by these sarcomeres, increases the rate of contraction, and decreases the change in length. These physiologic correlates of muscle structure suggest that myotendinous junctions are first preloaded by force generated by the terminal sarcomeres and then subjected to an increased load as the majority of the cell sarcomeres reach peak tension. Beyond this, any mechanical significance of shorter sarcomere length at myotendinous junctions is unknown. However, some experiments have dealt with mechanical failure of muscle cells in or within this zone of short sarcomeres. These are discussed below. Whether differences in cell mechanics attributable to differences in sarcomere length are responsible for these muscle tears is unknown.

Molecular Composition of Myotendinous Junctions

Subsarcolemmal Densities

For the contractile apparatus of muscle to act on extracellular structures such as tendon, there must be some mechanical coupling of the contractile proteins to extracellular structures (Fig. 5–8). Thin filaments have been observed to pass from terminal Z-disks of myofibrils to subsarcolemmal densities and they are, therefore, likely to be the myofibril component joined to the cell membrane.[6,7,9,10,13–15,18,22,24] These thin filaments bind heavy meromyosin, indicating that they are filamentous actin.[44] This suggests that molecular links between thin filaments and the membrane should have components with a high affinity for actin and an affinity for an integral or transmembrane protein.

Thus far, no technique has been developed for isolating the material that constitutes the subsarcolemmal densities of skeletal muscle. Investigations into the molecular composition of these sites have relied, instead, on comparisons with possibly analogous or homologous structures with better characterized molecular compositions.

Skeletal muscle Z-disks have been proposed as homologues to subsarcolemmal densities. The evidence supporting this proposal is essentially morphologic. First, several investigators have inferred from electron microscopic observations that Z-disks form as dense plaques subjacent to muscle cell membranes and then become separated from the membrane to become Z-disks.[45,46] Second, subsarcolemmal densities at myotendinous junctions are structurally continuous with Z-disks.[1,7,8,24]

Skeletal muscle subsarcolemmal densities and cardiac fascia adherens are analogues in that they are both sites at which force is transmitted across membranes. This suggests the possibility of molecular similarities in thin filament-membrane attachment. Molecular similarities between cardiac fascia adherens and Z-disks have also been demonstrated.[47,48] Both fascia adherens and Z-disks contain apparently comparable densities of the cytoskeletal protein α-actinin as seen in immonocytochemical studies using anti-α-actinin as a morphologic probe.[48] A major structural protein of Z-disks, α-actinin has been implicated in bundling actin filaments in vitro.[49-51]

The sites at which cultured cells form contacts with substrata provide a third possible molecular model of subsarcolemmal densities.[23] These sites, called focal contacts, resemble myotendinous junctions in that thin filaments converge on and are attached to the plasma membrane, and forces generated by the cell are transmitted to the substratum at these sites. In fact, observations and calculations suggest that the stress placed on the membrane by cultured cells are within an order of magnitude of those experienced by myotendinous junction membrane and that membrane stresses in both systems are primarily shear.[23] Thus, the mechanical functions of focal contacts and myotendinous junctions may be quantitatively as well as qualitatively similar.

Analysis of the utility of each of these systems as a molecular model of myotendinous junctions has used immunocytochemical procedures employing specific antibodies to cytoskeletal proteins located at Z-disks, fascia adherens, and focal contacts. A recent quantitative immunocytochemical study using affinity-purified (Figs. 5–9 and 5–10) polyclonal anti-α-actinin produced the unexpected finding that α-actinin, as recognized by this probe, was absent from myotendinous junctions and from the terminal sarcomere of Z-disks.[24] These findings indicate that thin filaments at myotendinous junctions and in terminal Z-disks are bundled either by some protein other than α-actinin or by some isoform of α-actinin that differs from the form found throughout nonterminal regions of the myofibril. These observations coincide with those of Small,[52] who similarly noted an absence of α-actinin from

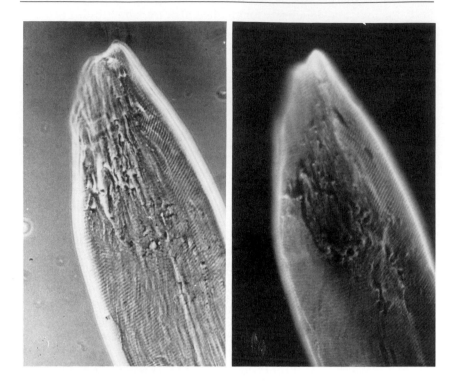

Fig. 5–9 *Light micrograph of single frog semitendinosus cell.* **Left:** *Cell seen by phase microscopy displays regular, transverse striations of sarcomeres.* **Right:** *Same cell after treatment with affinity-purified, anti-α-actinin and fluorescent second antibody and then viewed by epifluorescence microscopy. Note that α-actinin is located in transverse bands at the position of the Z-disk. Resolution is inadequate to determine whether α-actinin is found at sites other than Z-disks near myotendinous junctions (× 375). (Reproduced with permission from Tidball JG: Alpha-Actinin is absent from the terminal segments of myofibrils and from subsarcolemmal densities in frog skeletal muscle. Exp Cell Res 1987;170:469–482.)*

sites of thin filament-membrane attachment in smooth muscle. However, these findings are not consistent with those reported in an earlier, preliminary study that suggested that α-actinin was located at myotendinous junctions.[15] The experimental protocol was not reported in this earlier study,[15] however, so the data cannot be compared here.

Immunocytochemistry has also shown that vinculin, another cytoskeletal protein, is present at subsarcolemmal densities. Vinculin, a 130-kD molecular mass protein, has been implicated in mediating thin filament-membrane associations at focal contacts.[53,54] Vinculin has also been demonstrated at fascia adherens[55,56] but, consistent with its role in thin filament-membrane association, is not found at Z-disks.

A third cytoskeletal protein, called talin, is also a component of focal contacts and has been implicated in thin filament-membrane association.[57,58] Immunocytochemistry has been used to demonstrate that

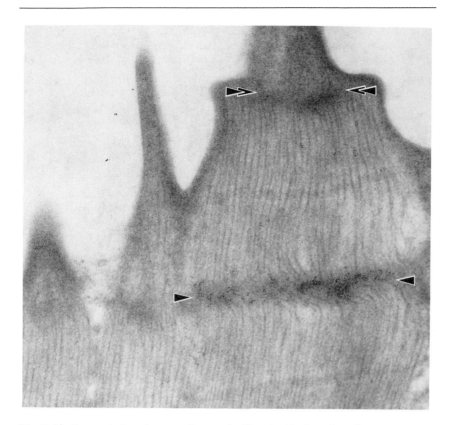

Fig. 5–10 *Transmission electron micrograph of longitudinal section of a myotendinous junction. The section was incubated with affinity-purified, rabbit anti-chick α-actinin and then a ferritin-conjugated, goat anti-rabbit IgG. Dense ferritin grains indicate the location of α-actinin. Note that the nonterminal Z-disk is rich in α-actinin (arrowhead), whereas terminal Z-disks (double arrowhead) and digit-like processes of the cell at myotendinous junctions contain little or no α-actinin (× 24,000).*

talin is present in the digit-like processes of the cell at myotendinous junctions[23] (Figs. 5–11 and 5–12) but absent from intracellular sites such as Z-disks and from fascia adherens.[55] Talin and vinculin have a high affinity for one another in vitro and are believed to be two links in a molecular chain between actin filaments and extracellular structural proteins.[59] The distribution of talin seems to be restricted to sites at which cells are attached to the extracellular matrix, whereas vinculin is found both at those sites and where cells are attached to other cells.[55] This indicates that there are different molecular interactions mediating force transmission from cell to substratum and from cell to cell.

There may also be differences in the molecular interactions mediating actin filament-membrane associations subjected to different loading conditions. In an interesting comparison of the molecular composition of subsarcolemmal densities of twitch and tonic muscle, Shear

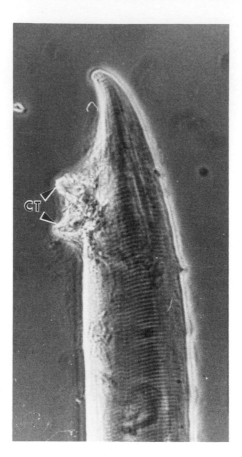
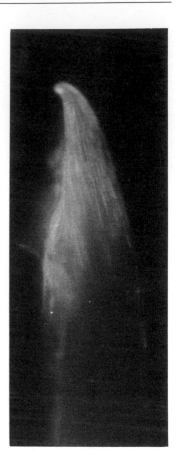

Fig. 5–11 *Single chick skeletal muscle cell demembranated with detergent and then treated with affinity-purified, rabbit anti-chick talin and a fluorescent goat anti-rabbit IgG.* **Left:** *Cell viewed by phase microscopy displaying connective tissue (CT) stuck onto side of cell.* **Right:** *Same cell viewed by epifluorescence. Note that labeling for talin is most apparent at the junctional region of the cell (× 400).*

and Bloch[60] found that the cytoskeletal protein vinculin was present in tonic-cell subsarcolemmal densities but not at twitch-cell subsarcolemmal densities other than neuromuscular junctions. This suggests that the distribution of these linking proteins is related to the physiologic type of the muscle and may reflect differences in the loading of myotendinous junctions and, therefore, in the loading of the linking proteins as well. Thus, myotendinous junctions may be modified in relation to differing physiologic stresses through morphologic changes at the junction, such as membrane folding, or through changes in the molecular composition of these force-transmitting sites.

Clearly, further investigations of the composition of this important region involved in linking cytoskeletal proteins to extracellular, struc-

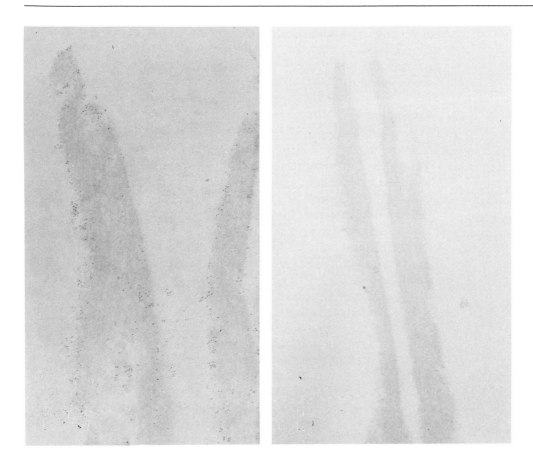

Fig. 5–12 *Single chicken skeletal muscle cell demembranated with detergent and then incubated with affinity-purified, rabbit anti-chick talin followed by ferritin-conjugated, goat anti-rabbit IgG.* **Left:** *Two digit-like processes at myotendinous junction display extensive labeling.* **Right:** *Two digit-like processes in a control cell treated with pre-immune rabbit serum and then the second antibody display little labeling (× 57,000).*

tural proteins would be informative. Only through a complete molecular inventory of linking molecules in this region and their interactions can precise, functional questions be asked. Which molecules are essential in forming these mechanical connections? Does either talin or vinculin interact with F-actin at these sites or are other linking molecules involved? Are there different physiologic functions for different subsarcolemmal proteins? For example, do some linking proteins effectively transmit force with little energy loss while others are more energy-dissipative? Do some loading patterns tend to cause mechanical failure of any of the subsarcolemmal constituents? Does a history of loading (for example, a training program) modify the concentration, distribution, or mechanical properties of any of the subsarcolemmal constituents?

The many unanswered questions relating to the structure and function of this region reflect the technical difficulty of isolating either chemically or physiologically the components of myotendinous junctions. These investigations are also hampered by reliance on molecular analogies to other systems that eliminate the possibility of identifying unique components at myotendinous junctions.

Myotendinous Junction Membrane

The junctional membrane separates contractile and other cytoskeletal proteins at myotendinous junctions from the extracellular structural proteins on which they act. This membrane, like other biologic membranes, is not composed entirely of a lipid bilayer. As Trotter and associates[13] demonstrated, non-ionic detergent extraction of skeletal muscle solubilizes the lipid bilayer at myotendinous junctions, but forces generated by the cell are transmitted to tendon even after lipid extraction. This indicates that some nonlipid structure, which is likely to be a transmembrane protein, connects intracellular to extracellular structural proteins.

Once again, focal contacts have provided a valuable molecular model for myotendinous junctions. Two important cytoskeletal proteins, vinculin and talin, may mediate actin-membrane associations at focal contacts and myotendinous junctions. External to both of these sites of cell-substratum attachment are layers of connective tissue, called basement membranes, that ensheathe and support the cells. These basement membranes are rich in large glycoproteins called fibronectin, laminin, and type IV collagen.[61,62] There must be some physical connections between basement membrane constituents and these cytoskeletal proteins mediating actin-membrane attachments. Explorations of these connections at focal contacts have revealed molecular interactions that may be relevant to other systems such as myotendinous junctions.

Investigators in several laboratories have identified a large transmembrane glycoprotein that may link intracellular and extracellular structural proteins. This glycoprotein, called integrin (also called CSAT antigen, cell-substratum attachment antigen, JG22, JG9, avian fibronectin receptor), comprises three glycoproteins with molecular masses of about 140 kD each.[63] Evidence supporting cell-substratum linkage involving this protein includes the following: (1) the antigen co-localizes with fibronectin along stress fibers and at the termini of stress fibers in cultured cells[64]; (2) the antigen binds laminin and fibronectin[65,66]; and (3) antibodies to integrin can inhibit adhesion to substrata by cultural cells.[67-70] Furthermore, integrin is co-distributed with and has binding affinity for talin.[71] Integrin is present not only at focal contacts[64,72,73] but also in smooth muscle[72] and at the ends of myotubes in vitro.[64] Also, preliminary immunohistochemical data show that integrin is distributed along the surface of adult skeletal muscle fibers, including at myotendinous junctions (Fig. 5–13). The location, transmembrane position, and affinity for both extracellular and intracellular structural proteins suggest that integrin may be an important

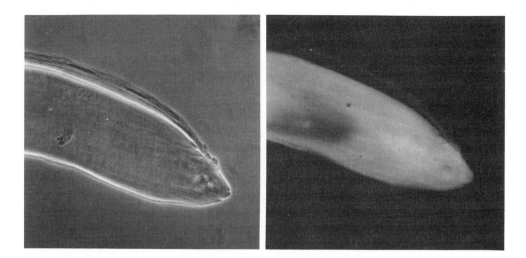

Fig. 5–13 *Single frog semitendinosus muscle cell incubated with rabbit, anti-chicken integrin followed by fluorescent, goat anti-rabbit IgG.* **Left:** *Phase image of junctional region of cell.* **Right:** *Same cell viewed by epifluorescence displays binding along surface of cell (× 350).*

transmembrane link between the cytoskeleton and extracellular substrata, at least in muscle and fibroblasts.[71] Other membrane proteins may have similar functions. Vitronectin, also called serum spreading factor, may also be involved in linkage of cytoskeletal to extracellular matrix proteins. Vitronectin, a 70-kD molecular weight protein,[74–77] was first localized to the surface of fibroblasts,[76] but indirect immunofluorescence studies using antivitronectin have revealed that vitronectin is also present at sites of cultured myotube-substratum attachment.[78] Amino acid sequencing reveals that the β-subunits of vitronectin and integrin share many similarities although the α-subunits display few structural similarities.[79] Hence, these two cell-substratum attachment molecules may be phylogenically or ontogenically related but appear as distinct molecules in adult vertebrates investigated thus far. No affinity of vitronectin for talin or any other cytoskeleton protein has been demonstrated.

Basement Membrane

Basement membrane lies external and parallel to the junctional plasma membrane and is separated from the membrane by a zone of low electron density, called the lamina lucida. The basement membrane includes the lamina densa (also called the external lamina)[15] and the reticular lamina. The basement membrane of the myotendinous junction resembles basement membranes at the nonjunctional regions of muscles as well as basement membranes associated with other cell types.[28] Because morphologic studies indicate that the basement membrane is an uninterrupted sheath enclosing the cell, contractile forces generated by the cell are presumed to be transmitted across the basement membrane and may involve basement membrane constituents as mechanical links.

Type IV collagen, laminin, and fibronectin are high-molecular-weight glycoproteins that constitute a major portion of all basement membranes studied thus far.[61,62] Immunocytochemical studies suggest that laminin and type IV collagen at least are components of the basement membrane of the myotendinous junction, although whether they act as force-transmitting structures is unknown.[15] The possible involvement of fibronectin and laminin as force-transmitting elements can be inferred from studies concerning their interactions with integrin.

Other extracellular, cell surface, or basement membrane molecules may also be involved in force transmission from muscle to substratum. Analysis of hybridomas that secrete antibodies against muscle connective tissue components has been used to identify putative muscle-substratum attachment molecules in the extracellular matrix.[80,81] One such monoclonal antibody recognizes an antigen called M1 or myotendinous antigen.[80] This antigen is present in early developing chick limb buds where differentiating tendon and muscle tissue make contact with one another.[80] The antigen is a major component of secreted glycoproteins in cultured muscle tissue and comprises several, disulfide-linked peptides composed in part of pepsin-resistant fragments of about 80 kD.[81] Indirect immunofluorescence has also shown the M1 antigen to be present in embryonic ligaments, cartilage, lens capsule, and surrounding notochord and aorta.[75] The proposed function of M1 antigen is linking muscle cells to tendon,[81] although its distribution suggests other functions as well. No immunocytochemical data exist to show whether M1 antigen is a component of the basement membrane of muscle or is co-distributed with the basement membrane.

The mechanical behavior of skeletal muscle basement membranes has been studied recently under conditions simulating the strain rates experienced by muscle cells during slow locomotion.[82] This study compared the dynamic mechanical behavior of single skeletal muscle cells with intact basement membranes to that of cells from which basement membranes were removed by enzymatic treatments. Under loading conditions chosen to simulate a stage of slow locomotion (0.7% to 2.7% strain at 1 Hz sinusoidal loading on cells at sarcomere lengths of 2.6 to 2.8 μ), intact cells were nearly twice as stiff as basement membrane-depleted cells (complex tensile modulus of 3.3 MPa vs 1.8 MPa). However, although intact cells were stiffer, they were much more energy-dissipative. For example, at strains of 0.7% to 2.7%, an intact cell transmits about 300 J/m^3 (energy storage $= E_s$) without loss and dissipates about 1,000 J/m^3 (energy loss $= E_d$) whereas a basement membrane-depleted cell transmits about 180 J/m^3 without loss while dissipating about 230 J/m^3.

These measurements[82] were used to calculate the mechanical properties of basement membranes under these loading conditions. These calculations indicate that the basement membrane is a stiff structure with a complex modulus of 161 MPa. It is also an energy-dissipative structure with a specific loss [defined as 0.5 (E_d/E_s)], of 2.9 compared with 1.6 for an intact cell.

What do these measurements of the mechanical behavior of basement membranes tell us about myotendinous junctions? First, we can predict that the junctional basement membrane will be a site of energy losses during muscle contraction. Both tendon and cross-bridged muscle have specific losses that are significantly lower than that of basement membrane. Considering only a simplified system comprising cell-basement membrane-tendon in series, the basement membrane will be the most energy-dissipative structure. This may seem functionally inappropriate. Why introduce a "loss" element in a chain of structures whose function is to transmit force? Recent theoretical work (T.L. Daniel, oral communication, 1987) has indicated that periodically strained systems such as the muscle cell or muscle-tendon system require an energy-dissipative element to impart stability to the system. If a periodically oscillated system is perfectly elastic, the magnitude of oscillations within the system can increase until a critical strain is reached, leading to mechanical failure of the system. Tendon and cross-bridged skeletal muscle without basement membrane are highly elastic biologic structures that may reach critical strain during periodic length changes if no viscous element is present. The basement membrane may provide for these functionally necessary viscous losses. Second, it is also noteworthy that junctional membrane folding permits basement membrane to extend tens of microns into the cell. Introduction of extensive basement membrane in parallel with myofibrils in the terminal regions of the cell modifies the mechanical properties of the cell near its ends. If junctional and nonjunctional basement membranes have similar mechanical properties, the ends of the cells containing extensive basement membrane in invaginations will be stiffer but more energy-dissipative than noninvaginated regions of the cells. This may pertain to factors determining the site of cell failure during some forms of eccentric loading (described below).

Muscle Strain Injury

The region of the myotendinous junction has assumed new importance as more is learned about indirect muscle injury or muscle strains. Indirect injury must be considered apart from lacerations, or direct injuries, or contusions to muscle which injure the muscle at the location of trauma. Indirect muscle injuries are caused by stretching or a combination of muscle activation and stretching. These injuries have a strong tendency to occur near the region of the myotendinous junction.

Delayed Muscle Soreness

Delayed muscle soreness is defined as muscular pain that generally occurs 24 to 72 hours after intense exercise. This should be distinguished from discomfort during exercise that is often associated with muscle fatigue. It should also be distinguished from painful involuntary cramps caused by strong contractions of susceptible muscles such as the gastrocnemius. Delayed muscle soreness is characterized by a variable sense of discomfort in the muscle beginning several hours after exercise and reaching a maximum after one to three days.

Many of the concepts of delayed muscle soreness were introduced by Hough[83] at the beginning of this century. He showed that forceful jerky movements produced muscle discomfort 24 to 48 hours after exercise, particularly in someone unaccustomed to such exercise. Hough further demonstrated that measurable muscle weakness accompanied the muscle soreness. Even after a brief conditioning period allowing for resolution of the soreness, the weakness persisted for several days. This was interpreted as the result of intramuscular damage to the structural elements of muscle. This injury was readily reversible. He also noted that the pain was most marked at the regions of the myotendinous junctions of the involved muscle.

Delayed muscle soreness has been the object of other investigations. The basic observations of Hough have been supported. The weakness accompanying delayed muscle soreness has been reported in several studies.[7,84,85] This weakness is transient and normal strength is recovered in a few days to weeks.[86] Eccentric exercises are defined as muscle activation involving lengthening of a muscle. The force applied to the activated muscle exceeds the muscle force and the muscle lengthens rather than shortens. Muscle is capable of generating more force in an eccentric mode than in isometric contraction. Eccentric exercise is much more likely to produce soreness than concentric exercise, defined as muscle activation combined with muscle shortening.[87,88]

Serologic studies have shown that exhausting exercise may be associated with increased levels of intramuscular enzymes in the serum. The increased levels of enzymes from skeletal muscle may correlate with the presence of soreness in the muscle.[89,90] In addition to serum markers for intramuscular enzymes, there are interesting indications that connective tissue breakdown is also a part of the syndrome of delayed muscle soreness. Urinary levels of hydroxyproline excretion are indicative of collagen or connective tissue breakdown. Increased levels of urinary hydroxyproline excretion have been associated with delayed muscle soreness.[91]

Muscle Strains

Indirect muscle injuries are caused by excessive force or stress on a muscle rather than by any direct trauma. This injury has been called a strain, muscle pull, muscle tear, and various other names. It is certainly a frequent injury and a major reason for time lost from athletic or occupational pursuits.[92–94] Interest in the basic pathophysiologic characteristics of this injury has been relatively slow to develop despite its clinical significance.

Complete Muscle Tears   One of the earliest studies of failure of the muscle-tendon unit in response to stretch was that of McMaster.[95] The

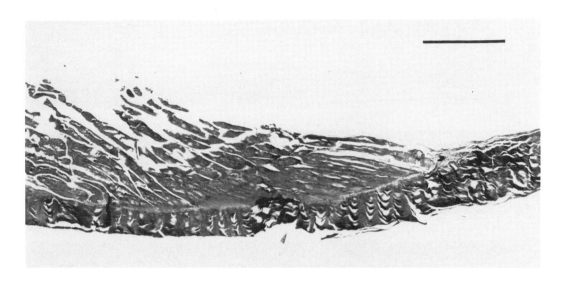

Fig. 5–14 *Complete avulsion of muscle fibers from myotendinous junction. The tendon is at the lower margin. Approximately 2 mm of muscle fiber remain attached to the tendon (bar gauge = 1 mm).*

gastrocnemius muscle-tendon unit of rabbits was strained to failure. Interestingly, the healthy tendon did not fail even after partial transection of the tendon. Failure occurred at the bone-tendon junction, the myotendinous junction, or within the muscle.

More recently, a number of studies have been performed on muscle-tendon preparations. Rabbit muscle-tendon units stretched to failure uniformly fail near the myotendinous junction.[96-98] The tendons of origin and insertion of most muscles extend well into the length of the muscle and the muscle fibers have an oblique angle of insertion into the tendons. The fiber architecture is quite variable in different muscles. Rabbit hind-limb muscles were strained to failure without any direct nerve or muscle activation.[99] For all muscles, failure consistently occurred near the myotendinous junction. Further, the rate of stretch does not appear to be a factor, as failure consistently occurred near the region of the myotendinous junction for extension rates between 1 and 100 cm/minute.

The exact site of failure has not been examined in great detail. The tendon avulses from the muscle, often with a short length of muscle fiber still attached, rather than at the precise junction of muscle and tendon. Fiber lengths of less than several millimeters can still be attached to the tendon after failure (Fig. 5–14). We do not know with certainty how this relates to the unfolding and junctional amplification noted previously.[20,22,34] It is also of interest that the terminal sarcomeres near the myotendinous junction are stiffer than the middle sarcomeres

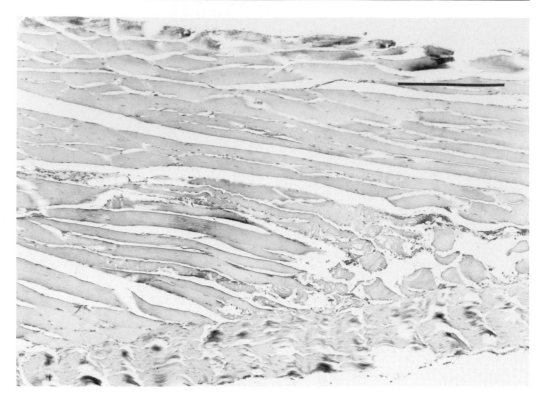

Fig. 5–15 *Incomplete tear of muscle at myotendinous junction immediately after injury. Tendon is at lower margin. Avulsion of fibers can be seen near the distal myotendinous junction. Less than 1 mm of muscle fiber remains attached to the tendon. Some fibers are uninjured (bar gauge = 0.5 mm).*

of a muscle fiber.[26,27] The injury to muscle occurs within this region of relatively limited extensibility. The reasons for the relative lack of extensibility in this area are unclear. There are structural differences in this region, but detailed studies of the region of injury are lacking. Research is needed to define precisely the ultrastructure of muscle at the site of failure and to relate it to the available data regarding the normal myotendinous junction.

Further, these studies demonstrated that tears occurred near the myotendinous junction in muscle stretched at different rates but without any stimulation of its active contractile properties. Investigation of the response of activated muscle to stretch has been done.[100] Rabbit anterior tibial muscles were stretched to failure and the effects of electrical stimulation of the muscle by the motor nerve were recorded. Tetanically stimulated muscle tore at the same increase in length as submaximally stimulated muscle or unstimulated muscle. Slightly greater forces were recorded at failure for the stimulated muscles. The major effect of stimulation was the ability of stimulated muscle to

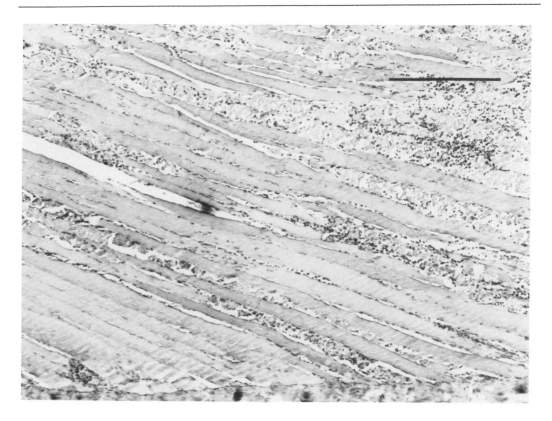

Fig. 5–16 *Incomplete tear of muscle 48 hours after strain. Tendon is at lower margin. An inflammatory reaction is pronounced near the myotendinous junction (bar gauge − 0.5 mm).*

maintain a higher force during stretching. This difference allowed a stimulated muscle to absorb much more energy than an unstimulated muscle. This is a significant finding in light of the ability of muscle to avoid injury to joints and to muscles themselves. Muscles must often contract eccentrically to absorb energy (for example, the quadriceps muscle contracts eccentrically on landing from a jump). Joint control and protection from injury may be better when muscles are able to absorb more kinetic energy.

Incomplete Muscle Tears Although complete muscle tears are seen clinically, incomplete tears are much more common. Only a few studies have addressed the pathophysiologic features of incomplete muscle injury. Nikolaou and associates[101] produced nondisruptive muscle injuries by stretching rabbit anterior tibial muscles well into the elastic region of the load-deformation relation, resulting in a characteristic lesion near the myotendinous junction. Fiber disruption and a small amount of hemorrhage were initially present (Fig. 5–15). During the

Fig. 5–17 *Incomplete tear of muscle seven days after strain. Tendon is at lower margin. A large increase in fibrous tissue and fibroblast can be seen near the junction. Few inflammatory cells are seen at this point (bar gauge = 0.2 mm).*

next one to four days a cellular inflammatory response occurred with the appearance of edema and granulation tissue with fibroblasts and inflammatory cells (Fig. 5–16). After one week a significant amount of scar tissue was present (Fig. 5–17). Physiologically these injured muscles had a consistent decrease in their ability to produce active tension for the first few days. After seven days the muscle approached normal tension production.

Clinical signs of recovery from muscle strain injuries might be expected to include an initial inflammatory and regenerative phase followed by scar formation near the region of the myotendinous junction. Although there are many clinical studies of muscle injuries, few have considered the pathophysiologic features of the injury. One recent computed tomographic study has shown that the early response of a muscle to a strain injury is the appearance of a region of low electron density within the muscle at the site of injury.[102] This study investigated hamstring strain injuries. Low electron density is indicative of an inflammatory process with edema formation. Acute intramuscular hematoma, on the other hand, should have been visible as an area of electron density higher than surrounding muscle. Although no biopsy specimens were obtained, anatomic dissections that demonstrated the lesions seen by computed tomography were near the myotendinous junction. The region of the myotendinous junction is quite extensive in the hamstring muscles. For example, the proximal tendon and myotendinous junctions of the semimembranosus and biceps femoris extend approximately 70% of the length of the entire muscle-tendon unit.

Effects of Immobilization   Stress-induced changes occur at the myotendinous junction. Length changes in muscle are associated with changes in sarcomere length. Long-term immobilization in a flexed or extended position results in readjustment of sarcomere number. Sarcomeres are added or deleted until sarcomere length at the position of immobilization equals the length before immobilization. When new sarcomeres are added, they are incorporated in the region of the myotendinous junction.[103,104] It may well be that a similar mechanism is responsible for the increased length of the developing muscle as it is stretched by growing bones. Stretch-induced growth is associated with the incorporation of new structural proteins and often with a new gain in muscle weight even though the muscle is immobilized.

Immobilization in a lengthened position not only leads to a reorganization of the active components of muscle tension, as shown by sarcomere number, but also leads to reorganization of the passive components of muscle tension.[105,106] The elastic components and connective tissue are rearranged so that the resting length of muscle readjusts to its position of immobilization. The specific changes occurring at the myotendinous junction have not been studied in detail. One biomechanical study investigating the passive properties of immobilized rabbit muscle found that muscle immobilized in a shortened position developed less force and stretched to a shorter length before tearing than did the nonimmobilized contralateral control muscle.[107] The tear occurred at the myotendinous junction of both muscle groups, indicating an alteration in the mechanical properties of this junction. Muscle immobilized in a lengthened position responded quite differently. More force and more change in length to tear were necessary than in the nonimmobilized control. The muscles immobilized in a stretched position also had a significant increase in their weight. Clearly, the position of immobilization is an important factor in determining the nature of the changes at the myotendinous junction.

Prevention and Treatment of Muscle Injury   Exercise regimens and stretching sequences have often been cited as an important way of avoiding injury to muscles, ligaments, tendons, and joints. Muscles immobilized in an extended position may be less susceptible to injury because longer stretches and more tension are required to disrupt the muscle-tendon unit at the junction of muscle and tendon.[107] Immobilization in extension, however, may not be similar to various stretching protocols employed or advised clinically.

A warm-up or conditioning period has recently been shown to be effective in altering the biomechanical properties of muscle in a way that may be effective in avoiding injury. Several different rabbit muscles subjected to only one maximal stimulation lasting ten to 15 seconds were strained to failure and compared to control unstimulated muscle. In all of the muscles the single stimulation caused the muscles to develop more force and to stretch farther when pulled to failure compared with control muscles.[108]

In a recent study evaluating the effects of nonsteroidal anti-inflammatory medication on the healing of experimental injures in the anterior tibial muscles of rats, Almekinders and Gilbert[109] stretched the muscles in the region of plastic deformation and then immobilized the limbs. The combination of injury and immobilization led to continued weakening of the muscle for several days as measured by maximum failure load of muscle passively strained to failure. Treatment with piroxicam slightly improved the maximum failure load. Histologic changes in the healing process were also noted when injured and immobilized muscles with and without piroxicam treatment were compared.

Summary

It is apparent from recent investigations that an understanding of the myotendinous junction is important in understanding the nature of indirect injuries or "strains" of muscle. Injured muscles most often fail in this region. It is also apparent that the properties of the myotendinous junction can be altered in ways that may influence the susceptibility of muscle to injury. This could be of enormous benefit in the prevention and treatment of the muscle-tendon injuries so common in sports medicine and occupational medicine.

It is obvious that there are few studies connecting injury in the region of the myotendinous junction to what we know about the fine structural detail of the normal myotendinous junction. Much greater structural detail is needed for the region of myotendinous junction failure. The site of tear should be related to the anatomic junction of muscle and tendon and to the region of the myotendinous junction, which has properties known to differ from those of most fibers separated from the junction.

Changes associated with biomechanical adaptations at the myotendinous junction should be analyzed in more structural detail. Only in this way can structure-function relationships be united with our pathophysiologic knowledge to provide an understanding of the myotendinous junction that can be useful in the prevention and management of injury.

Future Directions

It seems clear that much information about the basic science and pathologic characteristics of the myotendinous junction is missing.

More work is necessary to define these basic questions before the broad problems of injury, treatment, and prevention can be addressed.

At the basic level, the molecular composition of the myotendinous junction must be investigated further. The mechanism for attaching the cytoskeletal proteins to the extracellular structural proteins must be identified. It is necessary to determine which proteins transmit force and with what efficiency they do so. When the molecular interactions are understood, then the mechanism of failure can be addressed.

New data from strain injury studies also highlight the region of the muscle fibers near the myotendinous junction. The fibers there are stiffer than in the regions away from the junction. Muscle disruptions often occur through fibers near their termination rather than at the true myotendinous junction. Presumably there are cellular differences or differences in the basement membrane and extracellular space that cause differences in muscle fiber stiffness at that region. These differences should be determined. Structural studies detailing the region of increased stiffness in stretched muscle would be helpful. More detailed ultrastructural studies of this region are necessary. Biochemical studies demonstrating any differences in distribution of the structural and linking proteins are lacking.

That experimental muscle strain injuries occur near the myotendinous junction focuses attention on this area. However, it is apparent that very little basic science information exists regarding the nature of muscle injuries. Is the animal model adequate to predict that clinical muscle injuries will occur at the myotendinous junction? Can the muscle-tendon unit be injured or disrupted elsewhere than at the junction? Can injury be produced by a mechanism other than by the extreme stretches necessary in the present model?

The study of muscle injury is complex because of the many variables of muscle activity. Muscle can be studied when it is contracting or relaxed. There are actomyosin interactions even in relaxed muscle. Activated muscle can be shortening, lengthening, or isometric. There are active reflex mechanisms. Muscles have highly variable fiber architectural schemes. Strain rate and total amount of strain must be considered. This complexity has probably prevented relevant research in the past. However, progress in this field is now limited by the lack of basic studies isolating these variables and providing meaningful data regarding muscle injury.

Clinical studies of muscle injuries relating to the pathophysiologic characteristics of the injury are surprisingly scant. Computed tomography and ultrasound have been used sparingly for injuries of the myotendinous junction. New techniques such as magnetic resonance imaging may be very useful in defining the location and nature of the common diagnosis of "muscle strain." When one considers the extent of time lost because of muscle injuries, research into this area becomes imperative. One of the leading causes of disability is back strain in the cervical or lumbar area. Our lack of knowledge of muscle injuries does not allow us to conclude whether this injury involves muscles, ligaments, joints, or disk material.

Once the mechanisms of injury are determined, attention can then be focused on prevention. The number of sports-related injuries in this country is growing as more emphasis is placed on health and fitness through exercise. The medical profession has many more questions than answers when the proper techniques of training and injury prevention are discussed. Do stretching routines or warm-up routines really prevent joint or myotendinous junction injury? If so, how? Are the recommendations of medical and coaching professionals effective in preventing injury? Both clinical and basic studies are necessary. Prospective clinical trials with large numbers of participants over a relatively long period are needed. Present data suggest that muscle injury and disruption often occur near the myotendinous junction, although injuries in the muscle or tendon proper are certainly not excluded. It is surprising that at a time when molecular mechanisms of many disease processes are understood, so little is known about common injuries to muscle, tendon, and myotendinous junction.

Acknowledgments

Supported by a grant-in-aid from the American Heart Association, California Affiliate and Riverside and Orange County Chapters, and grant DCB–8417485 from the National Science Foundation (J.G.T.). Antisera to CSAT antigen was donated by Dr. A. Horwitz, University of Pennsylvania.

References

1. Gelber D, Moore DH, Ruska H: Observations of the myo-tendon junction in mammalian skeletal muscle. *Z Zellforsch Mikrosk Anat* 1960;52:396–400.
2. Auber J: Ultrastructure de la jonction myo-épidermique chez les Diptères. *J Microsc* 1963;2:325–336.
3. Lai-Fook J: The structure of developing muscle insertions in insects. *J Morphol* 1967;123:503–528.
4. Schippel K, Reissig D: Zur Feinstruktur des Muskel-Sehnenüberganges. *Z Microsk Anat Forschung* 1968;78:235–255.
5. Caveney S: Muscle attachment related to cuticle architecture in Apterygota. *J Cell Sci* 1969;4:541–559.
6. Mackay B, Harrop TJ, Muir AR: The fine structure of the muscle tendon junction in the rat. *Acta Anat* 1969;73:588–604.
7. Mair WGP, Tomé FMS: The ultrastructure of the adult and developing human myotendinous junction. *Acta Neuropathol* 1972;21:239–252.
8. Nakao T: Fine structure of the myotendinous junction and "terminal coupling" in the skeletal muscle of the lamprey, *Lampetra japonica. Anat Rec* 1975;182:321–338.
9. Nakao T: Some observations on the fine structure of the myotendinous junction in myotomal muscles of the tadpole tail. *Cell Tissue Res* 1976;166:241–254.
10. Wake K: Formation of myoneural and myotendinous junctions in the chick embryo: Role of acetylcholinesterase-rich granules in the developing muscle fibers. *Cell Tissue Res* 1976;173:383–400.
11. Ajiri T, Kimura T, Ito R, et al: Microfibrils in the myotendon junction. *Acta Anat* 1978;102:433–439.

12. Sweeny PR, Brown RG: Ultrastructural studies of the myotendinous junction of selenium-deficient ducklings. *Am J Pathol* 1980;100:481–496.

13. Trotter JA, Corbett K, Avner BP: Structure and function of the murine muscle-tendon junction. *Anat Rec* 1981;201:293–302.

14. Tidball JG: The geometry of actin filament-membrane associations can modify adhesive strength of the myotendinous junction. *Cell Motil* 1983;3:439–447.

15. Trotter JA, Eberhard S, Samora A: Structural connections of the muscle-tendon junction. *Cell Motil* 1983;3:431–438.

16. Hikida RS, Peterson WJ: Skeletal muscle-smooth muscle interaction: An unusual myoelastic system. *J Morphol* 1983;177:231–243.

17. Tidball JG: Myotendinous junction: Morphological changes and mechanical failure associated with muscle cell atrophy. *Exp Mol Pathol* 1984;40:1–12.

18. Eisenberg BR, Milton RL: Muscle fiber termination at the tendon in the frog's sartorius: A stereological study. *Am J Anat* 1984;171:273–284.

19. Trotter JA, Hsi K, Samora A, et al: A morphometric analysis of the muscle-tendon junction. *Anat Rec* 1985;213:26–32.

20. Trotter JA, Samora A, Baca J: Three-dimensional structure of the murine muscle-tendon junction. *Anat Rec* 1985;213:16–25.

21. Andreev DP, Wassilev WA: Specialized contacts between sarcolemma and sarcoplasmic reticulum at the ends of muscle fibers in the diaphragm of the rat. *Cell Tissue Res* 1986;243:415–420.

22. Tidball JG, Daniel TL: Myotendinous junctions of tonic muscle cells: Structure and loading. *Cell Tissue Res* 1986;245:315–322.

23. Tidball JG, O'Halloran T, Burridge K: Talin at myotendinous junctions. *J Cell Biol* 1986;103:1465–1472.

24. Tidball JG: Alpha-actinin is absent from the terminal segments of myofibrils and from subsarcolemmal densities in frog skeletal muscle. *Exp Cell Res* 1987;170:469–482.

25. Porter KR: The myo-tendon junction in larval forms of *Amblyostoma punctatum*, abstract. *Anat Rec* 1954;118:342.

26. Huxley AF, Peachey LD: The maximum length for contraction in vertebrate striated muscle. *J Physiol* 1961;156:150–165.

27. Gordon AM, Huxley AF, Julian FJ: Tension development in highly stretched vertebrate muscle fibres. *J Physiol* 1966;184:143–169.

28. Mauro A, Adams WR: The structure of the sarcolemma of the frog skeletal muscle fiber. *J Biophys Biochem Cytol* 1961;10:177–185.

29. Schmidt V: Die Histogenese der quergestreiften Muskelfaser und des Muskelsehnenüberganges. *Z Mikrosk Anat Forschung* 1927;8:97–184.

30. Speidel CC: Studies of living muscles: I. Growth, injury and repair of striated muscle, as revealed by prolonged observations of individual fibers in living frog tadpoles. *Am J Anat* 1938;62:179–236.

31. Crawford GNC: An experimental study of muscle growth in the rabbit. *J Bone Joint Surg* 1954;36B:294–303.

32. Mackay B, Harrop TJ: An experimental study of the longitudinal growth of skeletal muscle in the rat. *Acta Anat* 1969;72:38–49.

33. Williams PE, Goldspink G: Changes in sarcomere length and physiological properties in immobilized muscle. *J Anat* 1978;127:459–468.

34. Ishikawa H, Sawada H, Yamada E: Surface and internal morphology of skeletal muscle, in Peachey LD, Adrian RH, Geiger SR (eds): *Handbook of Physiology, Section 10. Skeletal Muscle.* Bethesda, American Physiological Society, 1983, pp 1–21.

35. Evans EA, Hochmuth RM: Mechanochemical properties of membranes. *Curr Top Membranes Transport* 1978;10:1–64.

36. Lubkin JL: The theory of adhesive scarf joints. *J Appl Mech* 1957;24:255–260.

37. Bikerman JJ: Stresses in proper adhints, in *The Science of Adhesive Joints.* New York, Academic Press, 1968, pp 192–263.

38. Ramsey RW, Street SF: The isometric length-tension diagram of isolated skeletal muscle fibers of the frog. *J Cell Comp Physiol* 1940;15:11–34.

39. Rand RP: Mechanical properties of the red cell membrane: II. Viscoelastic breakdown of the membrane. *Biophys J* 1964;4:303–316.

40. Rand RP: The structure of a model membrane in relation to the viscoelastic properties of the red cell membrane. *J Gen Physiol* 1968;52:173S–186S.

41. Burke RE, Levine DN, Zajac FE III, et al: Mammalian motor units: Physiological-histochemical correlation in three types in cat gastrocnemius. *Science* 1971;174:709–712.

42. Ewell RH, Eyring H: Theory of the viscosity of liquids as a function of temperature and pressure. *J Chem Phys* 1937;5:726–736.

43. Joly M: Non-Newtonian surface viscosity. *J Colloid Sci* 1956;11:519–531.

44. Maruyama K, Shimada Y: Fine structure of the myotendinous junction of lathyritic rat muscle with special reference to connectin, a muscle elastic protein. *Tissue Cell* 1978;10:741–748.

45. Heuson-Stiennon J-A: Morphogenèse de la cellule musculaire striée étudiée au microscope électronique: Formation des structures fibrillaires. *J Microsc* 1965;4:657–678.

46. Warren RH, Porter KR: An electron microscope study of differentiation of the molting muscles of *Rhodnius prolixus. Am J Anat* 1969;124:1–30.

47. Tokuyasu KT, Dutton AH, Geiger B, et al: Ultrastructure of chicken cardiac muscle as studied by double immunolabeling in electron microscopy. *Proc Natl Acad Sci USA* 1981;78:7619–7623.

48. Lemanski LF, Paulson DJ, Hill CS, et al: Immunoelectron microscopic localization of α-actinin on Lowicryl-embedded thin-sectioned tissues. *J Histochem Cytochem* 1985;33:515–522.

49. Zeece MG, Robson RM, Bechtel PJ: Interaction of α-actinin, filamin and tropomyosin with F-actin. *Biochim Biophys Acta* 1979;581:365–370.

50. Stromer MH, Goll DE: Studies on purified α-actinin: II. Electron microscopic studies on the competitive binding of α-actinin and tropomyosin to Z-line extracted myofibrils. *J Mol Biol* 1972;67:489–494.

51. Podlubnaya ZA, Tskhovrebova LA, Zaalishvili MM, et al: Electron microscopic study of α-actinin, letter. *J Mol Biol* 1975;92:357–359.

52. Small JV: Geometry of actin-membrane attachments in the smooth muscle cell: The localisations of vinculin and α-actinin. *EMBO J* 1985;4:45–49.

53. Geiger B: A 130K protein from chicken gizzard: Its localization at the termini of microfilament bundles in cultured chicken cells. *Cell* 1979;18:193–205.

54. Geiger B, Tokuyasu KT, Dutton AH, et al: Vinculin, an intracellular protein localized at specialized sites where microfilament bundles terminate at cell membranes. *Proc Natl Acad Sci USA* 1980;77:4127–4131.

55. Geiger B, Volk T, Volberg T: Molecular heterogeneity of adherens junctions. *J Cell Biol* 1985;101:1523–1531.

56. Volk T, Geiger B: A-CAM: A 135-kD receptor of intercellular adherens junctions: I. Immunoelectron microscopic localization and biochemical studies. *J Cell Biol* 1986;103:1441–1450.

57. Burridge K, Connell L: Talin: A cytoskeletal component concentrated in adhesion plaques and other sites of actin-membrane interaction. *Cell Motil* 1983;3:405–417.

58. Burridge K, Connell L: A new protein of adhesion plaques and ruffling membranes. *J Cell Biol* 1983;97:359–367.

59. Burridge K, Mangeat P: An interaction between vinculin and talin. *Nature* 1984;308:744–746.

60. Shear CR, Bloch RJ: Vinculin in subsarcolemmal densities in chicken skeletal muscle: Localization and relationship to intracellular and extracellular structures. *J Cell Biol* 1985;101:240–256.

61. Timpl RH, Rohde H, Robey PG, et al: Laminin—a glycoprotein from basement membranes. *J Biol Chem* 1979;254:9933–9937.

62. Bornstein P, Sage H: Structurally distinct collagen types. *Annu Rev Biochem* 1980;49:957–1003.

63. Knudsen KA, Horwitz AF: Tandem events in myoblast fusion. *Dev Biol* 1977;58:328–338.

64. Damsky CH, Knudsen KA, Bradley D, et al: Distribution of the cell substratum attachment (CSAT) antigen on myogenic and fibroblastic cells in culture. *J Cell Biol* 1985;100:1528–1539.

65. Horwitz A, Duggan K, Greggs R, et al: The cell substrate attachment (CSAT) antigen has properties of a receptor for laminin and fibronectin. *J Cell Biol* 1985;101:2134–2144.

66. Pytela R, Pierschbacher MD, Ruoslahti E: Identification and isolation of a 140 kd cell surface glycoprotein with properties expected of a fibronectin receptor. *Cell* 1985;40:191–198.

67. Decker C, Greggs R, Duggan K, et al: Adhesive multiplicity in the interaction of embryonic fibroblasts and myoblasts with extracellular matrices. *J Cell Biol* 1984;99:1398–1404.

68. Knudsen KA, Rao PE, Damsky CH, et al: Membrane glycoproteins involved in cell-substratum adhesion. *Proc Natl Acad Sci USA* 1981;78:6071–6075.

69. Greve JM, Gottlieb DI: Monoclonal antibodies which alter the morphology of cultured chick myogenic cells. *J Cell Biochem* 1982;18:221–229.

70. Chapman AE: Characterization of a 140Kd cell surface glycoprotein involved in myoblast adhesion. *J Cell Biochem* 1984;25:109–121.

71. Horwitz A, Duggan K, Buck C, et al: Interaction of plasma membrane fibronectin receptor with talin—a transmembrane linkage. *Nature* 1986;320:531–533.

72. Chen W-T, Greve JM, Gottlieb DI, et al: Immunocytochemical localization of 140 kD cell adhesion molecules in cultured chicken fibroblasts, and in chicken smooth muscle and intestinal epithelial tissues. *J Histochem Cytochem* 1985;33:576–586.

73. Chen W-T, Hasegawa E, Hasegawa T, et al: Development of cell surface linkage complexes in cultured fibroblasts. *J Cell Biol* 1985;100:1103–1114.

74. Barnes D, Wolfe R, Serrero G, et al: Effects of a serum spreading factor on growth and morphology of cells in serum-free medium. *J Supramol Struct* 1980;14:47–63.

75. Hayman EG, Engvall E, A'Hearn E, et al: Cell attachment on replicas of SDS polyacrylamide gels reveals two adhesive plasma proteins. *J Cell Biol* 1982;95:20–23.

76. Hayman EG, Pierschbacher MD, Öhgren Y, et al: Serum spreading factor (vitronectin) is present at the cell surface and in tissues. *Proc Natl Acad Sci USA* 1983;80:4003–4007.

77. Whateley JG, Knox P: Isolation of a serum component that stimulates the spreading of cells in culture. *Biochem J* 1980;185:349–354.

78. Baetscher M, Pumplin DW, Bloch RJ: Vitronectin at sites of cell-substrate contact in cultures of rat myotubes. *J Cell Biol* 1986;103:369–378.

79. DeSimone DW, Stepp MA, Fonda D, et al: Structural and functional analyses of integrin, a glycoprotein complex linking fibronectin to the cytoskeleton, abstract. *J Cell Biol* 1986;103:287a.

80. Chiquet M, Fambrough DM: Chick myotendinous antigen: I. A monoclonal antibody as a marker for tendon and muscle morphogenesis. *J Cell Biol* 1984;98:1926–1936.

81. Chiquet M, Fambrough DM: Chick myotendinous antigen: II. A novel extracellular glycoprotein complex consisting of large disulfide-linked subunits. *J Cell Biol* 1984;98:1937–1946.

82. Tidball JG: Energy stored and dissipated in skeletal muscle basement membranes during sinusoidal oscillations. *Biophys J* 1986;50:1127–1138.

83. Hough T: Ergographic studies in muscular soreness. *Am J Physiol* 1902;7:76–92.

84. Davies CTM, White MJ: Muscle weakness following eccentric work in man. *Pfluegers Arch* 1981;392:168–171.

85. Komi PV, Buskirk ER: Effect of eccentric and concentric muscle conditioning on tension and electrical activity of human muscle. *Ergonomics* 1972;15:417–434.

86. Armstrong RB, Ogilvie RW, Schwane JA: Eccentric exercise-induced injury to rat skeletal muscle. *J Appl Physiol* 1983;54:80–93.

87. Newham DJ, Mills KR, Quigley BM, et al: Pain and fatigue after concentric and eccentric muscle contractions. *Clin Sci* 1983;64:55–62.

88. Tiidus PM, Ianuzzo CD: Effects of intensity and duration of muscular exercise on delayed soreness and serum enzyme activities. *Med Sci Sports Exer* 1983;15:461–465.

89. Newham DJ, Jones DA, Edwards RHT: Large delayed plasma creatine kinase changes after stepping exercise. *Muscle Nerve* 1983;6:380–385.

90. Schwane JA, Johnson SR, Vandenakker CB, et al: Delayed-onset muscular soreness and plasma CPK and LDH activities after downhill running. *Med Sci Sports Exer* 1983;15:51–56.

91. Abraham WM: Factors in delayed muscle soreness. *Med Sci Sports* 1977;9:11–20.

92. Pritchett JW: High cost of high school football injuries. *Am J Sports Med* 1980;8:197–199.

93. Apple DV, O'Toole J, Annis C: Professional basketball injuries. *Phys Sports Med* 1982;10:81.

94. Garrick JG, Requa RK: Epidemiology of women's gymnastics injuries. *Am J Sports Med* 1980;8:261–264.

95. McMaster PE: Tendon and muscle ruptures: Clinical and experimental studies on the causes and location of subcutaneous ruptures. *J Bone Joint Surg* 1933;15:705–722.

96. Almekinders LC, Garrett WE Jr, and Seaber AV: Pathophysiologic response to muscle tears in stretching injuries. *Trans Orthop Res Soc* 1984;9:307.

97. Almekinders LC, Garrett WE Jr, Seaber AV: Histopathology of muscle tears in stretching injuries. *Trans Orthop Res Soc* 1984;9:306.

98. Garrett WE Jr, Almekinders LC, Seaber AV: Biomechanics of muscle tears in stretching injuries. *Trans Orthop Res Soc* 1984;9:384.

99. Garrett WE Jr, Nikolaou PK, Ribbeck BM, et al: The effect of muscle architecture on the biomechanical failure properties of skeletal muscle under passive extension. *Am J Sports Med*, in press.

100. Garrett WE Jr, Safran MR, Seaber AV, et al: Biomechanical comparison of stimulated and skeletal muscle pulled to failure. *Am J Sports Med*, in press.

101. Nikolaou PK, Macdonald BL, Glisson RR, et al: Biomechanical and histological evaluation of muscle after controlled strain injury. *Am J Sports Med* 1987;15:9–14.

102. Garrett WE Jr, Rich FR, Nikolaou PK, et al: Computed tomography of hamstring muscle strains. *Am J Sports Med*, in press.

103. Williams PE, Goldspink G: The effect of immobilization on the longitudinal growth of striated muscle fibres. *J Anat* 1973;116:45–55.

104. Griffin GE, Williams PE, Goldspink G: Region of longitudinal growth in striated muscle fibres. *Nature [New Biol]* 1971;232:28–29.

105. Tardieu C, Tabary J-C, Tabary C, et al: Adaptation of connective tissue length to immobilization in the lengthened and shortened positions in cat soleus muscle. *J Physiol [Paris]* 1982;78:214–220.

106. Williams PE, Goldspink G: Connective tissue changes in immobilised muscle. *J Anat* 1984;138:343–350.

107. Jones VT, Garrett WE Jr, Seaber AV: Biomechanical changes in muscle after immobilization at different lengths. *Trans Orthop Res Soc* 1985;10:6.

108. Safran MR, Garrett WE Jr, Seaber AV, et al: The role of warm-up in muscular injury prevention. *Am J Sports Med*, in press.

109. Almekinders LC, Gilbert JA: Healing of experimental muscle strains and the effects of nonsteroidal antiinflammatory medication. *Am J Sports Med* 1986;14:303–308.

Section Five
Skeletal Muscle

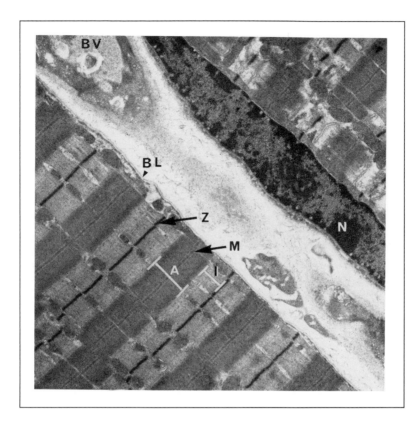

Group Leader

Arnold Caplan, PhD

Group Members

Joseph Buckwalter, MD
Bruce Carlson, MD, PhD
Bruce Caterson, PhD
Donald Fischman, MD

Group Participants

Richard Coutts, MD
William Garrett, Jr., MD, PhD
Ernst Hunziker, MD
Richard Lymn, PhD
Jerry Maynard, PhD
Helen Muir, PhD
Lawrence Rosenberg, MD
Ileen Stewart, MS
James Tidball, PhD

Synopsis

Acute injuries to skeletal muscle—including contusions, lacerations, ruptures, ischemia, compartment syndromes and denervation—occur frequently and cause significant disability. Blunt trauma can diminish muscle strength, limit joint motion, and lead to myositis ossificans. Muscle lacerations, surgical incisions, traumatic loss of muscle tissue, and denervation weaken muscles. Ruptures of muscle also cause weakness. Like the other injuries, they may result from direct trauma, but muscle contraction against resistance can also tear muscle. Acute muscle ischemia and compartment syndromes can cause extensive muscle necrosis. The many potential causes of compartment syndromes all result in increased pressure within a confined muscle compartment. Failure to relieve the pressure rapidly causes complications that range from weakness and decreased motion to loss of an entire limb. Despite the frequency and significance of acute muscle injuries, the long-term clinical results of treatment, the basic repair responses, and methods of facilitating the repair response in clinically significant injures have not been extensively studied.

Much of what is known about the potential for restoration of skeletal muscle function following injury is based on current understanding of muscle development combined with investigations of the cellular and molecular events during muscle regeneration following injuries. These investigations have shown that adult skeletal muscle has substantial capacity to repair itself. This repair process is in some respects a recapitulation of embryonic development but, unlike development, the potential of repair following a specific injury is influenced by the prior innervation pattern, vascularity, physical constraints of the surrounding tissues, the extent and condition of the remnant extracellular matrices (especially the basement membrane), the developmental lineage of the repair cells (fast vs slow lineage satellite cells), and the presence of non-myogenic lineage repair cells.

In general, the sequence of events following injury includes active degeneration and removal of injured muscle, proliferation of specialized myogenic repair cells, fusion to form myotubes, synthesis of contractile proteins, reinnervation, and activity-induced integration of the regenerated muscle into the unaffected muscle tissue. Repair not only follows acute traumatic injury, but proceeds continuously on a microscopic level. Individual myotubes degenerate, then regenerate to occupy exactly the same location as the predecessor myotube, although such replacement need not be by a myotube of the same lineage. For example, during strenuous exercise, it is clear that some individual myotubes become injured, are phagocytized, and replaced by dividing and fusing myoblasts and then myotubes. Molecular transitions in both the contractile proteins and extracellular matrix components are comparable to embryonic events but occur more rapidly. Rarely, connective tissue replaces these myotubes. In older individuals, the number of repair cells may be diminished and repair may not occur, resulting in muscle atrophy, but not fibrous tissue replacement or scarring.

Although the potential for regeneration of adult human skeletal muscle exists, it does not occur following clinically significant muscle loss. Thus, it is important to define the factors required for regeneration of skeletal muscle and to explore how these factors might be provided to restore functional adult muscle tissue following injury. At the level of cells and tissues, the major factors necessary for muscle regeneration include: (1) a source of myoblastic cells, (2) adequate vascularization, (3) an adequate number and distribution of motor and sensory nerves, (4) an appropriate mechanical environment, and (5) a space-occupying matrix that stimulates migration, proliferation, and differentiation of myoblastic cells and vascularization, innervation, and mechanical loading of the regenerating muscle.

Myoblastic cells survive many types of severe muscle injury and can form new muscle tissue, if they have an adequate vascular supply. When devascularized muscle grafts larger than a few hundred milligrams are grafted into vascularized muscle, the rate and extent of revascularization is a limiting factor. Ischemic satellite cells do not survive for much more than four hours, and multinucleated muscle fibers die even sooner. Freshly damaged, ischemic skeletal muscle does possess angiogenic properties, but the rate of vascular penetration into ischemic muscle is only 0.5 to 1.0 mm/day. Thus, devascularized segments of skeletal muscle larger than a few millimeters cannot rely on vascular ingrowth. In primates, collagen is deposited in areas of muscle that are not rapidly revascularized. Purified angiogenesis factors might help accelerate revascularization of muscle and thereby make possible regeneration of large volumes of muscle.

Although revascularization is critical for regeneration to begin and for survival of regenerated muscle fibers, vascularized muscle fibers are of little use if they are not appropriately innervated. Most mammalian muscle will regenerate to the stage of having cross-striated muscle fibers in the absence of innervation, but the final differentiation into fast and slow muscle fiber types and the necessary increase in cross-sectional area depend on motor innervation. In many types of muscle injury, remnants of the original myoneural junctions persist and are commonly reused by regenerating nerve fibers. The original synaptic sites are not required because ectopic motor end plates are common, and neuromuscular junctions regenerate readily in the absence of regional junctions. The problem with ingrowing sensory nerve fibers finding regenerating muscle spindles and Golgi tendon organs is much greater than that of motor nerve fibers finding their appropriate termination points. The most important limiting factor in the reinnervation of regenerating muscle is the pathway available to the regenerating nerve fiber. For example, in small rat muscles, the critical factor in attaining complete functional return is the quality of the neural pathways. Under any circumstance, the larger the mass of regenerating muscle, the greater the problem facing ingrowing nerves.

The influence of mechanical loading on muscle regeneration is one of the least studied areas. Incidental observations and anecdotal evidence suggest that the appropriate mechanical environment stimulates muscle regeneration and improves regenerating muscle function. Conversely, immobilization inhibits muscle regeneration.

In the presence of appropriate vascularity, potential nerve supply, and an appropriate mechanical environment, myoblasts need a space-occupying matrix that stimulates the migration, proliferation, and formation of normal muscle cells. Under normal circumstances muscle cells continually test and interact with their extracellular matrix. Some matrix components, such as hyaluronic acid, are appropriate substrates for cell migration and proliferation, whereas others promote or lock cells into differentiative activities. Some muscle proteoglycans may have this effect. The basal lamina that surrounds muscle fibers has a number of important functions in regeneration. It acts as a substrate for satellite-cell proliferation, as a selective cell filter, and possibly as a recognition site for nerves regenerating toward the muscle fibers. Future investigations undoubtedly will show that other matrix molecules have important roles in muscle regeneration.

Chapter 6
Skeletal Muscle

Arnold Caplan, PhD
Bruce Carlson, MD, PhD
John Faulkner, PhD
Donald Fischman, MD
William Garrett, Jr., MD, PhD

Chapter Outline

Introduction

Skeletal muscle is the most complex musculoskeletal soft tissue. It consists of muscle cells, elaborate, highly organized networks of nerves and vessels, and a muscle-specific extracellular matrix. In addition to the organelles found in other cells, muscle cells contain multiple nuclei, complex systems of intracellular membranes, and specific types of contractile proteins. Normal muscle function depends not only on the integrity of the muscle cells, but on the nerve and vessel networks and appropriate mechanical loading.

During embryonic and early postnatal development, muscles follow an encoded developmental program, responsive to both intrinsic and extrinsic cues, that brings the muscles to their normal mature form and function. Even in the adult, muscle structure and composition is not static; it constantly adapts to changing conditions. Muscles can become injured or diseased or their functional environment can change if the activity pattern of the individual changes. Muscles may respond to each of these circumstances, either by changing their functional characteristics, or by scarring. As the individual ages, the musculature undergoes a characteristic sequence of changes, some of which are also seen upon injury, adaptation, and repair. Understanding normal muscle structure and function, the response of muscle to training and injury, and potential methods of improving muscle repair depends on knowledge of muscle development.

Development

Origin of Muscle Progenitor Cells

Somites   Skeletal muscle is derived from a mesodermal cell population arising from the somite,[1] not as originally believed from the somatopleural (lateral plate) mesoderm. On the other hand, muscle connective tissues (epimysium, perimysium, and endomysium), cartilage, and bone arise from the somatopleural mesoderm.[1-3] Thus, skeletal muscle differs in embryologic origin from the other mesodermal tissues of the limb. Except for visceral striated muscle innervated by cranial nerves V, VII, IX, and XI, which originates from the branchial arches, all other limb, body-wall, and back musculature originates from a subdivision of the somite termed the myotome. It is not known when in embryonic development the precursors of the myogenic lineage become committed to forming muscle. Presumably, this occurs during condensation of the axial mesoderm to form the somites, but critical tests of this hypothesis are lacking.

Fiber Types   On the basis of studies with embryonic chicks, Miller and Stockdale[4-6] demonstrated that at least three clonal myoblast lineages exist in muscle primordia prior to any functional innervation of the muscle. These three lineages can be distinguished by the type of myosin heavy chain (MHC) expressed by the primary myotubes in

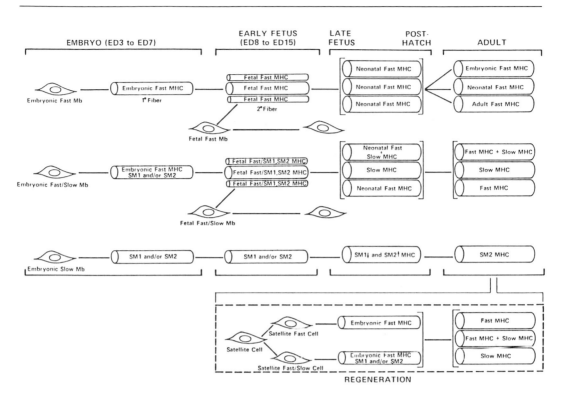

Fig. 6–1 *A model for avian myogenesis illustrating the formation of primary, secondary, and regenerating muscle fibers from myoblasts committed to distinct fast, fast/slow, and slow myogenic cell lineages and the sequential expression of fast- and/or slow-class MHCs within each of the distinct types of muscle fibers. The model postulates a cellular basis for restricting MHC expression within primary, secondary, and regenerating fibers to the fast, slow, or fast/slow classes of MHC. The MHC names are based on the developmental stage at which the isoform is first expressed. Thus the fetal fast MHC in this model is that fast MHC found in the fetus (d8-d15) and in regenerating avian muscle that others have designated "embryonic" MHC; whether or not a unique form of either fast- or slow-class MHC will be found in the embryonic period of development is not known. Only two isoforms have been indicated within the slow class of MHC because only two have been described; it is very likely that additional isoforms will be described within the slow class at earlier stages. (Reproduced with permission from Stockdale FE, Miller JB: The cellular basis of myosin heavy chain isoform expression during development of avian skeletal muscles. Dev Biol 1987;123:1–19.*

vivo and in clonal cell cultures.[4-6] Operationally, these have been termed "fast, slow, or fast/slow" myogenic lineages based on whether the myotubes contain fast-, slow- or both fast- and slow-type MHCs (Fig. 6–1). After innervation, at least one additional lineage can be distinguished that gives rise to secondary myotubes.[7] In this latter population, fiber type diversity apparently depends on innervation. Independent studies by Narusawa and associates[8] suggested that a different developmental pattern may exist in rats. These authors have observed that "in the 16-day fetus all primary myotubes in future fast and future slow muscles homogeneously express slow as well as embryonic myosin. Fiber heterogeneity arises owing to a developmentally regulated inhibition of slow MHC accumulation, as muscles are progressively assembled from successive orders of cells." Ho and associates[9] recently analyzed the development of the extensor digitorum longus, soleus, and temporalis muscles of the developing cat. These authors found strong evidence that different myoblast lineages exist in the temporalis and the limb muscles; their work can also be interpreted to support the concept of at least two different myoblast lines within the hindlimb muscle. These apparent differences between birds and mammals remain to be explained. However, studies with rat embryos have not examined stages as early in development as those of the chicken. Thus, current observational differences may be explained in the future. It should also be recognized that relatively few markers have been used to identify myogenic lineages, and more diversity is likely to be detected as additional markers, particularly cell surface markers, are developed. The specificity of neuromuscular connections in the limb suggests the existence of considerable informational content at the surface of embryonic muscle essential for functional innervation of skeletal muscle.[10,11]

Mitotic Properties of Myogenic Cells

Except for embryonic somites and some small muscle fibers of the middle ear and of the extraocular muscles, virtually all differentiated skeletal muscles are multinucleated. These nuclei, termed myonuclei, contain a 2C concentration of DNA and a diploid complement of chromosomes.[12] Myonuclei are inactive in both chromosome replication and karyokinesis.[13,14] Except for the ^{3}H-thymidine incorporation associated with repair DNA synthesis, normal myonuclei incorporate virtually no thymidine, and no validated reports have appeared in which DNA replication has been reinitiated within such nuclei. The concept has arisen that myonuclei are in a G_0 compartment of the mitotic cycle.[15] In contrast, mononucleated precursors of skeletal muscle are active in DNA synthesis and cell division. As noted above, this precursor population arises in the paraxial mesoderm of the somites from which the cells migrate to their definitive sites of differentiation. The terminology for the muscle precursors has been a point of dispute.[16,17] Since no good markers exist for the undifferentiated precursors of skeletal muscle, most workers have used the term "myoblast" to denote any mononucleated cell that can differentiate to form skeletal muscle. Since the mononucleated cells fall into two categories, those that replicate and those that have ceased division prior to cytoplasmic fusion and differentiation, Holtzer and Bischoff[6] limited the term myoblast to denote the postmitotic, mononucleated cell that is the immediate precursor of the myotube. Precursors of the myoblast have been termed myogenic-α, myogenic-β, and so forth, to allow for the existence of intermediate stages akin to those previously established for the erythropoietic lineage. Evidence for the existence of distinct myogenic precursor stages has been obtained.[18-20] In the discussion to follow, the term myoblast denotes mononucleated muscle precursors, without implying any particular stage of myogenesis.

Myoblasts respond to fibroblast growth factor (FGF) by synthesizing DNA and undergoing mitosis.[21] Removal of FGF causes myoblasts to withdraw from the cell cycle in G_1 and initiate terminal differentiation, as monitored by the expression of sarcomeric MHC, M-creatine kinase, and acetylcholine receptor (AchR). There is a small window in the G_1 compartment of the cycle when myoblasts remain FGF-sensitive, yet have begun synthesis of muscle-specific gene products. Re-exposure of the myoblasts to FGF during this time will stimulate DNA synthesis and down-regulate the three gene products listed above.[17] In addition, myogenic cells in culture are quite sensitive to transferrin-depleted media.[22] Both FGF and transferrin exhibit high affinity for proteoglycans of the extracellular matrix.[23] It is likely that these and other components of the extracellular matrix play an important role in regulating mitosis and differentiation of myoblasts. Depletion of matrix-associated mitogens prolongs the average length of stay of myoblasts in the G_1 compartment of the mitotic cycle.[24] Current evidence would support the concept that this prolongation in G_1 triggers muscle differentiation, presumably by initiating the transcription of regulatory genes required for the terminal differentiation of the myoblasts. Myoblasts exhibit characteristic shapes at specific compartments of the cell cycle (Fig. 6–2).

Fusion

Myotubes are formed by cytoplasmic fusion of myoblasts. The mechanism of this process is unknown. In the absence of a good genetic system, investigations of this phenomenon[25-30] have relied on the use of fusion inhibitors, such as phospholipase C, low levels of extracellular Ca^{2+}, inhibitors of metalloendoproteases, and tranquilizers such as diazepam. It is clear that the process is quite specific: myoblasts do not fuse with other cell types nor with other myoblasts still active in cell division. Thus, a cell-surface recognition process must exist, but membrane components putatively involved in the cell recognition have not been isolated. Myoblasts initiate fusion while in the G_1 compartment of the cell cycle. In addition to myoblast-myoblast fusion, myoblasts continue to fuse with myotubes as muscles elongate, both at the ends and along the shaft of the multinucleated cells.[31] In fact, growth of muscle depends on the continuing recruitment of myonuclei from a mononucleated population.

Cell fusion of myoblasts is not necessary for biochemical differentiation of muscle. In the chicken, blockade of fusion with EGTA, phospholipase C, or diazepam does not prevent the expression of the sarcomeric proteins nor of the AchR. In the rat, EGTA blockade of fusion does prevent contractile protein synthesis and myofibrillogenesis. However, transcription of the sarcomeric protein genes is not blocked.[32] Apparently, in the rat myoblast EGTA specifically inhibits translation of terminal muscle-specific gene products; the synthesis of "house-

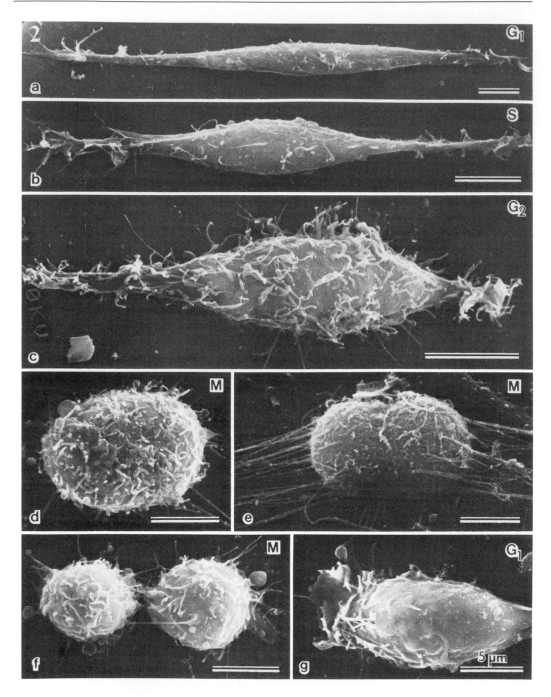

Fig. 6–2 *Scanning electron micrographs of cultured embryonic chick myoblasts at specific stages of the cell cycle: **a:** Spindle-shape in G_1; **b:** Spindle-shape in S; **c:** Large, partially retracted myoblast in late G_2; **d:** Rounded myoblast in M, probably metaphase; **e:** Myoblast in early anaphase of M (cytokinesis is beginning); **f:** Late anaphase of M (cytokinesis is virtually complete); **g:** Elongating myoblast in early G_1. (Reproduced with permission from Yutaka Shimada, Chiba, Japan.)*

keeping" gene products, such as proteins involved in nonmuscle motility and respiration, is not blocked. On the other hand, prevention of myoblast fusion by the inhibition of metalloendoproteases also prevents terminal muscle differentiation, but in this case a blockade seems to occur at the level of gene transcription.[32] Thus, the metalloendoproteases may act at an earlier stage of muscle differentiation than phospholipasae C or EGTA. Although mononucleated myoblasts held in G_1 in vitro can up-regulate most of the proteins used to identify mature skeletal muscle, in vivo expression of the muscle phenotype follows cytoplasmic fusion of the precursor population.

Contractile Proteins

Nonmuscle vs Muscle   Contractile proteins are ubiquitous components of all eukaryotic cells. Different sets of genes encoding these proteins are expressed in striated, smooth, and nonmuscle cells. These related proteins are termed isoforms to denote their structural and functional homologies. Contractile proteins underly nonmuscle-cell motility, and participate in diverse cellular functions such as cytokinesis, cytoplasmic streaming, and the stabilization and rearrangement of the cytoskeleton. Several sarcomeric proteins of striated muscle have counterparts in nonmuscle cells; these include actin, myosin (both heavy and light chains), tropomyosin, and α-actinin. However, there is a notable absence of troponin in smooth and nonmuscle cells. The most striking difference between muscle and nonmuscle contractile systems is at the morphologic level. In striated muscle cells the contractile proteins are arranged in a semicrystalline, double hexagonal lattice of interdigitating thick and thin myofilaments.[33] Bundles of these myofilaments constitute the myofibril. Within each myofibril the myofilaments are precisely arranged into cylindrical contractile units termed sarcomeres, approximately 2 to 2.5 μ long and 1 μ in diameter. Since the sarcomeres of adjacent myofibrils are transversely aligned, and each sarcomere contains characteristic A, I, and Z bands, the myofibers exhibit cross-striations that are the hallmark of skeletal and cardiac muscle.

Nonmuscle cells lack true sarcomeres and are nonstriated. However, their contractile proteins can be arranged in filamentous arrays that are primarily dependent on the polymerization and packing arrangement of actin. The assemblages of actin, in turn, are regulated by the actin-binding proteins expressed in specific cell types.[34] These assemblages of actin appear to participate in five major functions: (1) structural rigidity and growth of cytoplasmic extensions, as in stereocilia of cochlear hair cells, the brush border microvilli of epithelial cells, and the acrosomal process of sperm in several marine organisms; (2) the polarized adhesion of cells to a substrate, such as the adhesion of stress fibers along the ventral cytoplasm of cultured fibroblasts and within endothelial cells in vivo; (3) contractility, as evidenced by the contractile ring during cytokinesis and the terminal web of absorptive epithelia; (4) the reversible gel-like state of cytoplasm as demonstrated by the microfilament network at the periphery of many ameboid cells;

and (5) the linkage of organelles to one another and to the plasma membrane. A full discussion of contractile systems in nonmuscle cells can be found in recent reviews.[34,35]

Of importance to the present discussion is the fact that myoblasts express the panoply of nonmuscle contractile protein isoforms prior to terminal differentiation.[36] As noted above, the myoblasts must migrate to the limb buds from the myotomes. Upon withdrawal from the mitotic cycle within the limb, the myoblasts assume an elongated, spindle-shaped configuration, align and aggregate with other myoblasts or myotubes, and become anchorage-dependent. At this stage the myoblasts initiate the transcriptive program characteristic of muscle. The cells initiate cytoplasmic fusion, down-regulate the expression of nonmuscle contractile protein genes, and up-regulate the transcription of genes encoding proteins of the sarcomere, the sarcotubular system, the glycolytic pathway, and the membrane proteins necessary for the electrogenic properties of the myofibers and the reception of acetylcholine (Ach). Since the expression of the sarcomeric proteins begins while the myoblasts and early myotubes contain significant quantities of nonmuscle contractile proteins, the initial assembly of the myofibrils must take place in the context of a dual contractile system.

Coordinate Expression In tissue culture it is possible to establish experimental conditions in which myoblasts initiate relatively synchronous cytoplasmic fusion and myodifferentiation.[27,37,38] When such cultures were initially analyzed for the synthesis and accumulation of MHCs and myosin light chains, (MLCs), creatine kinase, the AchR, and other phenotypic markers of differentiated muscle, there appeared to be a temporal synchrony in the expression of each of these proteins. Furthermore, when mRNA levels for these proteins were assessed by either cell-free translation studies[39] or Northern blot analyses,[40] it was clear that regulation governing the expression of these proteins took place primarily within the nucleus, either at the transcriptional or early post-transcriptional levels. Although evidence has been obtained for translational control of MHC synthesis,[41] the majority of studies in this field have failed to identify significant pools of stored and untranslated mRNA encoding any of the sarcomeric proteins in developing muscle. Current evidence suggests that once mRNA exits from the myonucleus, it is rapidly translated. On the basis of these studies, the concept of coordinate expression of the contractile proteins arose.[42] The basic thrust of this concept is that a cassette, or group of muscle-specific genes, is regulated coordinately, possibly by a common trans-acting regulator. It is assumed that this regulator is the product of another gene or gene set that is activated when the myoblasts withdraw from the mitotic cycle and enter a G_0 state.

Although this concept has been profoundly influential, problems have arisen that have yet to be resolved. Most of the myofibrillar proteins exist as isoforms encoded by different genes or as the products of differential transcript splicing. Many of these transcripts exhibit tissue- and stage-specific expression that is not directly related to terminal myodifferentiation. For example, during skeletal muscle development the first fibril-specific isoform of actin that is expressed appears to be cardiac α-actin and not the skeletal muscle isoform.[43-45] Only later in the development of the myofiber is the skeletal muscle α-actin gene expressed and the cardiac α-actin gene down-regulated. Similarly, most skeletal muscles exhibit the sequential expression of several MHC isoforms during the course of development.[46,47] However, different muscle groups demonstrate unique patterns of MHC expression with different time schedules for isoform switching.[48] Thus, although there appears to be a coordinate activation of muscle-specific gene sets, individual gene products may differ in selected muscles or fiber types. These observations have important implications for any model of gene regulation, for they point to the existence not only of muscle-specific trans-acting regulatory molecules, but of tissue- and stage-specific regulators potentially responsive to neural and work-related parameters of muscle function.

Isoforms   Different isozymes of myosin, creatine kinase, lactic dehydrogenase, and aldolase have been recognized for some time in skeletal and cardiac muscles. With the advent of improved peptide fingerprinting techniques, the use of monoclonal antibodies, and the application of recombinant DNA procedures, isoform variants for virtually every myofibrillar protein have been recognized.[49] Although posttranslational modifications have been demonstrated for many of these proteins, isoform variants differ in their primary amino acid sequences. Thus, each of the different isoforms must be encoded by different mRNAs. The multiplicity of isoform mRNAs can arise from the transcription of different genes, the generation of two or more transcripts from a single gene based on several transcriptional initiation and termination sites, or the formation of several mRNAs from a single gene based on alternate post-transcriptional splicing.[50]

Most important to the present discussion, however, has been the recognition that virtually all of these isoforms exhibit developmental regulation. This has been demonstrated most dramatically with the myosin molecule, particularly its heavy chain.[47] In 1974, Masaki and Yoshizaki[51] demonstrated antigenic differences between adult and embryonic myosins in the chicken pectoralis muscle. Related evidence was subsequently obtained for embryonic mammalian myosin.[52] By peptide mapping, Whalen and associates[53] showed that fetal rat muscle contains an MHC differing in primary sequence from that in the same adult muscle. Comparable peptide mapping studies using the chicken model were soon published. Using in vitro translation of poly(A)RNA from embryonic and adult muscle, it was next proven that different mRNA transcripts, and not post-translational modifications of the MHCs, were responsible for the different MHC peptide maps. A third MHC isoform, termed neonatal MHC, was next detected in newborn rat muscle and at hatching stages of chicken development.[46] Several laboratories then generated monoclonal antibodies against the MHC, which have not only confirmed the peptide mapping experiments but permitted cytologic detection of the different isoforms in frozen sections of muscle and in cultured myotubes.[54-56] Each of these vertebrate MHCs appears to be a product of an independent structural gene[57-59]; there is no evidence that alternative splicing of single gene transcripts gives rise to more than one mRNA encoding the MHC. In *Drosophila*, however, a single MHC gene encodes three developmentally regulated mRNAs.[60,61]

Although sequential expression of embryonic, neonatal, and adult MHC isoforms has been demonstrated for rat hindlimb and chicken pectoralis muscles, there is equally good evidence that not all muscles have the same development programs. For example, the lateral head of gastrocnemius muscle of the chicken appears to express the neonatal MHC isoform in the adult. Other muscles continue to express the embryonic form of myosin throughout life.[48] In short, the identical MHC genes in different muscles appear to be under tissue-specific developmental regulation. A comparable example in mammals is the case of a slow MHC gene that is co-expressed in both cardiac and skeletal muscle.[62] In the heart, transcription of this gene is tightly regulated by circulating levels of thyroid hormone, whereas in slow, oxidative skeletal muscles of the hindlimb this gene is under less stringent regulation by this same hormone.

The discussion above has focused primarily on the MHC. However, at least three genes encode the protein subunits of myosin: one for the MHC, one for MLC1 and MLC3 (also termed the alkali or essential light chains), and one for MLC2 (also termed the DTNB or regulatory light chain).[63-65] For cardiac muscle, heterodimers of the MHC have been demonstrated; thus, a fourth gene would have to be invoked for the synthesis of those myosin molecules. Clear evidence for MHC heterodimer formation in skeletal muscle is lacking. However, there are clear examples in which both fast and slow MHCs are synthesized in the same myofibers.[4] Whether or not co-assembly of these two gene products can occur will have important bearing on potential models of myosin assembly. As with the MHCs, the MLCs appear to be developmentally regulated. At least in the chicken, early myotubes express both fast and slow MLC1 and MLC2. As the tissues mature, the fast muscles repress the expression of the slow MLCs while continuing fast MLC synthesis. MLC1 and MLC3 are encoded by the same gene but have different mRNAs, which arise from two separate transcripts.[64,66] Early in myogenesis only the MLC1 transcript is synthesized. In late embryonic development, while transcription of MLC1 continues, the alternative transcription unit is activated and MLC3 mRNA is formed. In addition to the stage-dependent regulation of the adult MLC isoforms, there is convincing evidence for the synthesis of embryonic-specific MLCs in both birds and mammals.[67,68]

Isoforms have how been detected for at least two other thick-filament proteins: C-protein[69] and myomesin,[70] both of which exhibit developmental regulation. At a minimum, I-band filaments contain the proteins actin, tropomyosin, and troponin. Although I-band filaments in skeletal muscle contain only the α isoform of actin, there are three α-

actin variants: skeletal, cardiac, and smooth.[71] In newly formed myotubes, the α-actin cardiac isoform appears first. Somewhat later in development, skeletal α-actin is synthesized, and cardiac α-actin is eventually down-regulated. For a significant time during muscle development, however, both α-actins are synthesized by the same myotubes. Each of these actin isoforms is encoded by different structural genes. A total of six actin genes are present in higher vertebrates, including humans.[71] Embryonic isoform transitions of tropomyosin and troponin have been discussed in detail recently.[72-75] The Z band contains predominantly α-actinin in its core, and desmin, filamin, synemin, and zeugmatin at its circumference.[76,77] Striated muscle, smooth muscle, and nonmuscle forms of α-actinin have been identified immunologically,[78] but no data are available about striated isoform transpositions during myogenesis. To date, only one form of desmin has been identified; there is only one desmin gene in vertebrates.[79] Two forms of filamin have been distinguished immunologically, and their expression does appear to change during myogenesis.[80] Neither filamin nor synemin has been cloned. No information currently exists about the polymorphism of zeugmatin. Finally, two very high molecular weight proteins, titin and nebulin, have been conclusively identified as significant components of all striated muscles.[81,82] No data are available about possible isoform variants of these two proteins.

Fast and Slow Fiber Types   Different striated muscles exhibit significant variations in structure at both the histologic and ultrastructural levels; as well as in innervation; physiology, as exemplified by speed of contraction or relaxation; biochemistry, as exemplified by glycogen and myoglobin contents and myosin adenosine triphosphatase (ATPase) activity; and in circulation. These profound variations are consonant with significant compositional differences in the myofibrils and the sarcotubular system. In birds, muscles of virtually homogeneous fiber type have been identified, for example, the pectoralis major and the anterior latissimus dorsi are fast-twitch and slow-tonic, respectively. Although certain fiber-type populations usually predominate in mammalian muscles, it is rare to identify muscles of homogeneous composition. Thus, biochemical studies of whole muscles must examine complex mixtures of the different cell populations and their constituent proteins. Morphometric, immunocytochemical, and microanalytical studies of single fibers are essential for defining isoform distributions in such heterogeneous systems. The following classification of fiber types[83] is the most widely used. Fibers are termed types I, IIA, IIB, IIC, or IIM. Type I fibers are slow-twitch fibers that have the slowest contraction time and the lowest content of glycogen and glycolytic enzymes. They are rich in mitochondria and myoglobin and are quite resistant to fatigue. The muscle fibers contain slow-type MHC. Morphologically, the sarcomeres contain a wide Z band. Type IIA are fast-twitch oxidative fibers. They have a faster contraction time than the type I fibers but have a higher content of mitochondria and myoglobin than type IIB fibers and are usually considered fast-twitch oxidative fibers. The Z bands are wide but slightly narrower than in the type I fibers. Physiologically, these are more fatigue-resistant than the type IIB fibers, which are also fast-contracting but rely more on glycolytic pathways for energy production and consequently contain fewer mitochondria than the IIA fibers. Both type IIA and type IIB fibers contain pure fast MHCs, but they are immunologically distinguishable myosin isoforms and are products of different genes. Operationally, the myosin from the IIA fibers is termed fast/red myosin while that in the IIB fibers is termed fast/white. Type IIC fibers contain both of these fast myosins and are thus termed intermediate fibers. Finally, most mammalian carnivores, including humans, possess a third fiber type termed the superfast fiber, or type IIM fiber, which is most prominent in the jaw muscles. This superfast fiber type contains a unique myosin that can be distinguished from both the slow and fast MHCs expressed in the type I and II fibers, respectively.[84] At a minimum, four different MHCs are found in these mammalian skeletal muscles. An additional MHC has been identified in the extraocular muscles.[85] If we include the embryonic and neonatal MHCs, seven MHC genes are required to encode the myosins found in skeletal muscle. At least one or two additional MHCs are found in the heart. Thus, nine genes encode the MHCs found in the striated muscle of humans.

The number of MLC genes is less certain, but it is thought to be approximately six. These do not include, however, the smooth muscle and nonmuscle components. Studies with chicken genomic DNA suggest the possibility of many more MHC genes than in the mammal.[86] The nomenclature of the different myosin isoforms is based on the fiber types in which the proteins are expressed. It is interesting that several of the slow-muscle isoforms are also expressed in the heart. This is well documented for slow MHC and has recently been established for slow troponin C (Tn-C). It is evident from both cDNA and genomic sequence analyses that these slow and cardiac-specific isoforms are products of the same genes. However, there must be tissue-specific regulation, for, as noted above, the slow MHC gene in the heart is under tight thyroid hormone regulation, while this is not true for slow skeletal muscle. Presumably, there are tissue-specific regulatory proteins under thyroid control, but this has not yet been proven. It is not known why slow skeletal and cardiac muscle co-expresses a common set of contractile protein genes.

At the chromosomal level, the MHC genes appear to be clustered. In the mouse the cardiac muscle genes are found on chromosome 14, while the skeletal muscle MHC genes are on chromosome 11.[87,88] In humans, the MHC genes appear to be clustered on chromosome 17.[89] In contrast, the genes for all other contractile proteins that have been mapped appear to be unlinked and widely dispersed on many chromosomes.[88]

Self-Assembly Properties   Many of the contractile proteins exhibit impressive self-assembly properties in vitro. This is especially true for actin and myosin, the major proteins of the sarcomere. For example, actin is traditionally solubilized in a low-ionic-strength solution containing only 2 to 5 mM of sodium adenosine triphosphate (ATP). This soluble form of actin is termed G-actin (globular actin). When the ionic strength is increased by adding either $MgCl_2$ or NaCl, rapid polymerization occurs, resulting in the formation of F-actin (fibrous actin), which is a double-stranded helical polymer of the G-actin subunits.[35] Such F-actin filaments are polarized; this is most readily demonstrated by binding subfragment–1 (S–1), or the myosin head, to the G-actin subunits of the filaments.[90] The decorated polymers exhibit characteristic arrowhead-like structures along their length, which point unidirectionally along each filament. Thus, the pointed end of each filament is shaped like the pointed end of an arrowhead. Conversely, the barbed end of each F-actin polymer is shaped like the base of an arrowhead. This polarity has functional significance because F-actin polymers add G-actin subunits at a much faster rate at the barbed than at the pointed ends of the filaments. Furthermore, there are different actin-binding proteins that cap the filaments at either the pointed or the barbed ends. For example, cytochalasin-B, an agent that disassembles actin microfilaments in nonmuscle cells, binds specifically at the barbed ends of the F-actin. A detailed discussion of actin protein-binding can be found in a recent review.[35] Of direct relevance to the present discussion is the fact that S–1 decoration of actin can be used for establishing the polarity of actin filaments either associated or not associated with myofibrils. In this manner it has been demonstrated that the pointed ends of all I-band filaments are oriented toward the A bands and away from the Z-disks.[90] S–1 fragments can also be used as a fluorescently labeled probe, akin to an antibody, for localizing F-actin by fluorescence microscopy.

Myosin, the predominant protein of the A band, also exhibits significant self-assembly properties. For example, myosin is usually dissolved in a high-ionic-strength solution, 0.6 M KCl, at neutral pH. When the KCl concentration is reduced to 0.1 M, myosin rapidly polymerizes to form filaments that possess many of the structural features of A-band filaments.[90,91] These include bipolarity and a central bare zone. The role of thick filament-associated proteins, C-protein, M-protein, and myomesin in polymerization remains uncertain.

Although both actin and myosin polymerize spontaneously when their concentrations at physiologic saline levels exceed the "critical" concentration, precise length regulation is lacking in vitro. Within the sarcomeres of vertebrate muscle, the lengths of thin and thick filaments are almost exactly 1.0 μ and 1.6 μ, respectively. Presumably, there are length-regulating mechanisms within muscle cells that have not been reproduced in cell-free systems.

Myofibrillogenesis

The registration of myofibrils in three-dimensional alignment generates the cross-striated appearance that distinguishes skeletal and cardiac muscle (Fig. 6–3). As noted above, the cells destined to form limb muscle arise in the axial, somitic mesoderm. The cells migrate from

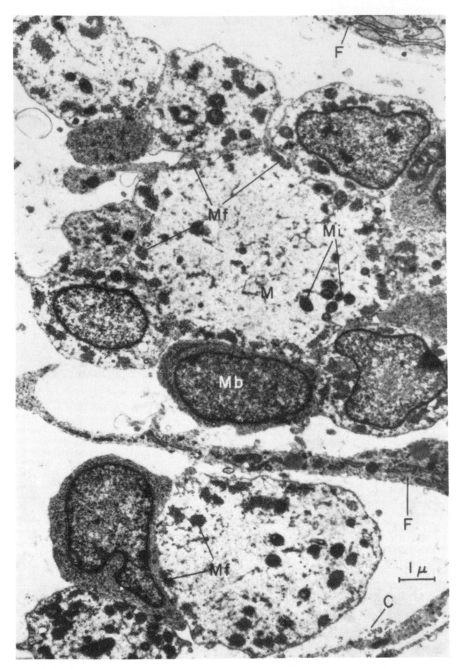

Fig. 6–3 *Low-power electron micrograph of a transverse section of 12-day embryonic chick limb muscle. Two clusters of myogenic cells are separated by a fibroblast (F). The large, more lightly stained primary myotubes (M) are surrounded by secondary myotubes and myoblasts (Mb). Myofibrils (Mf) and mitochondria (Mi) are indicated. Note the subsarcolemmal distribution of the newly formed myofibrils. (Reproduced with permission from Fischman DA: An electron microscope study of myofibril formation in embryonic chick skeletal muscle. J Cell Biol 1967;32:557–575, Rockefeller University Press.)*

the myotome by a process that is similar to the migration of neural-crest cells to peripheral sites in the organism. This migratory process involves the nonmuscle contractile proteins typical of most motile cells, such as platelets, macrophages, and neuritic growth cones. Once in the limb buds, the myoblasts divide, although the precise number of cell divisions is uncertain. Local concentrations of matrix-associated growth factors, such as FGF, are believed to regulate the replicative cycle.[17,21] Upon depletion of these growth factors, the cells enter a prolonged G_1 period and initiate terminal muscle differentiation. This phase is characterized by (1) the aggregation and alignment of the myoblasts; (2) the initiation of cytoplasmic fusion to form multinucleated syncytia termed myotubes; (3) the initiation of transcription and translation of proteins characteristic of muscle, such as the myofibrillar proteins, the membrane proteins of the sarcotubular system and plasmalemma, and proteins required for energy production by both aerobic and anaerobic processes; and (4) innervation. At this time the myotubes become anchored to the substrate and migration ceases. The cytoplasmic microfilaments form subplasmalemmal bundles or sheets that resemble stress fibers of cultured fibroblasts.

Morphologically, the earliest manifestation of myofibril synthesis appears to be the appearance of large, spiral-shaped polyribosomes, which are believed to synthesize MHCs.[92] Over the next few hours, thick myofilaments can be detected in the myotubes,[49,93] most of which are in proximity to the subplasmalemmal microfilament bundles.[94] If the myotubes are stained at this stage with antibodies specific for the different contractile protein isoforms, the accumulation of sarcomere-specific proteins can be observed. For example, α-actinin, α-actin, tropomyosin, and MHC can be detected. In addition, new proteins appear. These include troponins C, T, and I, myomesin, M-protein, C-protein, and titin. The mechanisms underlying sarcomere assembly are unknown. However, it is assumed that once the critical concentrations for polymerization of actin and myosin are exceeded, they assemble their respective filaments. The binding properties of the other proteins for actin and myosin lead to their association with the newly assembled thin and thick filaments. These events seem to occur beneath the plasma membrane in proximity to the microfilament bundles. Possibly, the microfilament bundles provide the organizing center for myofibril assembly, but this has not been proven directly. It has also been proposed that assembly of the contractile proteins is coupled to translation on the polyribosomes.[95]

As myosin polymerizes, thick filaments are formed that have axial threefold symmetry. Actin filaments, after binding to the myosin filaments, are thus positioned appropriately for the generation of the myofibril hexagonal lattice.[93] Presumably, additional filaments are added appositionally to increase the diameter of the nascent sarcomeres and longitudinally to increase myofibrillar length (Fig. 6–4). The definitive sequence of these events is still unknown. No successful in vitro models have been established to reproduce sarcomere assembly in cell-free systems; therefore, the process must be observed from elec-

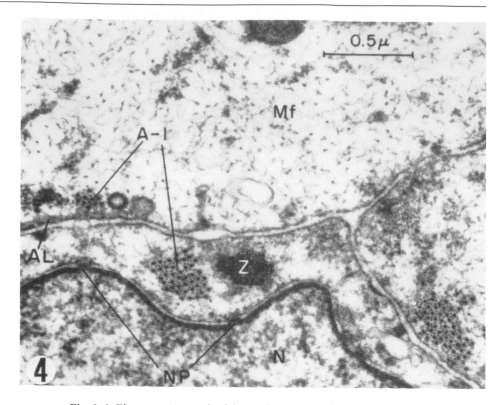

Fig. 6–4 *Electron micrograph of three adjacent myotubes containing newly formed myofibrils beneath the sarcolemmal membranes. Some of the sarcomeres have been cut at the region of thick and thin filament overlap (A-I) and others through the Z-disk (Z). Unassembled thin, thick, and intermediate filaments (Mf) are found near the myotube center. Amorphous dense (AL) material can be seen lining the sarcolemma beneath the myofibers. A nucleus (N) with nuclear pores (NP) is seen in the lower myotube. (Reproduced with permission from Fischman DA: An electron microscope study of myofibril formation in embryonic chick skeletal muscle. J Cell Biol 1967;32:557–575, Rockefeller University Press.)*

tron micrographs of fixed preparations. Methodologic advances are likely to come from a variety of techniques and models that use molecular genetics, traditional genetic systems (*Drosophila*), specific antibodies in combination with microinjection, improved video-enhanced image processing and three-dimensional reconstruction, and, finally, in vitro assembly systems for defining the components and rate-limiting steps in the assembly reactions.

Innervation

The neuromuscular junction is a chemical synapse that is organized both functionally and structurally to transmit a signal from a motor nerve terminal to a circumscribed region of a myofiber. The structure of the mature neuromuscular junction has been reviewed in detail[96,97];

Bennett[98] provided a thorough review of our knowledge of neuro-muscular development before 1983.

The structure of the neuromuscular junction can vary significantly between different species, between different fiber types of the same species, and during the course of development. As emphasized by Engel,[96] however, all neuromuscular junctions appear to contain five principal components: (1) a nerve terminal containing the neurotrans-mitter; (2) a Schwann cell overlying the nerve terminal; (3) a synaptic space lined with basement membrane; (4) a postsynaptic membrane containing the receptor for the neurotransmitter; and (5) a postjunc-tional sarcoplasm required for the structural and metabolic support of the postsynaptic membrane. In vertebrate voluntary skeletal muscle, the neurotransmitter is Ach, the receptor is the nicotinic AchR, and the synaptic space contains acetylcholinesterase (AchE).

Mature mammalian muscle receives three types of motor neurons: (1) α-motor neurons which innervate extrafusal muscle fibers; (2) β-motor neurons innervating both extrafusal and intrafusal fibers; and (3) γ-motor neurons innervating intrafusal fibers exclusively. Sensory nerves terminate on intrafusal muscle fibers and Golgi tendon organs. In adult muscle, each myofiber is innervated by only one motor neuron, but each motor neuron generally innervates more than one my-ofiber. A motor unit constitutes the motor neuron and the muscle fibers it innervates. In most extrafusal mammalian muscle, there is one neuromuscular junction per myofiber and, generally but not always, this is near the midportion of the muscle fibers. This focal pattern of neuromuscular junctions is characteristic of the phasic (twitch) muscle fibers of all vertebrates. In slow tonic muscle fibers of amphibians and birds, in mammalian intrafusal muscle fibers, and in certain extrafusal fibers of the larynx, and lingual and extraocular muscles, there are distributed neuromuscular junctions, with many end plates per muscle fiber.

In its preterminal region, the myelinated neuron is invested by two layers, a sheath of perineural epithelial cells (Henle's sheath) sur-rounded by connective tissue fibroblasts. Within a few microns of the neuromuscular junction, the myelin sheath ends abruptly, and between the last node of Ranvier and the neuromuscular junction the neuron is surrounded by a Schwann cell invested by Henle's sheath. The latter sheath does not cover the neuromuscular junction but ends just prior to the nerve terminal. From this point on, the Schwann cell covers that aspect of the neuron not in contact with the postsynaptic surface. Laterally, the basement membrane of the Schwann cell becomes con-tinuous with the extrasynaptic basement membrane of the myofiber, while medially it merges with synaptic basement membrane. At the synaptic region, only a basement membrane separates nerve and muscle.

Nerve terminals (Figs. 6–5 and 6–6) ending on skeletal muscle con-tain membrane-enclosed vesicles 50 to 60 nm in diameter that contain Ach, ATP, a vesicle-specific proteoglycan, and a membrane phos-phoprotein, synapsin I, characteristic of synapses.[99,100] Within the end-ing, clusters of the vesicles are found close to the synaptic zone

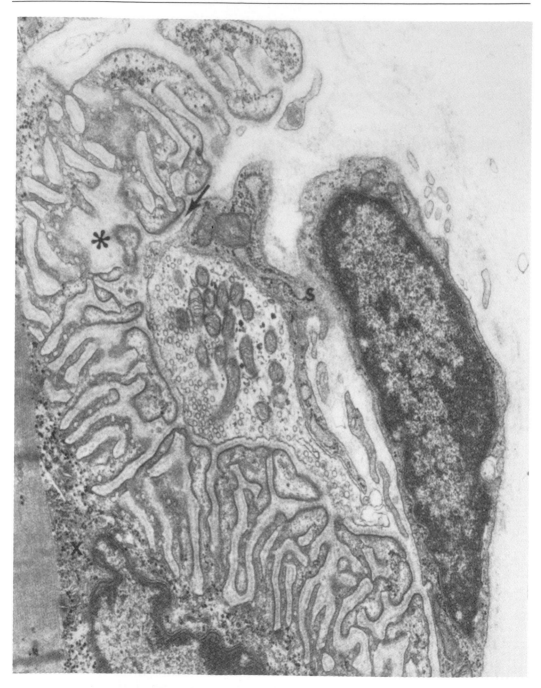

Fig. 6–5 *Electron micrograph of normal human neuromuscular junction. The right side of the nerve terminal is covered by a Schwann cell (s); the left side of the terminal faces the postsynaptic region of clefts and folds. Junctional sarcoplasm (X) contains glycogen granules, ribosomes, small tubular profiles, and nucleus. The arrow and asterisk mark primary and secondary synaptic clefts, respectively (× 30,600). (Reproduced with permission from Engel AG:* Handbook of Clinical Neurology. *Amsterdam, Elsevier Biomedical Division, 1979, vol 41, part 2, pp 99–145.)*

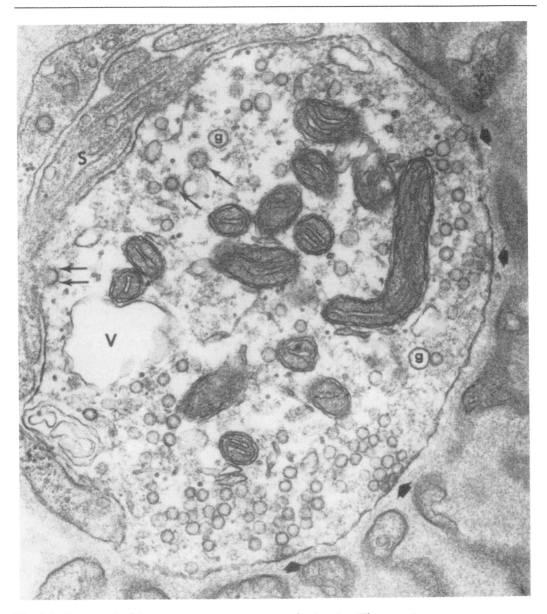

Fig. 6-6 *Nerve terminal in rat gastrocnemius neuromuscular junction. The synaptic vesicles are concentrated near the presynaptic membrane; the mitochondria cluster near the center of the nerve terminal. A few giant synaptic vesicles (g) and coated vesicles (arrows) are present. A coated pit (double arrows) is budding from axolemma covered by Schwann cell (S). The nerve terminal also contains glycogen granules, canaliculi, a small vacuole (V), and an amorphous or finely granular matrix. Solid arrows indicate four active zones that consist of dense spots on the inner surface of the presynaptic membrane and associated snyaptic vesicles. The active zones are in register with the secondary synaptic clefts. Also note dense membrane specializations on the terminal expansions of the junctional folds (× 63,000). (Reproduced with permission from Engel AG: The neuromuscular junction, in Engel AG, Banker BQ (eds):* Myology. *New York, McGraw-Hill, 1986, vol 1, pp 209–253.)*

whereas other organelles, particularly mitochondria, are distributed away from the synaptic surface. The clusters of synaptic vesicles lie over electron-dense strips of the synaptic plasma membrane, termed active zones.[101-103]

In addition to the 50-nm vesicles characteristic of cholinergic neuromuscular junctions, resting junctions contain giant synaptic vesicles 100 to 150 nm in diameter that appear to increase in number during recovery from tetanic stimulation.[104] The relationship of the two types of vesicles is uncertain, but the appearance of the large-sized vesicles correlates with giant/miniature end-plate potentials that are seen after prolonged stimulation. Thus, it is likely that the large vesicles also contain Ach.

By means of quick-freezing and deep-etch electron microscopy, clathrin-coated vesicles, associated with endocytic pits of the nerve ending, have been clearly described.[105,106] It is currently believed that these coated vesicles are involved in the recycling of exocytosed materials released from the nerve terminals during synaptic vesicle fusion with the plasmalemma.

Vesicles approximately 80 to 100 nm in diameter containing an electron-dense core have been observed in the motor nerve terminal. Their number appears to be higher in growth cones and regenerating nerve terminals.[107] The presence of such vesicles raises the possibility that other neurotransmitters are present at the neuromuscular junction.[108]

Approximately 15% of the nerve terminal volume is taken up by mitochondria, which are presumably involved in both ATP production and lipid metabolism.

Numerous microtubules are present at nerve endings often closely associated with synaptic vesicles. It is widely believed that the microtubules are involved in vesicle transport down the axon. Nonmuscle actin and neurofilament proteins have also been identified in motor axons, but their precise localization in the nerve terminal has not been established.

The presynaptic plasmalemma contains electron-dense stripes that mark the sites of synaptic vesicle release.[109] These regions can be seen most clearly by freeze-fracture electron microscopy. The active zones appear to correlate with the postjunctional folds of the muscle sarcolemma.

The synaptic space is approximately 70 nm wide. It is divided into primary and secondary clefts. The secondary clefts are produced by the junctional folds, and each secondary cleft communicates with the primary cleft at its apex. Basement membrane extends into both the primary and secondary clefts, always intervening between muscle and nerve cell membranes. The composition of the synaptic basement membrane is not identical to that of the extrajunctional basement membrane. Some proteins such as laminin, collagen type IV, and fibronectin are shared. Some unique components are AchE and a ligand for the lectin prepared from *Dolichos biflorus*. During the course of development and nerve regeneration, the basement membrane changes in composition and is believed to play central roles in directing the nerve terminal to its appropriate location on the muscle surface and in the morphogenesis of the neuromuscular junction.[110] AchE, a secretory product of muscle, is located in the junctional space. Except during myogenesis or regeneration, the extrajunctional basement membrane is unreactive for AchE. Presumably, the junctional basement membrane contains binding sites for the collagen-like tail of mature AchE that are lacking outside the neuromuscular junction.

The junctional folds are a synaptic specialization unique to the neuromuscular junction. This infolding of the postsynaptic surface increases the surface area for transmitter reception approximately eightfold to tenfold. Morphologic differences are seen in the development, complexity, and size of the junctional folds in different species, in different fibers, and at different stages of synaptogenesis. In general, junctional folds are not seen at early stages of neuromuscular junction formation; postsynaptic infoldings develop gradually as the neuromuscular junctions mature. Although well-developed junctional folds are evident at birth, the complexity and size of these structures increase postnatally. Fast-twitch muscle fibers (type IIB) exhibit better developed junctional folds than slow-twitch fibers (type I), while the junctional folds of type IIA fibers are intermediate between types I and IIA. Some intrafusal muscle fibers exhibit relatively shallow or absent junctional folds.

The AchRs are distributed predominantly along the apical crests of the junctional folds, and the plasma membrane in this region is more electron-dense than adjacent muscle plasma membrane as demonstrated with transmission electron microscopy (Fig. 6–6). By freeze-fracture electron microscopy and by potassium permanganate staining, this membrane region has been shown to contain granules of AchR 6 to 12 nm in diameter. The number of receptor molecules is approximately 10 to $20,000/\mu^2$.

The junctional folds contain cytoskeletal components believed to anchor the AchRs and to maintain the structural rigidity of the folds. The following proteins have been localized to this region: β-actin, vinculin, α-actinin, and filamin. By electron microscopy, microtubules, intermediate filaments, and microfilaments have been shown to be part of this cytoskeletal matrix. The AchRs are thought to be contained in compact arrays by cross-linking to a 43-kD external membrane protein that, in turn, is apparently coupled through the membrane to cytoskeletal elements.

Unmyelinated nerves penetrate muscle anlage shortly before the myoblasts initiate cytoplasmic fusion.[111] Nerve-muscle contacts can be detected shortly after myotube formation.[112] The appearance of functional neuromuscular junctions varies in different species. In humans, primitive neuromuscular junctions are detected between the sixth and tenth weeks of gestation,[113] but in the mouse and rat neuromuscular junctions have been seen between the 14th and 16th days of embryonic life.[114,115] Initial contacts usually appear at the midsection of the muscle, but there are many exceptions to this rule. Initially, each myotube is contacted by more than one neuron. Also, the same neuron may make more than one synapse with the same myotube. In the perinatal period, polyneuronal innervation is lost[98]; the mature myofiber is functionally innervated by only one motor nerve. The initial nerve endings are simple bulbous termini lying in a shallow depression on the muscle surface, which lacks junctional infoldings. The nerve terminals contain few synaptic vesicles. With progressive maturation of the neuromuscular junction, basement membranes form, junctional infoldings develop, synaptic vesicles increase in number, and the postsynaptic membrane becomes more electron-dense. In the rat, Schwann cells grow over the nerve endings during the perinatal period, and myelination occurs soon afterward. AchRs can be detected on myotubes in 14- to 15-day mouse or rat embryos. Initially, the AchRs are widely distributed over the muscle surface. About a week before birth, extrajunctional AchRs begin to decline and junctional AchRs increase significantly. AchE begins to accumulate at the neuromuscular junctions soon thereafter; its concentration at the synapse seems to correlate with the accumulation of basement membrane at that site. During the second postnatal week, junctional folds mature and polyaxonal innervation disappears. Physiologic measurements of end-plate potential, miniature end-plate potential, and quantum content of vesicle release indicate that maturation of functional parameters correlates well with morphologic changes of the neuromuscular junction during the second postnatal week.[98]

There has been considerable interest in the phenomenon of polyaxonal innervation of fetal muscle and the withdrawal of these collateral branches during the neonatal period. Our understanding of these processes is far from complete. Although motoneuron cell death accompanies withdrawal of incorrect projections from motoneuron pools, and loss of polyaxonal innervation of synaptic sites accompanies motoneuron death, a recent study indicates that neuronal cell death cannot account quantitatively for the withdrawal of polyinnervation in the postnatal period.[116] As Engel[96] emphasized, elimination of collateral sprouts rather than motoneuron cell death must be the major factor in the elimination of polyaxonal innervation in the postnatal period. Little is understood about the mechanisms underlying these phenomena.

Structure and Composition of Adult Muscle

Tissue Organization

The fundamental cellular unit of skeletal muscle is termed the myofiber or muscle fiber, and, except for some specialized muscles of the eye and ear, it is a multinucleated syncytium. Clusters of myofibers are grouped into bundles, termed fascicles, that aggregate to form a whole muscle. Each myofiber is surrounded by a basement membrane, sometimes termed an external lamina to distinguish this investment from that at the base of epithelia. This coat is usually 100 to 200 nm thick and can be seen on electron microscopy to consist of two layers: the inner, electron-lucent lamina rara and a more electron-dense outer layer, the lamina densa. The outer layer is usually thicker. The outermost edge of the lamina densa merges imperceptibly with a reticular layer, often termed the reticular lamina. At least four components of the basement membrane have been identified: collagen, laminin, fibronectin, and a muscle-specific proteoglycan. Carbohydrate residues of sarcolemmal proteins protrude into the lamina rara; this layer stains intensely with the PAS reagent. It is clear from recent studies in vitro that both fibroblasts and myofibers contribute to the basement membrane; in muscle cultures devoid of fibroblasts no basement membrane forms. This ensheathement of each myofiber by a basement membrane merging into irregularly arranged collagen fibrils embedded in extracellular matrix is termed the endomysium. These fine collagen fibrils aggregate into larger collagen fibers intermingled with elastic fibers. This denser connective sheath that covers the muscle fasciculi is termed the perimysium. The entire muscle is enclosed in a sheath of connective tissue, the epimysium, which is directly continuous with the deep fascia observed by blunt dissection of limb tissues. The muscle-tendon attachment, or myotendinous junction, is discussed in chapter 5.

The blood vessels supplying skeletal muscle run in the connective septa between muscle bundles, eventually reaching the endomysium where they form a rich capillary network around the individual myofibers. To accommodate the extensive changes in length of the muscle, the capillaries tend to be tortuous in contracted muscle and straight during muscle extension. Blood flow around the muscle fibers is regulated by terminal arterioles that can either shunt blood through arteriovenous anastomoses or direct it through the capillaries. These arterioles are richly supplied by vasomotor sympathetic nerves which, when stimulated, initiate contraction of the arterioles, resulting in greater blood flow through the capillary bed and a bypassing of the arteriovenous shunts. A thorough discussion of the vascular system in skeletal muscle has been presented by Jerusalem.[117]

Plasma Membrane

The structure of the muscle cell plasma membrane, when viewed by thin-section transmission electron microscopy, is quite similar to that seen in other nonmigratory cells of the body.[118] Specializations unique to skeletal muscle include the junctional folds at the neuromuscular junction, the extensive infoldings at the myotendinous junction, and the tubular invaginations of the transverse tubular system (T-tubules). Each of these specializations serves to increase the surface area of the myofiber, in the first case to facilitate Ach reception, in the second case to strengthen the adhesive interactions between muscle and tendon, and in the last case to expedite excitation-contraction coupling. During myogenesis, primary and secondary myotubes are coupled by an extensive number of low-resistance gap junctions. Upon innervation of the myotubes, these gap junctions disappear and the muscle fibers are not coupled to other muscle fibers or to neurons. Striated muscle also lacks maculae and zonulae adherentes, further distinguishing this tissue from cardiac muscle.

Transverse Tubules

Virtually all vertebrate skeletal muscle contains tubular invaginations of the plasma membrane that form an extensive anastomotic network within the myoplasm.[119] Since these tubules run mainly across the long axis of the myofibers, they are termed transverse or T-tubules. In amphibian skeletal muscle, the T-tubules invaginate from the plasma membrane and encircle the myofibrils at the level of the Z bands, whereas in mammalian muscle the T system aligns with the junction of the A and I bands. In birds, different muscles exhibit either the Z or A-I pattern of T-tubule distribution. At the circumference of each myofibril, the T-tubules abut with dilated sacs of the sarcoplasmic reticulum, termed terminal cisternae. Since each T-tubule comes into contact with two sacs of the sarcoplasmic reticulum, the complex of these three membrane components is termed a triad. As discussed below, the T-tubules are structurally linked to the sacs of the sarcoplasmic reticulum by bridging structures[120] usually termed "junctional feet,"[121] which play an important role in transduction of the electrical signal of the T-system, leading to Ca^{2+} release from the sarcoplasmic reticulum. While it is widely acknowledged that T-tubules in mammalian skeletal muscle are distributed at the A-I junctions of the sarcomeres, detailed stereoscopic analyses by high-voltage electron microscopy have revealed that the T-tubules often run obliquely in the sarcoplasm and that longitudinal connections between adjacent T-tubules in adjacent sarcomeres are rather common.[122]

During development, the T-tubules arise from caveolar invaginations of the myotube plasma membrane.[123] There is direct continuity of the T-tubule interior with the extracellular space. However, fluid within this space has no direct access to the lumen of the sarcoplasmic reticulum. Communication between these two compartments takes place across the triadic junctions, presumably at the junctional feet.

Sarcoplasmic Reticulum

All skeletal muscles contain a unique form of the endoplasmic reticulum termed the sarcoplasmic reticulum that surrounds the myofibrils and plays an essential role in both the storage and release of ionic calcium. The extent of sarcoplasmic reticulum development can vary widely in the animal kingdom depending on the need for differing rates of contraction-relaxation cycles.[124] For example, the bat cricothyroid muscle, required for emitting the high-frequency sound used in echo location, has a remarkably rich sarcoplasmic reticulum network. The converse is true for slowly contracting tonic muscles.

During development, the sarcoplasmic reticulum forms as an outgrowth of the nuclear envelope, in a manner akin to the endoplasmic reticulum of all eukaryotic cells.[125] Markers, such as lanthanum chloride, ferritin, or horseradish peroxidase, when placed in the extracellular space, will not enter the interior of the sarcoplasmic reticulum. If muscle is fixed in solutions containing sodium oxalate, calcium oxalate precipitates form in the sarcoplasmic reticulum because of the high concentrations of Ca^{2+} in this membrane compartment. Release of calcium ions from the sarcoplasmic reticulum is the final signal to the contractile proteins.

There are three morphologic components of the sarcoplasmic reticulum: the terminal cisternae (or junctional sarcoplasmic reticulum), the fenestrated sarcoplasmic reticulum, and the tubular sarcoplasmic reticulum. The terminal cisternae are found at two sites: alongside the T-tubules as part of the triadic or dyadic structures, and at peripheral couplings with the plasma membrane. It is currently believed that excitation of the sarcolemma and T-tubules will effect Ca^{2+} release from the cisternae at both of these sites. The junctional regions exhibit four characteristic features[119]: (1) They have one flat or regularly scalloped (dimpled) membrane surface. (2) This surface is covered by small, periodically arranged densities, the junctional feet. (3) Most of this surface faces a parallel surface of the T-tubules. (4) The interior of the sarcoplasmic reticulum in this region has a faint flocculent content. The fenestrated regions of sarcoplasmic reticulum are perforated by small round openings (fenestrae) that permit direct communication from one side of the double layer of sarcoplasmic reticulum membranes to the other. The size of the fenestrae vary; they are large in chicken slow fibers and the myocardium but small in human, rat, frog, and fish fast-twitch fibers. The number of fenestrae may vary with sarcomere length and type of fixation, but their distributional pattern is fairly constant for each fiber type. The function of the fenestrae is uncertain, but presumably they facilitate sarcoplasmic diffusion. The tubular (or longitudinal) sarcoplasmic reticulum is found between the fenestrated and junctional regions. In this region, the membranes of the sarcoplasmic reticulum are organized in parallel, longitudinal tubules. In some fibers, intermediate cisternae with flat surfaces and narrow lumina are found between tubular and junctional regions. Apparently, this is the only region of that contains cholesterol[123]; it is highly sensitive to freeze injury.

The major function of the sarcoplasmic reticulum is to pump calcium ions into its lumen against a steep concentration gradient and to maintain a relatively high internal concentration of Ca^{2+}. The sarcoplasmic reticulum membranes contain an intrinsic membrane component, the calcium ATPase. A second protein is calsequestrin, an extrinsic protein found on the luminal side of the sarcoplasmic reticulum. A third protein is a high-affinity calcium-binding glycoprotein, found in much lower concentrations.[124] When sarcoplasmic reticulum fractions are isolated experimentally, they are conventionally described as "heavy" and "light" sarcoplasmic reticulum. These terms reflect the varying contents of calsequestrin; the heavy fractions contain more calsequestrin but less calcium ATPase. Very little is known about the composition of the T-tubule-junctional sarcoplasmic reticulum complex, particularly in regard to signal transmission. A set of high-molecular-weight proteins that may be components of the junctional feet has been described.[125]

The Myofibril

The contractile proteins of striated muscle are organized into cylindrical organelles, termed myofibrils (Fig. 6–7). The diameter of the myofibrils is approximately 1 μ; length can be quite variable, often reaching many centimeters. Detailed reviews of their structure and composition have been prepared by Squire,[126] Huxley,[127] and Craig.[128] The myofibrils are encircled by membranes of the sarcoplasmic reticulum and the transverse tubules. The interfibrillar sarcoplasm also contains mitochondria, lysosomes, ribosomes, and glycogen. Each myofibril is organized into sarcomeres, the fundamental contractile unit of this organelle (Fig. 6–7). The transverse alignment of sarcomeres in adjacent, parallel myofibrils creates the cross-striated appearance of the muscle fiber. These striations result from the repetitive series of transverse bands in each sarcomere, the most prominent of which are the Z, A, and I bands. The distance between two Z bands is defined as the unit length of each sarcomere and will vary with the state of contraction or relaxation of the muscle cell. The A band, so called because of its anisotropy when viewed with polarization optics, contains thick myofilaments (Fig. 6–8). The length of the A band does not change during contraction. In its midregion, the A band contains a sub-band, termed the M band. The I bands (or isotropic bands) are located between Z and A bands and contain thin myofilaments. I-band length decreases upon sarcomere shortening. Huxley[127] demonstrated that thick myofilaments are arranged in an hexagonal lattice and that thin filaments interdigitate with the thick filaments at each trigonal point, producing what is now termed the double hexagonal lattice of myofilaments. Radiating out from each thick filament is a series of crossbridges that interconnect thick and thin filaments in the zone of filament overlap. Each thick filament appears to contain 300 crossbridges; three of these project from the axis of the filament every 14.3 nm.[129] Thus, there are 100 crossbridge-levels on each filament. The crossbridges appear to project from the filament shaft in a right-handed helix exhibiting a 60-degree rotation for every 14.3 nm of axial translation. The midregion of each thick filament, termed the bare zone, is devoid of crossbridges for a length of about 0.16 μ. A second series of transverse bridges found in the M band is located in the center of the bare zone, but in this case the bridges interconnect adjacent thick filaments. The detailed structure of the Z band is complex and remains unknown.[130] Basically, the Z band is built of two or more transverse plies of orthogonally arranged filamentous proteins conferring a square lattice-like appearance when viewed *en face* by electron microscopy.

Each of the myofilaments contains a characteristic set of contractile proteins. The thick myofilaments are composed predominantly of myosin, the major constituent of the sarcomere. It accounts for about 55% of the myofibril. Myosin has a total mass of approximately 480 kD.

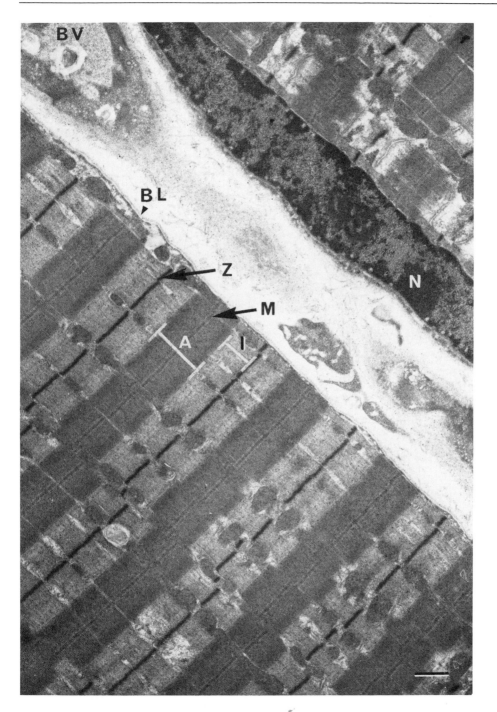

Fig. 6-7 *Longitudinal section of biopsy specimens from human vastus lateralis muscle. Two myofibers are seen with intervening blood vessels (BV). The nucleus (N) and basement membrane (BL) are indicated. Also labeled are the A, I, and Z bands of the myofibrils. (Reproduced with permission from Fischman DA, Meltzer AY, Poppei RW: Disruption of myofibrils in the skeletal muscle of psychotic patients.* **Arch Gen Psychiatry** *1970;23:503–515.)*

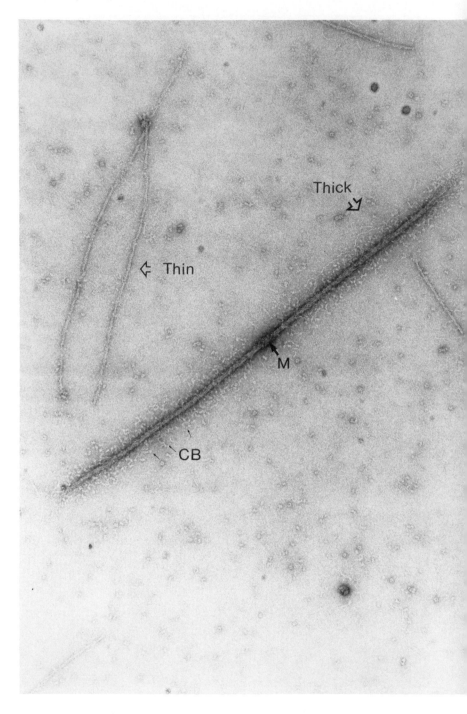

Fig. 6–8 *Negatively stained myofilaments isolated from adult chicken pectoralis major muscle. Both thick and thin myofilaments are visible. The M band and crossbridges (CB) are indicated on the thick filament. Note the double-stranded thin filament with actin subunits (Reproduced with permission from J. Dennis and D.A. Fischman.)*

It is composed of six subunits, two heavy chains (200 kD), and four light chains (approximately 20 kD). Within each myosin molecule, the two heavy chains are usually identical. There are three different light chains in fast-twitch muscle (MLC1f, MLC2f, MLC3f) and only two light chains in slow muscle (MLC1s, MLC2s). Each heavy chain is associated with two light chains. The most reliable data suggest that there are 300 myosin molecules per thick filament[128]; thus, there is one myosin molecule per crossbridge. Each myosin molecule is rod-shaped with two adjacent globular heads at one end of the rod. The molecules are approximately 15.5 nm long and 2 nm in diameter; the heads are oval structures about 19 nm long and 5 nm wide. The carboxyl terminus of each MHC subunit is at the tip of the rod, while the amino terminal end resides in the head. Proteolytic digestion of myosin with trypsin produces two fragments, termed heavy meromyosin (HMM) and light meromyosin (LMM). The HMM fragments contain two heads and a short piece of the rod. LMM, a purely α-helical fragment, contains about two thirds of the tail of the myosin rod. By selected digestion with two other enzymes, such as chymotrypsin or papain, it is possible to cleave the myosin molecule where each head joins the rod. This results in the generation of single-headed fragments, termed subfragment–1 (S–1) and the complete tail, termed the rod fragment.

In addition to myosin, all skeletal thick filaments of vertebrates contain at least five or six other proteins: C-protein,[131] M-protein,[132] myomesin,[70] end protein,[133] creatine kinase,[134] and titin.[81] C-protein is a single-chain protein of 140 kD[131]; by antibody labeling it has been localized to a series of stripes in the A band 43 nm apart. This region of the thick filaments has been termed the C-region. There appear to be between 14 and 18 C-protein stripes per A band. Based on the amount of C-protein isolated from myofibrils, it has been estimated that there are between two and four C-protein molecules per filament at each of these stripes or about 50 to 76 molecules per filament. The function of C-protein is unknown. Another protein associated with the C-region is found in rabbit slow muscle and termed X-protein.[135] Some controversy exists about whether X-protein is an isoform of C-protein or a totally distinct sarcomeric protein. The M-line contains three proteins: M-protein, myomesin, and M-creatine kinase. M-protein is a single-chain molecule of about 165 kD that may be located in the M-bridges. Myomesin has a subunit molecular weight of 185 kD. Approximately 10% of the creatine kinase of skeletal muscle is associated with the M-band region of the sarcomere. It is exclusively of the MM variety (that is, MM-creatine kinase). The functions of M- protein and myomesin are unknown; it is presumed that they participate in the stabilization of the A-lattice. The localization of MM-creatine kinase in the M-region is presumed to facilitate the rephosphorylation of adenosine diphosphate produced as a by-product of the actomyosin ATPase cycle. The last two components of the A band appear to be related. When native thick filaments are isolated from skeletal muscle and negatively stained or shadowed for electron microscopy, a periodic filamentous projection can be identified at the ends of the filaments.

These have been termed end filaments. Recent studies indicate that these filaments may contain titin, a gigantic elastic filament that may wrap around each thick filament and span the distance from A band to Z-disk. Titin has a subunit mass of 1×10^6 daltons and may exist as a dimer in the sarcomere. Titin may account for as much as 10% of total myofibrillar protein. Some workers have referred to the titin filaments as connecting filaments and the protein as connectin.[136] For a more thorough discussion of these newly discovered proteins, their localization and potential function, readers should consult the review by Wang.[137]

The thin filaments contain the following proteins: actin, tropomyosin, and the troponin complex. Other proteins have been suspected to cap the barbed and pointed ends of the thin filaments, but these remain to be purified to homogeneity. Actin is a highly conserved globular protein of 41.8 kD that polymerizes to form a double-chain, right-handed helical polymer, termed F-actin.[35] Thin myofilaments are almost exactly 1.0μ long and 8 nm in diameter in all vertebrate skeletal muscle. Each thin filament contains approximately 360 actin molecules. The axial spacing along each strand of the helix is 5.5 nm and there are 13 to 14 subunits in one turn. Thin filaments are polar structures; the barbed end of each filament is at the Z-disk, and the pointed ends of the filaments are directed toward the A band. In all vertebrates, the actin is associated with tropomyosin and three troponin subunits essential for Ca^{2+} regulation of muscle contraction. Tropomyosin is a two-stranded α-helical, coiled-coil molecule, each subunit of which has a mass of about 35 kD. Each tropomyosin molecule is 41 nm long and interacts with seven adjacent actin molecules within each strand of the thin filaments. Troponin, with a total mass of about 80 kD, is a noncovalent complex of three subunits, TN-I, TN-T, and TN-C, each of which has a distinct physiologic function in muscle. TN-I, the inhibitory subunit, has a molecular weight of 20 kD. When bound to F-actin, it inhibits the actin-myosin interaction, particularly in the presence of tropomyosin. TN-C is the Ca^{2+}-binding subunit; its molecular weight is 18 kD. In the presence of Ca^{2+}, TN-C can relieve the inhibition of the actin-myosin interaction imposed by TN-I. When the ionic calcium levels fall below 10^{-7} M, this effect of TN-C is reversed and the filament returns to its inhibited state. TN-T is the tropomyosin-binding component; its molecular weight is 30 kD. Its main function appears to be to facilitate the binding of TN-C to the actin-tropomyosin complex. The troponin molecule is lollipop-shaped; the entire complex is about 26 nm long and the tail is about 2 nm in diameter. The shape of troponin suggests that it interacts with tropomyosin over almost half of its 41-nm length.

The Z band (Z line or Z-disk) is composed of at least four proteins: α-actinin, desmin, filamin, and zeugmatin. The organization of the Z band is not understood; α-actinin appears to be in the core of the Z band, whereas the other proteins seem to be organized around the Z-band periphery. The Z band appears to account for about 6% of total myofibrillar mass. A thorough discussion of the contractile process as related to sarcomere structure and composition may be found in the reviews of Squire,[126] Huxley,[127] and Craig.[128]

Recent discussions of muscle cell lysosomal structure, distribution, and changes during development can be found in the review of Bird and Roisen.[138]

The distribution of ribosomes in skeletal muscle is not well understood. Most of the ribosomes in adult muscle are found in the intermyofibrillar sarcoplasm adjacent to the I-Z-I region and beneath the sarcolemma, especially near the nuclei. Glycogen granules tend to obscure the ribosomes when muscle is examined by routine thin-section transmission electron microscopy. Most of the rough-surfaced endoplasmic reticulum seems to be found near the nuclei, adjacent to the Golgi complexes, which are rather modest structures in skeletal muscle.

The Response of Muscle to Training

Even in adults, muscle structure changes in response to functional demands. An adaptive response indicates that a stimulus has been detected by a receptor and that appropriate signals have been transmitted to an effector. The stimulus for adaptation is described by its intensity, duration, frequency, and pattern, whereas the adaptive response is characterized by its nature, rate, magnitude, and duration. Negative feedback usually regulates adaptive responses. The stimuli required to produce a desired set of adaptive responses in skeletal muscles are organized into a training program. Training programs are based on the principles of overload, specificity, and reversibility.[139] Skeletal muscle fibers increase their structural or functional capability in response to overloading, that is, training that taxes muscle fibers beyond a critical level. Specificity of training means that a specific stimulus for adaptation elicits specific structural and functional changes in specific elements of skeletal muscle. As adaptation occurs in response to a specific training program, the rate of change will eventually plateau; no further adaptive change will occur unless the training stimulus is increased. The effects of training are reversible, that is, discontinuation of a training stimulus results in de-training and the adaptive changes regress.[139]

Low-tension high-repetition contractions of skeletal muscle fibers performed for 30 to 60 minutes on a regular basis result in endurance training. The tension is low only in a relative sense; the tension developed and the total duration of the activity must be great enough to produce an overload. "High-repetition" implies cyclic contractions of muscles as in walking, running, cycling, or swimming. The cycling of activity and rest periods provides time for recovery or for repositioning of a limb for the next stride, cycle, or stroke. The intensity, duration, and frequency of endurance training are designed to stimulate an increased capacity of fibers for sustained effort without inducing chronic fatigue. Chronic fatigue is defined operationally as an impairment in muscle function more than 24 hours after the exercise.

Endurance training usually involves large muscle masses that develop large metabolic loads and cause conditioning of some elements of the respiratory and circulatory transport systems.

Endurance training doubles the oxidative capacity of whole-muscle homogenates and increases the percentage of fibers classified as high oxidative in the limb muscles used in the training program.[140] These observations are supported by electron microscopic evidence that the number and size of mitochondria increase in trained skeletal muscles. The effect of endurance training on the capillary density of skeletal muscle is controversial. Some studies have shown no change; other studies show increases of 10% to 15%. The physiologic meaning of capillary density has not been clarified nor has the issue of adaptations following endurance training.

Muscle hypertrophy and increases in muscle strength are produced by high-tension low-frequency contractions of skeletal muscle.[139] Because recovery from a strength-training session requires up to 48 hours, sessions are usually performed on alternate days. Strength-training programs increase total muscle mass primarily by increasing the cross-sectional area of individual fibers (the number of myofibrils per fiber). Type II fibers hypertrophy more than type I fibers. Exercise programs designed to increase muscle strength do not increase the maximum muscle blood flow, the ratio of capillary to muscle fiber, oxidative capacity, or mitochondrial density of skeletal muscles.[139]

A major controversy concerning the response of skeletal muscle to strength-training concerns the role of hyperplasia and fiber splitting. Muscle hypertrophy of 7% to 34% has been reported for muscles in cat forelimbs following weightlifting. Following the hypertrophy, Gonyea and associates[141] reported an increase of 19% in the total number of fibers in a cross section through the belly of the muscle. This is not, however, a definitive method of demonstrating hyperplasia. Gollnick and associates[142] used ablation of the synergists and exercise to load the soleus, plantaris, and extensor digitorum longus of rats. Wet muscle masses increased 29% to 88% over control values. After nitric acid digestion, the total number of individual fibers was counted. As compared with muscles in control animals, differences in total number of fibers were observed in muscles that had been overloaded by ablation, or ablation plus exercise. In the ablation model as compared with the weight-training model, the stimulus for adaptation differed in intensity, duration, frequency, and variability. Thus, hypertrophy of muscle fibers is the predominant adaptation to strength-training, but the specific contributions of hyperplasia, fiber splitting, fiber branching, and fiber fusion to muscle enlargement have not been determined.

Stretch-adaptation in skeletal muscle has been studied in a number of different models, including the denervated hemidiaphragm, the denervated chicken wing, ablation of synergistic muscles, shortening of the distal tendon, stretched-immobilized ankle flexors, spring-loaded extension of the immobilized chicken wing, stretched myocytes in vitro, a bite-opening appliance, and hyperinflation of the diaphragm induced by emphysema.[143] The passive stretch results in a training stimulus. In innervated muscles, the stretch reflex results in increased tension development and subsequent hypertrophy. The stretch-induced growth is associated with an increased protein turnover. In several models, the number of sarcomeres in series increases or decreases, with commensurate changes in fiber length.

De-training of skeletal muscles has not been studied as intensively as has training. Upon cessation of endurance training, the number of oxidative fibers regresses back to control values in a matter of weeks and reductions occur in the oxidative capacity of muscle. De-training occurs after almost any surgical intervention.

In summary, three patterns of adaptive stimuli have been identified: low-tension high-repetition stimuli that result in an increased endurance; high-tension low-frequency stimuli that result in increased strength; and a stretch-stimulus that results primarily in increased strength. The effect of inactivity may be studied during de-training programs, by bedrest, through immobilization of limbs, or by reduction or elimination of weightbearing. These conditions produce specific adaptive changes in skeletal muscles and in the cardiovascular and respiratory systems.

Muscle Injury and Regeneration

Development, Regeneration, and Repair

Because of the complex structure and function of muscle, return of muscle function following injury requires restoration of essentially normal muscle tissue with appropriate innervation and vascular supply. Thus, unlike the other musculoskeletal soft tissues, with the exception or nerve, successful muscle repair requires regeneration, that is, formation of new essentially normal tissue.

It is essential to understand the molecular and cellular aspects of embryonic muscle development discussed previously before the process of regeneration can be fully appreciated. Moreover, the fundamental variables governing development serve as a framework for examining the controlling steps of the regeneration processes.

The process of development, maturation, and aging is a continuum of sequential cellular and molecular replacement events wherein tissues, cells, and molecules are replaced by structures of continually reduced developmental potential and increased degree of specialization.[144] Some of the new structures in this sequence are variants, or isoforms, of their predecessors; some of the new structures are unique. Three fundamental variables control both the developmental and regeneration processes:

(1) The genomic repertoire of the organism sets the limits to developmental and regenerative possibilities. For example, birds cannot form teeth, although their evolutionary ancestors had these structures.[145]

(2) The previous developmental decisions also set the limits; development is a progressive, irreversible process. There is no dediffer-

entiation to an earlier developmental state. For instance, although the mesoderm can give rise to either heart or skeletal muscle, once committed to one of these pathways, neither can be converted to the other.[146]

(3) The local molecular and cellular environment provides cues that affect the rate and extent of development and regeneration. This is summarily referred to as positional information and can profoundly affect the process. For example, the local site at which a mesenchymal stem cell finds itself can determine whether it will become an osteoblast or chondrocyte.[147]

Regeneration or repair cannot be an exact recapitulation of embryonic development, although it may be very similar. Repair or regeneration must take place on the fabric of existing adult structures in an ambiance of adult physiologic activity. Muscle is an especially useful tissue to investigate since a large body of information is known about its development at both the molecular and the cellular levels.[148,149] Importantly, muscle comes equipped with its own repair cells, the satellite cells, and thus, has a substantial and specific repair capacity.[150,151] This repair potential is controlled by the genomic repertoire, the previous developmental decisions, and the local milieu in which repair can take place. Clearly, we cannot currently manipulate the genomic repertoire or previous developmental decisions of human tissue, but we can affect the environment in which repair takes place. Alternatively, it may be possible in the future to manipulate the hosts so that mesenchymal stem cells that still have not made major developmental commitments can be introduced at a specific repair site.

A muscle is formed from the cooperative interplay between the cells of two unique and separate developmental lineages, one associated with the formation of individual muscle cells and the other accounting for the connective tissue "packaging" of individual muscles.[1,2] This latter lineage, the mesenchymal cell lineage, has the capacity to form bone, cartilage, ligament, tendon, and all the connective tissue associated with muscle and other mesenchymal structures. The cells of this mesenchymal lineage determine muscle shape and the pattern of muscle, including insertion sites and size.[3] Thus, the wrapping and confinement in which myogenesis takes place directly determines where the muscles are located, what they can do, and how they perform their site-specific functions.

All muscle is not alike. Not only are the final shapes and functions different, but the molecules within the muscle itself may also be different. Thus, the problem of restoring normal function is more complicated than in the other musculoskeletal connective tissues. The correct lineage cell has to get to the right place at the right time, has to be wrapped in the appropriate connective tissue package, has to be wired to the central nervous system to coordinate the activation of muscle contraction, and has to do all of this while a myriad of other changes are going on around this developing structure. Furthermore, it is probable that the type of muscle, fast vs slow, directly feeds back to determine the properties of the nerve that innervates it.[152] Once specified, the nerve is fixed in its electrical and impulse properties.

Thus, if regeneration occurs at that site, the nerve cannot revert to a different type, for instance from slow to fast, and thus the regeneration of that muscle must take place within the confinement of that fixed neural input.

Regeneration of muscle also requires production of the appropriate myofibrillar proteins. The myofibrillar proteins are products of multigene families of surprising complexity that undergo a sequence of developmentally and temporally controlled replacements of individual isoforms. That some of these molecular replacements normally occur as the muscle is actively functioning, as the animal is walking about and using the muscle in question, represents a major engineering feat in its own right.[46] During development, and presumably in successful repair, as these events are going on inside the muscle cell, there are dramatic replacement and rearrangement events occurring in the extracellular matrix that surrounds each muscle cell. These events involve the functional interconnections of nerves, ligaments, and tendons as well as the positioning of blood vessels.

The Morphology of Degeneration and Regeneration

The older literature on muscle pathology is replete with detailed descriptions of patterns of muscle fiber degeneration.[153] Many of these were worked out before it was recognized that skeletal muscle fibers can regenerate. Now, except for diagnostic purposes, different patterns of muscle fiber degeneration are less important than was once assumed. In most cases, there are two main phases in the degeneration of injured muscle fiber. In ischemic muscle, for instance, the first phase is one of intrinsic degeneration, characterized by membrane damage, disruption of sarcomeres at the Z bands, mitochondrial swelling, and nuclear pyknosis (Fig. 6–9).

Depending upon the model of muscle injury, the phase of intrinsic degeneration is followed by a phase of cell-mediated fragmentation of the damaged muscle fiber. The timing of the entry of phagocytic cells (mostly macrophages) into damaged muscle fibers depends on vascular relations. If muscle injury is produced without damage to the local blood supply, as occurs following injections of local anesthetics, then macrophages can be seen within the muscle fibers 12 hours after the injection. In contrast, after an ischemic lesion like a nonvascularized muscle graft, revascularization of a given area may not occur for weeks. Only after blood vessels have entered an area of formerly ischemic muscle do macrophages penetrate and fragment the necrotic muscle fibers. Little is known about what causes macrophages to penetrate the basal lamina and enter the necrotic muscle fiber. Using antibodies to a variety of complement antigens, Engel and Biesecker[154] demonstrated the activation of several complement components, leading to the formation of the membranolytic C5b–9 membrane attack complex and the attraction of macrophages and neutrophils to the damaged muscle fibers.

A relationship between macrophage invasion and the initiation of muscle fiber regeneration has been suspected for years, but the evidence

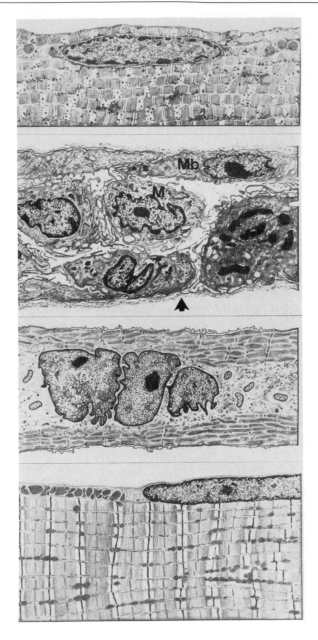

Fig. 6–9 *Summary drawings of major phases of the degeneration and regeneration of a single muscle fiber.* **Top:** *Early ischemic damage. The nucleus is becoming pyknotic, with chromatin clumping inside the nuclear membrane. The mitochondria are swollen, and the bundles of contractile filaments are breaking apart throughout the muscle fiber.* **Top center:** *Fragmentation phase. Macrophages (M) associated with ingrowing vasculature enter the degenerating muscle fiber and remove bundles of contractile filaments and other cytoplasmic debris. Beneath the basal lamina (arrow) spindle-shaped myoblasts (Mb) line up in preparation for the formation of new muscle fibers.* **Bottom center:** *Myotube. Beneath the original basal lamina, myoblasts have fused to form a multinucleated fiber with bundles of newly forming contractile filaments at the periphery.* **Bottom:** *Muscle fiber. The mature regenerated muscle fiber is in most respects indistinguishable from a normal muscle fiber.*

is contradictory. Typically, in vivo the activation of myogenic cells is not seen until macrophages have invaded the damaged muscle fiber. It has been postulated that by removing muscle fiber cytoplasm, a negative feedback influence on myogenic cells is lifted or that macrophages themselves provide a positive stimulus for myogenesis. There is no direct evidence that either the interruption of negative feedback or the macrophages initiate myogenesis, but neither of these two possibilities has been disproved. Bischoff,[155] however, cited several instances in which myogenic cells are activated in the absence of degeneration of the muscle fiber and has extracted a mitigen from crushed adult rat muscle that stimulates proliferation of myogenic cells. This evidence of a positive mitogenic stimulus for regenerative myogenesis seems quite convincing.

In the normal course of in vivo regeneration, a population of activated, spindle-shaped myogenic cells appears beneath the basal lamina of the original muscle fibers[156] as macrophage-mediated removal of the old muscle fiber is well underway (Fig. 6–9). These cells proliferate both in vivo and in vitro. At some point the myogenic cells withdraw from the cell cycle and can then be properly called myoblasts. The myoblasts then fuse with one another to form long syncytial myotubes, with characteristic chains of central nuclei. It has been very difficult to demonstrate directly the withdrawal from the cell cycle and fusion in vivo, and investigators have, to a large extent, extrapolated from the pattern of embryonic myogenesis in vitro. Recently, Bischoff[157] directly observed these events on single muscle fibers in culture.

More commonly than not, several early regenerating myotubes are found within the basal lamina tube of a single degenerated original muscle fiber. The structure of a single regenerating myotube is quite similar to that of its embryonic counterpart (Fig. 6–9). The cytologic features of myotubes are characteristic of cells producing large amounts of intracellular protein. The nuclei of the central chain are enlarged and show prominent nucleoli. The early myotubes contain helical polyribosomes upon which myosin is synthesized. The high content of cytoplasmic RNA is reflected in histologic preparations, in which early myotubes have an intensely basophilic cytoplasm. The contractile proteins are first assembled into well-organized bundles of filaments at the periphery of the myotube; later, myofibrils are added successively toward the center of the myotube. Construction of the sarcoplasmic reticulum and the triads, consisting of a T-tubule and two cisternal elements of the sarcoplasmic reticulum, keeps pace with the formation of the myofibrils. Pentads are commonly seen in early regenerating muscle fibers.

The transition from a myotube to a muscle fiber is a gradual one, with few distinct endpoints. One of the most conspicuous events is the breaking up of the central chains of nuclei and the migration of individual nuclei to the periphery of the muscle fiber. It is convenient to use the arrival of nuclei at the periphery as the point marking the differentiation of the myotube to a myofiber. In keeping with the accumulation of contractile proteins and the decrease in cytoplasmic

RNA, histologic preparations show a marked reduction in cytoplasmic basophilia and a correspondingly greater eosinophilia.

Final differentiation of a regenerating muscle fiber follows its innervation. After a functional neuromuscular junction has become established, the regenerating muscle fiber will assume a morphology characteristic of a fast or slow muscle fiber depending on whether it is innervated by a fast or slow nerve. Satellite cells establish a normal relationship with the regenerated muscle fiber. Schultz[158] has shown that the percentage of satellite cells in regenerated fast and slow muscles returns to normal levels.

Continuous Regeneration   The older literature on muscle pathology and regeneration commonly subdivided muscle regeneration into "continuous" and "discontinuous" categories. The description of muscle fiber regeneration given above applies to discontinuous regeneration, in which the regenerated muscle fiber is formed completely *de novo* with no connection to a pre-existing muscle fiber. Continuous muscle regeneration, or budding, occurs after transection of a muscle; it is characterized by the presence of an outgrowth of basophilic sarcoplasm containing varying numbers of nuclei with a morphology similar to that of myotube nuclei.[159-161] Bischoff[57] observed muscle budding in vitro and by analyzing creatine kinase isoenzymes provided evidence of dedifferentiation of the budded part of the muscle fiber. He suggested that budding may be a mechanism for bridging a gap in the continuity of a muscle fiber. The origin of the nuclei in the muscle buds has not been determined. Possible explanations are (1) that myonuclei can migrate into the muscle bud, (2) that myoblastic cells can fuse to the membrane of the bud, and (3) that both of the above can occur. The membrane of a muscle bud may have properties that allow myoblasts to fuse with it. In a simple transection lesion, the budding muscle fibers appear to become embedded in dense connective tissue, and further development ceases.

The Origin of Regenerating Muscle Fibers   With the discovery of the myosatellite cell by Mauro,[162] opinions were divided regarding the cellular origins of regenerating muscle. According to one, the myoblasts of regenerating muscle arise from myonuclei that separate off from dedifferentiating muscle fibers. The other considered that satellite cells are reserve cells with myogenic capacity and that these cells become activated after muscle damage. These arguments are well summarized by Mauro and associates.[163]

The concept of dedifferentiation to explain muscle regeneration arose mainly from studies on amphibian limb regeneration after amputation. One of the prominent phases in this process is a stage called dedifferentiation when the tissues at the distal end of the limb stump lose their adult structure. Shortly thereafter a regeneration blastema consisting of mesenchyme-like cells accumulates at the end of the limb.[164] Several descriptive studies at the light and electron microscopic levels have identified sequences of stages that were interpreted to show the release of mononucleated fragments from the ends of the transected muscle fibers and their ultimate incorporation into the regeneration blastema.[165-167] A more recent study by Kintner and Brockes[168] used a double labeling with antibodies assumed to be specific for blastemal cells (22/18) and muscle cells (12/101). Certain double-labeled mononucleated cells near the base of the blastema were suggested to be derived from myonuclei that fragmented off from damaged muscle fibers. Subsequent work,[169] however, has shown that the 22/18 antibody is not blastema-specific and that the double-labeled cells may be participating in tissue rather than epimorphic regeneration of muscle.

The muscle satellite cell is an undistinguished-looking mononucleated cell that lies between a muscle fiber and its surrounding basal lamina.[170,171] A mature satellite cell is a spindle-shaped cell that contains an ovoid nucleus with highly condensed peripheral heterochromatin and a thin rim of cytoplasm containing few and rudimentary cell organelles. By definition, basal lamina material should not be interposed between a satellite cell and its adjoining muscle fiber.

It is commonly believed that satellite cells represent myoblasts that did not fuse into multinucleated muscle fibers during embryonic myogenesis. Whether this is true or whether satellite cells are descended from a separate line of cells, typical satellite cells appear alongside the young muscle fibers late in the fetal period.

In rodents, satellite cells are numerous at birth, consisting of over 30% of the nuclei in or associated with the muscle fiber.[172,173] The percentage of satellite cell nuclei typically declines with increasing age until values in the range of 5% are attained. Quantitative studies have shown that changes in the percentage of satellite cell nuclei do not always reflect their total number within a muscle. Also, red and white muscles differ substantially in satellite cell content. Gibson and Schultz[174] showed that both the absolute number of satellite cell nuclei and the proportion of satellite cell nuclei to myonuclei are higher in the red soleus muscle than in the white extensor digitorum longus muscle in the rat. With increasing age, the satellite cell population of the extensor digitorum longus muscle declines much more dramatically than does that of the soleus muscle. Using a different means of quantitation to examine the soleus and extensor digitorum longus muscles in 2-month-old rats, Schmalbruch and Hellhammer[175] found 5,000 and 1,000 satellite cells/mm³, respectively. Schmalbruch[176] has estimated that the biceps brachii muscle of adult humans contains about 800 satellite cells/mm³. In vitro studies of satellite cells taken from rats of different ages have shown a decreased proliferative potential of satellite cells in older rats.[177]

Snow[178,179] demonstrated directly in vivo that during regeneration rat satellite cells are precursor cells of myoblasts. His experimental design was to expose muscles to ³H-thymidine so that only myonuclei or only satellite cells were labeled. He then minced the muscle, waited until regeneration reached the myotube stage, and looked for the presence of labeled myonuclei in the regenerating muscle fiber. After mincing, the myonuclei became obviously pyknotic after a few hours of ischemia, and regenerating muscle fibers arising from muscle in which myonuclei alone were labeled did not contain labeled nuclei within the myotubes. In contrast, minced muscles that contain labeled satellite cells produced regenerates with labeled myotubes. In vitro studies[157] have also shown directly that satellite cells can proliferate and fuse into myotubes. Although there is currently little evidence that myonuclei can participate in the regeneration of a muscle fiber, the possibility has not been disproved, especially in certain nonischemic lesions of muscle.

There have been periodic suggestions that circulating mononucleated cells might serve as precursors of regenerating muscle.[180] More recent work, however, has failed to support that hypothesis.[181]

The Extracellular Matrix The extracellular matrix that surrounds every myofiber physically isolates each contractile unit, the myofiber, from its neighbors by forming a basement membrane that completely surrounds each individual cell. This basement membrane serves as a specialized interface for attachment and integration of neural elements.[182] The basement membrane is used to isolate and sequester reparative cells, the satellite cells, which are dormantly packaged within the basement membrane and are liberated within this confinement when repair is required.[150,151] The extracellular matrix must merge to interdigitate and integrate with ligamentous and tendinous tissue.[183,184] The extracellular matrix also stores bioactive molecules, such as growth factors (some muscle-specific) that are anchored to its member molecules and released to play a central role in reparative events.[23,185]

With such a complex set of specialized end-functions, it is predictable that the formation of the basement membrane is a complex multistep phenomenon.[186] Like other developmental processes, the formation of the basement membrane is preceded by a series of programmed replacement events in which various matrix macromolecules are replaced by molecular variants or operationally similar molecules.[187] A specific class of matrix molecules having large amounts of associated polysaccharide are important components of these extracellular matrices and are involved in replacement phenomena.

Early myogenic cells produce a highly hydrated extracellular matrix. The predominant polysaccharide is hyaluronic acid.[188] It not only occupies positions in the extracellular matrices, but is also associated with the cell surfaces of myoblasts. These large polysaccharide-containing molecules are highly anionic and, thus, structure a large volume of water. The hyaluronic acid-rich matrix is replaced by a unique, huge, chondroitin sulfate, muscle-specific proteoglycan.[189,190] Like hyaluronic acid, these molecules are anchored at the cell surface and also occupy extracellular matrix sites. Their position in the extracellular matrix ensures that a sufficient space will be reserved by water to allow for the girth-wise expansion of myotubes. In addition, the cell surface chondroitin sulfate proteoglycan is involved in some way in inhibiting basement membrane formation and/or establishing the micro-environment in which basement membranes will eventually form. This chondroitin sulfate proteoglycan is replaced by heparan sulfate proteoglycans that insert into the cell membrane and form components of basement membranes. Eventually, neural elements integrate synaptic connections through the basement membrane and form a specialized interface structure in which AchE is anchored in the basement membrane at the formation of synapses.[182,191]

During muscle regeneration, the sequence of replacement events involving hyaluronic acid, chondroitin sulfate muscle proteoglycans, and basement membrane heparan sulfate proteoglycan is recapitulated to reform a functional muscle unit; although the overall sequence is comparable to the embryologic sequence, the timing and some details are unique to repair.

When viewed with the electron microscope, the basement membrane of the original muscle fibers can be identified throughout the degenerative period and for many days into the regenerative period. During the degenerative phase, macrophages penetrate the basal lamina and destroy the damaged muscle fiber, leaving the activated satellite cells to proliferate and then fuse into multinucleated myotubes on the inner border of the basement membrane. As they differentiate, the myotubes begin to produce their own basement membrane, sometimes directly apposed to the persisting original basement membrane. By conventional electron microscopy, it is difficult to follow the relationship between old and new basement membranes later in regeneration.

With the use of immunocytochemistry and lectin binding, Gulati and associates[192-194] looked at the more dynamic aspects of turnover of basement membrane components during muscle regeneration. During the degenerative phase, the pericellular fibronectin first took on a fragmented distribution and then disappeared altogether. Components of the basement membrane proper—laminin, types IV and V collagen, and heparan sulfate proteoglycan—persisted longer than fibronectin, but these also disappeared. Concanavalin A binding on the basement membrane was demonstrable throughout the degenerative period. Immunocytochemical methods have shown that as new basement membrane material is deposited around the regenerating myotubes, specific components are added. Fibronectin, which disappeared first from the surrounding matrix, reappeared last among the molecules that Gulati and associates tested.[194]

Myllylä and associates[195] noted increased immunocytochemical staining with antibodies to types III, IV, and V collagen in rat muscle that regenerated after exercise-induced injury; they concluded that collagen metabolism is stimulated during muscle regeneration. Some models of muscle regeneration, such as mincing, produce excessive amounts of collagen deposition, whereas other models produce minimal increases in muscle collagen content.

Two injury models using adult chicken muscle have recently been developed to explore the role of the extracellular matrix in directing muscle regeneration: cold injury and muscle excision. The large muscle-specific chondroitin sulfate proteoglycan (M-CSPG) pictured in Figure 6–10 is never synthesized by adult tissue and is found only in the newly forming embryonic muscle.[189,190] However, when muscle is

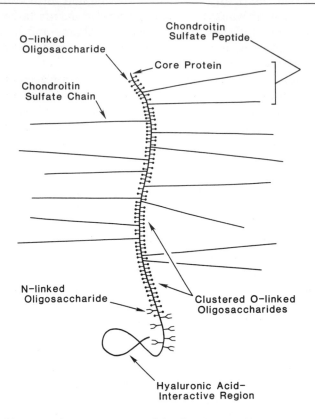

O-linked
Oligosaccharide

Chondroitin
Sulfate Peptide

Core Protein

Chondroitin
Sulfate Chain

N-linked
Oligosaccharide

Clustered O-linked
Oligosaccharides

Hyaluronic Acid-
Interactive Region

Fig. 6–10 *Diagrammatic representation of the chondroitin sulfate proteoglycan synthesized by chick embryonic skeletal muscle. The proteoglycan is organized around a core protein that is approximately 250 nm long and weighs approximately 300 to 400 kD. Covalently attached to the core protein are a large number of saccharide constituents, the major components being chondroitin sulfate glycosaminoglycans. These chondroitin sulfate chains are unusually large, 60 to 70 kD. About 40 to 60 chains bind to each proteoglycan. Chondroitin sulfate is the only glycosaminoglycan present in this proteoglycan, although small oligosaccharides of two types are also present. The more numerous of these oligosaccharides (10 to 15 per chondroitin sulfate chain) are the O-linked oligosaccharides, which are attached to serine or threonine residues of the core protein. The second class of oligosaccharides (one to two per chondroitin sulfate chain) are the N-linked oligosaccharides, which contain mannose and are attached to asparagine residues; these are similar to those found on other glycoproteins. The skeletal muscle chondroitin sulfate proteoglycan has the ability to interact in a link protein-stabilized fashion with hyaluronic acid similar to that of cartilage chondroitin sulfate proteoglycans and, therefore, its core protein presumably contains a functional hyaluronic acid-interactive region.*

injured by a cold (-90 C) probe or a piece of muscle is excised, this M-CSPG is synthesized by reparative cells as part of the sequence of reforming the muscle tissue (D.A. Carrino, U. Oron, D.G. Pechak, et al, unpublished data). The cold-injury model is of special interest because, although the myotube contents are phagocytically removed, the myotube basement membrane remains intact. Clearly then, the excision model can be considered the basement membrane-free variant. While the sequence of reparative events is roughly identical in these two models, repair is twice as fast in the cold-injury model. The formation of hyaluronic acid-rich basement membrane, followed by the formation of M-CSPG, and finally heparan sulfate-rich basement membrane is coordinated with cellular and isoformic replacement events. The result is the complete reformation of the damaged muscle sector. Failure of repair is usually associated with nonmuscle, fibrous connective tissue intrusion at the repair site. This is more likely to occur in the excision model than in the cold-injury model. The cold injury repairs faithfully and faster because all the reformation takes place inside the boundaries specified by the remnant basement membrane.[196] Still unanswered is the question of whether a totally new basement membrane eventually forms or whether the original structure is integrated into the newly forming structure. More importantly, it is not clear whether this newly formed basement membrane is synthesized by connective tissue cells or by the myotube. It is speculated that the myotube has a template for the self-assembly of the basement membrane, the components of which are mostly synthesized by fibroblasts.

The studies conducted to date on the extracellular matrix in regenerating muscle indicate that the matrix plays a dynamic role during the regenerative process. Major questions to be answered include (1) how individual components of the matrix are selectively removed; (2) what role the persisting components of the basement membrane play in the regeneration of muscle fibers; (3) how, or if, old and new molecular components of basement membranes are reintegrated during regeneration; and (4) how total levels of collagen synthesis are controlled after various types of muscle damage.

Regeneration of Entire Muscles The discussion thus far has concentrated on muscle regeneration at the cellular level, but entire muscles are also capable of regenerating. The regeneration of entire muscles involves a level of organizational complexity above that of the regeneration of individual muscle fibers. It is necessary to examine the degeneration and regeneration of muscle fibers in relation to the vascular supply, the nerve supply, and tendon connections.

In rodents entire muscles can regenerate from finely minced fragments.[197,198] The technique consists of removing a muscle from its bed, mincing it into cubic-millimeter fragments and placing the fragments in the original muscle bed at a different site. The muscle fragments are separated from a blood supply, a nerve supply, and tendon connections.

Within minutes after mincing, the muscle fragments fall into a profound state of ischemia, and intracellular degenerative changes occur. The myonuclei become pyknotic within hours, but the satellite cells are better able to withstand temporary ischemia.[178] Neither degeneration nor regeneration progresses significantly until vascular sprouts begin to grow into the minced mass. Numerous macrophages accumulate in the vicinity of the terminal vascular sprouts, and in these areas the macrophage-mediated fragmentation of muscle pieces begins. Concomitantly with the removal of the old muscle cytoplasm, the activation of satellite cells occurs.

As blood vessels grow farther into the minced tissue, a centripetal gradient of muscle-fiber breakdown and regeneration is set up. After several days, regeneration has proceeded to the myotube stage at the periphery of the minced area, whereas in an intermediate zone, muscle fragmentation and myoblast formation are the predominant processes. At the center of the minced tissue, the muscle fragments remain arrested in a state of ischemic necrosis until approached by blood vessels.

During the early stages of regeneration, tenuous connective tissue connections are established between the ends of the minced tissue and the tendon stumps of the original muscle. In this mechanical environment, the initially randomly oriented myotubes are reoriented into more parallel arrays. This reorientation first occurs at the periphery of the minced muscle and then proceeds toward the center. In the minced gastrocnemius muscle of the rat, original muscle fragments are replaced by regenerating muscle early in the second week after mincing, and the regenerating muscle differentiates from the stage of late myotubes to cross-striated muscle fibers. Fibroblasts are also active during this period, and in certain areas dense mats of collagen fibers are laid down. Reinnervation of minced muscle regenerates is slow, and neuromuscular junctions are not established until the third postoperative week.

Minced muscle regenerates are capable of contraction,[199] but their morphology is not identical to the external form or internal architecture of the original muscle. As a rule, they are considerably smaller. Muscles weighing over 1.5 g do not usually regenerate well after mincing.

An entire muscle can also regenerate after free nonvascularized grafting.[200–202] In this procedure a muscle is completely removed from its bed and then replaced, with the tendons of origin and insertion sutured to tendon stumps in the muscle bed. The main difference between a free muscle graft and minced muscle is that the internal architecture of the muscle is not mechanically disrupted. Nevertheless, the overall process of degeneration and regeneration is very similar (Fig. 6–11). The primary main difference is that at the periphery of the graft a thin rim of muscle fibers (2% to 5% of the total) survives the grafting procedure intact. The pattern of revascularization is like that in minced muscle except that segments of the original vascular bed of the muscle can be reused by the regenerating vessels. A centripetal gradient of muscle fiber degeneration and regeneration is set up (Fig. 6–11); the overall pattern and duration of muscle fiber regeneration and reinnervation differ little from those that occur after mincing. A major difference is that much less connective tissue forms in whole-muscle grafts. Free grafts are normally not successful in muscles weighing over 3 g, and a typical free muscle graft is restored to only about 35% to 50% of its original functional mass.[203]

The complete regeneration of an entire muscle occurs only in the

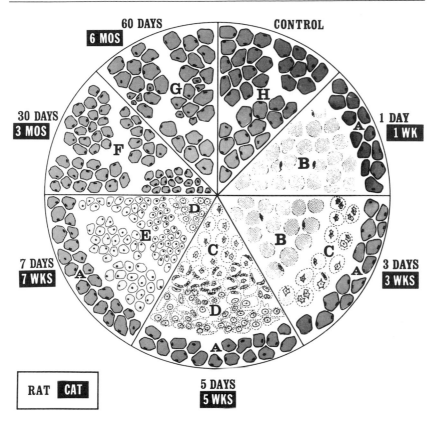

Fig. 6–11 *Schematic diagram showing major stages in the development of a free muscle graft. The days refer to times after grafting of extensor digitorum longus muscles in rats and cats. The diagram is divided into segments that represent the histologic appearance of the grafts in cross section. The letters refer to groups of cells showing similar histologic reactions.* **A:** *Surviving muscle fibers.* **B:** *Original muscle fibers in a state of ischemic necrosis.* **C:** *Muscle fibers invaded by macrophages, which are phagocytizing the necrotic cytoplasm.* **D:** *Myoblasts and early myotubes within the basal laminae of the original muscle fibers.* **E:** *Early cross-striated muscle fibers.* **F:** *Maturation of regenerating muscle fibers.* **G:** *Mature regenerated muscle fibers.* **H:** *Normal control muscle fibers.*

regenerating amphibian limb. Biologically, amphibian limb regeneration is different from the regeneration of individual mammalian muscles.[204,205] When a salamander's limb is amputated, interactions involving the wound epidermis, damaged mesoderm, and nerves[206] result in the loss of normal structure (dedifferentiation) of the tissues at the end of the limb stump and the formation of a mound of primitive-appearing cells (the regeneration blastema) beneath the wound epidermis. Despite considerable research over several decades, the cellular origins of the regeneration blastema are not known.

In many respects the properties of the regeneration blastema are similar to those of the embryonic limb bud,[207] and the pattern of muscle

257

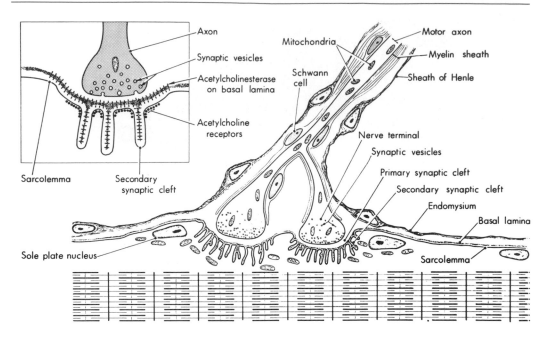

Fig. 6–12 *Schematic drawing of a neuromuscular junction. The inset shows the structural and molecular components involved in the release and uptake of acetylcholine from the nerve terminal to the muscle fiber.*

formation of the musculature in the regenerating limb duplicates that of the embryo.[208] The muscle anatomy of a regenerated amphibian limb is virtually indistinguishable from that of a normal limb.

Reinnervation of Regenerating Muscle The reinnervation of regenerating muscle has been investigated at two different levels: (1) the individual regenerating muscle fiber; and (2) the entire muscle. Studies on the latter have involved mainly the reinnervation of free nonvascularized muscle grafts and minced muscle. At the cellular and molecular level, investigators have studied the neuromuscular junction, and in particular the role of neuromuscular junction remnants of the original neuromuscular junction in guiding reinnervation of a regenerating muscle fiber. The basic structural features of the mammalian neuromuscular junction are illustrated in Figure 6–12.

Much of the recent research on the reinnervation of regenerating muscle has been conducted on frogs; it stems from reports by Rubin and McMahan[209] and Sanes and associates[191] that the basal lamina of the original neuromuscular junction is important as a cue for the localization of regenerating nerve terminals and for presynaptic and postsynaptic differentiation of the nerve and muscle. An experiment involving the inhibition of muscle fiber regeneration showed that nerve fibers regenerating down their original pathways terminated at the original end-plate zone of the persisting basal laminae of the muscle fibers.[191] Presynaptic differentiation of the nerve terminals occurs above the old end-plate region of the basal lamina, and, in the absence of innervation, postsynaptic folds form on the regenerating muscle fibers beneath the synaptic basal lamina.[210] More recent studies on regenerating mammalian muscle have also shown that the formation of postsynaptic folds and the localization of acetylcholine receptors do not depend upon the presence of the nerve.[211-213] A number of investigators have attempted to identify components of the synaptic basal lamina that cause the AchRs to cluster on the muscle fiber beneath it.[214-216] Attempts to discover the molecular basis for determining the site at which regenerating nerve fibers terminate have been less rewarding.

Regenerating nerve fibers spontaneously enter free nonvascularized muscle grafts or regenerating minced muscles. In the rat, nerve fibers are commonly first seen among the regenerating muscle fibers early or midway through the second postoperative week,[202] but functional neuromuscular transmission is usually not demonstrable at the whole muscle level until the end of the third week.[213,217] After the regenerated motor end plates have been established, the regenerating muscle fibers, which are histochemically homogeneous up to that time, differentiate further into fast and slow types.[218] If a fast muscle is grafted into the bed of a slow muscle and innervated by that nerve, or if a slow muscle is cross-transplanted into the bed of a fast muscle, it undergoes a fiber-type conversion so that the cross-transplanted fast muscle becomes slow and vice versa.[219,220] In cross-transplanted free muscle grafts, the extent of fiber-type conversion is slightly greater than that seen after cross-innervation of muscles that remain in their own beds.[221,222] The degree to which fast or slow mammalian muscle fibers are myogenically specified, as may be the case in avian muscle, remains to be determined. Such knowledge is important in the interpretation of cross-innervation experiments.

Mature free muscle grafts are typically hypoinnervated. Whether directly counting motor end plates[223] or estimating choline acetyltransferase activity,[224] investigators found that approximately 50% of the muscle fibers are innervated, although the total number of muscle fibers is normal.[225]

Côté and Faulkner[226] found that grafts of extensor digitorum longus muscle in the rat contain only about 50% of the normal number of motor units, but that the tetanic tension generated per motor unit was 80% of normal. The arrangement of the muscle fibers in regenerated motor units is not normal, and type grouping of muscle fibers is commonly seen in mature muscle grafts.[227]

Although reports of frog muscle regeneration suggest that the regenerating nerve fibers' termination points depend on information present at the synaptic basal lamina, studies of mammalian muscle regeneration have shown that ectopic reinnervation is common.[228] Womble[213] surgically removed the motor end-plate zone in grafted muscles and 60 days later found little difference between the number of ectopic end plates formed in the muscles without motor end plates and in standard muscle grafts with the end plates left intact.

Muscle spindles do not form in minced muscle regenerates.[229,230] Regenerated spindles are present in free muscle grafts,[227,231] but their numbers are only about half of normal.[232] Regenerated spindles are, as a rule, incompletely differentiated and, although both motor and sensory endings have been seen, they are abnormally innervated.[233,234] Nevertheless, some afferent signals have been recorded from mature free muscle grafts.[235]

In the absence of sensory innervation, muscle spindles do not form in the embryo,[236] but they do regenerate in the absence of nerves.[232] It appears that the presence of an intact spindle capsule, rather than innervation, is the key to permitting the regeneration of muscle spindles. In minced muscle regeneration, the spindle capsules are mechanically destroyed.

Muscle regeneration can begin in the absence of a nerve supply. By studying cases in which this occurs, one can determine whether or not muscle regeneration can occur in the absence of nerves and, if it can, what stages of the process are independent of neural input. In all species investigated to date, muscle regeneration begins in the absence of a nerve supply, but noninnervated regenerating muscle fibers of some species, such as frogs and mice, undergo a rapid atrophy in the late myotube stages and have disappeared by three weeks.[237,238] In rats, on the other hand, aneural muscle regeneration continues to the stage of cross-striated muscle fibers, but if the muscle fibers are not contacted by nerves, prolonged atrophy sets in.[239] Work by Zhenevskaya[240,241] showed that motor, not sensory, innervation prevents the atrophy of regenerating muscle.

Noninnervated regenerating muscles in rats develop full sets of thin cross-striated muscle fibers, but they do not undergo the normal conversion into histochemically fast or slow muscle fiber types.[242] Also, the contractile properties of noninnervated regenerating muscles do not undergo final maturation, resulting in slower-than-normal speeds for a regenerating fast muscle and faster-than-normal speeds for a regenerating slow muscle.

Although the reinnervation of previously denervated human muscle has been studied extensively,[243] the innervation of regenerating muscle has been studied only rarely, mainly in human muscle grafts. The reinnervation pattern is one of the main factors determining the success or failure of a muscle graft.[244,245]

A number of earlier muscle grafting models employed "muscular neurotization" to stimulate nerve ingrowth into a graft.[246,247] According to this principle, which was defined by Erlacher,[248] a noninnervated muscle graft is placed next to an innervated muscle, which is often "freshened" by scraping its surface. Nerves from the innervated muscle were assumed to sprout and then grow into the muscle graft. More recent research, however, has shown that muscular neurotization takes place as the result of regeneration, not sprouting, of damaged nerve fibers from the innervated "feeder" muscle into the graft.[249]

The reinnervation of regenerating human muscle is being attempted by guiding regenerating nerve fibers from a graft to the muscle and ensuring their distribution within the graft. This is accomplished primarily by anastomoses between nerve stumps of graft and host. The formation of individual neuromuscular junctions in regenerating human muscle has not yet been studied.

Revascularization of Regenerating Muscle There is an intimate relationship between the microvasculature and individual myotubes. Successful muscle regeneration depends at least in part on reestablishing this relationship. The details of how this relationship is established in development or regeneration are lacking, although some facts are available. The microvascular pattern within mammalian skeletal muscle has been studied in several different species by light and transmission electron microscopy.[250–254] Although variations exist, the basic muscle vasculature is arranged in a typical pattern. Principal arteries and companion veins enter a muscle and branch repeatedly into smaller vessels with numerous arcades. The distributing arterioles then course through the muscle perimysium, often at oblique or right angles to the muscle fibers. The branches of the terminal arterioles decrease in diameter and then course parallel to the muscle fibers as they become precapillary arterioles. The decrease in lumen size also involves the smooth muscle cells encircling the distributing arterioles, which exist as a single cell layer that becomes more dispersed and irregularly oriented. These smaller-diameter muscle capillaries course in close contact with the individual muscle fibers in a distinctive geometric pattern.[255] Small venules are characterized by the reappearance of perivascular cells (pericytes) that increase in density with increased vessel diameter. Larger venules have smooth muscle cells and often accompany a similar-sized arteriole.[256]

There is a growing awareness that capillaries are not homogeneous throughout the body.[257] Using histochemical staining for alkaline phosphatase in the arterial portion of the muscle capillary bed and staining for dipeptidyl-peptidase IV (DPPIV) in the venous portion, differences in the lengths of arterial and venous segments were established for fast- and slow-twitch muscles. For example, in the rat, the arterial portion of the "white" part of the sternomastoid was the longest, with the DPPIV-positive venous segment composing only 8% of the capillary length. In contrast, the "red" part of the sternomastoid had 60% of the total capillary length positive to DPPIV.[256] In addition, there is a clear correlation between the number of capillaries per muscle fiber. Importantly, when fast muscles of rabbits were chronically stimulated at rates occurring naturally in slow muscles, the number of capillaries increased weeks before any changes in muscle metabolism could be detected.

The intimacy of capillaries and myotubes is apparently not random in that an organized spatial relationship exists between capillaries and muscle fiber nuclei.[255] It has been pointed out that extracts of skeletal muscles (and other tissues) have chemotactic activity for vascular endothelial cells in vitro; this suggests that the myotube could produce agents that directly attract ingrowing vascular elements.[258] It may be that both organelle distribution and metabolic activity provide organizational cues for capillary density and composition. In muscle regeneration, the reformation of myotubes must be coordinated with restoration of the normal capillary arrangement and density.

Function in Regenerating Muscle   At the level of the individual muscle fiber, skeletal muscle has the ability to regenerate with almost complete restoration of function.[259] However, studies on the contractile physiology of regenerating muscle are commonly carried out on entire regenerating muscles, and it is then necessary to extrapolate from whole-muscle data to obtain information at the level of the muscle fiber.

Regenerating muscle is capable of contracting measurably by the time the first cross-striations can be seen at the light microscopic level.[199,242] The first contractions are very weak and slow. In the rat the contractile properties of early regenerating muscle closely resemble the properties of neonatal muscle. With time (roughly six weeks in the rat and six months in the cat), certain of the contractile properties return to near-normal values. These parameters are contraction time, half-relaxation time, and maximum velocity of shortening.[242,260-262] In mature cat-muscle regenerates, the maximum tetanic tension is 25.6 N/cm^2. This is 90% of the control value (28.5 N/cm^2).[261]

Other parameters, in particular those related to the contractile tension generated by the entire muscle, do not usually return to normal (Fig. 6–13).[259] In rodents, these deficits are attributable to the smaller-than-normal cross-sectional area of individual muscle,[263] whereas in the cat they are associated with regenerated muscles that have a smaller-than-normal number of muscle fibers. With the proper innervation, regenerating muscles in the rat can be restored to almost normal contractile functions (Fig. 6–13), but in the cat functional deficits are the rule in regenerated muscles.

Regenerated muscle in cats is much more fatigable than normal muscle.[261] This correlates well with a decrease in succinate oxidase activity of whole-muscle homogenates[264] and a large decrease (29% of control) in maximum blood flow within the muscle. Autografted cat extensor digitorum longus muscles are recruited less frequently and less intensely than normal control muscles.[265] In contrast to cat muscles, small muscle regenerates in the rat are substantially less fatigable.[203] The basis for muscle fatigue is still poorly understood.

Laboratory Models of Muscle Regeneration   There are numerous methods of eliciting skeletal muscle regeneration in laboratory animals. Each method has certain advantages and disadvantages, which should be considered before selecting an experimental model.

For obtaining the regeneration of entire muscles, mincing and free nonvascularized grafting are the most common methods.[197,198,200,201,266] Both of these models are good for studying integrative activities between the regenerate and the rest of the body. By briefly soaking muscles in bupivacaine (Marcaine) prior to grafting, it is possible to obtain

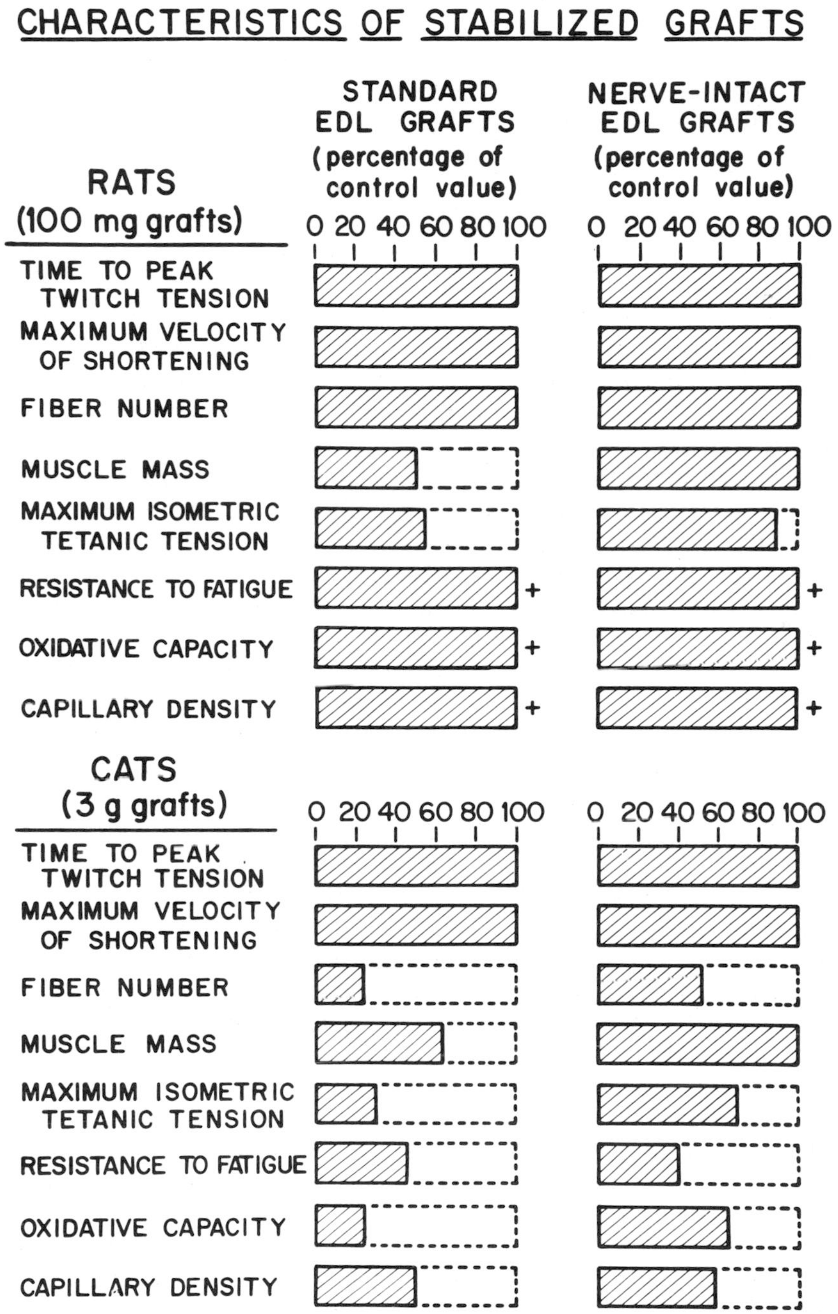

CHARACTERISTICS OF STABILIZED GRAFTS

Fig. 6–13 *Functional deficits in standard and nerve-intact grafts of extensor digitorum longus (EDL) muscles of rats and cats compared with values for control EDL muscles.*

pure populations of regenerating muscle fibers for studies of contractile physiology.[225] A disadvantage of both mincing and free grafts is that because of the initial ischemia, early stages of regeneration are asynchronous, as regeneration follows the ingrowth of blood vessels. In addition, free nonvascularized grafting and mincing are successful only on muscles weighing less than 3 g. A further disadvantage of mincing is the large amount of dense connective tissue that forms around the regenerate.

Nearly synchronous populations of degenerating muscle fibers can be obtained by injecting hot (60 to 70 C) Ringer's solution[267] or a variety of commonly used local anesthetics, such as bupivacaine, lidocaine or mepivacine.[268-272] Offsetting the advantage of synchrony is the fact that with a single injection of a local anesthetic it is very difficult to obtain lesions consistent in distribution and size. This reduces the value of the single-injection model for biochemical or physiologic studies.

The injection of certain snake and spider venoms results in widespread damage to muscle fibers in both humans and experimental animals.[273-276] These toxins act rapidly upon the membranes of muscle fibers and apparently spare the satellite cells. Unless the toxins also induce hemorrhage and subsequent necrosis, degeneration of damaged myofibers is rapid, and the uniform regeneration of new muscle occurs in a pattern reminiscent of that described for local anesthetic damage to muscle. The key for rapid and successful regeneration seems to be the preservation of reasonably intact basal laminae surrounding the infused muscle fibers. Certain organophosphate pesticides also produce muscle fiber necrosis,[277] but the lesions appear to be irregular and are more useful as an index of toxicity than as a regeneration model.

Localized heat and cold produce muscle degeneration and regeneration.[278-281] Extreme cold kills off all cells in a muscle and no regeneration ensues. An unusual model of muscle damage caused by a physical stimulus is the degeneration and regeneration of rat extraocular muscles after exposure to incandescent radiant energy.[282] Control experiments eliminated the heat component as the major effective agent, but the mechanism of damage is poorly understood.

Temporary ischemia, induced by either a tourniquet or by direct vascular ligation, has long been used as a model to induce muscle degeneration and regeneration.[283-285] Most investigators have found that muscle can tolerate ischemia for up to two hours without complete destruction of the muscle fibers, but certain ultrastructural changes, such as mitochondrial swelling and glycogen depletion, occur.[286] Ischemia lasting longer than two hours is followed by the complete muscle breakdown and regeneration. In smaller muscles the pattern of postischemic recovery is similar to that seen in a free muscle graft,[287] but larger muscles may fall into a state of liquifying necrosis.

Over the years, a number of investigators have used various types of mechanical trauma, such as crushing, muscle injury, or simple transections, to induce muscle regeneration.[288-292] These models are normally most useful for testing the destructive effects of a specific form of trauma that might be encountered by an individual, but normally they are too difficult to reproduce to provide consistent lesions for experimental analysis. Karpati and Carpenter[293] have used focal micropuncture of individual muscle fibers to advantage in studying the early intracellular response to trauma and later repair.

Exercise-induced muscle injury in laboratory animals has often been used to mimic human muscle injury in athletes. The experimental models can be broken down into several main types. Most common are treadmill models, leading to exhaustive exercise. Uphill running requires a long time to produce lesions.[195,294] Recently it has been suggested that downhill running involving lengthening (eccentric) contractions leads to more severe muscle injury than does uphill running.[295,296] Giddings and associates[297] presented evidence interpreted as showing muscle fiber necrosis and regeneration following prolonged weightlifting in the cat.

The regenerating amphibian limb offers a totally different model of muscle regeneration.[298,299] Here, the reconstitution of the entire complement of muscles in the adult limb can be studied, as the process recapitulates the process of muscle formation in the embryo. This epimorphic mode of regeneration differs in many respects from the regeneration of isolated muscles in the mammalian limb.[204,205]

Muscle Regeneration in Humans

Although the degeneration and regeneration of human muscle fibers were accurately described and illustrated as early as 1865,[300] the notion that human muscle does not regenerate crept into the English medical literature early in this century and has persisted with surprising tenacity. Human muscle regeneration follows a number of stimuli, which can be placed into several general categories.

A number of disease processes include muscle fiber breakdown and regeneration as part of the overall spectrum of pathology.[301] Myonecrosis followed by regeneration has been recognized for over a century in certain infectious diseases, such as typhus and diphtheria. The weakness and gradual recovery of muscle function in patients with polymyositis corresponds to the muscle breakdown and regeneration of muscle fibers.[302,303] A roughly similar picture is seen in patients with paroxysmal myoglobinuria.[176,304] Although muscle regeneration is a prominent process in the muscular dystrophies, it is difficult to study regeneration because of the chronic nature of the disease. This leads to nonsynchronous lesions and atypical regeneration, including complex branching patterns of the regenerated muscle fibers.[305-308] An interesting experiment for the study of regenerating human dystrophic muscle is to graft pieces into athymic nude mice.[309,310]

Trauma is a major source of muscle damage in humans, but there have been surprisingly few studies on the response of human skeletal muscle to direct mechanical trauma. If results of studies on experimental animals can be extrapolated to human beings,[311-313] one can assume that transection is followed by budding. If the two cut edges are allowed to spread, the likely result is the attenuation of any attempt at regeneration because the muscle buds become embedded in dense connective tissue that forms in the incision. On the other hand, if the edges of the transected muscle are closely apposed, intermingling and some fusion of buds from opposite sides of the wound can occur. Nevertheless, significant numbers of muscle fibers on the side of an incision opposite the motor end plate are likely to remain deinnervated. A major clinical problem in dealing with transversely incised muscles is providing a functional nerve supply to muscle fibers on both sides of the lesion.

Allbrook and associates[314] studied human muscle fiber degeneration and regeneration in patients with fractured limb bones, mainly the tibia and fibula. As early as six hours after injury, marked fragmentation of injured muscle fibers had already begun. During the first ten days after trauma, muscle fibers containing large numbers of phagocytic cells were noted. Regenerating myotubes were seen 13 days after injury, and cross-striated muscle fibers appeared by 18 days. Longterm biopsy specimens were difficult to interpret, but they did show excessive depositions of scar tissue.

Exhaustive exercise can produce localized muscle fiber necrosis in both experimental animals and humans. Hikida and associates[315] studied biopsy specimens of the gastrocnemius muscles of marathon runners for the first week after the race and found widespread structural signs of muscle-fiber degeneration. Their data, as well as those of Knochel,[316] suggested that muscle soreness, myoglobinuria, and rhabdomyolysis might be part of a complex of degenerative sequelae following strenuous exercise.

In humans, excessive amounts of corticosteroids, resulting from Cushing's disease or from high doses of exogenous cortisone, cause muscle weakness, atrophy, and necrosis of muscle fibers.[301] Studies on experimental animals have confirmed the muscle damage. Steiss[317] reported that dexamethasone administered over a period of one to three weeks interferes with muscle regeneration.

The myotoxicity of local anesthetics, although well-documented in experimental animals, has been little studied in humans. Yagiela and associates[318] examined specimens of human muscle shortly after injection of lidocaine and epinephrine and noted degenerative changes consistent with muscle-fiber breakdown. Carlson and associates examined local anesthetic effects on monkey thumb and extraocular muscles and found considerable degeneration followed by regeneration of muscle fibers (unpublished data).

Muscle Regeneration Following Specific Injuries

The principles of muscle regeneration discussed previously apply to all muscle injuries, but the results of attempted regeneration vary with the specific injury and the treatment. Contraction injuries, muscle lacerations, contusions, and ischemic injury are among the most common clinical problems.

Contraction-Induced Injury Recent studies of humans after eccentric exercise,[319] and rodents after downhill running,[320] indicate that exercise injures skeletal muscle fibers. McCully and Faulkner[321,322] studied contractions of whole muscles of mice isolated in situ and activated to 85% of maximum isometric tetanic force. They observed no evidence of overt injury during the five days after isometric or shortening contractions, but three to five days following the lengthening contractions, maximum force development decreased by 60%, and 37% of the fibers had degenerated completely. When animals perform strenuous eccentric exercise volitionally, less than 5% of the fibers in the contracting muscles show overt signs of injury when examined by histologic techniques.[319,320] Despite the small number of fibers showing overt injury, ultrastructural studies indicate that as many as 50% of the fibers in muscle biopsy specimens show evidence of disruption of myofilaments and other organelles.[319]

Some athletes have assumed that training is more effective if some injury to muscle fibers occurs. The role of injury and regeneration of muscle fibers in the adaptive process has been studied by moving muscles to an atypical site and innervating them by an atypical nerve, such as moving a slow muscle to a fast site and nerve.[323,324] Regenerating fibers adapt more rapidly than surviving fibers to the atypical site and innervation, but when the adaptive response stabilizes, the magnitude of the adaptation is not significantly different between regenerating and surviving fibers. The more rapid adaptation of regenerating fibers is observed for both slow-to-fast and fast-to-slow adaptations. Whether or not these observations can be applied to the adaptations associated with endurance and strength training is not clear. Furthermore, only a few variables have been studied and even these vary as to the effect of regeneration on the rate of adaptation. Much more research will be required to resolve these issues.

Muscle Laceration   Lacerations in skeletal muscle usually result from direct trauma by a sharp object. Surgical exposures also sometimes require division of muscles. For normal function following lacerations, muscle must regenerate across the repair site, and muscle tissue denervated by laceration of the muscle belly and separated from the intramuscular nerve supply must be reinnervated. Clinical experience has shown that functional recovery following muscle laceration is rarely complete although partial recovery is possible.

An investigation of complete and partial lacerations of rabbit skeletal muscles[325] demonstrated that following complete laceration and suture repair, the muscle fragments healed primarily by dense connective tissue scar. A small number of myotubes penetrated the scar tissue, but good regeneration of muscle tissue across the laceration was not seen. The muscle fragment isolated from the motor point had histologic findings of denervation. Recovery of muscle function was evaluated by measuring the isometric muscle tension following motor nerve stimulation. Muscles lacerated near the midbelly recovered approximately 50% of their ability to produce tension and could shorten about 80% of their normal amount. Partial lacerations recovered better function.

A recent clinical study[326] showed good correlation with these experiments. Muscle lacerations were repaired by connecting the proximal portion of the lacerated muscle to the distal portion by tendon grafts. At a mean follow-up period of 14 months, approximately 40% of the mean grip strength was recovered. Over half of the muscles achieved grade 4 or grade 5 strength by manual muscle testing.

Muscle Contusion   Muscle contusions usually result from nonpenetrating blunt injury. They occur frequently in accidents and sports and can cause significant disability and pain. The best method of treatment has not been clearly defined, but animal experiments provide some insight into the natural history of the injury. Järvinen and Sorvari[288] produced consistent lesions in rat gastrocnemius muscles. Soon after injury, an intense inflammatory reaction and hematoma formation occurred. Later, scar formation consisting of dense connective tissue with variable amounts of muscle regeneration was seen. Mobilized muscle was compared to immobilized muscle in the recovery process. More of an inflammatory reaction was seen in mobilized muscle, but it disappeared more rapidly with more scar formation than in immobilized muscle.[289] The speed of tissue repair was directly related to vascular ingrowth during the repair process.[327] Biomechanical testing showed faster recovery of tensile strength in mobilized muscles.[328] The quality of repair following blunt trauma varied with age. Young rats demonstrated a more intense inflammatory reaction, hematoma formation, and scar than older rats.[329] Incorporation of radioactively labeled proline demonstrated synthesis of extracellular connective tissue by two days following trauma, with most intensive synthesis occurring between days 5 and 21.[330] The appearance and distribution of collagen types I, III, IV, and V and of fibronectin have also been studied by histologic and immunofluorescent techniques.[331]

Severe blunt injury to muscle may result in bone formation within the muscle referred to as myositis ossificans. A prospective study of patients with quadriceps hematoma showed subsequent myositis ossificans in approximately 20% of cases.[332] The new bone formed after blunt trauma can be contiguous with normal bone (periosteal) or completely free of any connection with bone (heterotopic). Experimentally, some animals including rabbits appear to develop primarily periosteal new bone.[333] However, in other animals, such as sheep, both periosteal new bone and heterotopic bone occur.[334] Periosteal reactive new bone developed in all sheep receiving blunt trauma to the anterior thigh muscles, while heterotopic bone occurred in 17% of the sheep. Multiple episodes of injury and hematoma formation increased the likelihood of formation of heterotopic bone. Clinically, both types of new bone are present in humans, and it will be important in future studies to distinguish the specific types of posttraumatic bone formation.

Compartment Syndromes and Ischemia   There has been intense clinical and experimental interest in compartment syndromes. In these syndromes, increased pressure within a closed compartment bounded by bone and fascia damages the muscle. The increased pressure directly injures nerves and diminishes tissue perfusion, which may lead to an ischemic injury.[335] Common causes of increased compartment pressures include bleeding into a compartment, edema following partial or temporary ischemia, and crushing injuries.

A satisfactory method to measure tissue pressure has been central in the development of this field of study. Whitesides and associates[336] used a three-way stopcock connecting a manometer, a syringe, and an intramuscular line. Electronic pressure transducers are now in wide use. An intramuscular cannula or needle connects the fluid compartment to the transducer. Various techniques have been used to prevent muscle tissue from occluding the lumen of the cannula, including the use of a suture or thread used as a wick.[337] A cannula with multiple fenestrations or slits is also effective.[338]

The consequences of increased intracompartmental pressure are related to the magnitude and duration of the pressure. The deterioration of function in nerves passing through a compartment can be quantified as the nerve conduction velocity decreases.[339] The potential damage to muscle tissue is related to the pressure at which muscle capillary perfusion is prevented. The intracompartmental pressure can be significantly less than arterial pressure and still cause ischemia. Pulses distal to the compartment may still be present, in contrast to ischemia caused by arterial occlusion.

The critical pressure at which perfusion is compromised is an area of significant controversy. Determination of a critical pressure threshold is the central factor guiding surgical intervention. Release of the fascia around a compartment allows the compartment volume to expand and the pressure to fall, thereby preventing further ischemic damage, which often saves the function of the muscles or the entire limb. The normal capillary pressure is 30 mm Hg. It has been recommended that when compartment pressures higher than this level exist for a period of time, fasciotomy should be performed.[340] Higher critical pressures have been proposed by others, who believed that the capillary pressure should rise with the onset of increased compartment pressures and venous pressures. A clinical study suggested that only pressures greater than 45 mm Hg were associated with significant compartment syndromes.[341] It has also been argued that the critical pressure necessitating fasciotomy should be related to the diastolic blood pressure. Pressures within 10 to 30 mm Hg of diastolic pressure have been proposed as the threshold for surgery.[336]

Unfortunately, intramuscular pressure determination may not be an accurate predictor of tissue metabolic state or an accurate measure of the degree of muscle ischemia. Recently, the technique of phosphorus nuclear magnetic resonance spectroscopy has been applied to the study of normal muscle as well as to exercising and ischemic muscle. The technique is noninvasive and provides an excellent measure of the cellular metabolic state. This technique was applied to a study of compartment syndromes in the canine hindlimb.[342] The study supported the concept that the critical pressure threshold for metabolic derangement was related to the mean arterial blood pressure. At differences between compartment pressure and mean arterial blood pressure of less than 30 mm Hg in normal muscle and less than 40 mm Hg in traumatized muscle, the tissue began to employ anaerobic energy sources because blood flow and oxygen delivery were insufficient. However, at greater pressure differences there was no metabolic compromise even if the absolute compartment pressure exceeded 60 mm Hg.

This technique has also been used to study the effect of tourniquet application.[343] After varying periods of tourniquet ischemia up to three hours, little change was noted in cellular levels of ATP. However, phosphocreatine levels responded very quickly to ischemia and reperfusion. Reperfusion for five minutes allowed adequate time for restoration of the bioenergetic state of the muscle tissue. Cellular damage, as measured by creatine phosphokinase release and muscle biopsy, depended on the longest continuous period of ischemia.

Methods of Improving the Functional Mass of Regenerating Muscle

When larger masses of muscle are involved, it is the rule, rather than the exception, that the functional mass of the regenerate is less than that of the original muscle. Just as there are many possible reasons for the functional deficits, there are numerous ways in which investigators have been able to improve the functional mass of regenerating muscle. Because of the relatively greater ease of quantitation, most of the attempts at improvement have been made on grafted or minced muscles.

In situations involving the breaking of connections between the nerve supply and damaged muscle, deficiencies of reinnervation represent a major cause of a reduced functional mass of regenerating muscle. Reports of improvement have involved several approaches. Hall-Craggs and Brand[344] crushed the nerves in a muscle bed before a muscle graft was placed into that site in order to stimulate the faster ingrowth of nerve fibers into the graft. They reported some improvement in the quality of the regenerates. An experimental means by which small muscle grafts can be restored to normal mass and nearly normal function is the nerve-intact graft.[203,217] In this model the motor nerve to a muscle graft is not severed. Although the nerve terminals become ischemic and die back from the graft, functional motor end plates in the graft are reformed earlier (seven to eight days) than in standard free muscle grafts (21 to 23 days). Further testing, however, showed that the early return of nerve fibers into the regenerate is not the key to success; if functional reinnervation is delayed until four weeks by crushing of the sciatic nerve, the restoration of normal mass still occurs.[345] These results point to the broader distribution of regenerating nerve fibers throughout the graft as the most likely reason for the success of nerve-intact grafts. The results of nerve anastomosis experiments on cats also support this viewpoint. When the proximal nerve stump is anastomosed to a distal stump leading into a muscle graft, the regenerated muscle functions much better than it does if it is supplied by spontaneous reinnervation.[262]

Early in the history of transplantation, denervation of a muscle several weeks before it was grafted (predenervation) was considered essential for the success of the graft.[247,346] Further work has shown that although predenervation of a muscle produces substantial increases in the number of muscle fibers that survive free grafting,[225] predenervation produces no significant difference in the quality of the mature regenerate.

In dealing with larger masses of damaged muscle, innervation is not the sole factor determining the success of regeneration. The larger the mass of muscle, the more important revascularization becomes. The general rule is that spontaneous revascularization is not sufficient to support the regeneration of masses of muscle larger than 3 g. There have been few attempts to increase the normal rate or amount of spontaneous revascularization of regenerating muscle, although Phillips[347] has demonstrated the presence of angiogenic activity in ischemic muscle. Surgeons employing whole-muscle grafts to replace large masses of muscle now routinely use vascular anastomosis to maintain an uninterrupted blood supply to the graft. This eliminates the ischemic phase in an early graft and results in the survival of muscle fibers rather than their breakdown and subsequent regeneration.[348] Grafting of large vascular-anastomosed muscles has been used in the treatment of the deficit caused by Volkmann's ischemic contracture and severe soft-tissue trauma to limbs.[349]

The response of muscle grafts to various training regimens has been examined by White.[350] Chronic running exercise led to improvements in mass oxidative capacity, glycogen concentration, and total protein content of regenerating muscles.[351,352] Although not a practical model for human use, ablation of synergistic muscles resulted in an improvement of the functional mass of regenerating muscles.[323,353]

In some cases excessive deposition of connective tissue can interfere with muscle regeneration, especially when the muscle has been mechanically traumatized. Although surprisingly few studies have been directed at this problem, one study[354] used induced lathyrism to inhibit collagen formation in regenerating muscle.

Most somatic muscles appear to be minimally responsive to anabolic steroids during regeneration, but the hormonally sensitive levator ani muscle of the rat undergoes significant changes in response to the administration or deprivation of testosterone.[291] Transplantation of the levator ani into the leg has shown a myogenic hormonal responsiveness rather than an effect related to the bed or innervation of the levator ani.[224,355]

In 1979 both Jones[356] and Lipton and Schultz[357] reported the successful integration of implanted mononuclear myogenic cells into skeletal muscle fibers. This type of procedure has been used with some success to introduce cells into regenerating dystrophic muscles to increase their functional success.[358]

Although muscle in aged animals regenerates, the success of the

2. Delivering and maintaining the population of myoblasts in a region where regeneration is required.

3. Providing a proper substrate or matrix for the proliferation, orientation and differentiation of myoblasts.

4. Initiating and facilitating the growth of blood vessels into avascular regions or into areas of newly implanted material.

5. Connecting regenerating muscle to the appropriate origin and insertion points and providing a framework for the transmission of mechanical tension.

6. Providing adequate pathways for the regrowth of motor nerve fibers to all regenerating muscle fibers.

7. Ensuring appropriate proprioceptive feedback from muscle spindles and Golgi tendon organs.

Acknowledgments

Supported in part by NIH grants PO1 DE 07687, EY 05813, and AG 06157.

References

1. Chevallier A, Kieny M, Mauger A: Limb-somite relationship: Origin of the limb musculature. *J Embryol Exp Morphol* 1977;41:245–258.
2. Christ B, Jacob HJ, Jacob M: Experimental analysis of the origin of the wing musculature in avian embryos. *Anat Embryol* 1977;150:171–186.
3. Chevallier A: Role of the somitic mesoderm in the development of the thorax in bird embryos: II. Origin of thoracic and appendicular musculature. *J Embryol Exp Morphol* 1979;49:73–88.
4. Miller JB, Stockdale FE: Developmental origins of skeletal muscle fibers: Clonal analysis of myogenic cell lineages based on expression of fast and slow myosin heavy chains. *Proc Natl Acad Sci USA* 1986;83:3860–3864.
5. Miller JB, Stockdale FE: Developmental regulation of the multiple myogenic cell lineages of the avian embryo. *J Cell Biol* 1986;103:2197–2208.
6. Stockdale FE, Miller JB: The cellular basis of myosin heavy chain isoform expression during development of avian skeletal muscles. *Dev Biol* 1987;123:1–19.
7. Kelly AM, Zacks SI: The histogenesis of rat intercostal muscle. *J Cell Biol* 1969;42:135–153.
8. Narusawa M, Fitzsimons RB, Izumo S, et al: Slow myosin in developing rat skeletal muscle. *J Cell Biol* 1987;104:447–459.
9. Ho JFK, Hughes S, Chow C, et al: Immunocytochemical and electrophoretic analyses of changes in myosin gene expression in cat posterior temporalis muscle during postnatal development. *J Muscle Res Cell Motil,* in press.
10. Lance-Jones C, Landmesser L: Pathway selection by chick lumbosacral motoneurons during normal development. *Proc R Soc London [Biol]* 1981;214:1–18.
11. Lance-Jones C, Landmesser L: Pathway selection by embryonic chick motoneurons in an experimentally altered environment. *Proc R Soc London Ser B* 1981;214:19–52.
12. Lash JW, Holtzer H, Swift H: Regeneration of mature skeletal muscle. *Anat Rec* 1957;128:679–698.

13. Stockdale FE, Holtzer H: DNA synthesis and myogenesis. *Exp Cell Res* 1961;24:508–520.

14. Okazaki K, Holtzer H: An analysis of myogenesis *in vitro* using fluorescein-labeled antimyosin. *J Histochem Cytochem* 1965;13:726–739.

15. Holtzer H, Weintraub H, Mayne R, et al: The cell cycle, cell lineages, and cell differentiation. *Curr Top Dev Biol* 1972;7:229–256.

16. Holtzer H, Bischoff R: Mitosis and myogenesis, in Briskey E, Cassens R, Marsh B (eds): *The Physiology and Biochemistry of Muscle as a Food.* Madison, University of Wisconsin, 1970, vol 2, pp 29–51.

17. Konigsberg IR: The embryonic origin of muscle, in Engel AG, Banker BQ (eds): *Myology.* New York, McGraw-Hill, 1986, vol 1, pp 39–71.

18. White NK, Bonner PH, Nelson DR, et al: Clonal analysis of vertebrate myogenesis: IV. Medium-dependent classification of colony-forming cells. *Dev Biol* 1975;44:346–361.

19. Rutz R, Haney C, Hauschka S: Spatial analysis of limb bud myogenesis: A proximodistal gradient of muscle colony-forming cells in chick embryo leg buds. *Dev Biol* 1982;90:399–411.

20. Konieczny SF, Baldwin AS, Emerson CP: Myogenic determination and differentiation of 10T1/2 cell lineages: Evidence for simple regulatory system, in Emerson CP, Fischman DA, Nadal-Ginard B, et al (eds): *Molecular Biology of Muscle Development.* New York, Alan R Liss, 1986, pp 21–33.

21. Hauschka S, Lim R, Clegg C, et al: Trophic influences on developing muscle, in Strohman RC, Wolf S (eds): *Gene Expression in Muscle.* New York, Plenum Press, 1985, pp 113–122.

22. Ozawa E, Hagiwara Y: Degeneration of large myotubes following removal of transferrin from culture medium. *Biomed Res* 1982;3:16–23.

23. Matsuda R, Bandman E, Strohman RC: There is selective accumulation of growth factor chicken muscle: I. Transferrin accumulation in adult anterior latissimus dorsi. *Dev Biol* 1984;103:267–275.

24. Buckley PA, Konigsberg IR: Myogenic fusion and the duration of the post-mitotic gap (G_1). *Dev Biol* 1974;37:193–212.

25. Nameroff M, Munar E: Inhibition of cellular differentiation by phospholipase C: II. Separation of fusion and recognition among myogenic cells. *Dev Biol* 1976;49:288–293.

26. Shainberg A, Yagil G, Yaffe D: Alterations of enzymatic activities during muscle differentiation *in vitro*. *Dev Biol* 1971;25:1–29.

27. Emerson CP Jr, Beckner SK: Activation of myosin synthesis in fusing and mononucleated myoblasts. *J Mol Biol* 1975;93:431–447.

28. Vertel BM, Fischman DA: Myosin accumulation in mononucleated cells of chick muscle cultures. *Dev Biol* 1976;48:438–446.

29. Couch CB, Strittmatter WJ: Rat myoblast fusion requires metalloendoprotease activity. *Cell* 1983;32:257–265.

30. Bandman E, Walker CR, Strohman RC: Diazepam inhibits myoblast fusion and expression of muscle specific protein synthesis. *Science* 1978;200:559–561.

31. Ontell M, Kozeka K: Organogenesis of the mouse extensor digitorum longus muscle: A quantitative study. *Am J Anat* 1984;171:149–161.

32. Endo T, Nadal-Ginard B: Three types of muscle-specific gene expression in fusion-blocked rat skeletal muscle cells: Translational control in EGTA-treated cells. *Cell* 1987;49:515–526.

33. Huxley HE: The double array of filaments in cross-striated muscle. *J Biophys Biochem Cytol* 1957;3:631–648.

34. Stossel TP: Contribution of actin to the structure of the cytoplasmic matrix. *J Cell Biol* 1984;99:15s–21s.

35. Korn ED: Actin polymerization and its regulation by proteins from nonmuscle cells. *Physiol Rev* 1982;62:672–737.

36. Garrels JI: Changes in protein synthesis during myogenesis in a clonal cell line. *Dev Biol* 1979;73:134–152.

37. O'Neill MC, Stockdale FE: A kinetic analysis of myogenesis in vitro. *J Cell Biol* 1972;52:52–65.

38. Paterson B, Strohman RC: Myosin synthesis in cultures of differentiating chick embryo skeletal muscle. *Dev Biol* 1972;29:113–138.

39. Devlin RB, Emerson CP Jr: Coordinate regulation of contractile protein synthesis during myoblast differentiation. *Cell* 1978;13:599–611.

40. Yablonka Z, Yaffe D: Synthesis of polypeptides with the properties of myosin light chains directed by RNA extracted from muscle cultures. *Proc Natl Acad Sci USA* 1976;73:4599–4603.

41. Heywood SM, Thibault MC, Siegel E: Control of gene expression in muscle development, in Dowben RM, Shay JW (eds): *Cell and Muscle Motility*. New York, Plenum Press, 1983, vol 3, pp 157–193.

42. Devlin RB, Emerson CP Jr: Coordinate accumulation of contractile protein mRNAs during myoblast differentiation. *Dev Biol* 1979;69:202–216.

43. Gunning P, Ponte P, Blau H, et al: α-Skeletal and α-cardiac actin genes are coexpressed in adult human skeletal muscle and heart. *Mol Cell Biol* 1983;3:1985–1995.

44. Mayer Y, Czosnek H, Zeelon PE, et al: Expression of the genes coding for the skeletal muscle and cardiac actins in the heart. *Nucl Acids Res* 1984;12:1087–1100.

45. Paterson BM, Eldridge JD: α-Cardiac actin is the major sarcomeric isoform expressed in embryonic avian skeletal muscle. *Science* 1984;224:1436–1438.

46. Whalen RG, Sell SM, Butler-Browne GS, et al: Three myosin heavy-chain isozymes appear sequentially in rat muscle development. *Nature* 1981;292:805–809.

47. Bandman E: Myosin isoenzyme transitions in muscle development, maturation, and disease. *Int Rev Cytol* 1985;97:97–131.

48. Crow MT, Stockdale FE: The developmental program of fast myosin heavy chain expression in avian skeletal muscles. *Dev Biol* 1986;118:333–342.

49. Fischman DA: Myofibrillogenesis and the morphogenesis of skeletal muscle, in Engel AE, Banker BQ (eds): *Myology*. New York, McGraw-Hill, 1986, vol 1, pp 5–37.

50. Buckingham ME, Minty AJ: Contractile protein genes, in Maclean N, Gregory SP, Flavell RA (eds): *Eukaryotic Genes: Their Structure, Activity and Regulation*. London, Butterworths, 1983, pp 365–395.

51. Masaki T, Yoshizaki C: Differentiation of myosin in chick embryos. *J Biochem* 1974;76:123–131.

52. Lowey S, Benfield PA, LeBlanc DD, et al: Myosin isozymes in avian skeletal muscles: I. Sequential expression of myosin isoenzymes in developing chicken pectoralis muscles. *J Muscle Res Cell Motil* 1983;4:695–716.

53. Whalen RG, Schwartz K, Bouveret P, et al: Contractile protein enzymes in muscle development: Identification of an embryonic form of myosin heavy chain. *Proc Natl Acad Sci USA* 1979;76:5197–5201.

54. Bader D, Masaki T, Fischman DA: Immunochemical analysis of myosin heavy chain during avian myogenesis in vivo and in vitro. *J Cell Biol* 1982;95:763–770.

55. Winkelmann DA, Lowey S, Press JL: Monoclonal antibodies localize changes on myosin heavy chain isozymes during avian myogenesis. *Cell* 1983;34:295–306.

56. Bandman E: Continued expression of neonatal myosin heavy chain in adult dystrophic skeletal muscle. *Science* 1985;227:780–782.

57. Umeda PK, Sinha AM, Jakovcic S, et al: Molecular cloning of two fast myosin heavy chain cDNAs from chicken embryo skeletal muscle. *Proc Natl Acad Sci USA* 1981;78:2843–2847.

58. Kavinsky CJ, Umeda PK, Sinha AM, et al: Cloned mRNA sequences for two types of embryonic myosin heavy chains from chick skeletal muscle: I. DNA and derived amino acid sequence of light meromyosin. *J Biol Chem* 1983;258:5196–5205.

59. Wydro RM, Nguyen HT, Gubits RM, et al: Characterization of sarcomeric myosin heavy chain genes. *J Biol Chem* 1983;258:670–678.

60. Rozek CE, Davidson N: Drosophila has one myosin heavy-chain gene with three developmentally regulated transcripts. *Cell* 1983;32:23–34.

61. Davidson N, Falkenthal S, Rozek CE: Alternative splicing as a mechanism for generation of muscle protein diversity in *Drosophila*, in Emerson CP, Fischman DA, Nadal-Ginard B, et al (eds): *Molecular Biology of Muscle Development*. New York, Alan R Liss, 1986, pp 411–421.

62. Izumo S, Nadal-Ginard B, Mahdavi V: All members of the MHC multigene family respond to thyroid hormone in a highly tissue-specific manner. *Science* 1986;231:597–600.

63. Mahdavi V, Strehler EE, Periasamy D, et al: Sarcomeric myosin heavy chain gene family: Organization and pattern of expression, in Emerson CP, Fischman DA, Nadal-Ginard B, et al (eds): *Molecular Biology of Muscle Development*. New York, Alan R Liss, 1986, pp 345–361.

64. Nabeshima Y, Fujii-Kuriyama Y, Muramatsu M, et al: Alternative transcription and two modes of splicing results in two myosin light chains from one gene. *Nature* 1984;308:333–338.

65. Reinach FC, Fischman DA: Recombinant DNA approach for defining the primary structure of monoclonal antibody epitopes: The analysis of a conformation-specific antibody to myosin light chain 2. *J Mol Biol* 1985;181:411–422.

66. Periasamy M, Strehler EE, Garfinkel LI, et al: Fast skeletal muscle myosin light chains 1 and 3 are produced from a single gene by a combined process of differential RNA transcription and splicing. *J Biol Chem* 1984;259:13595–13604.

67. Whalen RG, Sell SM: Myosin from fetal hearts contains the skeletal muscle embryonic light chain. *Nature* 1980;286:731–733.

68. Obinata T, Masaki T, Takano H: Types of myosin light chains present during the development of fast skeletal muscle in chick embryo. *J Biochem* 1980;87:81–88.

69. Reinach FC, Masaki T, Shafiq S, et al: Isoforms of C-protein in adult chicken skeletal muscle: Detection with monoclonal antibodies. *J Cell Biol* 1982;95:78–84.

70. Grove BK, Cerny L, Perriard J-C, et al: Myomesin and M-protein: Expression of two M-band proteins in pectoral muscle and heart during development. *J Cell Biol* 1985;101:1413–1421.

71. Vanderkerckhove J, Weber K: The complete amino acid sequence of actins from bovine aorta, bovine heart, bovine fast skeletal muscle, and rabbit slow skeletal muscle: A protein-chemical analysis of muscle actin differentiation. *Differentiation* 1979;14:123–133.

72. Medford RM, Nguyen HT, Destree AT, et al: A novel mechanism of alternative RNA splicing for the developmentally regulated generation of troponin T isoforms from a single gene. *Cell* 1984;38:409–421.

73. Breitbart RE, Nguyen HT, Medford RM, et al: Intricate combinatorial patterns of exon splicing generate multiple regulated troponin T isoforms from a single gene. *Cell* 1985;41:67–82.

74. Abe H, Komiya T, Obinata T: Expression of multiple troponin T variants in neonatal chicken breast muscle. *Dev Biol* 1986;118:42–51.

75. Schachat FH, Bronson DD, McDonald OB: Heterogeneity of contractile proteins: A continuum of troponin-tropomyosin expression in mammalian skeletal muscle. *J Biol Chem* 1985;260:1108–1113.

76. Lazarides E, Capetanaki YG: The striated muscle cytoskeleton: Expression and assembly in development, in Emerson CP, Fischman DA, Nadal-Ginard B, et al (eds): *Molecular Biology of Muscle Development*. New York, Alan R Liss, 1986, pp 749–772.

77. Maher PA, Cox GF, Singer SJ: Zeugmatin: A new high molecular weight protein associated with Z lines in adult and early embryonic striated muscle. *J Cell Biol* 1985;101:1871–1883.

78. Endo T, Masaki T: Differential expression and distribution of chicken skeletal- and smooth-muscle-type α-actinins during myogenesis in culture. *J Cell Biol* 1984;99:2322–2332.

79. Capetanaki YG, Ngai J, Lazarides E: Characterization and regulation in the expression of a gene coding for the intermediate filament protein desmin. *Proc Natl Acad Sci USA* 1984;81:6909–6913.

80. Gomer RH, Lazarides E: The synthesis and development of filamin in chicken skeletal muscle. *Cell* 1981;23:524–532.

81. Wang K, McClure J, Tu A: Titin: Major myofibrillar components of striated muscle. *Proc Natl Acad Sci USA* 1979;76:3698–3702.

82. Wang K, Williamson CL: Identification of an N_2 line protein of striated muscle. *Proc Natl Acad Sci USA* 1980;77:3254–3258.

83. Gauthier GF: Skeletal muscle fiber types, in Engel AG, Banker BQ (eds): *Myology*. New York, McGraw-Hill, 1986, vol 1, pp 255–284.

84. Rowlerson A, Pope B, Murray J, et al: A novel myosin present in cat jaw-closing muscles. *J Muscle Res Cell Motil* 1981;2:415–438.

85. Wieczorek DF, Periasamy M, Butler-Browne GS, et al: Co-expression of multiple myosin heavy chain genes, in addition to a tissue-specific one, in extraocular musculature. *J Cell Biol* 1985;101:618–629.

86. Robbins J, Freyer GA, Chisholm D, et al: Isolation of multiple genomic sequences coding for chicken myosin heavy chain protein. *J Biol Chem* 1982;257:549–556.

87. Czosnek H, Nudel U, Shani M, et al: The genes coding for the muscle contractile proteins, myosin heavy chain, myosin light chain 2, and skeletal muscle actin are located on three different mouse chromosomes. *EMBO J* 1982;1:1299–1305.

88. Robert B, Barton P, Minty A, et al: Investigation of genetic linkage between myosin and actin genes using an interspecific mouse back-cross. *Nature* 1985;314:181–183.

89. Weydert A, Daubas P, Lazaridis I, et al: Genes for skeletal muscle myosin heavy chains are clustered and are not located on the same mouse chromosome as a cardiac myosin heavy chain gene. *Proc Natl Acad Sci USA* 1985;82:7183–7187.

90. Huxley HE: Electron microscope studies on the structure of natural and synthetic protein filaments from striated muscle. *J Mol Biol* 1963;7:281–308.

91. Pepe FA, Drucker B, Chowrashi PK: The myosin filament: XI. Filament assembly. *Prep Biochem* 1986;16:99–132.

92. Fischman DA: The synthesis and assembly of myofibrils in embryonic muscle. *Curr Top Dev Biol* 1970; 5:235–280.

93. Fischman DA: An electron microscope study of myofibril formation in embryonic chick skeletal muscle. *J Cell Biol* 1967;32:557–575.

94. Dlugosz AA, Antin PB, Nachmias VT, et al: The relationship between stress fiber-like structures and nascent myofibrils in cultured cardiac myocytes. *J Cell Biol* 1984;99:2268–2278.

95. Isaacs WB, Fulton AB: Cotranslational assembly of myosin heavy chain in developing skeletal muscle. *Proc Natl Acad Sci USA* 1987;84:6174–6178.

96. Engel AG: The neuromuscular junction, in Engel AG, Banker BQ (eds): *Myology*. New York, McGraw-Hill, 1986, vol 1, pp 209–253.

97. Salpeter MM: Development and neural control of the neuromuscular junction and of the junctional acetylcholine receptor, in Salpeter MM (ed): *The Vertebrate Neuromuscular Junction*. New York, Alan R Liss, 1987, pp 55–115.

98. Bennett MR: Development of neuromuscular synapses. *Physiol Rev* 1983;63:915–1048.

99. De Camilli P, Cameron R, Greengard P: Synapsin I (protein I), a nerve terminal-specific phosphoprotein: I. Its general distribution in synapses of the central and peripheral nervous system demonstrated by immunofluorescence in frozen and plastic sections. *J Cell Biol* 1983;96:1337–1354.

100. De Camilli P, Harris SM Jr, Huttner WB, et al: Synapsin I (protein I), a nerve terminal-specific phosphoprotein: II. Its specific association with synaptic vesicles demonstrated by immunocytochemistry in agarose-embedded synaptosomes. *J Cell Biol* 1983;96:1355–1373.

101. Couteaux R, Pécot-Dechavassine M: Vésicules synaptiques et poches au niveau des "zones actives" de la jonction neuromusculaire. *C R Acad Sci [D]* 1970;271:2346–2349.

102. Heuser JE, Reese TS, Landis DMD: Functional changes in frog neuromuscular junctions studied with freeze-fracture. *J Neurocytol* 1974;3:109–131.

103. Dreyer F, Peper K, Akert K, et al: Ultrastructure of the "active zone" in the frog neuromuscular junction. *Brain Res* 1973;62:373–380.

104. Heuser JE: A possible origin of the "giant" spontaneous potentials that occur after prolonged transmitter release at frog neuromuscular junctions. *J Physiol* 1974;239:106–108.

105. Heuser JE: Synaptic vesicle exocytosis and recycling during transmitter discharge from the neuromuscular junction, in Silverstein SC (ed): *Transport of Macromolecules in Cellular Systems*. Berlin, Dahlem Konferenzen, 1978, pp 445–464.

106. Heuser JE, Reese TS: Evidence for recycling of synaptic vesicle membrane during transmitter release at the frog neuromuscular junction. *J Cell Biol* 1973;57:315–344.

107. de Iraldi AP, de Robertis E: The neurotubular system of the axon and the origin of granulated and non-granulated vesicles in regenerating nerves. *Z Zellforsch Mikrosk Anat* 1968;87:330–344.

108. Chan-Palay V, Engel AG, Wu J-Y, et al: Coexistence in human and primate neuromuscular junctions of enzymes synthesizing acetylcholine, catecholamine, taurine, and γ-aminobutyric acid. *Proc Natl Acad Sci USA* 1982;79:7027–7030.

109. Birks R, Huxley HE, Katz B: The fine structure of the neuromuscular junction of the frog. *J Physiol* 1960;150:134–144.

110. Sanes JR: The extracellular matrix, in Engel AG, Banker BQ (eds): *Myology*. New York, McGraw-Hill, 1986, vol 1, pp 155–175.

111. Teräväinen H: Development of the myoneural junction in the rat. *Z Zellforsch Mikrosk Anat* 1968;87:249–265.

112. Kelly AM, Zacks SI: The fine structure of motor endplate morphogenesis. *J Cell Biol* 1969;42:154–169.

113. Juntunen J, Teräväinen H: Structural development of myoneural junctions in the human embryo. *Histochemie* 1972;32:107–112.

114. Matthews-Bellinger JA, Salpeter MM: Fine structural distribution of acetylcholine receptors at developing mouse neuromuscular junctions. *J Neurosci* 1983;3:644–657.

115. Dennis MJ, Ziskind-Conhaim L, Harris AJ: Development of neuromuscular junctions in rat embryos. *Dev Biol* 1981;81:266–279.

116. Banker BQ: Physiologic death of neurons in the developing anterior horn of the mouse, in Rowland LP (ed): *Human Motor Neuron Diseases*. New York, Raven Press, 1982, pp 473–485.

117. Jerusalem F: The microcirculation of muscle, in Engel AG, Banger BQ (eds): *Myology*. New York, McGraw-Hill, 1986, vol 1, pp 343–356.

118. Horowitz AF, Schotland DL: The plasma membrane of the muscle fiber, in Engel AG, Banker BQ (eds): *Myology*. New York, McGraw Hill, 1986, vol 1, pp 177–207.

119. Franzini-Armstrong C: The sarcoplasmic reticulum and the transverse tubules, in Engel AG, Banker BQ (eds): *Myology*. New York, McGraw-Hill, 1986, vol 1, pp 125–153.

120. Somlyo AV: Bridging structures spanning the junctional gap at the triad of skeletal muscle. *J Cell Biol* 1979;80:743–750.

121. Ferguson DG, Schwartz HW, Franzini-Armstrong C: Subunit structure of junctional feet in triads of skeletal muscle: A freeze-drying, rotary-shadowing study. *J Cell Biol* 1984;99:1735–1742.

122. Peachey LD, Eisenberg BR: Helicoids in the T system and striations of frog skeletal muscle fibers seen by high voltage electron microscopy. *Biophys J* 1978;22:145–154.

123. Sommer JR, Dolber PC, Taylor I: Filipin-cholesterol complexes in the sarcoplasmic reticulum of frog skeletal muscle. *J Ultrastruct Res* 1980;72:272–285.

124. MacLennan DH, Zubrzycka-Gaarn E, Jorgensen AO: Assembly of the sarcoplasmic reticulum during muscle development. *Curr Top Membranes Transport* 1985;24:337–368.

125. Cadwell JJS, Caswell AH: Identification of a constituent of the junctional feet linking terminal cisternae to transverse tubules in skeletal muscle. *J Cell Biol* 1982;93:543–550.

126. Squire J: *The Structural Basis of Muscular Contraction*. New York, Plenum Press, 1981.

127. Huxley HE: Molecular basis of contraction in cross-striated muscles and relevance to motile mechanisms in other cells, in Stracher A (ed): *Muscle and Nonmuscle Motility*. New York, Academic Press, 1983, pp 1–104.

128. Craig R: The structure of the contractile filaments, in Engel AG, Banker BQ (eds): *Myology*. New York, McGraw-Hill, 1986, vol 1, pp 79–124.

129. Kensler RW, Stewart M: Frog skeletal muscle thick filaments are three-stranded. *J Cell Biol* 1983;96:1797–1802.

130. Knappeis GG, Carlsen F: The ultrastructure of the Z disc in skeletal muscle. *J Cell Biol* 1962;13:323–335.

131. Offer G, Moos C, Starr R: A new protein of the thick filaments of vertebrate skeletal myofibrils: Extraction, purification and characterization. *J Mol Biol* 1973;74:653–676.

132. Masaki T, Takaiti O: M-protein. *J Biochem* 1974;75:367–380.

133. Trinick JA: End-filaments: A new structural element of vertebrate skeletal muscle thick filaments. *J Mol Biol* 1981;151:309–314.

134. Wallimann T, Turner DC, Eppenberger HM: Localization of creatine kinase isoenzymes in myofibrils: I. Chicken skeletal muscle. *J Cell Biol* 1977;75:297–317.

135. Starr R, Offer G: H-protein and X-protein: Two new components of the thick filaments of vertebrate skeletal muscle. *J Mol Biol* 1983;170:675–698.

136. Maruyama K, Kimura S, Ohashi K, et al: Connectin, an elastic protein of muscle: Identification of "titin" with connectin. *J Biochem* 1981;89:701–709.

137. Wang K: Sarcomere-associated cytoskeletal lattices in striated muscle, in Shay JW (ed): *Cell and Muscle Motility*. New York, Plenum Press, 1985, vol 6, pp 315–369.

138. Bird JWC, Roisen FJ: Lysosomes in muscle: Developmental aspects, enzyme activities, and role in protein turnover, in Engel AG, Banker BQ (eds): *Myology*. New York, McGraw-Hill, 1986, vol 1, pp 745–767.

139. Faulkner JA: New perspectives in training for maximum performance. *JAMA* 1986;205:741–746.

140. Saltin B, Gollnick P: Skeletal muscle adaptability: Significance for metabolism and performance, in Peachey LD, Adrian RH, Geiger SR (eds): *Handbook of Physiology: Section 10. Skeletal Muscle*. Bethesda, *American Physiological Society*, 1983, pp 555–631.

141. Gonyea W, Ericson GC, Bonde-Petersen F: Skeletal muscle fiber splitting induced by weight-lifting exercise in cats. *Acta Physiol Scand* 1977;99:105–109.

142. Gollnick PD, Timson BF, Moore RL, et al: Muscular enlargement and number of fibers in skeletal muscles of rats. *J Appl Physiol* 1981;50:936–943.

143. Faulkner JA: Structural and functional adaptations of skeletal muscle, in Roussos C, Macklem PT (eds): *The Thorax*. New York, Marcel Dekker, 1985, pp 1329–1351.

144. Caplan AI, Fiszman MY, Eppenberger HM: Molecular and cell isoforms during development. *Science* 1983;221:921–927.

145. Cummings EG, Bringas P Jr, Grodin MS, et al: Epithelial-directed mesenchyme differentiation in vitro: Model of murine odontoblast differentiation mediated by quail epithelia. *Differentiation* 1981;20:1–9.

146. DeHaan RL: Morphogenesis of the vertebrate heart, in DeHaan RL, Ursprung H (eds): *Organogenesis*. New York, Holt Rinehart & Winston, 1965, pp 377–419.

147. Osdoby P, Caplan AI: First bone formation in the developing chick limb. *Dev Biol* 1981;86:147–156.

148. Emerson CP, Fischman DA, Nadal-Ginard B, et al (eds): *Molecular Biology of Muscle Development*. New York, Alan R Liss, 1986.

149. Pearson ML, Epstein HF (eds): *Muscle Development: Molecular and Cellular Control*. Cold Spring Harbor Laboratory, 1982.

150. Bischoff R: The myogenic stem cell in development of skeletal muscle, in Mauro A, Shafiq S, Milhorat A (eds): *Regeneration of Striated Muscle, and Myogenesis*. Amsterdam, Excerpta Medica, 1970, pp 218–231.

151. Bischoff R: Enzymatic liberation of myogenic cells from adult rat muscle. *Anat Rec* 1974;180:645–662.

152. McLennan IS: The development of the pattern of innervation in chicken hindlimb muscles: Evidence for specification of nerve-muscle connections. *Dev Biol* 1983;97:229–238.

153. Hudgson P, Field EJ: Regeneration of muscle, in Bourne GH (ed): *The Structure and Function of Muscle*, ed 2. New York, Academic Press, 1973, vol 2, pp 311–363.

154. Engel AG, Biesecker G: Complement activation in muscle fiber necrosis: Demonstration of the membrane attack complex of complement in necrotic fibers. *Ann Neurol* 1982;12:289–296.

155. Bischoff R: A satellite cell mitogen from crushed adult muscle. *Dev Biol* 1986;115:140–147.

156. Vracko R, Benditt EP: Basal lamina: The scaffold for orderly cell replacement: Observations on regeneration of injured skeletal muscle fibers and capillaries. *J Cell Biol* 1972;55:406–419.

157. Bischoff R: Proliferation of muscle satellite cells on intact myofibers in culture. *Dev Biol* 1986;115:129–139.

158. Schultz E: A quantitative study of satellite cells in regenerated soleus and extensor digitorum longus muscles. *Anat Rec* 1984;208:501–506.

159. Ali MA: Myotube formation in skeletal muscle regeneration. *J Anat* 1979;128:553–562.

160. Hall-Craggs ECB: The regeneration of skeletal muscle fibres *per continuum*. *J Anat* 1974;117:171–178.

161. Shafiq SA, Gorycki MA: Regeneration in skeletal muscle of mouse: Some electron-microscope observations. *J Pathol Bacteriol* 1965;90:123–127.

162. Mauro A: Satellite cell of skeletal muscle fibers. *J Biophys Biochem Cytol* 1961;9:493–495.

163. Mauro A, Shafiq SA, Milhorat AT (eds): *Regeneration of Striated Muscle, and Myogenesis.* Amsterdam, Excerpta Medica, 1970.

164. Carlson BM: Factors controlling the initiation and cessation of early events in the regenerative process, in Sherbet GV (ed): *Neoplasia and Cell Differentiation.* Basel, S Karger, 1974, pp 60–105.

165. Thornton CS: The histogenesis of muscle in the regenerating fore limb of larval *Amblystoma punctatum. J Morphol* 1938;62:17–47.

166. Hay ED: Electron microscopic observations of muscle dedifferentiation in regenerating *Amblystoma* limbs. *Dev Biol* 1959;1:555–585.

167. Hay ED: Regeneration of muscle in the amputated amphibian limb, in Mauro A, Shafiq SA, Milhorat AT (eds): *Regeneration of Striated Muscle, and Myogenesis.* Amsterdam, Excerpta Medica, 1970, pp 3–24.

168. Kintner CR, Brockes JP: Monoclonal antibodies identify blastemal cells derived from dedifferentiating muscle in newt limb regeneration. *Nature* 1984;308:67–69.

169. Griffin KJP, Feketa DM, Carlson BM: A monoclonal antibody stains myogenic cells in regenerating new muscle. *Development*, in press.

170. Campion DR: The muscle satellite cell: A review. *Int Rev Cytol* 1984;87:225–251.

171. Mayr R: The muscle satellite cell and its role in muscle transplantation (a short review), in Freilinger G, Holle J, Carlson BM (eds): *Muscle Transplantation.* Vienna, Springer-Verlag, 1981, pp 19–27.

172. Allbrook DB, Han MF, Hellmuth AE: Population of muscle satellite cells in relation to age and mitotic activity. *Pathology* 1971;3:233–243.

173. Cardasis CA, Cooper GW: An analysis of nuclear numbers in individual muscle fibers during differentiation and growth: A satellite cell-muscle fiber growth unit. *J Exp Zool* 1975;191:347–358.

174. Gibson MC, Schultz E: Age-related differences in absolute numbers of skeletal muscle satellite cells. *Muscle Nerve* 1983;6:574–580.

175. Schmalbruch H, Hellhammer U: The number of nuclei in adult rat muscles with special reference to satellite cells. *Anat Rec* 1977;189:169–176.

176. Schmalbruch H: Muscle regeneration: Fetal myogenesis in a new setting. *Bibl Anat* 1986;29:126–153.

177. Schultz E, Lipton BH: Skeletal muscle satellite cells: Changes in proliferation potential as a function of age. *Mech Ageing Dev* 1982;20:377–383.

178. Snow MH: Myogenic cell formation in regenerating rat skeletal muscle injured by mincing: I. A fine structural study. *Anat Rec* 1977;188:181–200.

179. Snow MH: Myogenic cell formation in regenerating rat skeletal muscle injured by mincing: II. An autoradiographic study. *Anat Rec* 1977;188:200–218.

180. Partridge TA, Sloper JC: A host contribution to the regeneration of muscle grafts. *J Neurol Sci* 1977;33:425–435.

181. Grounds MD: Skeletal muscle precursors do not arise from bone marrow cells. *Cell Tissue Res* 1983;234:713–722.

182. Bayne EK, Anderson MJ, Fambrough DM: Extracellular matrix organization in developing muscle: Correlation with acetylcholine receptor aggregates. *J Cell Biol* 1984;99:1486–1501.

183. Kühl U, Timpl R, von der Mark K: Synthesis of type IV collagen and laminin in cultures of skeletal muscle cells and their assembly on the surface of myotubes. *Dev Biol* 1982;93:344–354.

184. Chiquet M, Fambrough DM: Chick myotendinous antigen: I. A monoclonal antibody as a marker for tendon and muscle morphogenesis. *J Cell Biol* 1984;98:1926–1936.

185. Markelonis G, Oh TH: A sciatic nerve protein has a trophic effect on development and maintenance of skeletal muscle cells in culture. *Proc Natl Acad Sci USA* 1979;76:2470–2474.

186. Bernfield M, Banerjee SD, Koda JE, et al: Remodelling of the basement membrane: Morphogenesis and maturation, in Porter R, Whalan J (eds): *Basement Membrane and Cell Movement*, Ciba Foundation Symposium 108. London, Pitman Medical, pp 179–196.

187. Caplan AI: The extracellular matrix is instructive. *BioEssays* 1986;5:129–132.

188. Toole BP: Glycosaminoglycans in morphogenesis, in Hay ED (ed): *Cell Biology of Extracellular Matrix*. New York, Plenum Press, 1981, pp 259–294.

189. Kosher RA, Savage MP, Walker KH: A gradation of hyaluronate accumulation along the proximodistal axis of the embryonic chick limb bud. *J Embryol Exp Morphol* 1981;63:85–98.

190. Pacifici M, Molinaro M: Developmental changes in glycosaminoglycans during skeletal muscle cell differentiation in culture. *Exp Cell Res* 1980;126:143–152.

191. Sanes JR, Marshall LM, McMahan UJ: Reinnervation of muscle fiber basal lamina after removal of myofibers: Differentiation of regenerating axons at original synaptic sites. *J Cell Biol* 1978;78:176–198.

192. Gulati AK: Basement membrane component changes in skeletal muscle transplants undergoing regeneration or rejection. *J Cell Biochem* 1985;27:337–346.

193. Gulati AK, Zalewski AA: An immunofluorescent analysis of lectin binding to normal and regenerating skeletal muscle of rat. *Anat Rec* 1985;212:113–117.

194. Gulati AK, Reddi AH, Zalewski AA: Changes in the basement membrane zone components during skeletal muscle fiber degeneration and regeneration. *J Cell Biol* 1983;97:957–962.

195. Myllylä R, Salminen A, Peltonen L, et al: Collagen metabolism of mouse skeletal muscle during the repair of exercise injuries. *Pflügers Arch* 1986;407:64–70.

196. Saad AD, Obinata T, Fischman DA: Immunochemical analysis of protein isoforms in thick myofilaments of regenerating skeletal muscle. *Dev Biol* 1987;119:336–349.

197. Studitsky AN: *The Experimental Surgery of Muscle*. Moscow, Izdatel Akad Nauk, SSSR, 1959, p 338.

198. Carlson BM: *The Regeneration of Minced Muscles*. Basel, S Karger, 1972.

199. Carlson BM, Gutmann E: Development of contractile properties of minced muscle regenerates in the rat. *Exp Neurol* 1972;36:239–249.

200. Studitsky AN: *Transplantation of Muscles in Animals*. Moscow, Meditsina, 1977, p 248.

201. Carlson BM: A review of muscle transplantation in mammals. *Physiol Bohemoslov* 1978;27:387–400.

202. Carlson BM, Hansen-Smith FM, Magon DK: The life history of a free muscle graft, in Mauro A (ed): *Muscle Regeneration*. New York, Raven Press, 1979, pp 493–507.

203. Faulkner JA, Carlson BM: Contractile properties of standard and nerve-intact muscle grafts in the rat. *Muscle Nerve* 1985;8:413–418.

204. Carlson BM: Relationship between the tissue and epimorphic regeneration of muscles. *Am Zool* 1970;10:175–186.

205. Carlson BM: Relationship between tissue and epimorphic regeneration of skeletal muscle, in Mauro A (ed): *Muscle Regeneration*. New York, Raven Press, 1979, pp 57–71.

206. Rose SM: Tissue-arc control of regeneration in the amphibian limb, in Rudnick D (ed): *Regeneration*. New York, Ronald Press, 1962, pp 153–176.

207. Muneoka K, Bryant S: Regeneration and development of vertebrate appendages. *Symp Zool Soc Lond* 1984;52:177–196.

208. Grim M, Carlson BM: The formation of muscles in regenerating limbs of the newt after denervation of the blastema. *J Embryol Exp Morphol* 1979;54:99–111.

209. Rubin LL, McMahan UJ: Regeneration of the neuromuscular junction: Steps toward defining the molecular basis of the interaction between nerve and muscle, in Schotland DL (ed): *Disorders of the Motor Unit*. New York, John Wiley & Sons, 1982, pp 187–196.

210. Burden SJ, Sargent PB, McMahan UJ: Acetylcholine receptors in regenerating muscle accumulate at original synaptic sites in the absence of the nerve. *J Cell Biol* 1979;82:412–425.

211. Bader D: Density and distribution of α-bungarotoxin-binding sites in postsynaptic structures of regenerated rat skeletal muscle. *J Cell Biol* 1981;88:338–345.

212. Hansen-Smith FM: Formation of acetylcholine receptor clusters in mammalian sternohyoid muscle regenerating in the absence of nerves. *Dev Biol* 1986;118:129–140.

213. Womble MD: The clustering of acetylcholine receptors and formation of neuromuscular junctions in regenerating mammalian muscle grafts. *Am J Anat* 1986;176:191–205.

214. Sanes JR, Cheney JM: Lectin binding reveals a synapse-specific carbohydrate in skeletal muscle. *Nature* 1982;300:646–647.

215. Barald KF, Phillips GD, Jay JC, et al: A component in mammalian muscle synaptic basal lamina induces clustering of acetylcholine receptors. *Prog Brain Res* 1987;71:397–408.

216. Magill C, Reist NE, Fallon JR, et al: Agrin. *Prog Brain Res* 1987;71:391–396.

217. Carlson BM, Hník P, Tuček S, et al: Comparison between grafts with intact nerves and standard free grafts of the rat extensor digitorum longus muscle. *Physiol Bohemoslov* 1981;30:505–513.

218. Carlson BM, Gutmann E: Transplantation and "cross-transplantation" of free muscle grafts in the rat. *Experientia* 1974;30:1292–1294.

219. Gutmann E, Carlson BM: Contractile and histochemical properties of regenerating cross-transplanted fast and slow muscles in the rat. *Pflügers Arch* 1975;353:227–239.

220. Vrbová G, Albani M: Regeneration of slow and fast muscles in rat hind limbs, in Frey M, Freilinger G (eds): *Proceedings of the Second Vienna Muscle Symposium*. Vienna, Facultas, 1986, pp 22–26.

221. Buller AJ, Eccles JC, Eccles RM: Interactions between motoneurones and muscles in respect of the characteristic speeds of their responses. *J Physiol* 1960;150:417–439.

222. Close R: Dynamic properties of fast and slow skeletal muscles of the rat after nerve cross-union. *J Physiol* 1969;204:331–346.

223. Bader D: Reinnervation of motor endplate-containing and motor endplate-less muscle grafts. *Dev Biol* 1980;77:315–327.

224. Carlson BM, Herbrychová A, Gutmann E: Retention of hormonal sensitivity in free grafts of the levator ani muscle. *Exp Neurol* 1979;63:94–107.

225. Carlson BM: A quantitative study of muscle fiber survival and regeneration in normal, predenervated, and Marcaine-treated free muscle grafts in the rat. *Exp Neurol* 1976;52:421–432.

226. Côté C, Faulkner JA: Motor unit function in skeletal muscle autografts of rats. *Exp Neurol* 1984;84:292–305.

227. Carlson BM, Gutmann E: Regeneration in free grafts of normal and denervated muscles in the rat: Morphology and histochemistry. *Anat Rec* 1975;183:47–62.

228. Landmesser LT: The generation of neuromuscular specificity. *Annu Rev Neurosci* 1980;3:279–302.

229. Zelená J, Sobotková M: Absence of muscle spindles in regenerated muscles of the rat. *Physiol Bohemoslov* 1971;20:433–439.

230. Zhenevskaya RP: *Neurotrophic Regulation of Plastic Activity of Muscle Tissue*. Moscow, Izdatel Akad Nauka, SSSR, 1974, p 239.

231. Schmalbruch H: Regeneration of soleus muscles of rat autografted in toto as studied by electron microscopy. *Cell Tissue Res* 1977;177:159–180.

232. Rogers SL, Carlson BM: A quantitative assessment of muscle spindle formation in reinnervated and non-reinnervated grafts of the rat extensor digitorum longus muscle. *Neuroscience* 1981;6:87–94.

233. Rogers SL: Muscle spindle formation and differentiation in regenerating rat muscle grafts. *Dev Biol* 1982;94:265–283.

234. Diwan FH, Milburn A: The effects of temporary ischaemia on rat muscle spindles. *J Embryol Exp Morphol* 1986;92:223–254.

235. Quick DC, Rogers SL: Stretch receptors in regenerated rat muscle. *Neuroscience* 1983;10:851–859.

236. Zelená J: The morphogenetic influence of innervation on the ontogenetic development of muscle-spindles. *J Embryol Exp Morphol* 1957;5:283–292.

237. Hsu L: The role of nerves in the regeneration of minced skeletal muscle in adult Anurans. *Anat Rec* 1974;179:119–136.

238. Mufti SA: Regeneration following denervation of minced gastrocnemius muscles in mice. *J Neurol Sci* 1977;33:251–266.

239. Mong FSF: Histological and histochemical studies on the nervous influence on minced muscle regeneration of triceps surae of the rat. *J Morphol* 1977;151:451–462.

240. Zhenevskaya RP: The influence of de-efferentation on the regeneration of skeletal muscle. *Ark Anat Gistol Embriol* 1960;39:42–50.

241. Zhenevskaya RP: The significance of sensory neurons for the structure and regeneration of skeletal musculature. *Arkh Anat Gistol Embriol* 1963;44:57–62.

242. Carlson BM, Gutmann E: Contractile and histochemical properties of sliced muscle grafts regenerating in normal and denervated rat limbs. *Exp Neurol* 1976;50:319–329.

243. Sunderland S: *Nerves and Nerve Injuries*, ed 2. Edinburgh, Churchill Livingstone, 1978, p 1046.

244. Freilinger G, Holle J, Carlson BM (eds): *Muscle Transplantation*. Vienna, Springer-Verlag, 1981.

245. Frey M, Freilinger G (eds): *Proceedings of the Second Vienna Muscle Symposium*. Vienna, Facultas, 1986.

246. Holle J: Die muskuläre neurotisation in der rekonstruktiven chirurgie. *Wien Klin Wochenschr* 1976;88(suppl 48):1–21.

247. Thompson N: A review of autogenous skeletal muscle grafts and their clinical applications. *Clin Plast Surg* 1974;1:349–403.

248. Erlacher P: Hyperneurotisation, muskuläre neurotisation; freie muskeltransplantation: Experimentelle untersuchungen. *Zentralbl Chir* 1914;41:625–627.

249. Must R: Muscular neurotization of rat free muscle grafts, in Frey M, Freilinger G (eds): *Proceedings of the Second Vienna Muscle Symposium.* Vienna, Facultas, 1986, pp 155–161.

250. Holley JA, Fahim MA: Scanning electron microscopy of mouse muscle microvasculature. *Anat Rec* 1983;205:109–117.

251. Grant RT, Wright HP: Further observations on the blood vessels of skeletal muscle (rat cremaster). *J Anat* 1968;103:553–565.

252. Baez S: Skeletal muscle and gastrointestinal microvascular morphology, in Kaley G, Altura BM (eds): *Microcirculation.* Baltimore, University Park Press, 1977, vol 1, pp 69–94.

253. Gorczynski RJ, Klitzman B, Duling BR: Interrelations between contracting striated muscle and precapillary microvessels. *Am J Physiol* 1978;235:494–504.

254. Myrhage R, Eriksson E: Vascular arrangements in hind limb muscles of the cat. *J Anat* 1980;131:1–17.

255. Brotzu G, Diaz G, Cherchi R, et al: A chemotactic relationship between skeletal muscle nuclei and capillaries. *Microvasc Res* 1982;23:87–92.

256. Mrázková O, Grim M, Carlson BM: Enzymatic heterogeneity of the capillary bed of rat skeletal muscles. *Am J Anat* 1986;177:141–148.

257. Plyley MJ, Sutherland GJ, Groom AC: Geometry of the capillary network in skeletal muscle. *Microvasc Res* 1976;11:161–173.

258. Glaser BM, D'Amore PA, Seppa H, et al: Adult tissues contain chemo-attractants for vascular endothelial cells. *Nature* 1980;288:483–484.

259. Faulkner JA, Côté C: Functional deficits in skeletal muscle grafts. *Fed Proc* 1986;45:1466–1469.

260. Carlson BM, Gutmann E: Regeneration in free grafts of normal and denervated rat muscles: Contractile properties. *Pflügers Arch* 1975;353:215–225.

261. Faulkner JA, Niemeyer JH, Maxwell LC, et al: Contractile properties of transplanted extensor digitorum longus muscles of cats. *Am J Physiol* 1980;238:C120-C126.

262. Faulkner JA, Markley JM Jr, McCully KK, et al: Characteristics of cat skeletal muscles grafted with intact nerves or with anastomosed nerves. *Exp Neurol* 1983;80:682–696.

263. Thomas D, Klueber K, Bourke D, et al: The size of the myofibers in mature grafts of the mouse extensor digitorum longus muscle. *Muscle Nerve* 1984;7:226–231.

264. Faulkner JA, Markley JM, White TP: Skeletal muscle transplantation in cats with and without nerve repair, in Freilinger G, Holle J, Carlson BM (eds): *Muscle Transplantation.* Vienna, Springer-Verlag, 1981, pp 47–54.

265. Gorniak GC, Gans C, Faulkner JA: Muscle fiber regeneration after transplantation: Prediction of structure and physiology from electromyograms. *Science* 1979;204:1085–1087.

266. Gutmann E: Transplantation of muscle. *Atti Accad Nazionale Lincei, Memorie* 1978;15(sezione 3):1–38.

267. Schmalbruch H: The morphology of regeneration of skeletal muscles in the rat. *Tissue Cell* 1976;8:673–692.

268. Basson MD, Carlson BM: Myotoxicity of single and repeated injections of mepivacaine (Carbocaine) in the rat. *Anesth Analg* 1980;59:275–282.

269. Benoit PW, Belt WD: Destruction and regeneration of skeletal muscle after treatment with a local anaesthetic, bupivacaine (Marcaine®). *J Anat* 1970;107:547–556.

270. Dolwick MF, Bush FM, Seibel HR: Regeneration of masseter muscle following lidocaine-induced degeneration: A histochemical study. *Acta Anat* 1977;98:325–333.

271. Foster AH, Carlson BM: Myotoxicity of local anesthetics and regeneration of the damaged muscle fibers. *Anesth Analg* 1980;59:727–736.

272. Jirmanová I, Thesleff S: Ultrastructural study of experimental muscle degeneration and regeneration in the adult rat. *Z Zellforsch Mikrosk Anat* 1972;131:77–97.

273. Maltin CA, Harris JB, Cullen MJ: Regeneration of mammalian skeletal muscle following the injection of the snake-venom toxin, taipoxin. *Cell Tissue Res* 1983;232:565–577.

274. Ownby CL, Odell GV: Pathogenesis of skeletal muscle necrosis induced by tarantula venom. *Exp Mol Pathol* 1983;38:283–296.

275. Queiroz LS, Santo Neto H, Rodrigues-Simioni L, et al: Muscle necrosis and regeneration after envenomation by *Bothrops jararacussu* snake venom. *Toxicon* 1984;22:339–346.

276. Tu AT: Local tissue damaging (hemorrhage and myonecrosis) toxins from rattlesnake and other pit viper venoms. *J Toxicol Toxin Rev* 1983;2:205–234.

277. Dettbarn W-D: Pesticide induced muscle necrosis: Mechanisms and prevention. *Fund Appl Toxicol* 1984;4:S18-S26.

278. Toader-Radu M: Dynamics of regeneration in skeletal muscle following localized heat injury. *Morphol Embryol* 1978;24:69–73.

279. Price HM, Howes EL Jr, Blumberg JM: Ultrastructural alterations in skeletal muscle fibers injured by cold: I. The acute degenerative changes. *Lab Invest* 1964;13:1264–1278.

280. Price HM, Howes EL Jr, Blumberg JM: Ultrastructural alterations in skeletal muscle fibers injured by cold: II. Cells of the sarcolemmal tube: Observations on "discontinuous" regeneration and myofibril formation. *Lab Invest* 1964;13:1279–1302.

281. Yanko L, Behar A, Yarom R: Regeneration of extraocular muscles following cold injury: A light and electron microscope study. *J Neurol Sci* 1974;21:23–32.

282. O'Steen WK, Shear CR, Anderson KV: Extraocular muscle degeneration and regeneration after exposure of rats to incandescent radiant energy. *J Cell Sci* 1975;18:157–177.

283. Le Gros Clark WE, Blomfield LB: The efficiency of intramuscular anastomoses, with observations on the regeneration of devascularized muscle. *J Anat* 1945;79:15–32.

284. Karpati G, Carpenter S, Melmed C, et al: Experimental ischemic myopathy. *J Neurol Sci* 1974;23:129–161.

285. Santavirta S, Luoma A, Arstila AU: Morphological and biochemical changes in striated muscle after experimental tourniquet ischaemia. *Res Exp Med* 1979;174:245–251.

286. Mäkitie J, Terävänen H: Ultrastructure of striated muscle of the rat after temporary ischemia. *Acta Neuropathol* 1977;37:237–245.

287. Hall-Craggs ECB: Ischemic muscle as a model of regeneration. *Exp Neurol* 1978;60:393–399.

288. Järvinen M, Sorvari T: Healing of a crush injury in rat striated muscle: 1. Description and testing of a new method of inducing a standard injury to the calf muscles. *Acta Pathol Microbiol Scand* 1975;83:259–265.

289. Järvinen M: Healing of a crush injury in rat striated muscle: 2. A histological study of the effect of early mobilization and immobilization on the repair processes. *Acta Pathol Microbiol Scand Sect* 1975:83:269–282.

290. Berlin RH: Missile injury to live muscle tissue. *Acta Chir Scand*, 1977, suppl 480, pp 1–450.

291. Gutmann E, Carlson BM: The regeneration of a hormone-sensitive muscle (levator ani) in the rat. *Exp Neurol* 1978;58:535–548.

292. Schultz E, Jaryszak DL, Valliere CR: Response of satellite cells to focal skeletal muscle injury. *Muscle Nerve* 1985;8:217–222.

293. Karpati G, Carpenter S: Micropuncture lesions of skeletal muscle cells: A new experimental model for the study of muscle cell damage, repair, and regeneration, in Schotland DL (ed): *Disorders of the Motor Unit*. New York, John Wiley & Sons, 1982, pp 517–533.

294. Salminen A: Effects of the protease inhibitor leupeptin on proteolytic activities and regeneration of mouse skeletal muscles after exercise injuries. *Am J Pathol* 1984;117:64–70.

295. Armstrong RB, Ogilvie RW, Schwane JA: Eccentric exercise-induced injury to rat skeletal muscle. *J Appl Physiol* 1983;54:80–93.

296. Kasperek GJ, Snider RD: The susceptibility to exercise-induced muscle damage increases as rats grow larger. *Experientia* 1985;41:616–617.

297. Giddings CJ, Neaves WB, Gonyea WJ: Muscle fiber necrosis and regeneration induced by prolonged weight-lifting exercise in the cat. *Anat Rec* 1985;211:133–141.

298. Schmidt AJ: *Cellular Biology of Vertebrate Regeneration and Repair*. Chicago, University of Chicago Press, 1968.

299. Wallace H: *Vertebrate Limb Regeneration*. Chichester, John Wiley & Sons, 1981.

300. Waldeyer W: Ueber die Veränderungen der quergestreiften Muskeln bei der Entzündung und dem Typhusprozess, sowie über die Regeneration derselben nach Substanz-defekten. *Arch Path Anat Physiol Klin Med* 1865;34:473–514.

301. Adams RD: Pathological reactions of the skeletal muscle fibre in man, in Walton JN (ed): *Disorders of Voluntary Muscle*, ed 3. Edinburgh, Churchill Livingstone, 1974, pp 168–233.

302. Mastaglia FL, Kakulas BA: A histological and histochemical study of skeletal muscle regeneration in polymyositis. *J Neurol Sci* 1970;10:471–487.

303. Mastaglia FL, Walton JN: An ultrastructural study of skeletal muscle in polymyositis. *J Neurol Sci* 1971;12:473–504.

304. Schutta HS, Kelly AM, Zacks SI: Necrosis and regeneration of muscle in paroxysmal idiopathic myoglobinuria: Electron microscopic observations. *Brain* 1969;92:191–202.

305. Mastaglia FL, Papadimitriou JM, Kakulas BA: Regeneration of muscle in Duchenne muscular dystrophy: An electron microscope study. *J Neurol Sci* 1970;11:425–444.

306. Ontell M, Feng KC, Klueber K, et al: Myosatellite cells, growth, and regeneration in murine dystrophic muscle: A quantitative study. *Anat Rec* 1984;208:159–174.

307. Schmalbruch H: Regenerated muscle fibers in Duchenne muscular dystrophy: A serial section study. *Neurology* 1984;34:60–65.

308. Terasawa K: Muscle regeneration and satellite cells in Fukuyama type congenital muscular dystrophy. *Muscle Nerve* 1986;9:465–470.

309. Wakayama Y, Schotland DL, Bonilla E: Transplantation of human skeletal muscle to nude mice: A sequential morphologic study. *Neurology* 1980;30:740–748.

310. Wakayama Y, Ohbu S: Light- and electron-microscopic studies of transplanted human dystrophic muscles to nude mice. *J Neurol Sci* 1982;55:59–77.

311. Millar WG: Regeneration of skeletal muscle in young rabbits. *J Pathol Bacteriol* 1934;38:145–151.

312. Jones DS: Healing of muscle tissue, in Williamson MB (ed): *The Healing of Wounds*. New York, McGraw-Hill, 1957, pp 149–167.

313. Gay AJ Jr, Hunt TE: Reuniting of skeletal muscle fibers after transection. *Anat Rec* 1954;120:853–871.

314. Allbrook D, Baker W deC, Kirkaldy-Willis WH: Muscle regeneration in experimental animals and in man: The cycle of tissue change that follows trauma in the injured limb syndrome. *J Bone Joint Surg* 1966;48B:153–169.

315. Hikida RS, Staron RS, Hagerman FC, et al: Muscle fiber necrosis associated with human marathon runners. *J Neurol Sci* 1983;59:185–203.

316. Knochel JP: Rhabdomyolysis and myoglobinuria. *Annu Rev Med* 1982;33:435–443.

317. Steiss JE: Effect of high- and low-dose dexamethesone on regeneration of minced skeletal muscle autografts in rats. *Exp Neurol* 1986;93:300–310.

318. Yagiela JA, Benoit PW, Buoncristiani RD, et al: Comparison of myotoxic effects of lidocaine with epinephrine in rats and humans. *Anesth Analg* 1981;60:471–480.

319. Fridén J, Sjöström M, Ekblom B: Myofibrillar damage following intense eccentric exercise in man. *Int J Sports Med* 1983;4:170–176.

320. Armstrong RB: Mechanisms of exercise-induced delayed onset muscular soreness: A brief review. *Med Sci Sports Exerc* 1984;16:529–538.

321. McCully KK, Faulkner JA: Injury to skeletal muscle fibers of mice following lengthening contractions. *J Appl Physiol* 1985;59:119–126.

322. McCully KK, Faulkner JA: Characteristics of lengthening contractions associated with injury to skeletal muscle fibers. *J Appl Physiol* 1986;61:293–299.

323. Donovan CM, Faulkner JA: Plasticity of skeletal mucle: Regenerating fibers adapt more rapidly than surviving fibers. *J Appl Physiol*, in press.

324. Faulkner JA, Sachdev N, Brooks SV, et al: Adaptation of soleus muscles to an atypical site and innervation: Comparison of regenerating and surviving fibers, abstract. *Fed Proc* 1987;46:639.

325. Garrett WE Jr, Seaber AV, Boswick J, et al: Recovery of skeletal muscle after laceration and repair. *J Hand Surg* 1984;9A:683–692.

326. Botte MJ, Gelberman RH, Smith DG, et al: Repair of severe muscle belly lacerations using a tendon graft. *J Hand Surg* 1987;12A:406–412.

327. Järvinen M: Healing of a crush injury in rat striated muscle: 3. A micro-angiographical study of the effect of early mobilization and immobilization on capillary ingrowth. *Acta Pathol Microbiol Scand* 1976;84:85–94.

328. Järvinen M: Healing of a crush injury in rat striated muscle. *Acta Pathol Microbiol Scand* 1976;142:47–56.

329. Järvinen M, Aho AJ, Lehto M, et al: Age dependent repair of muscle rupture: A histological and microangiographical study in rats. *Acta Orthop Scand* 1983;54:64–74.

330. Lehto M, Järvinen M, Nelimarkka O: Scar formation after skeletal muscle injury: A histological and autoradiographical study in rats. *Arch Orthop Trauma Surg* 1986;104:366–370.

331. Lehto M, Duance VC, Restall D: Collagen and fibronectin in a healing skeletal muscle injury: An immunohistological study of the effects of physical activity on the repair of injured gastrocnemius muscle in the rat. *J Bone Joint Surg* 1985;67B: 820–828.

332. Rothwell AG: Quadriceps hematoma: A prospective clinical study. *Clin Orthop* 1982;171:97–103.

333. Zaccalini PS, Urist MR: Traumatic periosteal proliferations in rabbits: The enigma of experimental myositis ossificans traumatica. *J Trauma* 1964;4:344–357.

334. Walton M, Rothwell AG: Reactions of thigh tissues of sheep to blunt trauma. *Clin Orthop* 1983;176:273–281.

335. Mubarak S, Hargens AR: *Compartment Syndromes and Volkmann's Contracture.* Philadelphia, WB Saunders, 1981, pp 106–118.

336. Whitesides TE Jr, Haney TC, Morimoto K, et al: Tissue pressure measurements as a determinant for the need of fasciotomy. *Clin Orthop* 1975;113:43–51.

337. Mubarak SJ, Hargens AR, Owen CA, et al: The wick catheter technique for measurement of intramuscular pressure: A new research and clinical tool. *J Bone Joint Surg* 1976:58A;1016–1020.

338. Rorabeck CH, Castle GSP, Hardie R, et al: Compartmental pressure measurements: An experimental investigation using the slit catheter. *J Trauma* 1981;21:446–449.

339. Matsen FA III, Mayo KA, Krugmire RB Jr, et al: A model compartmental syndrome in man with particular reference to the quantification of nerve function. *J Bone Joint Surg* 1977;59A:648–653.

340. Mubarak SJ, Owen CA, Hargens AR, et al: Acute compartment syndromes: Diagnosis and treatment with the aid of the wick catheter. *J Bone Joint Surg* 1978;60A:1091–1095.

341. Matsen FA III, Winquist RA, Krugmire RB Jr: Diagnosis and management of compartmental syndromes. *J Bone Joint Surg* 1980;62A:286–291.

342. Heppenstall RB, Sapega AA, Scott R, et al: The compartment syndrome: An experimental and clinical study of muscular energy metabolism utilizing phosphorus NMR spectroscopy. *Clin Orthop*, in press.

343. Sapega AA, Heppenstall RB, Osterman AL, et al: Noninvasive assessment of muscular metabolic state in human extremity compartment syndromes with phosphorus nuclear magnetic resonance spectroscopy (31P-NMR). *Trans Orthop Res Soc* 1985;10:368.

344. Hall-Craggs ECB, Brand P: Effect of previous nerve injury on the regeneration of free autogenous muscle grafts. *Exp Neurol* 1977;57:275–281.

345. Carlson BM, Foster AH, Bader DM, et al: Restoration of full mass in nerve-intact muscle grafts after delayed reinnervation. *Experientia* 1983;39:171–172.

346. Studitsky NN, Bosova NN: Development of atrophic muscular tissue in conditions of transplantation in place of mechanically damaged muscles. *Arch Anat Gist Embriol* 1960;39:18–32.

347. Phillips GD: Angiogenesis in injured skeletal muscle, abstract. *Anat Rec* 218:106A–107A.

348. Guelinckx P, Faulkner JA, Essig D: Rectus femoris muscles of rabbits autografted with microvascular repair with nerves intact or nerves anastomosed, in Frey M, Freilinger G (eds): *Proceedings of the Second Vienna Muscle Symposium*. Vienna, Facultas, 1986, pp 75–83.

349. Mantelow RT, Zuker RM: Functioning muscle transplantation to the extremities, in Frey M, Freilinger G (eds): *Proceedings of the Second Vienna Muscle Symposium*. Vienna, Facultas, 1986, pp 283–287.

350. White TP: Adaptations of skeletal muscle grafts to chronic changes of physical activity. *Fed Proc* 1986;45:1470–1473.

351. Van Handel PJ, Watson P, Troup J, et al: Effects of treadmill running on oxidative capacity of regenerated skeletal muscle. *Int J Sports Med* 1981;2:92–96.

352. White TP, Villanacci JF, Morales PG, et al: Exercise-induced adaptations of rat soleus muscle grafts. *J Appl Physiol* 1984;56:1325–1334.

353. Coan MR, Tomanek RJ: The growth of regenerating soleus muscle transplants after ablation of the gastrocnemius muscle. *Exp Neurol* 1981;71:278–294.

354. Gallucci V, Novello F, Margreth A, et al: Biochemical correlates of discontinuous muscle regeneration in the rat. *Br J Exp Pathol* 1966;47:215–227.

355. Hanzlíková V, Gutmann E: The absence of androgen-sensitivity in the grafted soleus muscle innervated by the pudendal nerve. *Cell Tissue Res* 1974;154:121–129.

356. Jones PH: Implantation of cultured regenerate muscle cells into adult rat muscle. *Exp Neurol* 1979;66:602–610.

357. Lipton BH, Schultz E: Developmental fate of skeletal muscle satellite cells. *Science* 1979;205:1292–1294.

358. Law PK: Beneficial effects of transplanting normal limb-bud mesenchyme into dystrophic mouse muscles. *Muscle Nerve* 1982;5:619–627.

359. Gutmann E, Hanzlíková V: Denervation, reinnervation and regeneration of senile muscle, in Cristofalo VJ, Holečková E (eds): *Cell Impairment in Aging and Development*. New York, Plenum Press, 1975, pp 431–440.

360. Gutmann E, Carlson BM: Regeneration and transplantation of muscles in old rats and between young and old rats. *Life Sci* 1976;18:109–114.

361. Carlson BM, Faulkner JA: Age of the host is an important factor in determining the success of skeletal muscle grafts between young and old rats. *Anat Rec*, in press.

Section Six
Peripheral Nerve

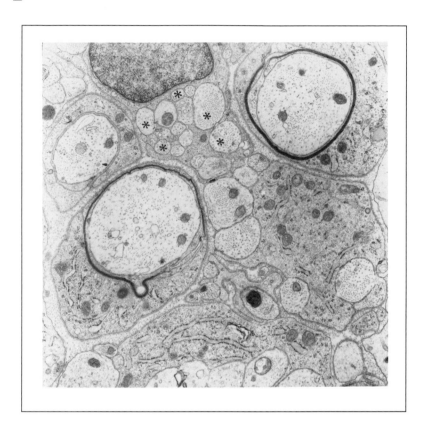

Group Leader

Göran Lundborg, MD, PhD

Group Members

Jack Lewis, PhD
Marston Manthorpe, PhD
Björn Rydevik, MD, PhD

Group Participants

Mark Adams, MD, FRCPC
Wayne Akeson, MD
Kai-Nan An, PhD
Albert Banes, PhD
Leo Furcht, MD
Richard Gelberman, MD
Van Mow, PhD
Theodore Oegema, PhD
Eric Radin, MD
A. Hari Reddi, PhD
Melvin Rosenwasser, MD
Lawrence Shulman, MD
Andrew Weiland, MD

Synopsis

Injuries to peripheral nerves still represent one of the most challenging surgical problems. Nerve injuries in the upper extremities result in considerable disability since these lesions may cause total or subtotal loss of sensory and/or motor function of the hand. When microsurgical techniques were introduced, hopes were raised that results of nerve repair would be much improved. As surgical techniques have reached a plateau and cannot be further refined, however, the overall recovery of nerve function remains unsatisfactory.

Successful regeneration of injured peripheral nerves involves a complicated sequence of events, the ultimate goal of which is to achieve a reconnection of peripheral targets with their corresponding original cortical projections in the brain. To achieve this goal, axons growing across the zone of injury should reinnervate their parent endoneurial pathways (endoneurial tubes) in the distal segment, and they should reach their peripheral targets before the end organs have irreversibly degenerated. Thus, axonal orientation of the site of injury and the growth rate in the distal segment are both crucial issues.

The peripheral nerve is a complex structure composed of several tissue elements. The primary functional elements, the axons, are long cellular processes deriving from their respective nerve cell bodies in the anterior horn of the spinal cord (motor neurons) or in the dorsal root ganglion (sensory neurons). The nerve fibers are located in fascicles embedded in the epineurium. Each fascicle is surrounded by a perineurium; the intrafascicular loose connective tissue is called the endoneurium. Peripheral nerves are also highly vascularized structures exhibiting well-developed vascular plexus in all tissue layers. A longitudinally oriented endoneurial capillary bed is present in the fascicles. The endothelial cells that form these vessels are connected by "tight junctions" and constitute a "blood-nerve barrier" corresponding to the "blood-brain barrier" of the central nervous system. A diffusion barrier to macromolecules is also present in the perineurium. These two barrier systems help to maintain a specialized endoneurial environment optimal for axonal function.

Inside the axons within the axoplasm, there is a system for bidirectional transport of substances. The so-called slow anterograde axonal transport takes place at a velocity of 1 to 6 mm/day and involves cytoskeletal elements such as microtubules and microfilaments. The so-called fast anterograde axonal transport takes place at a speed of up to about 400 mm/day and involves various enzymes and transmitter substances, vesicles, glycoproteins, and lipids. Retrograde axonal transport seems to provide recycling of neurotransmitter vesicles. Also, various target-derived substances of importance for survival and general growth as well as metabolic maintenance of the nerve cell body are probably transported by the retrograde axonal transport.

Impulse transmission and axonal transport depend on a local energy supply provided by the intraneural microvascular system. Moderate compression produces local ischemia in the nerve, thereby blocking impulse transmission as well as axonal transport. Such a "metabolic conduction block" is immediately reversible upon release of the compression. It has been demonstrated that extraneural pressures corresponding to those found inside the carpal canal of patients with carpal tunnel syndrome may induce a significant impairment of intraneural microcirculation.

Following compression at higher pressure levels, a local conduction block may be induced. This can persist for as long as several weeks after the compression ("neurapraxia"). In these cases the axons are still in continuity and nerve function will be restored when the local myelin has been repaired. In crush injuries, axonal continuity can be interrupted while the "endoneurial tubes," composed of Schwann cells and their basal lamina that normally surround axons, may remain intact ("axonotmesis"). In such cases wallerian degeneration occurs and axons have to regenerate through preserved endoneurial pathways in order to reach their peripheral previous cell targets. In more severe lesions, the connective tissue elements of the nerve trunk may be damaged ("neurotmesis"). In such cases considerable misdirection of axons will occur at the site of injury. The result is frequent reinnervation of inappropriate end organs.

With tension, intraneural microcirculation is compromised early. With increased stretching, the applied load increases correspondingly until a point is reached at which the load drops suddenly and the nerve fails structurally. Beyond this point, the nerve can carry very little tensile load. It has been shown that various components of nerve trunks have specific mechanical properties to resist loading—a fact that influences the stretching, compression, and other kinds of mechanical deformation of the nerve.

With severance of axons, the promixal portions start sending out sprouts and the regenerative process occurs. The mechanisms regulating axonal growth and guidance back to their target organ are not fully understood although experimental studies indicate that neuronotrophic factors as well as neurite-promoting factors in the local environment may act as molecular signals for axonal guidance. Clinically, nerve lesions are treated by approximating the proximal and distal nerve segments by suturing their ends or inserting nerve grafts if a segment of the nerve trunk is gone. Although microsurgical techniques have helped to improve the results of such procedures, there are indications that increased understanding and manipulation of microenvironmental factors influencing nerve growth will improve nerve repair in the future.

The final result of nerve repair depends on factors such as axonal growth, orientation of the apposing ends of the cut axons at the suture site, growth and maturation in the distal nerve segment, reinnervation of peripheral targets, and adaptation of the central nervous system to a modified pattern of afferent impulses caused by misdirected axons. Experimentally, the rate of axonal growth can be increased by inducing conditioning lesions in the nerve two weeks before the test lesion. In certain experimental systems local or general application of substances like cyclic adenosine monophosphate, triiodothyronine, and adrenocorticotropic hormones have been reported to improve regeneration. Although reports are conflicting, direct-current electric fields as well as pulsed electromagnetic fields may, under certain conditions, promote axonal regeneration.

Clinically, it is known that the functional restitution after nerve repair in children is superior to that in adults. This fact is probably based on the child's better propensity for cerebral adaptation after peripheral axonal misdirection. In adults, special sensory re-education programs have proved useful to facilitate such a central nervous adaptation after nerve repair.

Microsurgical techniques have helped to improve clinical outcomes in selected situations, but the results of nerve surgery are still unsatisfactory. Biochemical manipulation of the local microenvironment at the zone of injury and/or at the axonal and target cell area may help to improve nerve repair.

Chapter 7
Peripheral Nerve: The Physiology of Injury and Repair

Göran Lundborg, MD, PhD
Björn Rydevik, MD, PhD
Marston Manthorpe, PhD
Silvio Varon, MD
Jack Lewis, PhD

Chapter Outline

Chapter Outline   *continued*

Clinical Aspects of Nerve Repair
    Primary vs Secondary Repair
    Perineurial and Epineurial Suture Techniques
    Nerve Grafting
Sensory Re-education
Future Directions

Structure and Function

Microscopic Anatomy of the Nerve Trunk

The peripheral nerve is a complex structure composed of several tissue elements. The functional elements, the axons, are long cellular extensions from their respective nerve cell bodies in the anterior horn of the spinal cord (motor neurons) or in the dorsal root ganglion (sensory neurons) (Fig. 7–1). In the peripheral nerve, the nerve fibers are located in fascicles. The peripheral nerve trunk and its fascicles are held together by connective tissue layers named the epineurium, perineurium, and endoneurium[1,2] (Fig. 7–2). The epineurium is a loose connective tissue layer located superficially in the nerve and between the fascicles. Each fascicle is surrounded by a multilayered, condensed, mechanically strong membrane called the perineurium. The connective tissue in the intrafascicular space is called the endoneurium. This structural arrangement refers to the general anatomic principles of the peripheral nerves, and the relative amounts of epineurial and endo-

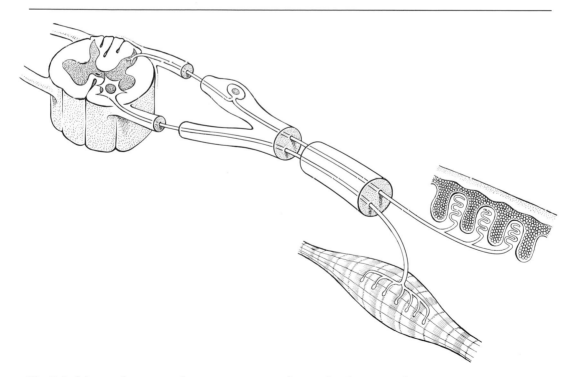

Fig. 7–1 *Schema of sensory and motor components of a peripheral nerve. Each axon represents a process from a nerve cell body situated in the dorsal root ganglion (sensory neuron) or the anterior horn of the spinal cord (motor neuron). (Reproduced with permission from Lundborg G: Nerve regeneration and repair.* Acta Orthop Scand *1987;58:145–169.)*

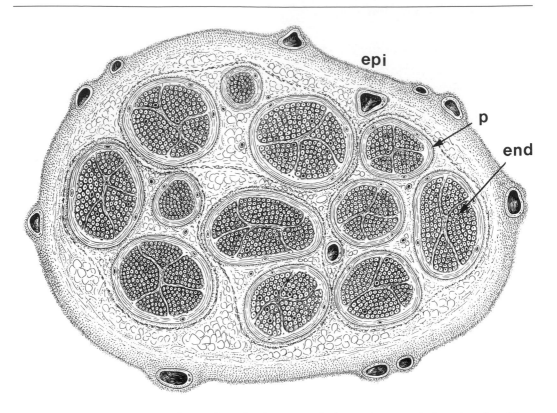

Fig. 7–2 *Schema of peripheral nerve trunk shows epineurium (epi), perineurium (p), and endoneurium (end).*

neurial tissue, for example, may vary among nerves. This relationship may also vary within the same nerve at various levels along its course.

Intraneural Microcirculation and Microvascular Permeability

Normal function of nerve fibers depends on a continuous supply of oxygen and other nutrients. This nutritional supply is provided by a rich intraneural microvascular network in all layers of the nerve (Figs. 7–3 to 7–5).[3–12] The nerve trunk receives a segmental vascular supply along its course. These segmental vessels usually approach the nerve in a coiled manner, reflecting their ability to adapt to physiologic movements of the nerve trunk. In the epineurium they divide into ascending and descending branches. The epineurial vascular bed anastomoses with vessels in the endoneurium via perineurial vessels, often piercing the innermost lamella at an oblique angle.[8–10]

When increased intrafascicular pressure is associated with endoneurial edema, these obliquely penetrating vessels may constitute a valve mechanism[9,13] that may reduce endoneurial blood flow.[14] Vessels in the endoneurium are primarily capillaries running parallel to the nerve

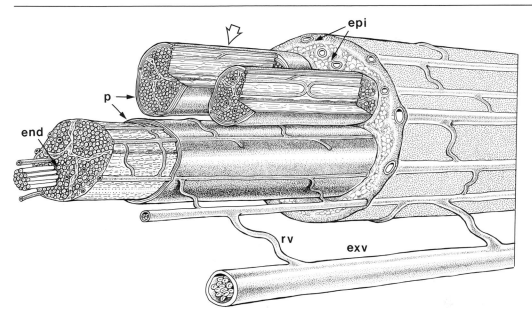

Fig. 7–3 *Schema of intraneural microcirculation. Note the well-developed vascular plexus in all layers of the nerve. Extrinsic vessels (exv) are linked via regional feeding vessels (rv) supporting vascular plexus in superficial and deep layers of the epineurium (epi), perineurium (p), and endoneurium (end). Vessels penetrate obliquely through the perineurium (arrows). (Reproduced with permission from Lundborg G:* Nerve Injury and Repair. *Edinburgh, Churchill Livingstone, in press.)*

fibers (Fig. 7–5). The peripheral nerve is a well-vascularized structure, one that is difficult to devascularize surgically: a nerve can be mobilized and its segmental vessels cut over a considerable length before intraneural blood flow is affected.[6,8,9,15] If a rabbit posterior tibial nerve is mobilized over a distance of 15 mm and cut distally, the intraneural blood flow is maintained to the very tip of the cut nerve.[8,9] However, if tension is applied to the cut end, intraneural blood flow, particularly venular flow in the epineurium, is already affected at about 8% elongation.[16] With gradually increasing tension, intraneural blood flow continues to decrease and at about 15% elongation there is complete ischemia of the nerve. Correspondingly, intraneural blood flow during compression can be impaired at low pressure levels. Local compression of a rabbit tibial nerve at 20 to 30 mm Hg reduces venular blood flow in the epineurium, and 60 to 80 mm Hg induces complete ischemia of the compressed nerve segment.[17] Thus, the peripheral nerve is well vascularized but its intraneural microcirculation is affected at relatively low stretching and compression levels. Such interference with the nutritive blood supply can lead to impairment of nerve function.[8,18–20]

The intraneural vessels respond to injury with increased permeability, resulting in edema (Fig. 7–6). The extent and localization of edema in various layers depend on the severity of the injury.[8,9,21,22] The

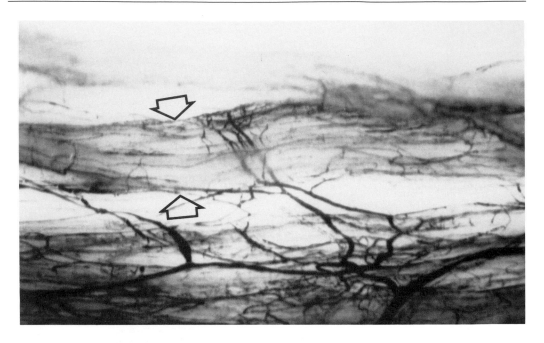

Fig. 7–4 *Microangiogram illustrating the intrinsic vascularization of the human median nerve from the carpal tunnel region. Arrows indicate the outline of one fascicle. (Reproduced with permission from Lundborg G: The intrinsic vascularization of human peripheral nerves: Structural and functional aspects. J Hand Surg 1979;4:34–41.)*

permeability properties of the nerve microvasculature have been analyzed by various tracer techniques such as albumin labeled with Evan's blue and horseradish peroxidase.[8,9,21-26] The studies demonstrated that the epineurial vessels normally allow the passage of small amounts of protein across their walls, unlike the endoneurial vessels, which normally do not allow any extravasation of the corresponding proteins. The walls of the endoneurial capillaries thus constitute a blood-nerve barrier analogous to the blood-brain barrier.[21,27] The anatomic basis for this barrier has been postulated to be the existence of "tight junctions" between the endothelial cells and the lack of pinocytotic vesicles in these cells.[21] A slight trauma to a nerve, such as ischemia of short duration or low-pressure compression, causes epineurial edema, whereas the endoneurial vascular bed is unaffected. Such epineurial edema is prevented from reaching the nerve fibers in the endoneurium by the perineurium, which acts as a diffusion barrier against several substances in both animals and humans.[21,28-30] More severe trauma may also injure the endoneurial vessels, resulting in endoneurial edema.[9,22]

The Fascicular Unit

The nerve fibers are located in fascicles that are often collected in groups of four or five. The fascicular arrangement varies along the

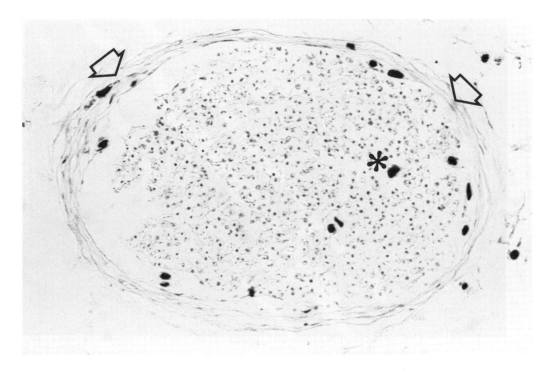

Fig. 7–5 *Microangiogram illustrating capillary pattern in one transversely cut fascicle from human median nerve. Arrows indicate outline of fascicle. Capillaries (*) are seen in the endoneurium and between individual lamellae of the perineurium. (Reproduced with permission from Lundborg G: Intraneural microcirculation and peripheral nerve barriers, in Omer G, Spinner M (eds): Management of Peripheral Nerve Problems. Philadelphia, WB Saunders, 1980, pp 903–916.)*

course of a nerve in both number of fascicles and size. This arrangement implies that there are anastomoses and branches between fascicles, called fascicular plexus formation.[2]

Structure and Barrier Properties of the Perineurium   Each fascicle is surrounded by a perineurium (perineurial sheath). This membrane consists of up to 15 lamellae, composed of flattened cells possessing basement membrane on both sides.[31-36] The cells in the lamellae are joined together by tight junctions about 9 mm wide.[36] It is well established that the perineurium acts as a diffusion barrier against several substances such as ferritin,[37,38] horseradish peroxidase,[30,39-41] and albumin.[9,23,28,30,42] The barrier function of the perineurium seems to be located at the innermost layers of the perineurial lamellae, where the tight junctions between the lamellar cells prevent diffusion across the perineurium. This barrier function has been shown to work both from the epineurium to the endoneurium and in the opposite direction.[43] This bidirectional perineurial barrier function may be related not only

cut is made in the perineurium, the nerve fibers herniate out through the "perineurium window."[55]

The slightly elevated EFP in normal cases can be further increased by various disorders and injuries (for example, the experimental neuropathies induced by chronic intoxication, substances such as hexachlorophene and lead, and metabolic diseases such as diabetes).[56] Nerve transection increases permeability in the distal nerve segment during the degeneration process.[57] It has also been shown that EFP may increase after cryogenic injuries,[56] as well as after local compression of the peripheral nerve.[13] Such increases in EFP, resulting from endoneurial edema formation, may have pronounced, long-lasting effects on nerve function. The edema may not be as easily drained as it is in other tissues because of such factors as the lack of lymphatics in the endoneurial tissue[2] and the barrier function of the perineurium. This perineurial barrier function, together with the mechanically unyielding properties of the perineurium, may result in a localized increase in EFP. This situation can be regarded as a "miniature compartment syndrome."[13] This pathophysiologic condition may be significant as the vessels that penetrate obliquely through the perineurium could be occluded by such a limited increase in EFP.[14]

The Nerve Fiber

Ultrastructure of Myelinated and Unmyelinated Fibers   The fibers in the peripheral nerves are of two basic types, myelinated and unmyelinated (Figs. 7–7 to 7–9). In both, the main component of each fiber is a central axon. The nomenclature is not perfectly defined, but a nerve fiber is usually defined as one axon and one Schwann cell (myelinated) or several axons and a Schwann cell (unmyelinated). In myelinated fibers, the axon is surrounded by one Schwann cell. During the developmental period, the membrane of the Schwann cell rotates in a spiral manner around the axon, thereby forming a sheath of lipid and protein layers—the myelin sheath. In unmyelinated nerve fibers, one Schwann cell surrounds several axons.

In myelinated nerve fibers, the Schwann cells are arranged longitudinally on the surface of the axon. Opposing Schwann cells come in close contact with each other at the nodes of Ranvier. At this location cellular processes from the Schwann cells interdigitate, forming a space between the processes, which allows extracellular ions to reach the axons. This is believed to be a structural organization important for the saltatory conduction of nerve impulses from one node of Ranvier to another. Thus, the nerve fiber at this location is exposed to its surroundings, whereas the portion between two nearby nodes of Ranvier is covered by myelin sheaths and the cytoplasm of the Schwann cell. These latter structures are believed to isolate the axon electrically from its surroundings.[58,59]

Conduction velocity is generally higher in myelinated fibers than in unmyelinated fibers because of saltatory conduction in the former. In unmyelinated fibers, impulse propagation takes place in a continuous

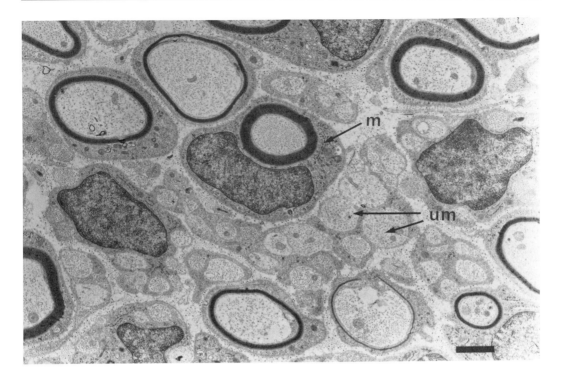

Fig. 7-7 *Electron micrograph showing myelinated (m) and unmyelinated (um) nerve fibers (bar gauge = 1 μ).*

manner along the axon and at a lower speed. It should also be noted that in myelinated nerve fibers the conduction velocity is directly proportional to total fiber diameter.

Erlanger and Gasser[60] defined three groups of nerve fibers: A, B, and C based on their dimensions. Group A contains the largest myelinated somatic afferent and efferent fibers, which have the highest conduction velocities. Group B contains myelinated autonomic and preganglionic fibers. Group C contains the thinnest, unmyelinated visceral and somatic afferent pain fibers, which have the lowest conduction velocities, as well as postganglionic autonomic efferent fibers. Group A can be further subdivided into three subgroups according to nerve fiber diameter: A-α (diameter, 15 to 20 μ), A-β (diameter, 10 to 15 μ), and A-δ (diameter, 2 to 5 μ).

The propagation of impulses along each axon is based on its action potential. This is an all-or-none phenomenon elicited in the axon once the resting membrane potential is changed enough to initiate this process. The propagation of the action potential along the axon is an energy-dependent process and, consequently, is blocked by such phenomena as ischemia. The action potential that can be recorded from a nerve trunk is called compound action potential, and it represents

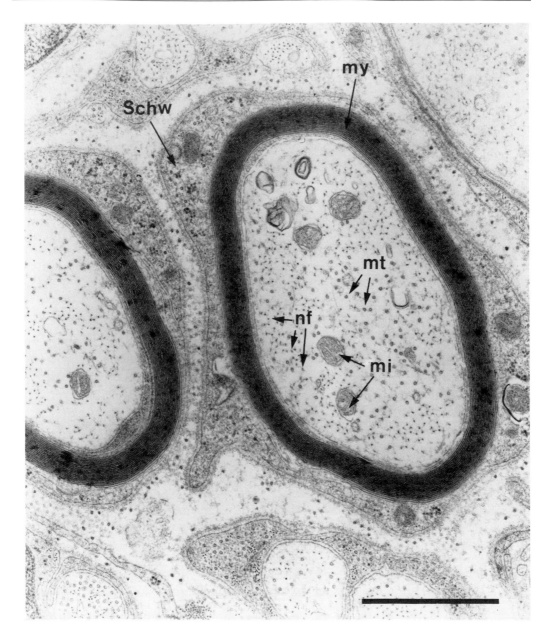

Fig. 7–8 *Electron micrograph showing myelinated fibers containing Schwann cell cytoplasm (Schw), myelin (my), mitochondrium (mi), microtubule (mt), and neurofilament (nf) (bar gauge = 1 μ).*

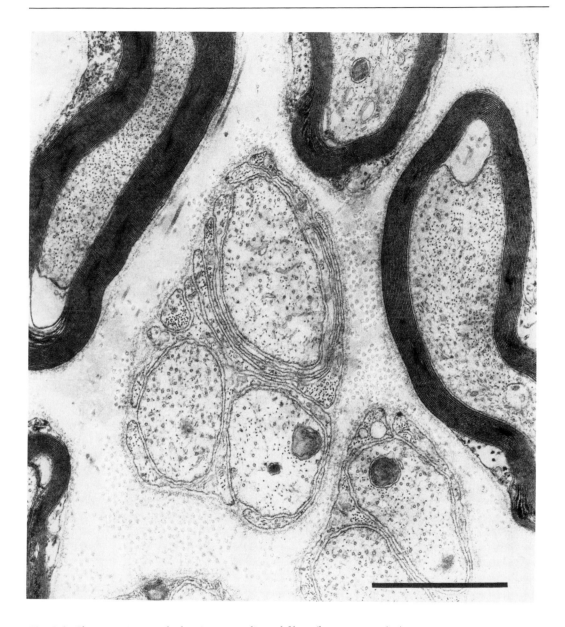

Fig. 7–9 *Electron micrograph showing unmyelinated fibers (bar gauge = 1 μ).*

a summary of the individual action potentials in all fibers conducting during recording.

Anterograde and Retrograde Axonal Transport   Because of the anatomy of the neuron, which has a very long process (the axon) extending from the nerve cell body, there is a need for efficient communication systems between various parts of the nerve cell. The distances between nerve cells are remarkable. The length of an axon can be 10,000 to 15,000 times the diameter of the nerve cell body.

Most substances necessary for the membrane integrity of the axon and the nerve terminal are synthesized in the nerve cell body. These essential products are transported from the centrally located place of synthesis to the periphery, and also in the opposite direction, by the axonal transport systems. These systems carry several neuronal components such as organelles (for example, mitochondria) and various proteins. Axonal transport from the centrally located nerve cell body to the periphery is called anterograde axonal transport; the process in the opposite direction is called retrograde axonal transport. Two main phases of anterograde axonal transport have been identified.[61-65] Slow anterograde axonal transport takes place at a velocity of 1 to 6 mm/day and involves cytoskeletal elements such as microtubules and microfilaments. Fast anterograde axonal transport takes place at speeds of as much as 400 mm/day and involves various enzymes, transmitter substance vesicles, and glycoproteins as well as lipid.[66-74]

Slow axonal transport was first described by Weiss and Hiscoe[75] in 1948. It is believed that slow axonal transport normally serves to replace the axoplasm along the axon continuously. When an injury disrupts axonal continuity, slow axonal transport is involved in reconstituting the new axoplasm. The exact mechanism of slow axonal transport is not completely known but it has been compared to the "protoplasmic streaming" of ameboid motion. There is evidence that microtubules are involved in both slow and fast axonal transport.[76,77]

Different theories have been proposed regarding the mechanism of fast axonal transport. Droz and associates[65,78] suggested that the smooth endoplasmic reticulum is involved in the rapid transport of synthesized material from the nerve cell body to the peripheral parts of the axons. However, there is increasing evidence that fast axonal transport is associated primarily with microtubules.[70,79,80] The microtubules constitute a continuous pathway on which transmitter substance vesicles, organelles, and proteins can be moved along the axons by "transport filaments." This mechanism is energy-dependent. Supporting evidence for the important role of microtubules in fast axonal transport is the fact that colchicine, which is known to disrupt microtubules, can block axonal transport.

A retrograde axonal transport system has also been identified in several neuronal systems.[81-89] Retrograde axonal transport seems to provide recycling of neurotransmitter vesicles. Also, various substances such as herpesvirus and rabies virus may be taken up at the nerve terminals and transported in a retrograde direction.[90] Nerve growth factor may be taken up by sensory and autonomic nerve terminals and transported by retrograde axonal transport.[91] Such phenomena in the retrograde axonal transport systems can take place at speeds ranging from 1 to 2 mm/day up to almost 300 mm/day.[92]

The exact functions and physiologic significance of axonal transport are not completely understood. The anterograde axonal transport systems may play various roles in the normal state and in injuries. It has been postulated that retrograde axonal transport systems may involve a transfer of information to the nerve cell body regarding the status of the axon and the local environment at the nerve terminals.[84,93] The interaction between the nerve cell body and its peripheral connections is illustrated by the cell body response after nerve injury, including the chromatolytic reaction. This reaction in the nerve cell body may be based on interruption of the retrograde axonal transport systems.

Generally, the axonal transport systems require a local energy supply along the axon to function properly. Thus, anoxia and ischemia both in vitro and in vivo block fast and slow anterograde axonal transport and retrograde axonal transport.[94–97] Although both axonal transport and impulse propagation seem to depend on a continuous energy supply, they are not based on the same mechanisms. This is illustrated by the fact that axonal transport can be maintained over a nerve segment that does not conduct impulses after inhibition of membrane excitability by a blocking substance such as procaine.[98,99]

Mechanical Properties of Peripheral Nerves

The peripheral nerves have the ability to adapt to limited tensile loads by exhibiting a nonlinear relationship between the load and deformation (Fig. 7–10). This load-deformation relationship, beyond a small amount of initial deformation, appears to be linear up to the elastic limit. With further stretching, at the point of failure, the load drops suddenly without complete rupture of the nerve trunk.[2,100,101] The ultimate load for the rabbit tibial nerve is about 5 N. Other reported values are 73 to 220 N for the human median nerve and 65 to 155 N for the human ulnar nerve.[2] The elastic limit, as reported for the human nerves, occurs at strain levels between 11% and 17%.[2] Before the elastic limit is reached, the nerve can regain its original length and retain its elastic properties if the tension is released. However, in the post-elastic region, the nerve behaves in a viscous inelastic fashion to the point of complete structural failure. In human nerves, complete structural failure occurs at strain levels between 15% and 23%.[2] However, intraneural structural damage and functional changes gradually occur before complete failure of the nerve trunk. A discussion of the mechanical properties of the nerve trunk, however, is complicated by the fact that nerves are not homogeneous but composite structures in which each tissue component has its own mechanical properties. If a peripheral nerve is cut in vivo, it retracts about 10% to 20% of its length; the exact amount of retraction depends on the local anatomic situation. This phenomenon indicates that nerves are under some tension in vivo.

There is a disagreement in the literature as to which tissue constituent in the nerve is responsible for its tensile properties. According to Sunderland and Bradley[101] and Sunderland,[2] elasticity and tensile strength reside in the fascicular tissue and in particular the perineurium. These authors argued that both the elasticity and the strength of a nerve are retained as long as the perineurium remains intact.

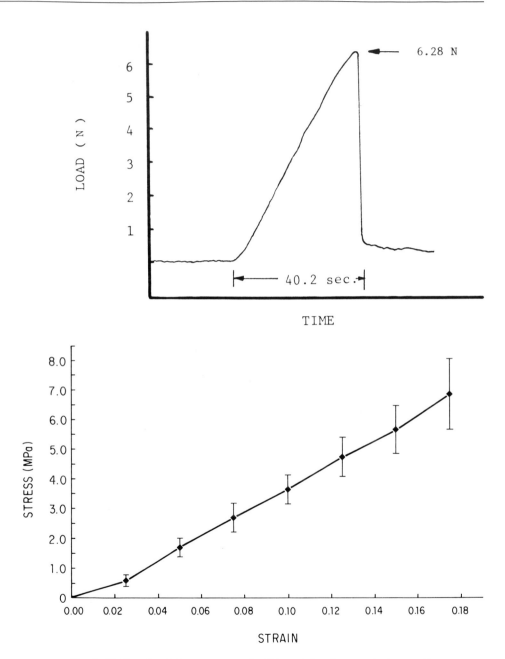

Fig. 7–10 ***Top:*** *Load-time response of rabbit peripheral nerves during tensile experiment. At the point of failure, although the tissues looked grossly intact, a sudden load drop was seen.* ***Bottom:*** *Stress-strain behavior of rabbit peripheral nerves. Linear stiffness was reached at 3% beyond in vivo length.*

Haftek,[100] however, argued that the elasticity and strength of the nerve trunk depend primarily on the epineurium.

The nerve fibers themselves do not seem to contribute to the tensile properties, based on the observation that the distal segment of a severed nerve, which has undergone wallerian degeneration, has the same mechanical properties as a normal nerve. However, the endoneurial connective tissue seems to contribute to the mechanical properties of the nerve trunk, as indicated by the tensile behavior of spinal nerve roots. The nerve roots lack epineurium and perineurium but still have some elasticity and tensile strength. Nonetheless, the nerve fibers in the nerve roots are more susceptible to mechanical damage because they lack the outer connective tissue layers.

The Response of Peripheral Nerves to Trauma

Peripheral nerves may be subjected to various kinds of mechanical deformation but, within limits, inherent protective mechanisms act to maintain normal nerve function. However, if the trauma exceeds a certain magnitude or duration, various degrees of structural and functional damage to the nerve will result. The three levels of nerve injury are low-pressure compression injury, high-pressure compression injury, and interruption of axonal continuity. It should be emphasized that this is an oversimplification. In any given nerve injury, there is usually a variety of damage to different nerve fiber groups. Of importance is the intraneural fascicular topography and the extent to which the fascicles are embedded in protective epineurium.[2] Large nerve fascicles in a small amount of epineurium are more susceptible to compression than are small fascicles embedded in an abundant epineurium. The degree of injury to various nerve fibers may vary depending on their location within each individual nerve fascicle: superficially located nerve fibers undergo more damage than centrally located fibers.[102 105] The diameter of the nerve fibers has also been shown to be significant. Large nerve fibers undergo more deformation than small-diameter fibers at a given pressure.[106] Also, ischemia affects large nerve fibers before thinner nerve fibers.[8]

Classification of Nerve Injuries

If a nerve is moderately compressed, conduction may be impaired by ischemia of the compressed nerve segment. This kind of nerve injury can be called local metabolic conduction block.[107,108] This term can also be used for a more prolonged functional impairment caused by endoneurial edema without structural nerve fiber abnormalities.[107] In more severe trauma, however, there may be a demyelinating conduction block,[108-113] which is characterized by blocked conduction at the level of compression along with preserved axonal continuity and, thus, also normal excitability distal to the lesion. This degree of nerve injury probably depends on the presence of both demyelinization and endoneurial edema. Recovery is usually complete within a few weeks or months. This degree of nerve injury corresponds to the first of the five degrees described by Sunderland[2] and to Seddon's[114] neurapraxia. More severe nerve injury corresponds to Seddon's axonotmesis or Sunderland's second-degree lesion. This lesion is characterized by interruption of nerve fibers with preservation of the Schwann cell sheaths. Recovery after this type of injury is usually complete but requires outgrowth of regenerating axons.

Increasingly severe trauma may induce loss of continuity of nerve fibers and their Schwann cell sheaths within intact fascicles (Sunderland's third-degree injury) or disruption of the perineurium (Sunderland's fourth-degree injury). Sunderland's fifth-degree injury is complete severance of the nerve trunk. Third-, fourth-, and fifth-degree nerve injuries all correspond to Seddon's neurotmesis. Recovery from these more severe lesions is usually incomplete. It should be noted that both these classification systems are qualitative.

Compression Injuries

Low-Pressure Compression   The first sign of impairment of the microcirculation during compression is reduced blood flow in epineurial venules at about 20 to 30 mm Hg of compression.[17] Such impaired blood flow in the venules of a peripheral nerve for a prolonged period is considered important in, for example, the pathophysiology of the carpal tunnel syndrome.[10,115] Impairment of the venular circulation in the epineurium may result in retrograde effects on the capillary endoneurial circulation, which in turn may lead to a depletion of oxygen to the endothelial cells of these capillaries. Anoxic injury to these cells may increase vascular permeability, leading to subsequent leakage of fluid and proteins into the endoneurial space and causing edema. Because there are no lymphatic vessels in the endoneurium,[2] resorption of such edema may be hindered.

Another important factor is impaired blood flow in endoneurial vessels caused by edema resulting in increased EFP. It has been shown that after low-pressure compression (30 to 80 mm Hg) for two to four hours, the EFP may be increased to more than three times the baseline value.[13] This increased EFP can occlude the vessels that penetrate the perineurium at an oblique angle,[9,10,107] thus reducing the intraneural blood flow.[14]

The earliest sign of injury to a nerve is edema formation in the epineurium.[8,9] More pronounced trauma may cause increased vascular permeability with edema formation in the endoneurium as well. Edema in the epineurium can probably be drained if the trauma is not prolonged or repeated. However, the occurrence of endoneurial edema has been shown to be associated with pronounced changes in nerve function.[8,9] Thus, a conduction block may occur in a nerve segment as a result of depletion of oxygen. If the period of compression-induced ischemia is limited to a few hours, and if the applied pressure is not too high, nerve function recovers as intraneural blood flow is restored. However, after prolonged periods of ischemia and/or compression, endoneurial edema formation with subsequent functional changes may occur. In both cases, these functional changes do not necessarily produce pathologic changes in the nerve fibers and the impairment may be considered a physiologic metabolic conduction block.

It should be noted that the axonal transport systems may be inhibited at a compressed nerve segment during pressure levels of 30 to 50 mm Hg.[96,116,117] Long-lasting inhibition of axonal transport may be consistent with depletion of essential substances in the distal parts of the axon. Similarly, retrograde effects may be induced in the nerve cell body in connection with compression lesions peripherally.[118]

High-Pressure Compression Peripheral nerves can be subjected to compression at higher pressure levels in association with acute injuries to extremities and by tourniquets. In such situations, the nerve suffers not only local vascular obliteration but also mechanical deformation of all the tissue components in the nerve trunk, including blood vessels and nerve fibers. As a result, a long-lasting conduction block may be induced. Important factors include local myelin changes and intraneural edema, both of which occur primarily at the edges of the compressed nerve segment. The functional recovery of such a lesion has been shown to be related to remyelination and drainage of endoneurial edema.[9,108,109] If the compressive pressure is very high, axonal continuity can be disrupted.

The local conduction block induced by external high pressure without loss of axonal continuity has been called neurapraxia by Seddon.[114] The pathophysiologic basis for such a lesion has been analyzed by means of compression induced by tourniquet application around the hind limbs of baboons.[108-110] The paranodal myelin is stretched on one side of the node and invaginated on the other. This nodal displacement occurs at both distal and proximal edges of the tourniquet and is followed by segmental demyelination and conduction block. Ochoa and associates[109] reported that the damage to the myelin occurs in large-diameter nerve fibers with sparing of smaller fibers with a total diameter of less than 5 μ.

It has also been shown that intraneural microvascular injury with subsequent edema formation occurs primarily at the edges of the compressed nerve segment.[22] Structural damage to both nerve fibers and intraneural microvessels and functional changes were indeed maximal. The pressure gradient is obviously the largest in these regions.[109,113,119] Experimental data confirmed that the nerve fiber injury may be related to the longitudinal displacement, and the microvascular injury at the edges may depend on the shear stresses at these locations.[119] Mathematical modeling of the tissue deformation associated with nerve compression has demonstrated that longitudinal displacement is maximized at the edges of the compressed segment (Fig. 7–11). Shear stresses are maximized superficially at the edges. The model analyses also indicate that the magnitude of intraneural shear stress is related to the shape of the compression, that is, whether the edge of the compression device is sharp or rounded. A sharp edge produces a more pronounced shear stress concentration than does a rounded edge.

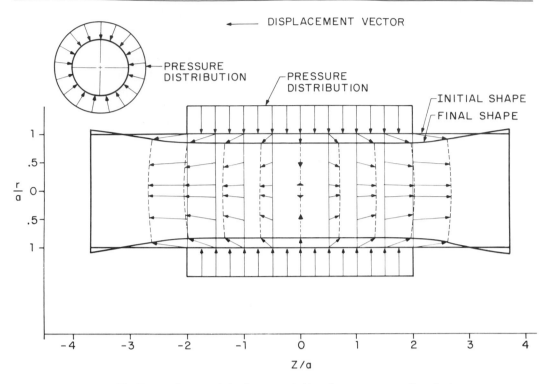

Fig. 7–11 *Theoretical displacement field under a pressure cuff applied to a model cylindrical nerve of radius "a." Displacement in radial (r) and longitudinal (za) directions are computed on the basis of isotropic material properties and elastic series so the deformation is proportional to the pressure. (Reproduced with permission from Rydevik B, Lundborg G, Skalak R: Biomechanics of peripheral nerves, in Frankel VH, Nordin M (eds): Basic Biomechanics of the Musculoskeletal System, ed 2. Philadelphia, Lea and Febiger, in press.)*

Basically, high-pressure compression injuries entail both an ischemic component, because of obliteration of intraneural vessels, and mechanical deformation of the compressed nerve segment. Therefore, it is not yet possible to separate these factors and the outcome of the injury is likely to depend on both.

Interruption of Axonal Continuity

Severe nerve trauma may interrupt axonal continuity and be followed by degeneration of the distal axonal segment. In severe compression or crush lesions, the tubes formed by Schwann cells and their basal laminae (endoneurial tubes) are usually preserved and can facilitate regrowth of axons to their correct peripheral targets. Complete severance of the nerve trunk, on the other hand, produces an extremely complicated situation in which the growth and orientation of axons in the space between the separated proximal and distal segments of the nerve trunk constitute a major problem. Thus, the prognosis depends on the type of nerve lesion. A crush lesion is followed by a very short "initial delay" and a "scar delay"[2] before axons start growing and before they have traversed the injury zone. The correct target organs are reinnervated[120] and there is a minimal blockage of intra-axonally transported material at the site of the lesion when regeneration has occurred.[121] After transection and repair of the nerve trunk, however, the initial and scar delays are extended, axonal misdirection is a major problem, and there is a significant damming of axonally transported material at the site of the lesion because of misdirected and/or permanently constricted axons.[121,122]

Proximal Nerve Segment After an axon is severed, its corresponding nerve cell body undergoes characteristic structural and functional changes.[123-125] The cell body is the major site for synthesis of proteins and other materials required for axonal growth, and the amputation of the distal part of the axon induces changes in the cell body, including metabolic preparation for replacement of lost axoplasmic volume. The typical cell body response to severance of its axon includes increasing cell body volume, displacement of the nucleus to the periphery, and an apparent disappearance of basophilic material from the cytoplasm, called chromatolysis. Ultrastructurally, such chromatolysis represents a reorganization of the rough endoplasmic reticulum, reflecting the increased RNA and protein contents of the regenerating nerve cell.[126] Cell body reaction may vary with age, species, and proximity and nature of the lesion.[127] A proximal lesion may lead to the death of the cell body.

Various explanations of the cell body response have been offered. Exclusion of axonal transport has been regarded as a significant factor. Interestingly, some of the changes produced by axotomy in sympathetic neurons can be reversed by local application of nerve growth factor.[128,129] It is believed that this substance is normally supplied to the sympathetic neurons by their target cells and transported to the cell body by retrograde transport to promote neuronal metabolism.[130,131] Axotomy, therefore, would prevent nerve growth factor from exerting such trophic effects. Changes in the cell body, resembling those elicited by axotomy, can also be elicited by local application to the axon of substances, such as colchicine or vinca alkaloids, that block axonal transport.

Distal Nerve Segment Severance of a nerve trunk progressively increases the collagen contents of the distal segment and produces extensive changes in collagen types and fibronectin.[2,132] Production of collagen in the endoneurium depends largely on fibroblast activity, but it has been suggested that this phenomenon may, to some extent, depend on Schwann cell activity.[133] The collagen becomes progressively denser and the diameter of the endoneurial tube decreases with time.

After axotomy, the distal parts of the fibers undergo wallerian degeneration. Initially, the myelin in the area of axon fragmentation swells into droplets; and over time it is phagocytized by Schwann cells and macrophages. The microtubules and neurofilaments of the axon itself undergo granular disintegration caused by protolytic activity.[134,135] There is accumulating evidence that this enzymatic disintegration of proteins is calcium-dependent and that neurofilament disruption in mammalian peripheral nerves is mediated by a calcium-activated enzyme in the axoplasm.[136] Degeneration of axons has been inhibited and delayed in vitro by incubation in calcium-free media[136] and in vivo by local application of leupeptin—a protease inhibitor.[137-139] After axotomy, the Schwann cells in the distal segment rapidly undergo mitosis, forming Schwann cell columns (band of Büngner). Schwann cell proliferation starts one to five days after axotomy, with the activity usually peaking after about three days.[140-143] The bands of Büngner are regarded as important pathways for the regenerating axons, representing guidelines and probably also trophic sources for regenerating fibers. The basal lamina of the Schwann cell contains the glycoproteins fibronectin and laminin; the latter is known to promote the growth of neurites.

Schwann cell proliferation is associated with synthesis of specific proteins in the distal nerve segment.[144,145] After transection, cells in the distal segment begin to synthesize and secrete an acidic 37-kD protein that accumulates in the extracellular space. According to Müller and associates,[146] the protein is synthesized only in those segments of the nerve in which axons regenerate. Kanje and associates[147] emphasized the significance of synthesized proteins for the growth of axons, showing that inhibition of Schwann cell proliferation or protein synthesis by local drug administration after a crush injury interferes with axonal growth in rat sciatic nerve. Increased synthesis of IGF-1 (somatomedin C) distally was observed by Hansson and associates[148] in crush injuries.

Stretching Injuries of Peripheral Nerves

Stretching injuries of peripheral nerve trunks are usually associated with severe accidents, for example, when high-energy traction is applied to the brachial plexus. Such plexus injuries may result in partial or total functional deficits in terms of sensory and motor loss. The outcome depends on which tissue components of the nerve trunks are damaged as well as on the extent of tissue injury. Plexus injuries represent one extreme type of stretching lesion caused by a sudden, violent trauma. A completely different stretching injury situation occurs when both ends of a cut nerve are sutured under moderate tension. This situation happens, for example, when there is a substantial loss in the continuity of a nerve trunk, and restoration of continuity requires the application of tension to the nerve ends. In such cases, the moderate gradual tension may be sufficient to reduce the transverse fascicular cross-sectional area, possibly resulting in impairment of intraneural nutritive capillary flow. Plexus injuries, however, involve larger forces that may break some intraneural tissue component. Another stretch injury involves more gradual extension applied over a long time, such as that caused by the growth of intraneural tumors.

According to Sunderland,[2] there are several stages of change occurring in the nerve under progressively increasing tension. At first the nerve elongates rapidly and the slack in the fascicles and nerve trunk is taken up. A point may be reached at which conduction is blocked before any significant morphologic changes can be detected. With increased tension, axons begin rupturing inside their as yet intact endoneurial tubes and a second-degree lesion is established. With continued stretching, the endoneurial tubes also rupture and a general disorganization of the endoneurial contents occurs inside the intact perineurium that gives elasticity to the nerve trunk. Such a third-degree injury is transformed into a fourth-degree injury when the perineurium breaks. Haftek[100] presented a different view regarding this sequence. He found that with increased tension the epineurium ruptures first while the perineurium is still intact and the endoneurial tubes are still in continuity.

Regeneration

Cellular Background: The in Vitro Approach

The peripheral nerve is a complex assembly of several discrete cell types although these components are interdependent. Schwann cells, for example, monopolize all axonal surfaces but not other cell surfaces; perineurial and epineurial cells do not invade the center of the nerve; and vascular and fibroblastic elements seem able to connect with any cell surface except that of the neuron. This organization serves to isolate the different cell types so that selected cell-to-cell communications are facilitated. When the peripheral nerve is damaged, for example, by crushing or transection, axons are interrupted, leading to degeneration of their distal segments. This axonal loss leaves the distal Schwann cells without their principal cellular contact and deprives the neurons of target contributions as well as part of their normal allotment of Schwann cells.

As shown in Figure 7–12, successful peripheral nerve regeneration requires a coordinated sequence of neuronal cell behaviors. The damaged neuron must remain viable and healthy. The healthy neuron must regenerate a new growth tip (growth cone) from its proximal nerve stump or, if the proximal axon completely degenerates, from the parent somata. The growing nerve tip must then extend distally, cross the transection site, invade and grow through the degenerated distal nerve toward its original innervation area, make contact, reinnervate, and restore functional interaction with a target cell. During all this, Schwann cells must reestablish their specific pre-injury relationships with the regenerated axon.

Each of these cellular behaviors is likely to be controlled by specific physical and/or chemical cues. For example, for nerve regeneration to proceed, the regeneration environment must contain specific molecules that (1) maintain the survival and general growth capabilities of the damaged neuron, (2) stimulate the initiation of a growth tip and elongation of axons, (3) guide the growing axon toward the periphery, (4)

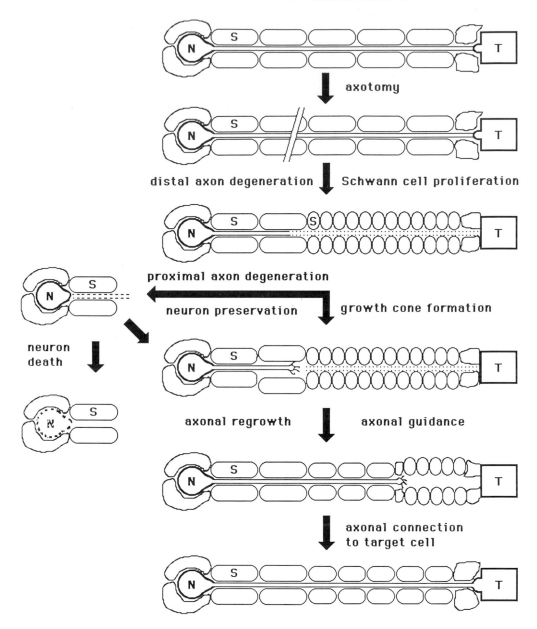

Fig. 7–12 *Neuronal and Schwann cell changes induced by peripheral nerve damage. In normal peripheral nerve, the neuron (N) is associated with Schwann cells (S) and is connected by a long axon to its target cell (T). When an axon is interrupted by an injury, the distal portion, deprived of its cell body, degenerates and the associated Schwann cells proliferate in the degenerating area. The proximal portion of the axon may then degenerate back to its soma before a new growth tip (growth cone) forms. Because the axotomized neurons are cut off from all potential survival-promoting contributions from their target cells and their Schwann cells, some of the damaged neurons may die. The regenerating growth cones from surviving neurons must extend across the transection site and be guided back to their target areas where they must recognize and reconnect with a target cell. During this axonal regrowth, Schwann cells reestablish an association with the new axon.*

enable the axon to reconnect with a target cell, and (5) stimulate the Schwann cell to re-ensheath and, where appropriate, remyelinate the regenerated axon. In vitro techniques remain the best ways to characterize substances with potential influence on regenerative nerve cell behaviors and to examine their modes of action.

Sensory, sympathetic, and motor neurons—the three neuronal types contributing axons to sciatic nerves—can be removed from the developing organism (and thus experimentally axotomized) and purified of the Schwann and other nonneuronal cell types present in their source tissues. When such neurons are cultured on a solid, adhesive substratum and in a standard nutrient-rich medium, it becomes possible to demonstrate unequivocally that these axotomized neurons do, in fact, require specific molecular signals for survival, growth-cone formation, axonal elongation, and axonal guidance. Figure 7–13 illustrates such varied responses from cultured parasympathetic chick embryo ciliary ganglion motor neurons when presented with selected growth-promoting signals. Ciliary ganglion neurons cultured in the absence of added growth factors (Fig. 7–13, *top left*) attach to the culture substratum but degenerate in place and die within one day. However, cultured ciliary ganglion neurons survive for prolonged periods in the presence of extracts from intraocular muscle (the normal innervation target tissue of ciliary ganglion in vivo) or from peripheral nerve (Fig. 7–13, *center left*). Thus, target or nerve tissue extracts contain a neuronotrophic factor that supports ciliary ganglion neuronal survival. Under the influence of neuronotrophic factor alone, the neurons do not readily form growth cones or extend axons. However, when the neuronotrophic factor is supplied to neurons cultured on a substratum previously exposed to medium collected from Schwann cell cultures or to extracts of basal lamina, the neurons form growth cones very rapidly (Fig. 7–13, *bottom left*) and with time extend long randomly oriented axons (Fig. 7–13, *top right*). Thus, the Schwann cell "conditioned medium" and basal lamina extracts contain a neurite-promoting factor that stimulates the regrowth of axons. If ciliary ganglion neurons are cultured with neuronotrophic factor on a substratum containing only restricted pathways of bound neurite-promoting factor or basal lamina containing neurite-promoting factor, axonal growth is confined almost exclusively along those pathways (Fig. 7–13, *bottom right*).

Survival and growth responses to these substances are dose-dependent. Figure 7–14 illustrates the survival response of ciliary ganglion neurons to their neuronotrophic factor supplied at twofold serial dilutions. Similar dose-response relationships have been shown for other cultured neurons and their neuronotrophic factors [149-155] as well. Also, dose-response relationships have been demonstrated for substances that specifically elicit axon initiation,[156-160] an increased rate of axon elongation,[161] axonal maintenance,[162] and axonal retraction.[163] Such in vitro bioassays are useful in determining the concentration of biologic activity (Fig. 7–14) and are used routinely as the method of choice to monitor such bioactivity during factor purification and characterization.

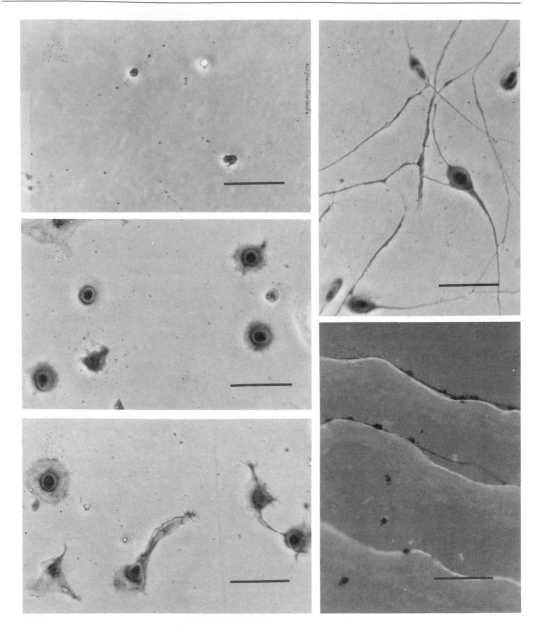

Fig. 7–13 *Influences of target tissue-derived substances on cultures of purified motor neurons. Embryonic chick parasympathetic motor neurons were cultured for 24 hours under different conditions.* **Top left:** *In the presence of a basal culture medium but absence of culture supplements, all neurons attach to the substratum but die in place, leaving only vestiges (bar gauge = 100 μ).* **Center left:** *In the presence of small amounts of extracts from peripheral nerve, neurons survive but do not extend axons (bar gauge = 100 μ).* **Bottom left:** *In the presence of small amounts of culture medium exposed to Schwann cells, the neurons form growth cones (bar gauge = 100 μ).* **Top right:** *In the presence of high levels of culture medium exposed to Schwann cells and after a longer period in culture, neurons extend long fibers (bar gauge = 100 μ).* **Bottom right:** *When these motor neurons are cultured in the presence of nerve extract on immobilized cryostat sections of basal lamina, the neurons (stained dark with antineurofilament antibody) extend axons that follow the contour of the basal lamina (stained light with antilaminin antibody) (bar gauge = 500 μ).*

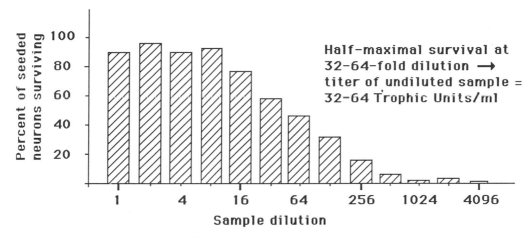

Fig. 7–14 *Dose-response relationships between neuronal survival in cultured ciliary ganglion motor neurons and purified rat sciatic nerve ciliary neuronotrophic factor. In the example, maximal neuronal survival is maintained when the sample is diluted 1:8 and survival declines when the sample is further diluted. Survival is only 50% of the maximum between 1:32 and 1:64. Thus, the sample "titer" in trophic units/ml is 32 to 64/ml.*

Thus far, a number of neuronotrophic and neurite-promoting factors have been isolated and characterized, largely because of the development of quantitative in vitro assays for their biologic activities. A list of the specific factors described to date is presented in Table 7–1 together with their reported occurrence in peripheral nerve. Particularly pertinent to this review is that peripheral nerves are the richest known source of ciliary neuronotrophic factor, a 28-kD protein supporting the in vitro survival of certain sensory, sympathetic, and parasympathetic motor neurons.[151,164] Also, the peripheral nerve is particularly rich in Schwann cell-associated basal laminae that contain high levels of laminin, a protein that is the most potent nonspecific in vitro promoter of axonal growth yet described.[159,165,173]

Certain aspects of the biologic activities of these factors are potentially relevant to peripheral axon regeneration:

(1) All of the neuronotrophic factors listed in Table 7–1 can express their biologic activities when presented in soluble form to the cultured neurons and most of them are capable of expressing their activities even after being bound to a solid substratum.[152,164,177] When nerve growth factor[166] and ciliary neuronotrophic factor[164] were bound only to selected areas of a culture substratum, neurons cultured without neuronotrophic factors in the medium survived only on those areas containing the bound factor. Thus, neuronal survival may be mediated in vivo by released, surface-bound, or extracellular matrix-bound factors.

(2) When nerve growth factor is presented in vitro from a micro-

Table 7–1 Isolated neuronotrophic and neurite-promoting factors in peripheral nerves

Factor	Reported in Peripheral Nerves	Investigators
Neuronotrophic factors		
Ciliary neuronotrophic factor	Yes	Barbin et al[149]; Manthorpe et al[151]; Carnow et al[164]; Rudge et al[169]
Brain-derived neuronotrophic factor	No	Barde et al[150]
Fibroblast growth factor	No	Walicke et al[154]; Morrison et al[166]
Nerve growth factor	Yes	Varon et al[174]; Korsching and Theonen[176]
Neuroleukin	No	Gurney et al[171]
Purpurin	No	Schubert and LaCorbiere[152]
Neurite-promoting factors		
Collagen	Yes	Adler et al[175]
Fibronectin	Yes	Rogers et al[172]
Laminin	Yes	Manthorpe et al[159]; Wewer et al[160]; Engvall et al[170]; Rogers et al[172]
Laminin-proteoglycan complexes	Yes	Lander et al[158]; Davis et al[167, 168]
Neurite extension factor (−S100)	No	Kligman[157]

pipette to an adjacent advancing sensory neuronal growth cone, the growing tip turns and grows toward the source of nerve growth factor.[178] Thus, axonal growth in vivo may be directed toward cellular sources (for example, target or distal Schwann cells) of nerve growth factor.

(3) Sympathetic neurons have been cultured in a multicompartment chamber in which cell bodies and their axons can exist in separate media environments.[162,179] This multicompartment technique allowed the investigator to establish that nerve growth factor supplied only to the axonal compartment supported both the axons and, presumably via retrograde axonal transport, the soma. However, factor supplied only to the soma supported only the soma. Thus, without a local supply of nerve growth factor, axons degenerate. Variations of these studies also established that axons supplied with nerve growth factor do not grow into a compartment free of nerve growth factor.

(4) Ciliary motor neurons require ciliary neuronotrophic factor for their in vitro survival but respond to nerve growth factor (in the presence of the ciliary neuronotrophic factor) with an increased rate of axonal elongation.[161] Thus, a given neuronal cell can respond differently to different factors. Also, certain sensory and sympathetic neurons can be supported by either ciliary neuronotrophic or nerve growth factors.[149,151] Thus, some neuronal survival could be assured in vivo by more than one factor.

(5) Basic fibroblast growth factor, a known mitogen for Schwann[180] and fibroblastic cells,[181] also supports the survival of different types of brain neurons[154,166] and is localized in brain neurons in vivo.[182] Thus,

a given factor may express biologic activity toward both neuronal and nonneuronal cells.

(6) No neurite-promoting factor supports neuronal survival by itself and all but one (B–5100) exert their axonal growth-promoting effects only when immobilized on the substratum.[159,167,168,183] Laminin is by far the most potent.[165] Localized pathways of immobilized neurite-promoting factors can serve as local pathways for axonal growth.[183,184]

It should be clear from the above discussion that in vitro techniques offer ways to detect, measure, and characterize the biologic activities of substances promoting neuronal survival and axonal growth. Several specific substances have been identified, many of which may be required in vivo for proper peripheral nerve regeneration.

Sprouting

After severance of a nerve trunk, each axon in the proximal stump produces a large number of sprouts that advance distally. Terminal sprouts arise from the tip of the axons and collateral sprouts from more proximal nodes of Ranvier. The sprouting process starts within the first days after transection. Regeneration units, representing clusters of many nonmyelinated sprouts originating from the same parent axon, appear at an early stage (Fig. 7–15). At the distal-most part of each sprout there is a growth cone, a swelling from which microspikes or filopodia arise. The filopodia palpate and explore their environment with constant movements. The average number of axonal sprouts decreases continuously as some of them (pioneer fibers) make peripheral connections and then mature at the expense of those sprouts that fail to establish peripheral contact.[185]

During the first weeks after nerve severance, endoneurial fibroblasts and perhaps Schwann cells undergo circumferential elongation to surround groups of axons, thus coming to resemble a perineurial cell (Fig. 7–16). This process is called compartmentation and is probably stimulated by the disturbance of the endoneurial environment resulting from damage to the normal perineurium. The formation of many miniature fascicles probably results from a need for rapid restoration of normal endoneurial environment by creating a new "perineurium." The compartmentation phenomenon occurs only during the first months after transection; by one year there are few or no signs of such compartmentation.

Axonal Regrowth Across the Interstump Gap

When axons in the nerve are transected by trauma, major issues requiring a better understanding are (1) whether all the axotomized neurons actually survive the transection, (2) the sequence of cellular events permitting the regenerating axons to cross the transection site, (3) the specific molecules required for the sequence of cellular events to occur, and (4) which manipulations of the nerve reformation environment best produce an approximation of the normal nerve structure. Thus, models are needed in which the regenerating nerve can be

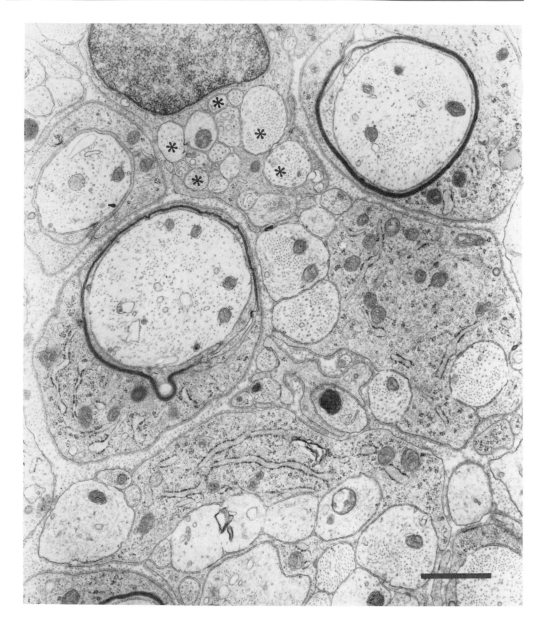

Fig. 7–15 *Regenerating sprouts in the silicone chamber model. Numerous sprouts (*) originating from one parent axon constitute a regeneration unit (bar gauge = 1 μ).*

studied in a defined and experimentally alterable environment. One such model, an impermeable "nerve regeneration chamber," has been useful in addressing some of these issues.

The Regeneration Chamber In 1979, Lundborg and Hansson[186] described regeneration of adult rat sciatic nerve within a preformed,

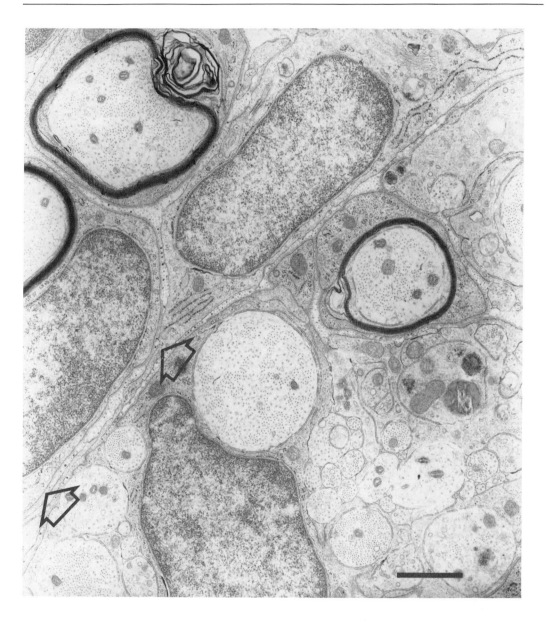

Fig. 7–16 *"Early compartmentation." Endoneurial cells extend long processes (arrows) encircling groups of axons to form minifascicles (bar gauge = 1 μ).*

autologous tissue (mesothelial) chamber implanted in vivo. Lundborg and associates[187,188] and Varon and Williams[189,190] described similar studies using impermeable silicone chambers, a modification allowing better control over the fluid environment surrounding the regenerating nerve structure.

327

Figure 7–17 illustrates the salient features of the silicone chamber system. Briefly, a 2-mm segment of adult rat sciatic nerve is resected from the midthigh level and the proximal and distal nerve stumps inserted with minimal compression into the opposing ends of a cylindrical silicone tube, leaving a10-mm interstump gap. A series of studies[188,191–194] has shown that within one day the chamber fills with fluid exuded from both stumps. Within one week, a fibrous, coaxial matrix has formed to connect the opposing stumps physically. By two weeks, the coaxial matrix has been invaded from both stumps by Schwann cells and other cells, although axons in typical "regeneration units" associated with Schwann cells are only occasionally seen midway along the interstump gap. After one month the nerve structure has reformed and displays endoneurial mass (including blood vessels, Schwann cells, myelinated and unmyelinated axons) surrounded by a perineurium-like outer cell layer.

The advantage of this system is that the chambers can be collected periodically and analyzed histologically to gain information on the sequence of cellular events occurring during nerve reformation. Since the reforming nerve structure is bathed in a fluid, it is possible to collect and test the fluid periodically for agents whose presence or absence might correlate with defined cellular events. Also, the contents of the fluid can be altered, that is, specific components can be removed or added and their impact on subsequent progression of matrix assembly and nerve reformation assessed.

In vitro studies indicated that embryonic neuronal cells at least require neuronotrophic and neurite-promoting factors to survive and regenerate their axons. If such requirements also apply to adult rat neurons contributing to the sciatic nerve (that is sensory, sympathetic, and spinal motor neurons), one must hypothesize (1) that such neuronotrophic and neurite-promoting factors are present within the implanted silicone chamber, (2) that the relevant adult rat neurons can respond to the neuronotrophic and neurite-promoting factors, (3) that axonal maintenance and elongation within the chamber correlate temporally with the presence of the neuronotrophic and neurite-promoting factors, respectively, and finally, (4) that changes in the levels of neuronotrophic factor and/or neurite-promoting factor within the chamber at the critical period or periods alter the survival of neurons contributing axons to the chamber or alter axonal maintenance and growth through the chamber. The following discussion presents evidence supporting the above hypothesis.

Neuronotrophic and Neurite-Promoting Factors Within Chamber Fluid In a series of studies,[187,192,194,195] fluid was collected from separate silicone chambers at various times after implantation and assayed quantitatively for neuronotrophic factor activity toward perinatal sensory, sympathetic, and spinal cord motor neurons. Neuronotrophic factor activities for all the test neurons were detectable in fluids collected at the earliest times (one hour), reached a peak within the first few days, decreased to much lower levels thereafter, and persisted at

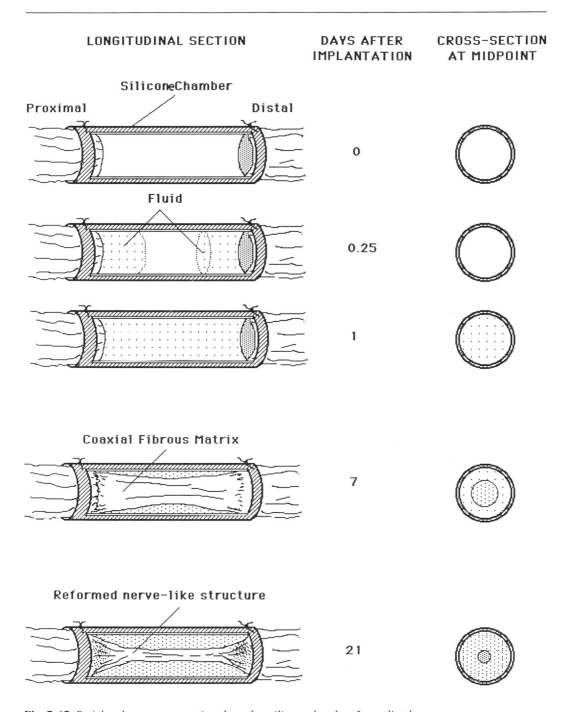

Fig. 7-17 *Peripheral nerve regeneration through a silicone chamber. Immediately after the two stumps of an experimentally resected adult rat sciatic nerve are sutured into the opposing ends of a silicone tube, the tube begins to fill with fluid from both ends, filling it by one day. By seven days, a coaxial matrix (stroma) has traversed the interstump gap and become disconnected from the tube walls. Histologic studies show that most of the center regions of this matrix contain fibrous material free of Schwann cells and connective tissue elements. By one month the matrix has reduced in diameter.*

relatively low levels for at least one month. The times at which peak activity was expressed differed for different neuronal cell types (with sensory activity peaking at three hours, sympathetic activity at 24 hours, and spinal motor activity at 72 hours), suggesting that the activities depend on distinct molecules released into the fluid under separate regulation. It remains to be determined if these neuronotrophic factors are related to the neuronotrophic factors measured in extracts of undamaged peripheral nerve (Table 7–1).

Chamber fluid also shows neurite-promoting factor activity. This was almost undetectable for the first few days, appeared at about day 6, and gradually increased thereafter for up to three weeks. This substance can be bound and removed from solution with immobilized anti-rat laminin antibodies but the activity is not inhibited by soluble anti-rat laminin antisera. Thus, fluid neurite-promoting factor activity may depend on a laminin-proteoglycan complex with similar properties that was recently purified from schwannoma tumor cell conditioned media.[166,167]

Response of Adult Rat Neurons to Neuronotrophic and Neurite-Promoting Factors   There are few reports on the culture behavior of dissociated adult neurons because of difficulties in disaggregating adult neural tissue. Thus far, only neurons from adult rat sensory neurons have been obtained and tested for their in vitro response to neurite-promoting factors.[196] Purified rat laminin, schwannoma cell-derived laminin, and the schwannoma cell-derived laminin-proteoglycan complex dramatically stimulate axonal regeneration from these cultured adult sensory neurons. However, the adult neurons are somewhat delayed in their response compared with their neonatal counterparts. The influences of chamber-fluid neuronotrophic and neurite-promoting factors have yet to be tested using this neuronal system.

Correlation of Neuronotrophic and Neurite-Promoting Factor Activities With Axonal Entry, Maintenance, and Growth in the Chamber   All neuronotrophic factor activities can be detected in the chamber by three hours (that is, as the chambers become filled with fluid) and all are present thereafter, although at different relative levels.[192] Thus, if the freshly transected axons within the proximal stump require these same neuronotrophic factors for maintenance and subsequent regenerative growth into the chamber, such axons should, from the start, be continuously bathed in fluid containing neuronotrophic factors.

As mentioned above, axons are seen in the chamber by two weeks after implantation. Williams and associates[191] reported that Schwann cells first emerge from the proximal and distal stumps at about one week and proceed along the fibrous coaxial matrix toward one another to meet near the center by about two weeks; thereafter they accumulate along the entire length. Axons seem to enter the chamber from the proximal stump within the second week. They are numerous and extensively myelinated by the fourth week. Neurite-promoting factor activity does not appear in the fluid surrounding the matrix before the first week (a time preceding substantial entry of cells) but gradually increases over the next two weeks—the time during which a large number of Schwann and other cells have entered—but few axons are present. Thus, the increase in fluid neurite-promoting factor activity correlates temporally with the accumulation of Schwann and other cells within the chamber and is at relatively high levels when axons enter the chamber. One interpretation of this correlation is that the neurite-promoting factors in the fluid reflect their accumulation on the surface of the nonneuronal cells and that this cell-surface neurite-promoting factor is the required growth substratum for entering axons.

Because of the reactivity of the fluid neurite-promoting factors to antilaminin antibodies, Longo and associates[193] examined the development with time of laminin immunoreactivity within the coaxial matrix. This study showed that laminin was not detectable in the matrix at one week, became minimally detectable at two weeks, and was extensive at three weeks. Thus, laminin, a potent in vitro axon-promoting agent, appeared bound to the coaxial structure at about the same time as did the nonneuronal cells and was present there at even higher levels at a time when axons were regenerating into the chamber.

Alteration of Neuronotrophic and Neurite-Promoting Factors in the Chamber Environment   One general modification of the chamber that does not allow nerve regeneration occurs when no distal nerve insert is provided. This can be accomplished either by sealing one end of the chamber or by leaving the distal end open without the nerve insert. In both cases, the fluid filling these chambers contained high levels of the same neuronotrophic factors measured in normal chambers.[197] Thus, the presence of these neuronotrophic factors does not assure a regenerative response. An important result of the lack of the distal nerve insert in open chambers is that the coaxial matrix either does not form or becomes populated by cells other than Schwann cells from the distal end. Therefore, the lack of regeneration in these modified chambers may be related to the lack of solid growth substratum and the presence of matrix-associated or cell-associated neurite-promoting factors.

The evidence thus far encourages the idea that peripheral nerve regeneration requires neuronotrophic and neurite-promoting factors, but remains insufficient to validate such a notion unequivocally. Current information suggests that appropriate application of neuronotrophic and neurite-promoting factor technology in vivo might improve the regenerative response.[198] Model systems such as the silicone chamber offer unique opportunities to define and fulfill the requirements for successful peripheral nerve regeneration.

Axonal Growth in the Distal Segment

Sprouts, growing toward the distal nerve segment, may either approach a Schwann cell column or grow at random into the connective tissue layer of the nerve. Schwann cell columns may be invaded by numerous sprouts coming from the parent axon or other axons in the proximal segment. Because an excess number of sprouts invade the distal segment, the total number of axons in the distal segment exceeds the number in the proximal segment. With time, the number of axons decreases distally as the sprouts that fail to make peripheral connections atrophy and disappear.

The Schwann cells are instructed by some signal from the advancing axons to develop into myelinated or unmyelinated fibers.[199,200] Apparently each neuron is coded to be myelinated or unmyelinated; the instruction to the Schwann cell to form myelin requires direct contact between the axon and the Schwann cell. Unmyelinated fibers remain unmyelinated even when they regenerate into a distal nerve segment that originally contained myelinated fibers.[140] In animal experiments the rate of axonal outgrowth in the distal segment is 2 to 3.5 mm/day after transection and 3 to 4.5 mm/day after a crush injury.[121,201,202] In humans, the average outgrowth rate has been estimated to be at most 1 to 2 mm/day.[203]

Factors Influencing Axonal Orientation and Growth

It has long been known that nerve fibers tend to grow toward a distal nerve segment rather than at random.[204,205] The term neurotropism implies preferential growth toward a distant target of nerve tissue. However, Weiss and Taylor[206] believed that this phenomenon represented randomized growth only in that some axons reached the peripheral nerve stump by accident, whereas others grew in wrong directions and disappeared. In media lacking oriented stroma, it was observed that nerve processes followed random courses, whereas an oriented interphase offered "contact guidance."[207] When nerve fibers regenerate into a blood plasma clot under some tension, the fibers follow the stretch lines, thus presenting directed and polarized growth.

Weiss and Taylor[206] studied the growth of axons in Y-shaped chamber systems (transplanted aorta bifurcation from a donor rat) in which a piece of nerve or tendon was introduced into their respective distal outlets. They observed no preferential growth toward the nerve segment compared with the tendon. However, the use of other materials in the chamber and modern methods of assessing nerve growth have produced indications that axons grow preferentially toward nerve tissue rather than other types of tissue.[208–211] Topographic specificity, that is, preferential growth of axons belonging to a certain fiber component of the nerve toward the corresponding distal parts, has been shown.[208,212,213] These authors cut the peroneal stumps into the proximal inlet of a Y-chamber system and peroneal and tibial nerve segments into both outlets, and found preferential growth of peroneal axons to the outlet associated with the distal stumps of peroneal rather than tibial nerves.

It has been demonstrated that a lesion created two weeks before a second test lesion produces more rapid axonal outgrowth.[214–216] Two weeks after creation of a conditioning lesion, the most rapidly growing sensory axons in the sciatic nerve of rat demonstrated a 23% increase in outgrowth rate.[215] Acceleration of axonal outgrowth has also been reported after a conditioning lesion created only two days before the test lesion.[217] The effects of the conditioning lesion probably result from increased protein synthesis in the cell body, reflected by the cellular response after injury.

Hormonal influence on nerve regeneration occurs in certain circumstances. Triiodothyronine (T_3) administration has been reported to stimulate protein synthesis in the nerve cell body. It increases and improves maturation of regenerating axons.[218–220] Treatment with thyroxine (T_4) for three months increases the ability of the axons to pass extended gaps.[221] Also, adrenocorticotropic hormone has been reported to increase the number of outgrowing axons after a crush injury, although no increase in outgrowth rate was observed.[222–224]

There are reports that direct-current electric fields and pulsed electromagnetic fields (PEMF) may, under certain conditions, promote axonal regeneration. Kort and Bassett,[225] studying the effects of PEMF on bone healing, reported improved nerve regeneration in the corresponding limbs. Published reports of the effects of electromagnetic fields, however, conflict. In vitro, there seems to be a preferential neurite outgrowth from nerve explants towards the cathode of extracellularly applied electric fields.[226–228] Radji and Bowden[229] reported that PEMF had a dramatic effect on axonal outgrowth after experimental nerve repair, but Orgel and associates[230] could not verify such effects. Orgel and associates did, however, verify that an increased number of motor neurons reestablished appropriate connections to the periphery as a result of improved axonal orientation at the suture site under the influence of the field.

In vitro, cyclic adenosine monophosphate (cAMP) has been found to stimulate neurite outgrowth.[231] In vivo, stimulatory effect of cAMP was reported by Pichichero and associates,[232] who observed a more rapid recovery of reflex function in the hind limb of rats after crushing the sciatic nerve. Others, however, have not been able to demonstrate corresponding effects.[215]

Clinical Aspects of Nerve Repair

Primary vs Secondary Repair

In light of the response of the nerve cell body after injury and the effects of a conditioning lesion, there may be a theoretical basis for delaying the repair of severed nerves. However, there are many reasons for immediate, primary repair of a sharply cut nerve. In the early phase it is generally possible to achieve an exact orientation of the cut nerve end because of well-preserved landmarks such as the fascicular pattern in the transected surfaces and the longitudinal epineurial vessels. The ends can be coapted without tension. If a delayed, secondary repair is performed, the nerve segments may be retracted and there may be considerable difficulty in suturing the nerve ends without tension.

Also, experimental data support the view that nerves should be sutured primarily.[233–235] Denervated muscles undergo rapid atrophy and a delay of 18 to 24 months may result in irreversible degenerative

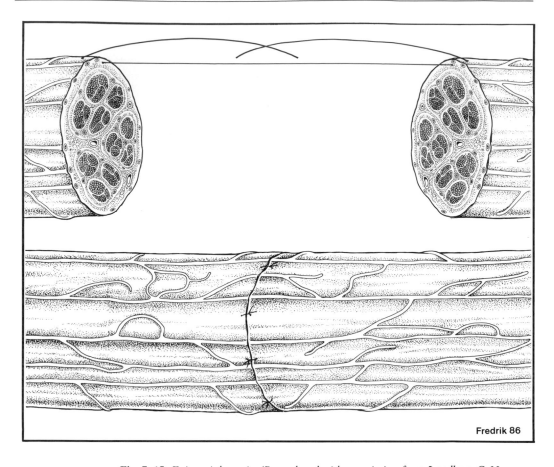

Fredrik 86

Fig. 7–18 *Epineurial repair. (Reproduced with permission from Lundborg G:* Nerve Injury and Repair. *Edinburgh, Churchill Livingstone, in press.)*

changes in muscles. Sensory organs seem to be more resistant to denervation, and it has been reported that there is little correlation between recovery of sensory nerves and the time between injury and repair.[236] However, poor results in some cases can be explained by atrophied skin receptors despite good reinnervation of digital nerves.[237]

Perineurial and Epineurial Suture Techniques

The epineurial suture technique (Fig. 7–18) represents a comparatively simple method of nerve repair. It does not, however, ensure correct matching of the fascicular internal structures.[4,238,239] The development of microsurgical techniques introduced the concept of group fascicular repair to achieve optimal orientation and adaptation of groups of fascicles.[240–243] With this technique, epineurial tissue is resected over a short distance, fascicular groups are dissected under high magnifi-

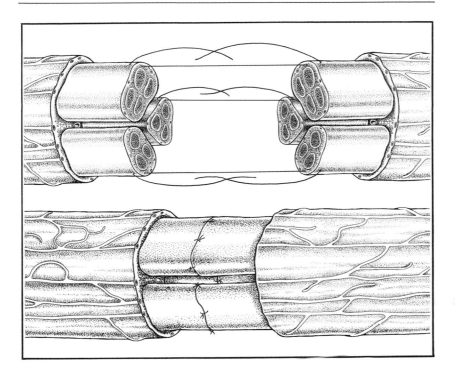

Fig. 7–19 *Group fascicular repair. (Reproduced with permission from Lundborg G:* Nerve Injury and Repair. *Edinburgh, Churchill Livingstone, in press.)*

cation, and corresponding fascicular structures in both nerve ends are sutured individually to achieve optimal matching of corresponding nerve elements (Fig. 7–19). Fascicular sutures do not resist tension and can usually be performed only as a primary procedure. The advantage of this method is improved matching of fascicular components, but the procedure includes a considerable amount of surgical dissection with potential risks in terms of microhemorrhage, edema, and scarring.

Despite many attempts to compare perineurial and epineurial suture techniques experimentally, no consistent superiority of one technique over the other has been demonstrated.[234,244-249] However, there are reports of improved specificity of muscle reinnervation after fascicular suture of rat sciatic nerve.[250,251] Tracer techniques (retrograde transport of horseradish peroxidase) demonstrated that inappropriate reinnervation of peroneal muscle by tibial motor neurons in the rat was minimized by individual fascicular sutures.

It is generally agreed that there is a place for both techniques and that each nerve lesion should be judged individually. A simple epineurial suture may be used to treat a clean, fresh nerve injury, especially at levels where fascicles are closely packed with minimal interfascicular epineurial tissue. However, at levels where the transectional area of the nerve is dominated by epineurial tissue, such tissue should be resected and fascicular groups formed by individual stitches. At distal levels of an extremity, nerve branches representing unique functions are usually well defined in separate fascicles within the main nerve trunk. In such situations (for instance, the median and ulnar nerves of the wrist at a level where motor and sensory components are separated and well defined), a microsurgical approach is indicated.

Nerve Grafting

It has been shown experimentally that tension at the suture line increases scar tissue formation and prohibits axonal regeneration.[252-257] Even slight tension may compromise intraneural microcirculation.[9,16] When it is not possible to suture the nerve ends together without tension, interposition of a nerve graft has been proposed.[252,253] A nerve graft not only serves as a guide for advancing axons, but also provides an optimal microenvironment of Schwann cell columns. Most transplanted Schwann cells in a thin graft survive, multiply, and form Büngner bands.[258] The grafts survive initially by diffusion from surrounding tissues; this is followed by revascularization within the first few days.[259]

In nerve grafting, fascicular groups in both nerve ends are dissected and isolated, and the gaps within these groups are bridged by separate nerve grafts (Fig. 7–20). The technical details of this procedure have been described elsewhere.[252,253] Most often the sural nerve from the lower limb is used as a graft because of its accessibility and appropriate thickness.

Free vascularized nerve grafts have been suggested in selected situations, especially when the recipient bed is badly scarred and conventional nerve grafts will not become optimally vascularized to act as vascular carriers of nonvascularized nerve grafts.[260-264]

A primary problem in nerve grafting is to match the corresponding components of the proximal and distal ends properly. Because intraneural fascicular topography rapidly changes from one level to another along the course of a peripheral nerve trunk,[2] such matching may be difficult to achieve. More recent studies[265] have shown that the fascicular topography in certain instances is more constant than had been realized previously, and tracer techniques[266] have shown that individual axons can be traced over long distances within the same quadrant of nerves. Thus, it appears that individual nerve fibers remain within the same sector of the nerve trunks for long distances and that a grafting procedure should make possible a reasonable matching of the corresponding axons in different levels of the nerve.

Sensory Re-education

After a crush injury, axons usually grow out to reinnervate the correct peripheral targets: in these cases the endoneurial tubes are unin-

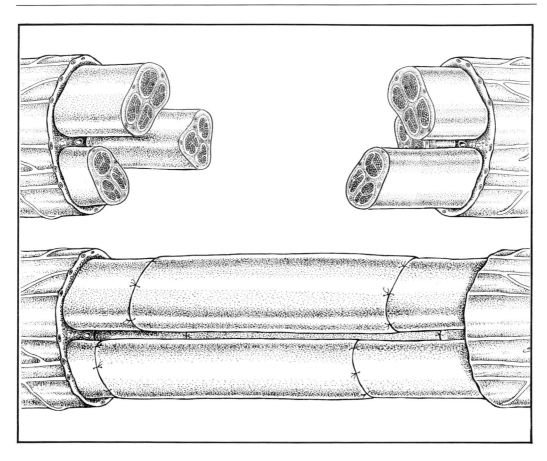

Fig. 7–20 *Nerve grafting. (Reproduced with permission from Lundborg G:* Nerve Injury and Repair. *Edinburgh, Churchill Livingstone, in press.)*

terrupted and the peripheral pathways guiding the axons remain intact. However, severance of a nerve trunk changes the situation dramatically. Because of axonal misdirection at the suture site, peripheral cutaneous receptor fields project into new cortical areas after axonal outgrowth. To restore discriminative sensibility the brain has to adapt to a new pattern of impulses in afferent fibers, a process that is not much of a problem for children but that presents considerable difficulties for adults. The situation is analogous to that of learning to speak a new language.

Special rehabilitation programs—sensory re-education—designed to teach the patient to recognize the form and shape of item and surface structures have been proposed.[267] Such programs have facilitated recovery of useful sensibility in the hand. This emphasizes that the problem involves not only the peripheral but also the central nervous components.

Future Directions

The following points represent a summary of future directions for research into nerve injury and repair:

For increased understanding of the reaction of peripheral nerves in traumatic conditions not entailing disruption of axonal continuity, more work should be directed to the mechanisms underlying various degrees of clinically relevant functional changes resulting from nerve compression and stretching.

In acute compression injuries, more knowledge is needed regarding the roles of ischemic and mechanical nerve fiber deformation. The detailed nature of the structural damage to nerve fibers and intraneural blood vessels should be analyzed and related to functional changes. In this respect, the events occurring during compression and recovery should be separated. The modes and rates of compression should also be investigated.

Chronic nerve compression is poorly understood in terms of the intraneural changes underlying various clinical symptoms. The work in this area should focus on developing improved experimental models for chronic nerve compression and analyses of the correlations between intraneural tissue reactions and functional changes. The neurophysiologic evaluation should include techniques comparable to those used clinically.

Endoneurial edema is a significant result of nerve injuries and neuropathies in which the occurrence of such edema leads to physiologic changes in the nerve. Future work should include efforts to improve the elimination of such edema from nerve fascicles.

Stretching injuries are not well understood, partly because of lack of knowledge regarding the mechanical properties of the various tissue components of the nerve ends. Also, the interactions of these components and structural models of the nerve trunk are not available. Further studies should include analyses of both short-term and long-term effects of controlled stretching on nerve structure and function.

Complete functional restitution, the ideal after peripheral nerve injury repair, will require a better understanding of the cell biology of both the normal and the regenerating nerve structure. Such an understanding will permit the development of a better rationale for intervention therapies aimed at achieving the ideal recovery. In general, nerve regeneration must involve at least four regenerative behaviors: (1) the axotomized neuronal cell body must remain alive and be supplied with the appropriate microenvironment that permits the neuron to respond to axonal growth-promoting factors; (2) the neuronal cell body or transected proximal axonal stump must form a growth cone and extend an axon; (3) the regenerating axon must be presented with a permissive or growth-promoting terrain through which to grow and, with appropriate physical and/or chemical cues, a terrain that guides the axon back toward the target organs; and (4) the target organ itself must be maintained in a receptive state and in an appropriate configuration to encourage functional contacts with the regenerating axonal growth tip. Thus, future improvements in nerve regeneration should be directed toward optimizing the success of these four regenerative behaviors. There is also a need for refined techniques to define, in quantitative terms, the various degrees of regeneration in both experimental and clinical situations.

The following experimental approaches aimed at achieving complete functional restitution should be investigated. They are generally categorized as mechanical or chemical approaches.

The primary mechanical approach is to create conditions that encourage regenerating fibers to grow along the very same pathways on which they were located before interruption. Although this may be realized to a significant extent after nerve crush or compression injuries, once a transection occurs, even the best surgical techniques can presently achieve only a minimally correct apposition of the proximal fibers with their corresponding distal counterparts. There are indications that attempts to refine the techniques used for apposition of nerve ends in experimental animals may improve the results.[268,269] Future research should aim at finding techniques to optimize the orientation of nerve ends at the site of injury. Refined in vivo histochemical techniques that differentiate between motor and sensory components may prove valuable in this context. Neurophysiologic recording techniques based on the use of microchips may be helpful in defining the electrical characteristics of fascicular domains. In such cases, presuming that the neuron survives and reextends a growth cone, the growing tip must still extend through spaces provided by degenerating distal cells. Attempts should be made to improve axonal outgrowth rate. Chemical, hormonal, and bioelectrical approaches may be helpful for this purpose.

Another suggested approach is to apply cell fusion techniques similar to those successfully used to reconnect severed invertebrate axons. Important considerations in applying this approach would be to allow or to stimulate regrowth of the proximal axon toward the distal part of the axon even though both may retract after transection. The application of this technique would require a delay of distal axonal degeneration and local application of membrane-fusing agents at the right moment. The mechanical approaches assume that enough chemical cues are available to achieve nerve regeneration, providing that the regenerating fibers can be redirected accurately back toward the target organ.

The chemical approach assumes that nerve regeneration fails because one or more molecular cues are missing and that once such cues are recognized and available they can be administered to improve regeneration. The molecules that promote neuronal survival and axonal growth and guidance have been recognized in vitro but animal models must be developed so that such molecules can be tested in vivo. More basic research must be directed toward recognizing and purifying such factors, and determining their neuronal specificity (motor, sensory, sympathetic), as well as toward developing techniques allowing their clinical use. Recent advances in biomaterial technology suggest that the use of biocompatible or bioresorbable chambers may be possible. Also, more information is required on the nature of molecules that regulate the proliferation and differentiation of the non-neuronal cells in the nerve (Schwann cells, fibroblasts, and other "neurial" cells as well as endothelial cells) and maintenance of receptive target cells (muscle, sensory receptors). Future experiments should allow the construction of chemically defined nerve channels that can replace a damaged nerve segment and promote neuronal growth and guided regeneration.

Acknowledgments

This work was supported by grants 5188 and 7651 from the Swedish Medical Research Council (G.L and B.R.), grants BNS85–01 (S.V.) and BNS86–17034 from the National Science Foundation, and a grant from the Swedish Work Environmental Fund (G.L.).

References

1. Key A, Retzius G: *Studien in der Anatomie des Nervensystems und des Bindegewebes.* Stockholm, Samson & Wallin, 1876.
2. Sunderland S: *Nerves and Nerve Injuries,* ed 2. Edinburgh, Churchill Livingstone, 1978.
3. Blunt MJ: Functional and clinical implications of the vascular anatomy of nerves. *Postgrad Med J* 1957;33:68–72.
4. Edshage S: Peripheral nerve suture: A technique for improved intraneural topography evaluation of some suture materials. *Acta Chir Scand* 1964; 331(suppl):1–104.
5. Smith JW: Factors influencing nerve repair: I. Blood supply of peripheral nerves. *Arch Surg* 1966;93:335–341.
6. Smith JW: Factors influencing nerve repair: II. Collateral circulation of peripheral nerves. *Arch Surg* 1966;93:433–437.
7. Lundborg G, Brånemark P-I: Microvascular structure and function of peripheral nerves: Vital microscopic studies of the tibial nerve in the rabbit. *Adv Microcirc* 1968;1:66–88.
8. Lundborg G: Ischemic nerve injury: Experimental studies on intraneural microvascular pathophysiology and nerve function in a limb subjected to temporary circulatory arrest. *Scand J Plast Reconstr Surg* 1970;6(suppl):1–113.
9. Lundborg G: Structure and function of the intraneural microvessels as related to trauma, edema formation, and nerve function. *J Bone Joint Surg* 1975;57A:938–948.
10. Lundborg G: The intrinsic vascularization of human peripheral nerves: Structural and functional aspects. *J Hand Surg* 1979;4:34–41.
11. Bell MA, Weddell AGM: A morphometric study of intrafascicular vessels of mammalian sciatic nerve. *Muscle Nerve* 1984;7:524–534.
12. Bell MA, Weddell AGM: A descriptive study of the blood vessels of the sciatic nerve in the rat, man and other mammals. *Brain* 1984;107:871–898.
13. Lundborg G, Myers R, Powell H: Nerve compression injury and increased endoneurial fluid pressure: A "miniature compartment syndrome." *J Neurol Neurosurg Psychiatry* 1983;46:1119–1124.
14. Myers RR, Murakami H, Powell HC: Reduced nerve blood flow in edematous neuropathies: A biomechanical mechanism. *Microvasc Res* 1986;32:145–151.
15. Smith DR, Kobrine AI, Rizzoli HV: Blood flow in peripheral nerves: Normal and post severance flow rates. *J Neurol Sci* 1977;33:341–346.

16. Lundborg G, Rydevik B: Effects of stretching the tibial nerve of the rabbit: A preliminary study of the intraneural circulation and the barrier function of the perineurium. *J Bone Joint Surg* 1973;55B:390–401.

17. Rydevik B, Lundborg G, Bagge U: Effects of graded compression on intraneural blood flow: An in vivo study on rabbit tibial nerve. *J Hand Surg* 1981;6:3–12.

18. Gerard RW: The response of nerve to oxygen lack. *Am J Physiol* 1930;92:498–541.

19. Lehmann JE: The effects of asphyxia on mammalian A nerve fibers. *Am J Physiol* 1937;119:111–120.

20. Eiken O, Nabseth DC, Mayer RF, et al: Limb replantation: II. The pathophysiological effects. *Arch Surg* 1964;88:54–65.

21. Olsson Y, Kristensson K: Permeability of blood vessels and connective tissue sheaths in the peripheral nervous system to exogenous proteins. *Acta Neuropathol* 1971;5(suppl):61–69.

22. Rydevik B, Lundborg G: Permeability of intraneural microvessels and perineurium following acute, graded experimental nerve compression. *Scand J Plast Reconstr Surg* 1977;11:179–187.

23. Olsson Y: Studies on vascular permeability in peripheral nerves: I. Distribution of circulating fluorescent serum albumin in normal, crushed and sectioned rat sciatic nerve. *Acta Neuropathol* 1966;7:1–15.

24. Olsson Y: Studies on vascular permeability in peripheral nerves: 2. Distribution of circulating fluorescent serum albumin in rat sciatic nerve after local injection of 5-hydroxytryptamine, histamine and compound 48/80. *Acta Physiol Scand* 1966;69(suppl 284):1–22.

25. Olsson Y: The effect of the histamine liberator, compound 48/80 on mast cells in normal peripheral nerves. *Acta Pathol Microbiol Scand* 1966;68:563–574.

26. Olsson Y, Reese TS: Permeability of vasa nervorum and perineurium in mouse sciatic nerve studied by fluorescence and electron microscopy. *J Neuropathol Exp Neurol* 1971;30:105–119.

27. Waksman BH: Experimental study of diphtheritic polyneuritis in the rabbit and guinea pig: III. The blood-nerve barrier in the rabbit. *J Neuropathol Exp Neurol* 1961;20:35–77.

28. Martin KH: Untersuchungen über die perineurale Diffusionsbarriere an gefriergetrockneten Nerven. *Z Zellforsch Mikrosk Anat* 1964;64:404–428.

29. Shantha TR, Bourne GH: The perineural epithelium—a new concept, in Bourne GH (ed): *The Structure and Function of Nervous Tissue*. New York, Academic Press, 1968, vol 1, pp 379–459.

30. Söderfeldt B, Olsson Y, Kristensson K: The perineurium as a diffusion barrier to protein tracers in human peripheral nerve. *Acta Neuropathol* 1973;25:120–126.

31. Thomas PK: The connective tissue of peripheral nerve: An electron microscope study. *J Anat* 1963;97:35–44.

32. Shanthaveerappa TR, Bourne GH: The "perineural epithelium," a metabolically active, continuous, protoplasmic cell barrier surrounding peripheral nerve fasciculi. *J Anat* 1962;96:527–537.

33. Shanthaveerappa TR, Bourne GH: The perineural epithelium of sympathetic nerves and ganglia and its relation to the pia arachnoid of the central nervous system and perineural epithelium of the peripheral nervous system. *Z Zellforsch Mikrosk Anat* 1964;61:742–753.

34. Shanthaveerappa TR, Bourne GH: The effects of transection of the nerve trunk on the perineural epithelium with special reference to its role in nerve degeneration and regeneration. *Anat Rec* 1964;150:35–50.

35. Shanthaveerappa TR, Bourne GH: Perineural epithelium: A new concept of its role in the integrity of the peripheral nervous system. *Science* 1966;154:1464–1467.

36. Thomas PK, Jones DG: The cellular response to nerve injury: 2. Regeneration of the perineurium after nerve section. *J Anat* 1967;101:45–55.

37. Waggener JD, Bunn SM, Beggs J: The diffusion of ferritin within the peripheral nerve sheath: An electron microscopy study. *J Neuropathol Exp Neurol* 1965;24:430–443.

38. Oldfors A: Permeability of the perineurium of small nerve fascicles: An ultrastructural study using ferritin in rats. *Neuropathol Appl Neurobiol* 1981;7:183–194.

39. Klemm H: Das Perineurium als Diffusionsbarriere gegenüber Peroxydase bei epi- und endoneuraler Applikation. *Z Zellforsch Mikrosk Anat* 1970;108:431–445.

40. Olsson Y, Reese TS: Inaccessibility of the endoneurium of mouse sciatic nerve to exogenous proteins. *Anat Rec* 1969;63:722–728.

41. Oldfors A, Sourander P: Barriers of peripheral nerve towards exogenous peroxidase in normal and protein deprived rats. *Acta Neuropathol* 1978;43:129–134.

42. Thomas PK, Olsson Y: Microscopic anatomy and function of the connective tissue components of peripheral nerve, in Dyck PJ, Thomas PK, Lambert EH (eds): *Peripheral Neuropathy.* Philadelphia, WB Saunders, 1973, pp 97–120.

43. Lundborg G, Nordborg C, Rydevik B, et al: The effect of ischemia on the permeability of the perineurium to protein tracers in rabbit tibial nerve. *Acta Neurol Scand* 1973;49:287–294.

44. Novikoff AB, Quintana N, Villaverde H, et al: Nucleoside phosphatase and cholinesterase activities in dorsal root ganglia and peripheral nerve. *J Cell Biol* 1966;29:525–545.

45. Schlaepfer WW, La Valle MC, Torack RM: Cytochemical demonstration of nucleoside phosphatase activity in myelinated nerve fibers of the rat. *Histochemie* 1969;18:281–292.

46. Rydevik B, Lundborg G, Nordborg C: Intraneural tissue reactions induced by internal neurolysis: An experimental study on the blood-nerve barrier, connective tissues and nerve fibres of rabbit tibial nerve. *Scand J Plast Reconstr Surg* 1976;10:3–8.

47. Low PA, Dyck PJ: Increased endoneurial fluid pressure in experimental lead neuropathy. *Nature* 1977;269:427–428.

48. Low P, Marchand G, Knox F, et al: Measurement of endoneurial fluid pressure with polyethylene matrix capsules. *Brain Res* 1977;122:373–377.

49. Myers RR, Powell HC, Costello ML, et al: Endoneurial fluid pressure: Direct measurement with micropipettes. *Brain Res* 1978;148:510–515.

50. Myers RR, Powell HC: Endoneurial fluid pressure in peripheral neuropathies, in Hargens AR (ed): *Tissue Fluid Pressure and Composition.* Baltimore, Williams & Wilkins, 1981, pp 193–207.

51. Myers RR, Costello ML, Powell HC: Increased endoneurial fluid pressure in galactose neuropathy. *Muscle Nerve* 1979;2:299–303.

52. Chen HI, Granger HJ, Taylor AE: Interaction of capillary, interstitial, and lymphatic forces in the canine hindpaw. *Circ Res* 1976;39:245–254.

53. Hargens AR, Akeson WH, Mubarak SJ, et al: Fluid balance within the canine anterolateral compartment and its relationship to compartment syndromes. *J Bone Joint Surg* 1978;60A:499–505.

54. Myers RR, Heckman HM, Powell HC: Endoneurial fluid is hypertonic: Results of microanalysis and its significance in neuropathy. *J Neuropathol Exp Neurol* 1983;42:217–224.

55. Spencer PS, Weinberg HJ, Raine CS, et al: The perineurial window—a new model of focal demyelination and remyelination. *Brain Res* 1975;96:323–329.

56. Powell HC, Costello ML, Myers RR: Endoneurial fluid pressure in experimental models of diabetic neuropathy. *J Neuropathol Exp Neurol* 1981;40:613–624.

57. Powell HC, Myers RR, Costello ML, et al: Endoneurial fluid pressure in wallerian degeneration. *Ann Neurol* 1979;5:550–557.

58. Ranvier LA: *Leçons sur l'histologie du système nerveux.* Paris, Savy, 1878.

59. Landon DN, Halle S: The myelinated nerve fibre, in Landon DN (ed): *The Peripheral Nerve.* London, Chapman and Hall, 1976, pp 1–105.

60. Erlanger J, Gasser HS: *Electrical Signs of Nervous Activity.* Philadelphia, University of Pennsylvania Press, 1937.

61. Black MM, Lasek RJ: Slow components of axonal transport: Two cytoskeletal networks. *J Cell Biol* 1980;86:616–623.

62. Brady ST, Lasek RJ: The slow components of axonal transport movements, compositions and organization, in Weiss DG (ed): *Axoplasmic Transport.* Berlin, Springer-Verlag, 1982, pp 206–217.

63. Grafstein B, Forman DS: Intracellular transport in neurons. *Physiol Rev* 1980;60:1167–1283.

64. McLean WG, McKay AL, Sjöstrand J: Electrophoretic analysis of axonally transported proteins in rabbit vagus nerve. *J Neurobiol* 1983;14:227–236.

65. Droz B: Synthetic machinery and axoplasmic transport: Maintenance of neuronal connectivity, in Tower DB (ed): *The Nervous System: Vol 1. The Basic Neurosciences.* New York, Raven Press, 1975, pp 111–127.

66. Dahlström A: Axoplasmic transport (with particular respect to adrenergic neurons). *Philos Trans R Soc Lond [Biol]* 1971;261:325–358.

67. Lasek RJ: Protein transport in neurons. *Int Rev Neurobiol* 1970;13:289–324.

68. Lubińska L: Axoplasmic streaming in regenerating and in normal nerve fibres. *Prog Brain Res* 1964;13:1–71.

69. Griffin JW, Price DL, Drachman DB, et al: Incorporation of axonally transported glycoproteins into axolemma during nerve regeneration. *J Cell Biol* 1981;88:205–214.

70. Ochs S: Axoplasmic transport, in Tower D (ed): *The Nervous System: Vol 1. The Basic Neurosciences.* New York, Raven Press, 1975, pp 137–146.

71. Sjöstrand J, McLean WG, Frizell M: The application of axonal transport studies to peripheral nerve problems, in Omer GE Jr, Spinner M (eds): *Management of Peripheral Nerve Problems.* Philadelphia, WB Saunders, 1980, pp 917–927.

72. Weiss DG: General properties of axoplasmic transport, in Weiss DG (ed): *Axoplasmic Transport.* Berlin, Springer-Verlag, 1982, pp 1–14.

73. Brady ST: Microtubules and the mechanisms of fast axonal transport, in Weiss DG (ed): *Axoplasmic Transport.* Berlin, Springer-Verlag, 1982, pp 301–306.

74. Alvarez J, Torres JC: Slow axoplasmic transport: A fiction? *J Theor Biol* 1985;112:627–651.

75. Weiss P, Hiscoe HB: Experiments on the mechanism of nerve growth. *J Exp Zool* 1948;107:315–395.

76. Bamburg JR, Shooter EM, Wilson L: Developmental changes in microtubule protein of chick brain. *Biochemistry* 1973;12:1476–1482.

77. Pleasure D: The structural proteins of peripheral nerve, in Dyck PJ, Thomas PK, Lambert EH (eds): *Peripheral Neuropathy.* Philadelphia, WB Saunders, 1975, pp 231–252.

78. Droz B, Di Giamberardino L, Koenig HL: Transports axonaux de macromolécules présynaptiques. *Actual Neurophysiol* 1974;10:260.

79. Ochs S: Systems of material transport in nerve fibres: (axoplasmic transport) related to nerve function and trophic control. *Ann NY Acad Sci* 1974;228:202–223.

80. Ochs S: Characteristics and a model for fast axoplasmic transport in nerve. *J Neurobiol* 1971;2:331–345.

81. Lubińska L: On axoplasmic flow. *Int Rev Neurobiol* 1975;17:241–296.

82. Bisby MA: Orthograde and retrograde axonal transport of labeled protein in motoneurons. *Exp Neurol* 1976;50:628–640.

83. Bisby MA: Retrograde axonal transport, in Hertz L, Fedoroff S, Hertz L (eds): *Advances in Cellular Neurobiology*. New York, Academic Press, 1980, vol 1, pp 69–117.

84. Bisby MA: Functions of retrograde axonal transport. *Fed Proc* 1982;41:2307–2311.

85. Kristensson K, Olsson Y: Retrograde transport of horseradish peroxidase in transected axons: 3. Entry into injured axons and subsequent localization in perikaryon. *Brain Res* 1976;115:201–213.

86. de Vito JL, Clausing KW, Smith OA: Uptake and transport of horseradish peroxidase by cut end of the vagus nerve. *Brain Res* 1974;82:269–271.

87. Olsson TP, Forsberg I, Kristensson K: Uptake and retrograde axonal transport of horseradish peroxidase in regenerating facial motor neurons of the mouse. *J Neurocytol* 1978;7:323–336.

88. Kristensson K, Sjöstrand J: Retrograde transport of protein tracer in the rabbit hypoglossal nerve during regeneration. *Brain Res* 1972;45:175–181.

89. Dahlin LB, Danielsen N, Ehira T, et al: Mechanical effects of compression of peripheral nerves. *J Biomech Eng* 1986;108:120–122.

90. Kristensson K: Implications of axoplasmic transport for the spread of virus infections in the nervous system, in Weiss DG, Bono A (eds): *Axoplasmic Transport in Physiology and Pathology*. Berlin, Springer-Verlag, 1982, pp 152–158.

91. Pleasure D: Axoplasmic transport, in Sumner AJ (ed): *The Physiology of Peripheral Nerve Disease*. Philadelphia, WB Saunders, 1980, pp 221–237.

92. Gainer H, Fink DJ: Covalent labelling techniques and axonal transport, in Weiss DG (ed): *Axoplasmic Transport*. Berlin, Springer-Verlag, 1982, pp 464–470.

93. Bisby MA: Retrograde axonal transport and nerve regeneration, in Elam JS, Cancalon P (eds): *Axonal Transport in Neuronal Growth and Regeneration*. New York, Plenum Press, 1984, pp 45–67.

94. Leone J, Ochs S: Anoxic block and recovery of axoplasmic transport and electrical excitability of nerve. *J Neurobiol* 1978;9:229–245.

95. Rydevik B, McLean WG, Sjöstrand J, et al: Blockage of axonal transport induced by acute, graded compression of the rabbit vagus nerve. *J Neurol Neurosurg Psychiatry* 1980;43:690–698.

96. Dahlin LB, Rydevik B, McLean WG, et al: Changes in fast axonal transport during experimental nerve compression at low pressures. *Exp Neurol* 1984;84:29–36.

97. Dahlin LB, McLean WG: Effects of graded experimental compression on slow and fast axonal transport in rabbit vagus nerve. *J Neurol Sci* 1986;72:19–30.

98. Schnapp BJ, Reese TS: New developments in understanding rapid axonal transport. *Trends Neurosci* 1986;9:155–162.

99. Ochs S: Energy metabolism and supply of ∼P to the fast axoplasmic transport mechanism in nerve. *Fed Proc* 1974;33:1049–1058.

100. Haftek J: Stretch injury of peripheral nerve: Acute effects of stretching on rabbit nerve. *J Bone Joint Surg* 1970;52B:354–365.

101. Sunderland S, Bradley KC: Stress-strain phenomena in human peripheral nerve trunks. *Brain* 1961;84:102–119.

102. Aguayo A, Nair CPV, Midgley R: Experimental progressive compression neuropathy in the rabbit: Histologic and electrophysiologic studies. *Arch Neurol* 1971;24:358–364.

103. Ochoa J: Nerve fiber pathology in acute and chronic compression, in Omer GE Jr, Spinner M (eds): *Management of Peripheral Nerve Problems*. Philadelphia, WB Saunders, 1980, pp 487–501.

104. Spinner M, Spencer PS: Nerve compression lesions of the upper extremity. *Clin Orthop* 1974;104:46–67.

105. Powell HC, Myers RR: Pathology of experimental nerve compression. *Lab Invest* 1986;55:91–100.

106. MacGregor RJ, Sharpless SK, Luttges MW: A pressure vessel model for nerve compression. *J Neurol Sci* 1975;24:299–304.

107. Lundborg G: *Nerve Injury and Repair*. Edinburgh, Churchill Livingstone, in press.

108. Gilliatt RW: Acute compression block, in Sumner AJ (ed): *The Physiology of Peripheral Nerve Disease*. Philadelphia, WB Saunders, 1980, pp 287–315.

109. Ochoa J, Fowler TJ, Gilliatt RW: Anatomical changes in peripheral nerves compressed by a pneumatic tourniquet. *J Anat* 1972;113:433–455.

110. Fowler TJ, Danta G, Gilliatt RW: Recovery of nerve conduction after a pneumatic tourniquet: Observations on the hind-limb of the baboon. *J Neurol Neurosurg Psychiatry* 1972;35:638–647.

111. Denny-Brown DE, Brenner C: Lesion in peripheral nerve resulting from compression by spring clip. *Arch Neurol Psychiatry* 1944;52:1–19.

112. Denny-Brown D, Brenner C: The effect of percussion of nerve. *J Neurol Psychiatry* 1944;7:76–95.

113. Denny-Brown D, Brenner C: Paralysis of nerve induced by direct pressure and by tourniquet. *Arch Neurol Psychiatry* 1944;51:1–26.

114. Seddon H: *Surgical Disorders of the Peripheral Nerves*, ed 2. Edinburgh, Churchill Livingstone, 1975.

115. Sunderland S: The nerve lesion in the carpal tunnel syndrome. *J Neurol Neurosurg Psychiatry* 1976;39:615–626.

116. Rydevik B, Nordborg C: Changes in nerve function and nerve fibre structure induced by acute, graded compression. *J Neurol Neurosurg Psychiatry* 1980;43:1070–1082.

117. Dahlin LB, Sjöstrand J, McLean WG: Graded inhibition of retrograde axonal transport by compression of rabbit vargus nerve. *J Neurol Sci*, in press.

118. Dahlin LB, Nordborg C, Lundborg G: Morphological changes in nerve cell bodies induced by experimental graded compression. *J Neurol Sci*, in press.

119. Rydevik B, Lundborg G, Skalak R: Biomechanics of peripheral nerves, in Frankel VH, Nordin M (eds): *Basic Biomechanics of the Musculoskeletal System*, ed 2. Philadelphia, Lea and Febiger, in press.

120. Horch K: Guidance of regrowing sensory axons after cutaneous nerve lesions in the cat. *J Neurophysiol* 1979;42:1437–1449.

121. Danielsen N, Lundborg G, Frizell M: Nerve repair and axonal transport: Outgrowth delay and regeneration rate after transection and repair of rabbit hypoglossal nerve. *Brain Res* 1986;376:125–132.

122. Danielsen N. Lundborg G, Frizell M: Nerve repair and axonal transport: Distribution of axonally transported proteins during maturation period in regenerating rabbit hypoglossal nerve. *J Neurol Sci* 1986;73:269–277.

123. Lieberman AR: The axon reaction: A review of the principal features of perikaryal responses to axon injury. *Int Rev Neurobiol* 1971;14:49–124.

124. Grafstein B: The nerve cell body response to axotomy. *Exp Neurol* 1975;48:32–51.

125. Grafstein B, McQuarrie IG: Role of the nerve cell body in axonal regeneration, in Cotman CW (ed): *Neuronal Plasticity*. New York, Raven Press, 1978, pp 155–195.

126. Brattgård S-O, Edstrom JE, Hydén H: The chemical changes in regenerating neurons. *J Neurochem* 1957;1:316–325.

127. Lieberman HR: Some factors affecting retrograde neuronal responses to axonal lesions, in Bellaris R, Gray EG (eds): *Essays on the Nervous System.* Oxford, Clarendon Press, 1974, pp 71–105.

128. Hendry IA: The response of adrenergic neurones to axotomy and nerve growth factor. *Brain Res* 1975;94:87–97.

129. Purves D, Njå A: Effect of nerve growth factor on synaptic depression after axotomy. *Nature* 1976;260:535–536.

130. Angeletti PU, Levi-Montalcini R, Caramia F: Ultrastructural changes in sympathetic neurons of newborn and adult mice treated with nerve growth factor. *J Ultrastruct Res* 1971;36:24–36.

131. Angeletti P, Vigneti E: Assay of nerve growth factor (NGF) in subcellular fractions of peripheral tissues by micro complement fixation. *Brain Res* 1971;33:601–604.

132. Salonen V, Lehto M, Vaheri A, et al: Endoneurial fibrosis following nerve transection: An immunohistological study of collagen types and fibronectin in the rat. *Acta Neuropathol* 1985;67:315–321.

133. Thomas PK: Changes in the endoneurial sheaths of peripheral myelinated nerve fibres during wallerian degeneration. *J Anat* 1964;98:175–182.

134. Schlaepfer WW: Calcium-induced degeneration of axoplasm in isolated segments of rat peripheral nerve. *Brain Res* 1974;69:203–215.

135. Schlaepfer WW, Micko S: Calcium-dependent alterations of neurofilament proteins of rat peripheral nerve. *J Neurochem* 1979;32:211–219.

136. Schlaepfer WW, Hasler MB: Characterization of the calcium-induced disruption of neurofilaments in rat peripheral nerve. *Brain Res* 1979;168:299–309.

137. Badalamente MA, Hurst LC, Makowski J, et al: Inhibition of neural and muscle degeneration after microsurgical repair of severed peripheral nerve. *Ann NY Acad Sci* 1984;435:380–381.

138. Badalamente MA, Hurst LC, Stracher A: Calcium-induced degeneration of the cytoskeleton in monkey and human peripheral nerves. *J Hand Surg* 1986;11B:337–340.

139. Badalamente MA, Hurst LC, Stracher A, et al: The effects of leupeptin on nerve repair in primates. *Ann NY Acad Sci* 1986;463:253–255.

140. Weinberg HJ, Spencer PS: Studies on the control of myelinogenesis: II. Evidence for neuronal regulation of myelin production. *Brain Res* 1976;113:363–378.

141. Weinberg HJ, Spencer PS: The fate of Schwann cells isolated from axonal contact. *J Neurocytol* 1978;7:555–569.

142. Aguayo AJ, Charron L, Bray GM: Potential of Schwann cells from unmyelinated nerves to produce myelin: A quantitative ultrastructural and radiographic study. *J Neurocytol* 1976;5:565–573.

143. Aguayo AJ, Epps J, Charron L, et al: Multipotentiality of Schwann cells in cross-anastomosed and grafted myelinated and unmyelinated nerves: Quantitative microscopy and radioautography. *Brain Res* 1976;104:1–20.

144. Skene JHP: Growth-associated proteins and the curious dichotomies of nerve regeneration. *Cell* 1984;37:697–700.

145. Skene JHP, Shooter EM: Denervated sheath cells secrete a new protein after nerve injury. *Proc Natl Acad Sci USA* 1983;80:4169–4173.

146. Müller HW, Gebicke-Harter PJ, Hangen DH, et al: A specific 37,000-dalton protein that accumulates in regenerating but not in nonregenerating mammalian nerves. *Science* 1985;228:499–501.

147. Kanje M, Lundborg G, Edström A, et al: Insulin-like growth factor (IGF–1) stimulates regeneration in rat sciatic nerve. Presented at the Axonal Transport Satellite Symposium, XXX International Congress of the International Union of Physiological Sciences, Calgary, 1986.

148. Hansson HA, Dahlin LB, Danielsen N, et al: Evidence indicating trophic importance of IGF-1 in regenerating peripheral nerves. *Acta Physiol Scand* 1986;126:609–614.

149. Barbin G, Manthorpe M, Varon S: Purification of the chick eye ciliary neuronotrophic factor. *J Neurochem* 1984;43:1468–1478.

150. Barde Y-A, Edgar D, Thoenen H: Purification of a new neurotrophic factor from mammalian brain. *EMBO J* 1982:549–553.

151. Manthorpe M, Skaper SD, Williams LR, et al: Purification of adult rat sciatic nerve ciliary neuronotrophic factor. *Brain Res* 1986;367:282–286.

152. Schubert D, LaCorbiere M: Isolation of an adhesion mediating protein from chick neural retina adherons. *J Cell Biol* 1985;101:1071–1077.

153. Selak I, Skaper SD, Varon S: Pyruvate participation in the low molecular weight trophic activity for central nervous system neurons in glia-conditioned media. *J Neurosci* 1985;5:23–28.

154. Walicke P, Cowan WM, Ueno N, et al: Fibroblast growth factor promotes survival of dissociated hippocampal neurons and enhances neurite extension. *Proc Natl Acad Sci USA* 1986;83:3012–3016.

155. Walicke P, Varon S, Manthorpe M: Purification of a human red blood cell protein supporting the survival of cultured CNS neurons, and its identification as catalase. *J Neurosci* 1986;6:1114–1121.

156. Davis GE, Manthorpe M, Varon S: Parameters of neuritic growth from ciliary ganglion neurons in vitro: Influence of laminin, schwannoma polyornithine-binding neurite promoting factor and ciliary neuronotrophic factor. *Dev Brain Res* 1985;17:75–84.

157. Kligman D: Isolation of a protein from bovine brain which promotes neurite extension from chick embryo cerebral cortex neurons in defined medium. *Brain Res* 1982;250:93–100.

158. Lander AD, Fujii DK, Reichardt LF: Purification of a factor that promotes neurite outgrowth: Isolation of laminin and associated molecules. *J Cell Biol* 1985;101:898–913.

159. Manthorpe M, Engvall E, Ruoslahti E, et al: Laminin promotes neuritic regeneration from cultured peripheral and central neurons. *J Cell Biol* 1983;97:1882–1890.

160. Wewer U, Albrechtsen R, Manthorpe M, et al: Human laminin isolated in a nearly intact, biologically active form from placenta by limited proteolysis. *J Biol Chem* 1983;258:12654–12660.

161. Collins F, Dawson A: An effect of nerve growth factor on parasympathetic neurite outgrowth. *Proc Natl Acad Sci USA* 1983;80:2091–2094.

162. Campenot RB: Development of sympathetic neurons in compartmentalized cultures: II. Local control of neurite survival by nerve growth factor. *Dev Biol* 1982;93:13–21.

163. Davis GE, Skaper SD, Manthorpe M, et al: Fetal calf serum-mediated inhibition of neurite growth from ciliary ganglion neurons in vitro. *J Neurosci Res* 1984;12:29–40.

164. Carnow TB, Manthorpe M, Davis GE, et al: Localized survival of ciliary ganglionic neurons identifies neuronotrophic factor bands on nitrocellulose blots. *J Neurosci* 1985;5:1965–1971.

165. Davis GE, Varon S, Engvall E, et al: Substratum-binding neurite-promoting factors: Relationships to laminin. *Trends Neurosci* 1985;8:528–532.

166. Morrison RS, Sharma A, de Vellis J, et al: Basic fibroblast growth factor supports the survival of cerebral cortical neurons in primary culture. *Proc Natl Acad Sci USA* 1986;83:7537–7541.

167. Davis GE, Klier FG, Engvall E, et al: Association of laminin with heparan and chondroitin sulfate-bearing proteoglycans in neurite-promoting complexes from rat schwannoma cells. *Neurochem Res*, in press.

168. Davis GE, Manthorpe M, Engvall E, et al: Isolation and characterization of rat schwannoma neurite-promoting factor: Evidence that the factor contains laminin. *J Neurosci* 1985;5:2662–2671.

169. Rudge JS, Davis GE, Manthorpe M, et al: An examination of ciliary neuronotrophic factors from avian and rodent tissue extracts using a blot and culture technique. *Dev Brain Res* 1987;32:103–110.

170. Engvall E, Krusius T, Wewer U, et al: Laminin from rat yolk sac tumor: Isolation, partial characterization and comparison with mouse laminin. *Arch Biochem Biophys* 1983;222:649–656.

171. Gurney ME, Heinrich SP, Lee MR, et al: Molecular cloning and expression of neuroleukin, a neurotrophic factor for spinal and sensory neurons. *Science* 1986;234:566–574.

172. Rogers SL, Letourneau PC, Palm SL, et al: Neurite extension by peripheral and central nervous system neurons in response to substratum-bound fibronectin and laminin. *Dev Biol* 1983;98:212–220.

173. Timpl R, Rohde H, Robey PG, et al: Laminin—a glycoprotein from basement membranes. *J Biol Chem* 1979;254:9933–9937.

174. Varon S, Nomura J, Perez-Polo JR, et al: The isolation and assay of the nerve growth factor proteins, in Fried R (ed): *Methods and Techniques of Neurochemistry.* New York, Marcel Dekker, 1972, vol 3, pp 203–229.

175. Adler R, Manthorpe M, Varon S: Separation of neuronal and nonneuronal cells in monolayer cultures from chick embryo optic lobe. *Dev Biol* 1979;69;424–435.

176. Korsching S, Thoenen H: Quantitative demonstration of the retrograde axonal transport of endogenous nerve growth factor. *Neurosci Lett* 1983;39:1–4.

177. Gundersen RW: Sensory neurite growth cone guidance by substrate adsorbed nerve growth factor. *J Neurosci Res* 1985;13:199–2121.

178. Gundersen RW, Barrett JN: Characterization of the turning response of dorsal root neurites toward nerve growth factor. *J Cell Biol* 1980;87:546–554.

179. Campenot RB: Development of sympathetic neurons in compartmentalized cultures: I. Local control of neurite growth by nerve growth factor. *Dev Biol* 1982;983:1–12.

180. Krikorian D, Manthorpe M, Varon S: Purified mouse Schwann cells: Mitogenic effects of fetal calf serum and fibroblast growth factor. *Dev Neurosci* 1982;5:77–91.

181. Gospodarowicz D, Neufeld G, Schweigerer L: Fibroblast growth factor. *Mol Cell Endocrinol* 1986;46:187–204.

182. Pettmann B, Labourdette G, Weibel M, et al: The brain fibroblast growth factor (FGF) is localized in neurons. *Neurosci Lett* 1986;68:175–180.

183. Collins F, Garrett JE Jr: Elongating nerve fibers are guided by a pathway of material released from embryonic nonneuronal cells. *Proc Natl Acad Sci USA* 1980;77:6226–6228.

184. Davis GE, Engvall E, Varon S, et al: Human amnion membrane as a substratum for cultured peripheral and central nervous system neurons. *Dev Brain Res* 1987;33:1–10.

185. Sanders FK, Young JZ: The influence of peripheral connexion on the diameter of regenerating nerve fibres. *J Exp Biol* 1946;22:203–212.

186. Lundborg G, Hansson H-A: Nerve regeneration through preformed pseudosynovial tubes: A preliminary report of a new experimental model for studying the regeneration and reorganization capacity of peripheral nerve tissue. *J Hand Surg* 1980;5:35–38.

187. Lundborg G, Longo FM, Varon S: Nerve regeneration model and trophic factors in vivo. *Brain Res* 1982;232:157–161.

188. Lundborg G, Dahlin LB, Danielsen N, et al: Nerve regeneration in silicone chambers: Influence of gap length and of distal stump components. *Exp Neurol* 1982;76:361–375.

189. Varon S, Williams LR: Peripheral nerve regeneration in a silicone model chamber; cellular and molecular aspects. *Periph Nerve Rep Regen* 1986;1:9–25.

190. Williams LR, Varon S: Experimental manipulations of the microenvironment within a nerve regeneration chamber, in Ruben RJ, Van De Water TR, Rubel EW (eds): *The Biology of Change in Otolaryngology.* Amsterdam, Excerpta Medica, 1986, pp 275–289.

191. Williams LR, Longo FM, Powell HC, et al: Spatial-temporal progress of peripheral nerve regeneration within a silicone chamber: Parameters for a bioassay. *J Comp Neurol* 1983;218:460–470.

192. Longo FM, Skaper SD, Manthorpe M, et al: Temporal changes of neuronotrophic activities accumulating *in vivo* within nerve regeneration chambers. *Exp Neurol* 1983;81:756–769.

193. Longo FM, Hayman EG, Davis GE, et al: Neurite-promoting factors and extracellular matrix components accumulating in vivo within nerve regeneration chambers. *Brain Res* 1984;309:105–117.

194. Longo FM, Manthorpe M, Skaper SD, et al: Neuronotrophic activities accumulate in vivo within silicone nerve regeneration chambers. *Brain Res* 1983;261:109–117.

195. Manthorpe M, Longo FM, Varon S: Comparative features of spinal neuronotrophic factors in fluids collected in vitro and in vivo. *J Neurosci Res* 1982;8:241–250.

196. Unsicker K, Skaper SD, Davis GE, et al: Comparison of the effects of laminin and the polyornithine-binding neurite promoting factor from RN22 schwannoma cells on neurite regeneration from cultured newborn and adult rat dorsal root ganglion neurons. *Dev Brain Res* 1985;17:304–308.

197. Williams LR, Powell HC, Lundborg G, et al: Competence of nerve tissue as distal insert promoting nerve regeneration in a silicone chamber. *Brain Res* 1984;293:201–211.

198. Varon S, Manthorpe M, Williams LR: Neuronotrophic and neurite-promoting factors and their clinical potentials. *Dev Neurosci* 1983/84;6:73–100.

199. Spencer PS: Neuronal regulation of myelinating cell function. *Soc Neurosci Symp 1979;4:275–321.*

200. Politis MJ, Spencer PS: A method to separate spatially the temporal sequence of axon-Schwann cell interaction during nerve regeneration. *J Neurocytol* 1981;10:221–232.

201. Gutmann E: Factors affecting recovery of motor function after nerve lesions. *J Neurol Psychiatry* 1942;5:81–95.

202. Black MM, Lasek RJ: The use of axonal transport to measure axonal regeneration in rat ventral motor neurons, abstract. *Anat Rec* 1976;184:360–361.

203. Buchthal F, Kühl V: Nerve conduction, tactile sensibility, and the electromyogram after suture or compression of peripheral nerve: A longitudinal study in man. *J Neurol Neurosurg Psychiatry* 1979;42:436–451.

204. Forssman J: Ueber die Ursachen welche die Wachsthumsrichtung der peripheren Nervenfasern bei der Regeneration bestimmen. *Beitr Pathol Anat* 1898;24:56–100.

205. Ramon y Cajal S: *Degeneration and Regeneration of the Nervous System.* London, Oxford University Press, 1928.

206. Weiss P, Taylor AC: Further experimental evidence against "neurotropism" in nerve regeneration. *J Exp Zool* 1944;95:233–257.

207. Weiss P, Taylor AC: Guides for nerve regeneration across gaps. *J Neurosurg* 1946;3:375–389.

208. Politis MJ, Ederle K, Spencer PS: Tropism in nerve regeneration in vivo: Attraction of regenerating axons by diffusible factors derived from cells in distal nerve stumps of transected peripheral nerves. *Brain Res* 1982;253:1–12.

209. Ochi M: Experimental study on orientation of regenerating fibers in the severed peripheral nerve. *Hiroshima J Med Sci* 1983;32:389–406.

210. Lundborg G, Dahlin LB, Danielsen N, et al: Tissue specificity in nerve regeneration. *Scand J Plast Reconstr Surg* 1986;20:279–283.

211. Mackinnon SE, Dellon AL, Lundborg G, et al: A study of neurotrophism in a primate model. *J Hand Surg* 1986;11A:888–894.

212. Politis MJ: Specificity in mammalian peripheral nerve regeneration at the level of the nerve trunk. *Brain Res* 1985;328:271–276.

213. Seckel BR, Ryan SE, Gagne RG, et al: Target-specific nerve regeneration through a nerve guide in the rat. *Plast Reconstr Surg* 1978;78:793–798.

214. McQuarrie IG, Grafstein B: Axon outgrowth enhanced by a previous nerve injury. *Arch Neurol* 1973;29:53–55.

215. McQuarrie IG, Grafstein B, Gershon MD: Axonal regeneration in the rat sciatic nerve: Effect of a conditioning lesion and of dbcAMP. *Brain Res* 1977;132:443–453.

216. McQuarrie IG: Acceleration of axonal regeneration in somatic rat motoneurons by using a conditioning lesion, in Gorio A, Millesi H, Mingrino S (eds): *Posttraumatic Peripheral Nerve Regeneration: Experimental Basis and Clinical Implications*. New York, Raven Press, 1981, pp 49–58.

217. Forman DS, McQuarrie IG, Labore FW, et al: Time course of the conditioning lesion effect on axonal regeneration. *Brain Res* 1980;182:180–185.

218. Cook RA, Kiernan JA: Effects of triiodothyronine on protein synthesis in regenerating peripheral neurons. *Exp Neurol* 1976;52:515–524.

219. Stelmack BM, Kiernan JA: Effects of triiodothyronine on the normal and regenerating facial nerve of the rat. *Acta Neuropathol* 1977;40:151–155.

220. Berenberg RA, Forman DS, Wood DK, et al: Recovery of peripheral nerve function after axotomy: Effect of triiodothyronine. *Exp Neurol* 1977;57:349–363.

221. Danielsen N, Dahlin LB, Ericson LE, et al: Experimental hyperthyroidism stimulates axonal growth in mesothelial chambers. *Exp Neurol* 1986;94:54–65.

222. Bijlsma WA, Jennekens FGI, Schotman P, et al: Stimulation by $ACTH_{4-10}$ of nerve fiber regeneration following sciatic nerve crush. *Muscle Nerve* 1983;6:104–112.

223. Bijlsma WA, van Asselt E, Veldman H, et al: Ultrastructural study of effect of $ACTH_{4-10}$ on nerve regeneration; axons become larger in number and smaller in diameter. *Acta Neuropathol* 1983;62:24–30.

224. Verghese JP, Bradley WG, Mitsumoto H, et al: A blind controlled trial of adrenocorticotropin and cerebral gangliosides in nerve regeneration in the rat. *Exp Neurol* 1982;77:455–458.

225. Kort J, Bassett CAL: Effects of pulsing electromagnetic fields (PEMF's) on peripheral nerve regeneration. *Trans Orthop Res Soc* 1980;5:132.

226. Marsh G, Beams HW: In vitro control of growing chick nerve fibers by applied electric currents. *J Cell Comp Physiol* 1946;27:139–157.

227. Sisken BF, Smith SD: The effects of minute direct electrical currents on cultured chick embryo trigeminal ganglia. *J Embryol Exp Morphol* 1975;33:29–35.

228. Jaffe LF, Poo MM: Neurites grow faster towards the cathode than the anode in a steady field. *J Exp Zool* 1979;209:115–128.

229. Radji ARM, Bowden REM: Effects of high-peak pulsed electromagnetic field on the degeneration and regeneration of the common peroneal nerve in rats. *J Bone Joint Surg* 1983;65B:478–492.

230. Orgel MG, O'Brien WJ, Murray HM: Pulsing electromagnetic field therapy in nerve regeneration: An experimental study in the cat. *Plast Reconstr Surg* 1984;73:173–182.

231. Roisen FJ, Murphy RA, Pichichero ME, et al: Cyclic adenosine monophosphate stimulation of axonal elongation. *Science* 1972;175:73–74.

232. Pichichero M, Beer B, Clody DE: Effects of dibutyryl cyclic AMP on restoration of function of damaged sciatic nerve in rats. *Science* 1973;182:724–725.

233. Grabb WC: Median and ulnar nerve suture: An experimental study comparing primary and secondary repair in monkeys. *J Bone Joint Surg* 1968;50A:964–972.

234. Grabb WC, Bement SL, Koepke GH, et al: Comparison of methods of peripheral nerve suturing in monkeys. *Plast Reconstr Surg* 1970;46:31–38.

235. Müller H, Grubel G: Long-term results of peripheral nerve sutures—a comparison of micro- and macrosurgical techniques. *Adv Neurosurg* 1981;9:381–387.

236. Önne L: Recovery of sensibility and sudomotor activity in the hand after nerve suture. *Acta Chir Scand* 1962;300(suppl):1–69.

237. Carlstedt T, Lugnegard H, Andersson M: Pacinian corpuscles after nerve repair in humans. *Periph Nerve Rep Regen* 1986;1:37–40.

238. Moberg E: Evaluation and management of nerve injuries in the hand. *Surg Clin North Am* 1964;44:1019–1029.

239. Wilgis EFS: Nerve repair and grafting, in Green DP (ed): *Operative Hand Surgery.* New York, Churchill Livingstone, 1982, vol 2, pp 915–938.

240. Millesi H: Microsurgery of peripheral nerves. *Hand* 1973;5:157–160.

241. Ito T, Hirotani H, Yamamoto K: Peripheral nerve repairs by the funicular suture technique. *Acta Orthop Scand* 1976;47:283–289.

242. Orgel MG, Terzis JK: Epineurial vs. perineurial repair: An ultrastructural and electrophysiological study of nerve regeneration. *Plast Reconstr Surg* 1977;60:80–91.

243. Terzis JK: Clinical microsurgery of the peripheral nerve: The state of the art. *Clin Plast Surg* 1979;6:247–267.

244. Wise AJ Jr, Topuzlu C, Davis P, et al: A comparative analysis of macro- and microsurgical neurorrhaphy technics. *Am J Surg* 1969;117:566–572.

245. Yamamoto K: A comparative analysis of the process of nerve regeneration following funicular and epineurial suture for peripheral nerve repair. *Arch Jpn Chir* 1974;43:276–301.

246. Bora FW Jr, Pleasure DE, Didizian NA: A study of nerve regeneration and neuroma formation after nerve suture by various techniques. *J Hand Surg* 1976;1:138–143.

247. Cabaud HE, Rodkey WG, McCaroll HR, et al: Epineurial and perineurial fascicular nerve repairs: A critical comparison. *J Hand Surg* 1976;1:131–137.

248. Cabaud HE, Rodkey WG, McCaroll HR: Peripheral nerve injuries: Studies in higher nonhuman primates. *J Hand Surg* 1980;5:201–206.

249. Kline DG, Hudson AR, Bratton BR: Experimental study of fascicular nerve repair with and without epineurial closure. *J Neurosurg* 1981;54:513–520.

250. Brushart TM, Henry EW, Mesulam MM: Reorganization of muscle afferent projections accompanies peripheral nerve regeneration. *Neuroscience* 1981;6:2053–2061.

251. Brushart TM, Tarlov EC, Mesulam MM: Specificity of reinnervation after epineurial and individual fascicular suture of the rat sciatic nerve. *J Hand Surg* 1983;8:248–253.

252. Millesi H, Meissl G, Berger A: The interfascicular nerve-grafting of the median and ulnar nerves. *J Bone Joint Surg* 1972;54A:727–750.

253. Millesi H, Meissl G, Berger A: Further experience with interfascicular grafting of the median, ulnar, and radial nerves. *J Bone Joint Surg* 1976;58A:209–218.

254. Miyamoto Y: End to end co-aptation under tension on repair of peripheral nerves, in Gorio A, Millesi H, Mingrino S (eds): *Posttraumatic Peripheral Nerve Regeneration: Experimental Basis and Clinical Implications.* New York, Raven Press, 1981, pp 281–286.

255. Miyamoto Y, Tsuge K: Effects of tension on intraneural microcirculation in end to end neurorrhaphy, in Gorio A, Millesi H, Mingrino S (eds): *Posttraumatic Peripheral Nerve Regeneration: Experimental Basis and Clinical Implications.* New York, Raven Press, 1981, pp 81–91.

256. Miyamoto Y, Tsuge K: Grafting versus end to end co-aptation of nerves, in Gorio A, Millesi H, Mingrino S (eds): *Posttraumatic Peripheral Nerve Regeneration: Experimental Basis and Clinical Implications.* New York, Raven Press, 1981, pp 351–356.

257. Millesi H, Meissl G: Consequences of tension at the suture line, in Gorio A, Millesi H, Mingrino S (eds): *Posttraumatic Peripheral Nerve Regeneration: Experimental Basis and Clinical Implications.* New York, Raven Press, 1981, pp 277–279.

258. Aguayo AJ, Bray GM: Experimental nerve grafts, in Jewett DL, McCarroll HR Jr (eds): *Nerve Repair and Regeneration: Its Clinical and Experimental Basis.* St. Louis, CV Mosby Co, 1980, pp 68–76.

259. Almgren KG: Revascularization of free peripheral nerve grafts: An experimental study in the rabbit. *Acta Orthop Scand* 1974;154(suppl):1–104.

260. Koshima I, Okabe K, Harii K: Comparative study of free and vascularized nerve grafts transplanted into scar tissue in rats, abstract. *J Microsurg* 1981;3:126.

261. Taylor GI, Ham FT: The free vascularized nerve graft: A further experimental and clinical application of microvascular techniques. *Plast Reconstr Surg* 1976;57:413–426.

262. Breidenbach W, Terzis JK: The anatomy of free vascularized nerve grafts. *Clin Plast Surg* 1984;11:65–71.

263. Bonney G, Birch R, Jamieson AM, et al: Experience with vascularized nerve grafts. *Clin Plast Surg* 1984;11:137–142.

264. Rose EH, Kowalski TA: Restoration of sensibility to anesthetic scarred digits with free vascularized nerve grafts from the dorsum of the foot. *J Hand Surg* 1985;10A:514–521.

265. Jabaley ME, Wallace WH, Heckler FR: Internal topography of major nerves of the forearm and hand: A current view. *J Hand Surg* 1980;5:1–18.

266. Brushart TM: The central course of primate digital nerve axons, abstract. *J Hand Surg* 1985;10A:426.

267. Dellon AL: *Evaluation of Sensibility and Re-education of Sensation in the Hand.* Baltimore, Williams & Wilkins, 1981.

268. de Medinaceli L, Freed WJ: Peripheral nerve reconnection: Immediate histological consequences of distributed mechanical support. *Exp Neurol* 1983;81:459–468.

269. de Medinaceli L, Wyatt RJ, Freed WJ: Peripheral nerve reconnection: Mechanical, thermal and ionic conditions that promote the return of function. *Exp Neurol* 1983;81:469–487.

Section Seven
Peripheral Blood Vessel

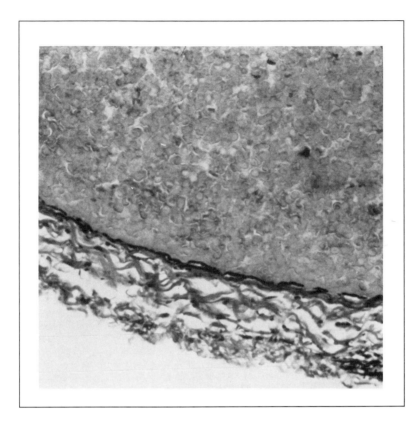

Group Leader

Andrew Weiland, MD

Group Members

Kai-Nan An, PhD
Leo Furcht, MD
Theodore Oegema, PhD

Group Participants

Thomas Andriacchi, PhD
Joseph Buckwalter, MD
Richard Coutts, MD
Bruce Carlson, MD, PhD
Victor Goldberg, MD
Göran Lundborg, MD, PhD
Marston Manthorpe, PhD
Eric Radin, MD
A. Hari Reddi, PhD
Björn Rydevik, MD, PhD

Synopsis

Many authors overlook the peripheral vessels in their discussions of the musculoskeletal soft tissues. Yet restoration of normal function following injury to tendons, ligaments, tendon and ligament insertions, myotendinous junctions, peripheral nerves, and skeletal muscles depends largely on the vascular response to injury.

Although the vascular system of each of these musculoskeletal tissues is anatomically and functionally unique, all systems share certain essential common features. Perhaps most important, they are living systems of cells and matrix, not fixed networks of rigid tubes. The volume of the vessels, the rate of blood flow, and the numerical density of the small vessels change following injury or in response to the demands placed on the tissues. Failure to restore normal blood flow after obstruction or mechanical disruption of the vessels causes necrosis or permanent damage to tissue. The success of free tissue transfer depends initially on restoration of blood flow through vessels anastomosed surgically, and eventually on growth of vessels into the transplanted tissue from the recipient site. Injury to the musculoskeletal soft tissues, with the exception of cartilage or the inner regions of the menisci, initiates a vascular response that forms a fibrin clot, allows migration of cells into the injury site, helps remove necrotic tissue, forms new blood vessels, and re-establishes blood flow. Lack of this vascular response is one of the factors that limits repair of some cartilage and meniscal injuries.

Despite the importance of the musculoskeletal soft-tissue vascular systems, surprisingly little is known about the details of their normal function, responses to changes in demands on the tissues, roles in repair, methods of improving their

response to injury, and growth into tissues transferred from one part of the musculoskeletal system to another. Thus, much of this chapter must be based on studies of vessels in other tissues and experience with surgical repair of transected vessels.

The peripheral vascular system consists of a continuous network of branching vessels that assume specific anatomic arrangements in each tissue. Vessels consist of three components: endothelial cells, smooth muscle cells, and connective tissue. The connective tissue component includes fibroblast-like cells and an extracellular matrix formed from collagens, elastin, proteoglycans, and noncollagenous proteins. The walls of larger vessels contain nerves and small blood vessels. Generally, the components of vessel walls are arranged in concentric layers: the innermost tunica intima (intima), the tunica media (media), and the outermost tunica adventitia (adventitia).

Endothelial cells line the inner surface of all blood vessels, actively help maintain the integrity of the vascular system, and have significant roles in repair. The composition and organization of media and adventitia vary considerably in relation to the mechanical demands placed on the vessels and the metabolic needs of the tissue. The large arteries have been the most extensively studied. In these vessels about 20% of the dry weight consists of elastin. Several types of collagen make up the bulk of the connective tissue with proteoglycans contributing approximately 0.5% to 1.0% of the dry weight. In addition, structural glycoproteins, including laminin and fibronectin, have been identified in vessels.

Vascular injury invokes the cell proliferation necessary for repair, but the proliferative response may be so extensive as to compromise or even occlude the lumen of the injured vessel. Release of a growth factor, either platelet-derived or endothelium-produced, appears to stimulate this smooth-muscle cell response. A human platelet-derived growth factor has been shown to initiate cell migration; heparin, heparan sulfate, and analogues of heparin derived from endothelium may also be regulators of smooth-muscle cell growth. Smooth-muscle cell proliferation appears to cease once an overlying layer of endothelium is established in a previously injured vessel.

Different mechanisms of injury can affect the vessel walls and the fluid movement in a variety of ways. For example, the mechanical characteristics of lacerations, avulsions, injections, compressions, and penetrating injuries differ significantly. Different techniques of vessel repair—including end-to-end, sleeve, and end-to-side repairs—use of vessel and prosthetic grafts, and specific suture materials produce anastomoses with different mechanical characteristics, rates of blood flow, and patency rates.

Formation of new vessels (angiogenesis) accompanies injury, inflammation, surgery, and tumor growth. Modulation of angiogenesis could help control inflammation, improve repair of musculoskeletal injuries, and accelerate the revascularization of tissues transferred from one part of the body to another. Cells, matrix components, and soluble factors all may help to control angiogenesis.

Future investigations of the peripheral vascular system that seem to have particular promise include studies of the specific structural and functional features of the musculoskeletal soft-tissue vascular systems, the mechanisms of mechanical damage to the peripheral vessels, methods of rapidly evaluating the nature and extent of injury to vessels, methods of facilitating the repair of vessels such as tissue adhesives or improved methods of anastomosis of severed vessels, the mechanisms that control the development of collateral circulation following injury, methods of decreasing tissue and vessel damage following injury and, perhaps most important, methods of promoting and controlling angiogenesis.

Chapter 8
Peripheral Blood Vessel

Theodore Oegema, PhD
Kai-Nan An, PhD
Andrew Weiland, MD
Leo Furcht, MD

Chapter Outline

Introduction

The peripheral vascular tree distributes blood to all tissues and organs.[1] Resting cardiac output is about 6 liters/min with an initial velocity in the aorta of about 33 cm/min. As the vascular area increases, the rate of flow gradually decreases, to as low as 0.1 to 0.5 mm/sec in the capillaries[2] where the surface area of the vessel wall expands to about 700 m[2]. Arteries and arterioles branch repeatedly, giving rise to tubes with progressively smaller diameters. For example, a 0.1-mm arteriole in the rat cremaster muscle undergoes four orders of branching. In the first branching, four second-order vessels arise. These, in turn, give rise to five to ten daughter branches each, so that as many as 14,000 to 70,000 capillaries come from the single 0.1-mm arteriole. With this branching, there is a 100- to 500-fold increase in cross-sectional area and a drop in velocity from between 30 and 50 mm/sec to between 0.1 and 0.5 mm/sec in the capillary. Thus, in the average capillary with a length of 0.3 mm, only six erythrocytes pass a specific point every second.[2,3] At any given time, about 5% of the total amount of blood is in the capillaries from which the blood gradually returns to the heart through veins of increasing diameter. Differences in the structure, composition, and mechanical properties of arteries, capillaries, and veins reflect differences in their function. The remarkable capacity of peripheral vessels to repair injuries, combined with advances in surgical techniques, has made it possible to restore blood supply to the musculoskeletal soft tissues after acute vascular damage. In addition, new vessels form after injury in most tissues.

Vessel Structure, Composition, and Function

Arteries

Generally, three layers of an arterial wall can be identified: The inner layer (or tunica intima) consists of endothelial cells, their basement membrane, and subendothelial connective tissue. The intermediate layer (or tunica media) consists of connective tissue and smooth muscle cells usually aligned with their long axis perpendicular to the flow or overlapping in a well-organized spiral orientation. The outer layer (or tunica adventitia) consists of connective tissue and transitional fibroblasts and merges with the loose surrounding connective tissue. The internal elastic lamina forms a distinct boundary between the tunica intima and the tunica media in many large arteries. Between the tunica media and the tunica adventitia an external elastic lamina is also present. Sympathetic nerves innervate most arterial walls; however, adrenergic and cholinergic nerves are found in the arteriovenous anastomoses. Both the large conducting arteries and the smaller distributing arteries, such as the common carotid, innominant, subclavian, and common iliac, contain many fenestrated layers of elastin. There may be 40 to 70 fenestrated lamina of elastin, 5 to 15 μ apart, in a large vessel. The outer walls of these large vessels may have microvasculature of their own. The innermost layers, however, are avascular. These arteries function as distensible conduits with an elastic recoil that helps maintain blood pressure between heart beats.

Elastic arteries give way to the distributing arteries (such as the femoral or radial) where the tunica media consists mainly of smooth muscle cells that adjust blood flow by controlling the diameter of the vessel lumen. This group contains the majority of the vessels down to those 0.5 mm in diameter. The intima is thinner and lacks smooth muscle cells. There is a well-developed elastica interna. The endothelium has well-established myoendothelial junctions with the innermost smooth muscle cells of the tunica media through the numerous fenestrations of the elastica interna. The thickness of the media varies from 40 helically oriented smooth muscle cells to as few as three or four in small arteries. A relatively thick basement membrane surrounds the smooth muscle cells. The elastica externa is an interrupted irregular sheet of elastin, thinner than in the conducting arteries and containing numerous fasicles of unmyelinated nerve axons. The smooth muscle cells are well-interconnected with gap junctions between the cells providing a low-resistance cell contact network for rapid passage of small ions.

In the transitional vessels, the tunica adventitia is thicker than the media and is only loosely connected with surrounding tissues, allowing adjustment of the artery as it contracts to regulate flow. The smallest arteries and arterioles form the principal component of resistance to flow that regulates blood pressure. In the arterioles, which range from 30 to 50 μ in diameter, the endothelium is still continuous, and there is now a thicker basal lamina with few reticulin and elastic fibers. The elastica interna is thin and windowed, and is absent in the terminal arterioles. The tunica media may consist of up to two layers of smooth muscle cells, but many of these vessels contain a single layer surrounding the endothelial cells in a hexagonal array.

During growth and development in elastic arteries, there is a progressive thickening of elastic lamina and an increase in thickness in the media, but there is little or no addition of elastin in the muscular arteries. With aging, large arteries become less pliant; connective tissue matrix and smooth muscle cells progressively increase as many of the smooth muscle cells transform from a contractile to a synthetic form.

Between the arterioles and capillaries there may be meta-arterioles with single smooth muscle cells spaced at intervals along a single layer of endothelial cells and providing contraction or a sphincter-like function. In some regions there are direct arteriovenous anastomoses that provide mechanisms for shunting blood around capillary beds. These are present in particularly high concentrations in skin[4] and largely absent in other tissues such as muscle, except as shunts in the surrounding fascia.[3]

Capillaries

At the level of the terminal arterioles, the vessels become true capillaries lined with a single, attenuated endothelial cell with a single basal lamina. Often, there is a single cell encompassing the entire circumference of the capillary. Occasionally, this cell is a contractile paracapillary pericyte that may help regulate the capillary diameter. Endothelial cells will also contract on mechanical stimulation.

Typically, ten to 40 capillaries may arise from a single terminal arteriole. In most of the musculoskeletal tissues (muscle, connective tissue, and nerve), the endothelium forms an uninterrupted layer or is nonfenestrated, whereas in some of the endocrine organs (pancreas, digestive tract), circular fenestrations interrupt the endothelial cell layer.

Veins

On the venous side, three layers of the vessel wall can generally be identified: the tunica intima, the tunica media, and the tunica adventitia. However, they are much less distinct than those seen on the arterial side. The muscular and elastin components are less developed than in arteries, and the connective tissue component is more prominent. When several capillaries unite, they form a venule of 15 to 20 μ in diameter consisting of a layer of endothelial cells with occasional fibroblasts. This layer of cells is still permeable to larger molecules, and it is also a site of continued exchange between the blood and the tissue. These cells are major regulatory sites for changes during inflammation and are particularly susceptible to compounds such as histamine and serotonin. In 50-μ vessels, partially differentiated smooth muscle cells gradually appear; as the vessels increase in diameter, the muscle cells increase in number to form well-established layers. Within vessels of 2 to 9 mm in diameter, the smooth muscle layer gradually becomes surrounded by elastic fibers; the tunica intima is still poorly developed and the tunica media is thinner and consists mainly of circular smooth muscle cells with many collagen fibers and few fibroblasts. The tunica adventitia is usually thicker and may still contain bundles of longitudinal collagen fibers within elastic networks and numerous smooth muscle cells.

In larger veins, the connective tissue layer of the tunica intima is thicker, the media is still poorly developed, and the tunica adventitia still makes up a greater part of the vessel wall. It contains elastic fibers and longitudinally oriented collagen fibers with many predominant smooth muscle cells and elastic networks. Valves are present in many of the medium-diameter blood vessels as part of the venous return system, particularly those in the extremities. These valves help regulate back-flow away from the heart.

Extracellular Matrix Composition

There are large regional differences in the composition of blood vessels.[5] In the larger arteries and the human aorta, about 20% of the dry weight consists of elastin, the biochemistry of which has recently been reviewed.[6] This component varies greatly in content throughout the vascular structure. Several types of collagens make up the bulk of the tissue and approximately 0.5% to 1% of the dry weight is proteoglycans. Numerous structural glycoproteins have also been described.

Collagens   Type I and type III collagens are the major collagenous components of human and bovine aortas. The proportion of type III to type I varies, but values for type III range from 42% to 21%. The higher values are found in the aorta and the lower values in the vena cava.[5]

Type IV collagen has been isolated from the human aorta. Numerous immunochemical studies have suggested localization in the basement membranes of the smooth muscle cells; these cells probably contain the largest amount of type IV collagen, with the endothelial-cell basement membrane also containing some. As much as 5% of the total pepsin-solubilized tissue from bovine aorta is type IV. Small amounts of type V have been isolated from both human and bovine aorta after pepsin digestion. There is still some controversy about the chain composition; according to Mayne and associates,[7] the αI (V) and αII (V) were not present in the typical 2:1 ratio and α3 (V) was not detected in human or bovine aorta; it may, however, be present in small amounts in porcine aorta.[8]

Type VI collagen was first isolated from the human tunica intima.[9] It has subsequently been isolated from a wide variety of connective tissues. Type VI contains both collagenous and noncollagenous domains stabilized by disulfide bonds and three genetically distinct chains. It has been suggested that type VI may form some of the microfibrils found around the elastin fibers and may be related to the 100-nm banded structures seen in many tissues and in tissue culture.[5]

Within blood vessels, the following collagen types have been localized by means of specific antibodies: In human aorta, type I tends to localize in close proximity to smooth muscle cells in the tunica media and to stain extensively in the tunica adventitia. Type III is predominant in the subendothelial layer close to the elastic lamina of the tunica media. In the tunica intima in both human and fetal bovine aorta, type III was observed throughout the entire layer and not just in the immediate subendothelium.[10,11] Another study suggests that neither type I nor type III is present in the subendothelium.[12]

Type IV has been localized in the basement membranes of smooth muscle cells and in the subendothelium of bovine aorta. In capillaries there is strong staining for type IV in the basement membrane, including strong staining in the basement membranes around capillary beds in muscle fibers.[5]

Studies of type V collagen have produced conflicting results, but it may be localized in the pericellular matrix surrounding smooth muscle cells underlying the capillary endothelial cells and largely localized within the interstitial collagen fibers. It is diffusely located throughout the tunica media and subendothelium of calf aorta. Type VI tends to be found throughout the matrix.[5]

Elastin   In many arteries, elastin fibers create an easily deformed network and are responsible for the early resistance in tension and the elastic recoil of the tissue. Elastin functions at relatively small distensions before the collagen network comes into tension and resists further deformation. As much as 90% of the elastin fiber is composed of an amorphous fraction with no banding pattern in association with small-diameter beaded microfibrils. The latter are chemically unrelated to elastin. Elastin fibers are assembled from soluble 70- to 72-kD tropho-elastin molecules and are extensively cross-linked with lysine-derived isodesmosine and desmosine. Elastin is one third glycine residues and has a high concentration of hydrophobic amino acids. The elastin peptides assume a highly mobile random coil conformation. When the cross-linked random coils are stretched, they assume a more ordered structure with fewer degrees of freedom and the hydrophobic residues are also brought into contact with water, causing an increased ordering of the water structure. Thus, energy is stored in the elastin network as increased available entropy and allows this network to recoil to a lower energy state with more than 90% efficiency.[6] Elastin is present as well-defined lamellae whose numbers vary depending on the pressure in the blood vessel.

Proteoglycans Blood-vessel proteoglycans have been only partially characterized. Initial studies indicated the presence of chondroitin 4- and 6-sulfate, dermatan sulfate, hyaluronic acid, and heparan sulfate. They are present in highest concentration in the intima with decreasing amounts in the media and adventitia of the aorta. The native proteoglycans have been partially characterized. A small amount of large aggregating proteoglycans similar to those found in cartilage,[13] which may be stabilized by link protein,[14] have been found in aorta extracts and preliminarily localized by immunochemical techniques in the highest concentration in the intima region of the aorta.[15] Several small dermatan sulfate and chondroitin sulfate proteoglycans have been partially resolved and characterized.[16,17] Staining of proteoglycans with cationic dyes such as Alcian blue or ruthenium red suggests networks of proteoglycans between collagen and elastin fibers connected by thin filaments and in association with a 60-nm periodic structure of collagen fibers.[18,19] The majority of these are sensitive to testicular hyaluronidase and chondroitinase ABC. Small granules were also noted in association with the basement membrane and with the endothelial cells and smooth muscle cell membranes. There is also considerable staining on the luminal surface of the endothelial cells. How much the anionic reactions on the endothelial luminal surfaces can be attributed to proteoglycans is unclear. However, at least part of the endothelial surface molecules are removed by heparitinase. By means of antibodies to purified aorta, heparan sulfate proteoglycan has been localized in the subendothelial basement membranes, at the surface of the smooth muscle cells, and between the elastin fibers of the media.[20] This finding is largely in agreement with the work using antibodies prepared against Engelbreth-Holm-Swarm (EHS) heparan sulfate proteoglycan where subendothelial staining was noted.[11]

Laminin and Fibronectin Laminin, the large glycoprotein isolated from a number of tissues including the transplantable mouse EHS tumor, is present in all basement membranes localized around smooth muscle cells of the aortic media and in the subendothelium of arteries and capillaries.[21,22]

Fibronectin, a well-characterized connective tissue molecule[23] with multiple substrate specificities, including binding to heparan sulfate, cell surfaces, and collagen, has been localized in the subendothelial layers of both capillaries and arteries and around the smooth muscle cells of the media, particularly muscular arteries and throughout the adventitia. It appears in high concentrations in the basement membrane of the endothelium.[24]

Endothelial and Smooth Muscle Cells

Endothelial cells are squamous epithelial cells that line the entire vascular system. They are generally nonmobile, adhere to the subendothelial basal lamina or basement membrane, and are slow to renew themselves.[25] The underlying smooth muscle cells are surrounded by a basement membrane and synthesize the collagen, proteoglycan, and elastin network. Recent advances in tissue culture have allowed study of smooth muscle[26] and endothelial cells[27] in vitro. However, very few comparisons, except by electron microscopic or immunochemical methods, have been made between their in vivo and in vitro responses.

The endothelium plays a central part in maintaining the integrity of the cardiovascular tree. It controls the exchange between the blood and the surrounding tissue. For small molecules measuring 0.5 to 1.5 nm, permeability is directly related to the free diffusion coefficient of the molecules.[28] This would be analogous to having an effective pore size of about 5.0 to 7.5 nm and is believed to be anatomically related to the spaces between the cells. A second pore size equivalent to 20 to 25 nm is thought to be directly related to vesicle transport. Endothelial cell vesicles measuring 60 to 70 nm actively transport molecules 20 nm or greater in size between the blood and basement-membrane surfaces of the cell. In muscle, microvesicles occupy up to 20% of the cytoplasmic volume of the attenuated periphery of endothelial cells. Active transport of molecules across endothelial cells by vesicles reaches a maximum density at the junction of the venules and the venous ends of the capillaries. In skeletal muscle, this forms a gradient of permeability toward the venous side of the microvasculature.[29] The magnitude of this transport mechanism can be appreciated when we realize that in the skeletal muscle of a 70-kg man, this rate of exchange is calculated to be 2 mg of protein per minute.

The permeability of small venule walls is regulated by local factors such as the release of histamine or serotonin, which rapidly causes, particularly in the postcapillary venules, retraction of the endothelial cells and an efflux of serum proteins into the underlying tissue. This leakage is aggravated in trauma or anoxia in which there is transient or permanent loss of endothelial cells, and plays a major role in the initial attempts at repair after trauma.

As reviewed recently by Simionescu and Simionescu,[30] the endothelial cell surface is in contact with the blood and plays a major role in the function of the blood vessel. Numerous specific functions have been attributed to molecules associated as either peripheral or integral components of the luminal surface plasma membrane. These luminal membrane proteins include a number of enzymes: adenosine triphosphatase (ATPase), adenosine diphosphatase (ADPase), and 5'-nucleotidase, and nonspecific esterase. These enzymes are thought to play a major role in limiting platelet aggregation after the platelet release of ADP. Carboxypeptidase N degrades bradykinin and converts the complement-derived anaphylactoid C5a peptide to the less potent C5a des Arg by removing the carboxyterminal arginine. Angiotensin I converting enzyme is a carboxyterminal dipeptidase that releases histidyl-leucine from angiotensin I to form the more potent angiotensin II. It also inactivates kinins. Two lipolytic esterase enzymes involved in degrading triglycerides and phospholipids are present. Lipoprotein lipase releases triglycerides from chymomicrons and very-low-density lipoproteins and is found in high concentration in the aorta.[31] These molecules are localized on the cell surface by interactions with the heparan sulfate chains of the proteoglycans and can be displaced by heparin.[31-34]

Endothelial cells can produce both coagulant and anticoagulant factors. The tissue factor thromboplastin, essential for rapid coagulation, is located on the endothelial cell surface. Endothelial cells produce platelet factor VIII-related antigen, which plays an important role in intravascular coagulation and platelet aggregation. It is localized in the endothelial cell basement membrane in association with fibronectin. Endothelial cells may also have plasminogen activator present in the cell surface membrane. Histamine receptors, particularly the H2 receptors in venules, have been localized within the vascular endothelium, which may explain some of the special involvement of the venules in inflammation.[30]

Endothelial cells both in situ and in vivo take up low-density lipoproteins by high-affinity receptor sites in coated pits and by low-affinity sites in the plasmalemmal vesicles. Certain endothelial cells also preferentially express a scavenger receptor pathway for charge-modified lipoproteins and nonlipoprotein ligands.[30]

Endothelial cells have relatively few specific hormone-binding sites. The best characterized are the insulin receptor sites, which are distinct from the insulin-like growth-factor receptor sites. It has been suggested that insulin is trancytosed across the endothelial cell by a specific receptor-mediated process. However, numerous transport proteins have been found in endothelial cells, including those for transferrin and albumin. The albumin receptor is involved in transcytosis and would also move other molecules known to be bound to albumin, such as fatty acids and testosterone. Endothelial cells may also mediate some immune reactions. Within specific regions, called the high endothelium of postcapillary venules of certain lymphoid organs, the endothelium is involved in homing of specific lymphocyte populations. In addition, since endothelial cells carry the blood group ABO and H antigens as well as the histocompatibility antigens, they can be targets after transplantation.[30]

Several molecules are associated with the luminal glycocalyx as peripheral molecules such as α_2-macroglobulin, a potent anti-protease that inhibits a wide variety of proteases, including those involved in fibrolysis, coagulation, and the kallikrein system. Fibrinogen-fibrin has also been localized in blood vessel walls as an extended thin network.[30]

Endothelial cells provide an antithrombogenic surface through several mechanisms. They synthesize and bind to their luminal sides molecules with anti-thrombin III binding activity. This has been shown to be present in the form of heparan sulfate proteoglycans[35] that contain the high-affinity antithrombin III modifications that have been described for heparin.[36,37] About 1% of the heparan sulfate chains present and synthesized by endothelial cells in culture contain a high-affinity anti-thrombin III site. The number of sites may be increased tenfold in the peripheral capillaries as judged by in vivo titration studies.[35] In addition, dermatan sulfate proteoglycans, which could use the alternate heparin cofactor II to inactivate thrombin,[38] have been found to be present in blood vessels.[13]

Additional limitations to the proliferation of a thrombus formed by aggregating platelets are provided by several mechanisms.[39] The endothelial cells contain specific thrombin-binding sites that rapidly remove thrombin from the newly formed clot. Also, the cells contain a large amount of membrane-bound ADPase that inactivates the ADP released by platelets, further limiting thrombus formation and platelet aggregation.[30]

Physical factors such as blood flow through the vessels are thought to play a major role in limiting the size of the thrombus in that there is a rapid dilutional effect of the activated intermediates, which are then quickly inactivated either by serum or cellular mechanisms. Thus, when endothelial cells are removed the highly thrombogenic surfaces of the basement membrane and the vessel collagens are exposed and these rapidly activate both platelet aggregation and the extrinsic pathway for clotting. However, the surrounding healthy endothelial cells then seek to limit the size of clot formation through the mechanisms discussed above and proceed to dissolve the forming clot through such processes as plasminogen activation to plasmin at their cell surfaces.[39]

Metabolites of arachidonic acid also play a role in the balance between thrombus formation and platelet aggregation, as reviewed by Needleman and associates.[40] In blood platelets, prostaglandin G_2 (PGG$_2$) and prostaglandin H_2 (PGH$_2$) are metabolized primarily to thromboxane A_2 (TxA$_2$). This metabolite is short-lived (half-life = 30 seconds) and spontaneously hydrolyzes to thromboxane B_2 (TxB$_2$). The contraction of vascular smooth muscle, platelet aggregation, and serotonin release are induced by a concentration of 20 nM or less of TxA$_2$ but not TxB$_2$. This is counterbalanced when vascular endothelial cells convert PGH$_2$ to prostacyclin (PGI$_2$), which spontaneously hydrolyzes to 6-keto-PGIx with a half-life of about ten minutes at pH 7.4. PGI$_2$ causes vasodilation and inhibits the aggregation of platelets. While vascular and nonvascular smooth muscle cells also synthesize PGI$_2$, the major source is the endothelial cell because of the high concentration of enzyme present. Additionally, there is a regional variation in PGI_2 synthesis; not all endothelial cells synthesize PGI_2. The PGI_2 synthetase enzyme undergoes rapid self-destruction during the catalytic cycle but can be partially protected by radical scavengers. Thrombin in vitro and ADP in rabbit aorta organ culture stimulates PGI_2 synthesis, and the cells may even use platelet-derived PGH_2 to form PGI_2. This response may limit the area of platelet deposition. Because of the self-inactivation, the cells become unresponsive to a continued or second challenge, suggesting a burst-like sequence of synthesis in response to repeated challenges. The response can also be mediated by histamine and bradykinin or even thrombin via thrombin receptors. In culture the response is maximal for nonconfluent cells and declines with confluence. Some platelet-derived PGH_2 can also be isomerized by a serum protein to prostaglandin D_2, which also inhibits platelet aggregation.

Leukotrienes may also play modulatory roles. They are synthesized by the 5-lipoxygenase pathway by neutrophils, eosinophils, monocytes, and mast cells. As peptioleukotrienes, they contract vascular smooth muscle cells and cause arteriolar constriction, dilation of venules, and plasma exudation and are chemotaxic for human polymorphonuclear leukocytes.

In injured tissue, the metabolism of arachidonic acid is often dramatically altered either by the direct invasion of inflammatory cells or by their interaction with resident fibroblasts and smooth muscle cells and these cells may play major roles in the long-term survival of repaired or grafted vessels.[40]

Endothelial cells synthesize and maintain the basement membrane. The composition of the basement membrane may vary with different sites, but it always contains type IV collagen, laminin, and a heparan sulfate proteoglycan,[41] along with other proteins such as fibronectin, nidogen, and, in some cases, pymphagoid antigen.[23] Models for the self-assembly of these structures have been proposed.[42]

In tissue culture, the response of the endothelial cells depends on the nature of their surrounding matrix.[43] Endothelial cells in culture rapidly synthesize type IV collagen[44] and various proteoglycans. The proteoglycans under normal circumstances are largely heparan sulfate proteoglycans. The heparan sulfate proteoglycans may be secreted immediately into the media, or they may become anchored in the matrix or on cell surfaces and subsequently released[45] as a family of disulfide-bonded dimers of heparan sulfate proteoglycans.[46] Several small dermatan sulfate-containing proteoglycans are also synthesized.[45] Smooth muscle cells predominantly synthesize a large aggregating dermatan sulfate molecule, as well as small heparan sulfate and dermatan sulfate proteoglycans[47–49] and type I and type III collagens.[50]

In wounding experiments both in vivo and in vitro, endothelial cells synthesize dermatan sulfate molecules when they migrate and divide to fill in injuries. Contact inhibition turns synthesis off when confluence is regained.[45] While migrating, the cells are very sensitive to the underlying matrix.[51] In small injuries, endothelial cells can migrate without proliferating, and early blood vessel formation does not require proliferation of the cells.[52] However, in larger injuries the cells must divide and migrate to form new vascular networks. From electron microscopic studies, it has been shown that the newly formed endothelial subsurface may contain increased levels of proteoglycan and more deposits of lipoprotein, which may make the wounded areas more vulnerable to atherosclerotic plaque formation.[45]

The endothelial cells overlying smooth muscle cells seem to help regulate maintenance of the smooth muscle cell-endothelial cell relationship, although the actual magnitude of this effect in vivo is unclear.[53] However, they have been shown both in vivo and in vitro to produce heparan sulfate fragments, which inhibit smooth muscle cell growth[41] and influence synthesis of matrix molecules by smooth muscle cells in vitro.[54] This mechanism of regulating smooth-muscle cell proliferation might be upset by endothelial cell damage. If endothelial cell loss is extensive, the denuded region becomes much more vulnerable to subsequent proliferation of smooth muscle cells and possible fibrotic replacement.

Mechanical Properties of Normal Vessels

To understand the hemodynamics of the vascular system, the mechanical properties of blood vessels, including the compliance and, more specifically, the viscoelastic properties of the vessel walls, must be known.[55] Knowing the force elongation relationship of the normal vessel under uniaxial tensile conditions provides information necessary to estimate existing tension in the anastomosed vessels of shorter segments. The most commonly used methods for characterizing these properties are tensile tests of either the entire vessel or a strip, infusion tests, or volume/pressure tests for compliance measurement. Vessels demonstrate nonlinearity of the force-deformation or stress-strain relationship and existing hysteresis. Vessels also demonstrate viscoelastic characteristics in terms of creep and stress relaxation.

Variations in the elasticity and viscoelasticity of vessels can be explained by differences in the composition and structure of their walls. Elastin is soft, elastic, and relaxes very little. While collagen is also elastic, it has a larger elastic modulus and relaxes more. The precise quantitative relationships between the composition of vessels and their mechanical properties remain to be studied. Although the mechanical properties of vessels change with age, disease, and medical treatment, these changes have not been well defined. Because of their biologic and mechanical properties, normal autogenous vessels are preferred for reestablishment of blood flow where grafts are needed. The value of irradiated graft vessels is controversial. Some hold the risk insignificant,[56] while others warn against their indiscriminate use.[57,58] In addition, the risk of arterial thrombosis increases following the use of previously irradiated tissue.[59,60] Currently available prosthetic vessels also do not duplicate the properties of normal vessels.

The Response of Vessels to Injury

Almost any injury to a blood vessel stimulates a cellular proliferative response. This response may be self-limiting and result in functionally insignificant internal thickening, or it may reduce or even occlude the lumen in which it occurs. As noted previously, control of endothelial- and smooth-muscle cell proliferation is complex and depends to some extent on interaction between endothelial cells and smooth muscle cells. Better understanding of this part of the injury and repair response of vessels is essential since it plays a significant role in arterial endarterectomy,[61,62] intra-arterial passage of a balloon catheter,[63] autologous vein grafting,[64] creation of a prosthesis, repair of lacerated vessels,[65] and the injuries that stretch or contuse vessels without rupturing them.

Using a canine model, Bush and associates[61] recently investigated the effect of endothelial-cell seeding (ECS) or antiplatelet therapy with aspirin on the inhibition of neointimal hyperplasia of the arterial wall following carotid endarterectomy. Endarterectomies were performed in 160 carotid arteries; 46 endarterectomies were treated perioperatively with aspirin, 34 were seeded with a high-density (3×10^6) of autogenous endothelial cells, and 80 were untreated carotid arteries. At selected time intervals, the patent arteries were perfusion-fixed, and the cross-sectional areas of neointimal hyperplasia were measured by means of digital planimetry. At six weeks, patency of the endarterectomized carotid artery was 88% in the aspirin and ECS groups in contrast to 35% in the control group (P $<$ .01). The cross-sectional area of neointimal hyperplasia was not significantly different in the aspirin and the control groups at six weeks. However, the ECS group showed a marked reduction in neointimal hyperplasia at six weeks (P $<$ .01). This inhibition of neointimal hyperplasia after carotid endarterectomy by ECS may reflect accelerated luminal healing or a direct inhibition of smooth-muscle cell proliferation in the injured arterial wall.

Schwarcz and associates[62] reviewed lesions from 20 patients with recurrent carotid artery stenosis in an attempt to characterize the histologic findings in these recurrences and provide additional data for understanding their pathogenesis. Endothelial cells were found lining the luminal surface of both early and late recurrent lesions—a finding not previously described. The presence of these cells implies that a phase of arterial healing after endarterectomy involves re-endothelialization. Contrary to previous reports, neither thrombosis nor intraplaque hemorrhage could be found to contribute significantly to the development of recurrent stenosis. The early presence of smooth muscle cells followed by their later disappearance suggests modification of recurrent stenosis over time. The authors suggested that recurrent stenosis is a dynamic process of arterial healing that, if amenable to treatment, could improve the results of carotid endarterectomy in the future.

Clowes and associates[63] studied the kinetics of cellular proliferation after arterial injury introduced by the intraluminal passage of a 2F balloon catheter introduced in the external carotid artery of Sprague-Dawley rats. The endothelial layer was regenerated from the ends of the denuded segment but failed to cover the central third of the artery by 12 weeks. Smooth-muscle cell proliferation reached a maximum at 48 hours in the media and 96 hours in the intima. Smooth-muscle cell proliferation persisted at a high level at the surface of the intima that lacked endothelium even at 12 weeks. These results support the concept that intimal smooth-muscle cell proliferation after arterial injury is an acute event related to the initial injury process.

Postoperative changes in autologous vein grafts have been reported by Fuchs and associates[64] who evaluated the short- and long-term alterations in saphenous bypass grafts. The most common and least understood change observed was intimal hyperplasia.

Chronic endothelial and smooth-muscle cell proliferation has also been observed in healing polytetrafluoroethylene (PTFE) prostheses, as reported by Clowes and associates.[65] At six and 12 months after grafts were placed, endothelial coverage by ingrowth from the anastomoses was more advanced than at three months, and by 12 months, 60% of grafts 7 to 9 cm long were covered. Endothelial cells proliferated in association with the growing edge. Underlying smooth muscle cells proliferated in the region of the growing edge of the endothelial cells and also at the anastomoses. Intimal cross-sectional area was greatest at anastomoses and at late times was principally the result of an increase in connective tissue. These results demonstrated slow but progressive healing of the grafts by ingrowth of endothelium. There was also an increased turnover rate of smooth muscle cells and endothelial cells in established intima at late times. In this experiment, 4-cm PTFE grafts were employed; thrombosis occurred in only two of 20 grafts. The use of smaller-diameter (< 2 mm) PTFE grafts has been associated with a higher rate of thrombosis, and these prostheses have not been employed in microvascular surgery.

The histologic characteristics of this vascular proliferation have been well delineated.[66] Although it is called intimal hyperplasia, the proliferating cells are smooth muscle cells that originate either in the media or in the subintimal spaces. In both arteriosclerosis and hypertension, the two major forms of vascular disease, the central cellular feature is smooth-muscle cell proliferation.[67] The common role of accumulation of smooth muscle cells suggests that control of cell proliferation may be critical to both diseases.

The pivotal element in the origin of this smooth muscle response appears to be the presence of growth factors either derived from platelets[68] or produced by the endothelium.[69] Antoniades and Williams[68] defined the structure and function of human platelet-derived growth factor (PDGF). It is a heat-stable cationic polypeptide transported in blood in the alpha granules of platelets and released from platelets during blood clotting. PDGF has been resolved into at least two closely related active polypeptides, PDGF-I and PDGF-II, each consisting of two inactive chains linked together by disulfide bonds. PDGF stimulates the growth of normal cells in culture, including fibroblasts, ar-

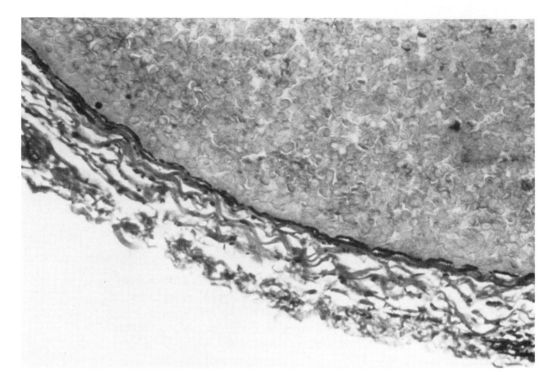

Fig. 8–1 *Normal rat vein (× 250).*

terial smooth muscle cells, and glial cells. In addition, PDGF has been shown to stimulate cell migration and many diverse metabolic functions such as amino acid transport, protein synthesis, cholesterol ester synthesis, phospholipid turnover, and prostacyclin synthesis. It modulates receptor binding of other active components such as epidermal growth factor, luteinizing hormone, low-density lipoprotein, and somatomedin-C. Specific cell membrane receptors for PDGF have been demonstrated in arterial smooth muscle cells and fibroblasts.

Experiments performed by Clowes[69] suggest that heparin, heparan sulfate, and analogues of heparin derived from endothelium may be regulators of smooth-muscle cell growth. Under certain conditions, endothelium may synthesize a smooth-muscle cell growth factor—endothelial cell-derived growth factor—and endothelial membranes may contain an inhibitor of endothelial growth. Heparin has been shown to be associated with a reduction of intimal thickening in a rat injury model, and smooth-muscle cell proliferation appears to cease once an overlying layer of endothelium is established in a previously injured vessel.

A number of specific questions remain unanswered concerning the process of intimal hyperplasia following repair of lacerated vessels.

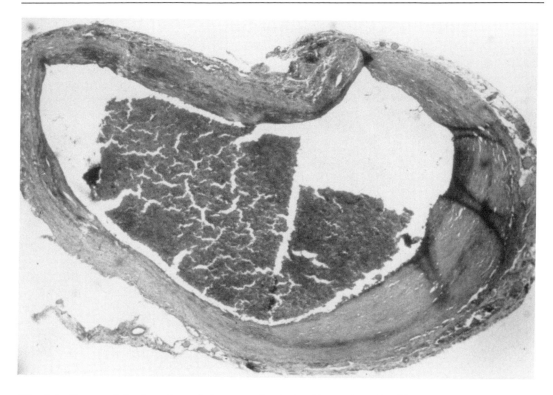

Fig. 8–2 *Five-month in situ vein graft demonstrating intimal hyperplasia (× 40).*

(1) Is artery or vein more susceptible to this process? This question is relevant to the clinical situation in which either artery or vein can be used to bypass diseased vessels, such as the choice between internal mammary artery and autologous vein bypass in coronary artery disease.[70]

(2) Does the amount of perivascular dissection and subsequent disruption of vasovasorum influence the degree of intimal hyperplasia seen in grafted artery and vein? This question is relevant to the issue of possible advantages of in situ as opposed to free, reversed autologous vein graft techniques.[71]

(3) Can drugs that alter platelet function (possibly reducing the production of PDGF) alter this process in artery and vein segments that have undergone surgery?[72] Clinical advantage has been demonstrated with the use of antiplatelet drugs in vascular bypass procedures.[73,74]

To begin to answer some of these questions, the following vascular procedures were performed on Sprague-Dawley rats.

Artery in artery in situ: A segment of iliac artery was left in place but divided twice and reanastomosed end-to-end in situ.

Artery in artery reversed: This arterial segment was fully dissected, reversed, and reanastomosed end-to-end into the arterial defect.

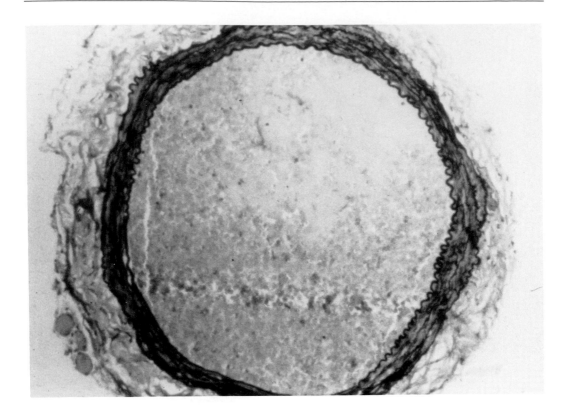

Fig. 8–3 *Normal rat artery (× 100).*

Vein in vein in situ: A segment of iliac vein was divided twice with minimal dissection in between and reanastomosed end-to-end in situ.

Vein in vein reversed: The vein segment was fully dissected, reversed, and anastomosed end-to-end into the venous defect.

Vein in artery in situ: The iliac artery was divided and anastomosed in an end-to-side technique to a segment of in situ iliac vein.

Vein in artery reversed: The segment of iliac vein was fully dissected, removed, and anastomosed end-to-end into a defect in the iliac artery.

The segments of vessels undergoing operation were removed and examined by light microscopy with hematoxylin and eosin and Verhoeff-Van Gieson staining techniques. Photomicrographs were made from the light microscopic sections, and the percentages of the areas occupied by lumen, intima, and media were quantified[74] to ascertain the degree of intimal hyperplastic response.

Light microscopy revealed the most profound intimal hyperplastic changes in those groups in which veins carried blood at arterial pressure. In this situation, the normally thin-walled vein showed evidence of irregular but consistent intimal hyperplasia as well as hyalinization and necrosis in the media and adventitia. These responses appeared equal in both in situ and completely dissected and reversed vein grafts.

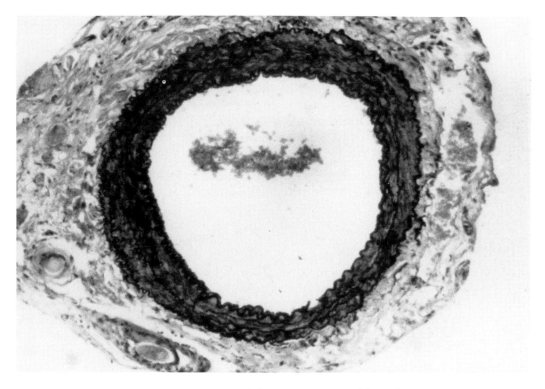

Fig. 8–4 *Five-month interpositional arterial graft demonstrating intimal hyperplasia but to a lesser degree than the vein graft shown in Figure 8–2 (× 100).*

Normal rat artery has a well-developed media with prominent elastic fibers. An internal elastic membrane defines the monocellular intima lining the lumen. When these vessels are grafted, they demonstrate a degree of adventitial and medial hyalinization and necrosis in both the in situ and free graft segments. In each of these situations, there is only a minimal degree of intimal thickening with a thin layer of tissue being demonstrated between the internal elastic membrane and the lumen.

These preliminary studies demonstrate consistent changes in the media and adventitia of vessels undergoing grafting maneuvers. The potentially significant intimal hyperplastic response is more consistent in vein grafts than in arterial grafts (Figs. 8–1 to 8–4). The degree of intimal hyperplasia and the influence of platelet-suppression drugs in this process should be studied further.

Mechanical Aspects of Vessel Injury and Repair

Mechanisms of Injury

Knowing the mechanism of vessel injury is important in planning the surgical repair and in the long-term results of the repair. In general, for sharp lacerations and local crush injuries, there has been a success rate of 80% or more in surgical revascularization. On the other hand, generalized crush and avulsion injuries cause segmental vessel disruption and diffuse tissue damage and thrombosis; thus, lower success rates are expected.

The only known nonthrombogenic surface is the normal vascular endothelium. Its nonthrombogenicity requires compatibility with both platelets and the plasma coagulation system. Various injuries of the vessel wall destroy these properties.[75] Injured vascular tissue must be resected back to the normal intima, and, when necessary, a vein graft should be used to bypass the local crush area.

Avulsion Injury Clinically, it has been noticed that in avulsion injuries, soft-tissue injury often occurs more proximally than skeletal injury.[76]

In a recent experimental study of avulsion injury by Mitchell and associates[77] using rabbits, venous and arterial injuries were noted both proximally and distally by operating microscope as well as by light and electron microscope. Deep clefts as well as circumferential skip lesions, intimal tears, holes, bruising, sleeving, and dilation were noted at vessel bifurcations through all three tunicae. With light and electron microscopic examination, the arterial and venous lesions were often noted to be as far as 4 cm from the rupture site both proximally and distally. This damage will affect the pattern of blood flow and therefore the healing response.

In Mitchell and associates' study, avulsion injury was less damaging in the venous wall than in the arterial wall. Veins are avulsed more easily than arteries. The prolonged stretching of the relatively rigid arteries probably leads to more severe damage. The extensive areas of endothelial and internal elastic lamina denudation in avulsed arteries, where damaged smooth muscle cells are exposed to the lumen, create probable sites of thrombus formation and subsequent occlusion.

High-Pressure Injection Injury The damage commonly associated with high-pressure injection injury is thought to result from the necrosis that occurs from the high-speed mechanical impact and the vascular ischemia.[78] However, according to Ramos and associates,[79] injection of normal saline under high pressure does not cause significant tissue reaction; thus, injection injury may be more closely related to the material and to the extent to which it is distributed by the high pressure rather than to high-pressure necrosis itself.[80] The mechanism of such injury deserves more investigation.

Compression Injury Compression injury caused by occlusive vascular clamps is commonly encountered clinically. Most microvascular clips exert high closure forces and produce endothelial damage,[81-85] which may cause platelet aggregation and thrombogenesis. The compressive response of the artery is caused by viscoelastic deformation of smooth muscle and collagen.[86] It is therefore desirable to ensure the minimal occlusive force with the minimal damage threshold during temporary vascular occlusion.[81]

According to Poiseuille's law, internal irregularity at the clip site drastically reduces the flow by a fourth-power function. In addition, the turbulence and vortex formation may further enhance thrombogenesis by the release of additional thrombogenic ADP.[87,88] However, the exact mechanism and threshold for lesion onset and the extent of damage caused by the microvascular clamp have not been defined.

Penetration Injury Penetrating injuries such as gunshot wounds have been reviewed and described extensively.[89] The more common effects of needle trauma to the vessel wall during microvascular anastomosis have been quantitated.[90] Using end-to-end anastomosis, there was significant thickening of the arterial wall two weeks postoperatively, but there was no relation between the amount of wall-thickness change and needle diameter. Patency rates of less than 100% of the femoral arteries were found in all groups of rats in which the needles used were greater than 100 μ in diameter; otherwise, patency was 100%.

The Micro Edge Taper (MET) pointed needle has been compared with the conventional taper pointed needle. The leading point of the MET needle is much sharper than that of a conventional taper needle and may cause less damage to the vessel wall. The breaking point of anastomoses created with the MET needle is significantly higher than with the other conventional method. At one week following anastomosis using the MET, the breaking point is the highest (0.8 N) and even comparable to the control value of 0.75 N. However, the strength decreases after three weeks.[91]

Comparison of Repair Techniques

Testing Methods The tensile test is the most commonly used method for quantification of mechanical strength and compliance of normal vessels and vascular repair. Lauritzen and Bagge[92] used the tensile strength test at a rate of 0.1 mm/sec for comparison of the end-to-end and end-in-end (sleeve) methods of anastomosis. A bursting and leakage test was also used by Lauritzen and Bagge to study leakage at the suture line. A static pressure of 240 mm Hg was used for one minute.

The compliance under bursting pressure is determined by distending the specimens with water at a constant infusion rate. Diameters are measured by a shadow technique.[93] The tangential tension and stress are calculated by using Laplace's law and the Lame approximation.

In vitro measurement of blood flow rate at the anastomosis has been made by measuring the time required for 50 ml of saline to pass the anastomosis at pressures of 120 and 240 mm Hg.[92] In vivo, electromagnetic flowmetry[94] and ultrasound and laser Doppler methods have been used. In addition, an isotope labeling technique has been used to study the accumulation of platelets at the repair sites.[95]

End-to-End Repair The tensile strength of the end-to-end repair technique has been studied by many investigators using slightly different techniques and animal models.

Lauritzen and Bagge,[92] using 1-mm rat femoral arteries, found that the tensile strength was maintained at the level of 75 g from the first hour to the first week mainly by the strength of the suture (seven to ten stitches of monofilament 10–0 suture) and increased up to 100 g progressively from the second week up to the eighth week. Thereafter, no statistical difference between the sleeve and end-to-end methods was noted. There was no leakage after three days under 240 mm Hg but, three hours postoperatively, three out of four arteries leaked.

The strength of the end-to-end anastomosis has been studied in rat vessels[96] repaired with six interrupted sutures of 10–0 nylon. The cross-sectional area and pull-out force were measured immediately and for up to 150 days. Among these anastomoses, 54 of 56 were patent at the time of harvest. The immediate anastomosis strength was 44% of the intact control vessel.

An end-to-end repair using wrap-around anastomosis with three stitches has been described. It functioned well when examined after four weeks with 100% patency.[97] However, the strength of such an anastomosis has not been defined.

End-in-End (Sleeve) Repair Lauritzen and Bagge[92] used three equally spaced stitches for the sleeve anastomosis. The tensile strength was in the range of 50 g within the first week and increased progressively from the second week to more than 100 g. All anastomoses three days or older, including the two aneurysms, withstood a static pressure of 240 mm Hg for one minute without leakage. Within the first three postoperative hours, however, the proximal end of one of the sleeve anastomoses slid out and three of the end-to-end anastomoses began to leak.

Marked stenosis was noted in end-in-end anastomoses of rabbit arteries.[94] Stenosis following end-to-end and end-in-end anastomoses was analyzed using angiography. The end-in-end anastomosis had an average luminal area of 22% of the intact vessel one hour postoperatively and maintained up to 63% after 90 days. Less stenosis was found in larger vessels (1.7 to 2.0 mm) than in smaller ones (0.8 to 1.2 mm).[98]

Meier's technique of suturing the distal end of the proximal segment of an end-in-end anastomosis is reported to produce less stenosis than Lauritzen and Bagge's technique of suturing the proximal end of the distal segment.

With the stenosis, the flow rate may be reduced by the end-in-end technique. The ratio between wall thickness and internal diameter of a blood vessel is fairly constant, particularly in arteries.[99] The flow rate is usually higher with the end-to-end than with the end-in-end technique within the first three hours following operation. The differences decrease with time, and by three days there were no differences.[92] The response and adaptation of the system to the stenosis in terms of the fluid flow deserve more study.

By using electromagnetic flowmetry, it was found that the mean blood flow in small arteries (0.8 to 1.2 mm) was almost unchanged after end-to-end anastomosis but was reduced to approximately 50% of the preoperative value following end-in-end anastomosis.

End-to-Side Repair   End-to-side anastomosis is often used in re-plantation and free-tissue transfers. Several laboratory studies have been performed to examine the effect of the type of arteriotomy and the angle of the anastomosis on the hemodynamics and the suture stress.

Suture line stress for end-to-side anastomosis in the synthetic graft caused by hemodynamic stress has been theoretically analyzed under steady-state and pulsatile flows.[100–102] Using the principles of fluid mechanics, the equation of continuity, the equation of energy (Navier-Stoke's equation and Bernoulli's equation), and the momentum equation, the equation on the suture line at the anastomosis as the result of blood flow can be obtained. The forces are caused by the kinetic energy of the flow and by the pressure energy of the flow, which is more dominant. The force is influenced by the ratio of the cross-sectional area between the anastomotic ends and the angle of end-to-side anastomosis. For an arterial pressure of 180 mm Hg, the hoop stress and axial stress are in the range of 10 to 20×10^5 dyn/cm^2, and the force in the suture line is in the range of 20 to 50×10^3 dyn.

In a recent study, the effects of the type of arteriotomy and the angle of the anastomosis on vessel patency and aneurysm formation were examined.[103] There was a decreased rate of aneurysm formation in the group that had an elliptical arteriotomy where the anastomosis was oriented at 90 degrees. The significant decrease in aneurysm formation seen in the 90-degree group suggests that an uneven stress was placed on the suture line in the 45-degree group. The difference in the patency rate was not statistically significant and suggests that in the high-pressure artery, the angle of orientation and type of arteriotomy do not play a role.

Ring-Pin Stapler   With the ring-pin stapler, the vessel end is passed through the ring and everted and fixed on the pins. A similar procedure is performed on the other end. The two rings are brought together so that the pin of one ring passes through the holes of the other. The patency of vein-to-vein anastomosis is increased by using such a stapler, which gives a smooth intima-to-intima junction.[104,105] The advantages of such a method are the absence of foreign material in the vascular lumen, atraumatic handling of the vessel ends, avoidance of intimal damage, and near-perfect coaptation of the intimal surface in the line of the anastomosis.

Absorbable Anastomotic Coupler   Any device that could achieve acceptable patency rates in a rapid fashion could be of value in reconstructive microsurgery. By combining the classic cuffing principle with the modern technology of modeling the polymer polyglactin, a coupler has been developed. It is absorbed by hydrolysis in 50 to 70 days. It is composed of two cuffs and an interconnecting collar. The patency rate was found to be more than 94% for short, intermediate, and long terms (up to one year). An average of 3.3 mm of vessel length is required by this technique, and may increase the tension at the repair site. All vessels were patent at two weeks.[106]

Arterial regeneration with an absorbable prosthesis has been studied by Greisler.[107] Currently, the shortcomings of small-vessel replacement include high thrombogenicity and low compliance. The ideal prosthesis would support a functioning thromboresistant endothelialized intima with a media of organized smooth muscle and elastin and have adequate strength to withstand the various hemodynamic stresses.[107]

Although the utilization of a coupler device may be attractive, more study is required before additional clinical application. Factors responsible for the initiation and regulation of arterial regeneration must be evaluated. The timing of the decrease in material strength is critical to the incidence of aneurysmal formation. A modular system to accommodate size discrepancies in the repaired vessels would also be helpful.

Adhesive Techniques Since any suture and needle used in microvascular anastomosis may inflict damage on the vessel wall, much effort has been spent in developing methods of anastomosis in which the number of sutures is reduced to a minimum.

Better understanding of blood coagulation and hemostasis has led to the trial of various fibrinogen and thrombin mixtures with experimental and clinical uses. By mimicking the end stage of the coagulation cascade, an adhesive clot may be formed (fibrin glue).[108-111] Increasing the concentration of fibrinogen increases the strength and adhesion of the glue. Increasing the thrombin concentration increases the rate of glue formation.

A combined suture-adhesive technique, using fibrinogen adhesive, was used by Dinges and associates.[112] The use of adhesive materials was analyzed by Karl and associates.[113] More studies to document their strengths and the healing process are needed.

CO_2 Laser-Assisted Repair Recently, laser-assisted anastomosis of vessels has been attempted. This technique uses three equally spaced sutures and multiple CO_2 laser pulses with a spot size of 150 μ.[114] The tensile strength of the anastomoses ranged from 10% to 25% of the untreated control over a period of 0 to 14 weeks and then continuously decreased to 15%. The laser apparently damaged the tissue so that the pull-out strength of the tissue required a longer time for recovery.

Mechanical Properties of Vessel Repair

Changes in vessel-repair strength with time following surgery are important to the characterization of the process of repair. Maximum breaking strength using end-to-end anastomosis was studied for a period of up to three weeks.[91] The strength is highest at week 1. Maximum strength at week 1 can be attributed to residual thrombus from the time of injury. The lower three-week value may be the result of a lack of medial regeneration following thrombus dispersion. Acland and Trachtenberg[57] have shown with serial histologic studies in the rat that there is little or no medial regeneration for up to three weeks.

The failure strength decreased to a minimum (23% to 50%) at four months and then returned to control values at 12 months.[115] Healing is slow, usually requiring 12 months. However, the stretch ratio at failure is almost constant. A detailed analysis of failure mode would be worthwhile in order to document the relationship between the extent and organization of the collagen network and failure, as well as the likelihood of suture failure as opposed to tearing-out stress at various times.

Bursting pressures did not change significantly in the first two weeks after transplantation of a vein graft but increased significantly after three to four weeks. This increase is related to the increase in collagen concentration at three to four weeks; the decrease in collagen concentration in the first two weeks may be compensated for by the increase in other components.[116] The tensile strength of a wound depends not only on the absolute quantity of collagen formed but also on its morphologic character and polymerization.[117] It is possible that the hemodynamic stress elicits the formation of more mature and stronger collagen in the graft in the same way as the tensile strength of wound collagen can be improved by moderate distraction forces.[118]

The healing process of a normal sutured arterial anastomosis[119] can be characterized by (1) loss of endothelium at the anastomotic site with reestablishment of complete coverage by 14 days; (2) hyaline degeneration of the media occurring within the suture area; (3) subintimal hyperplasia occurring to a degree compensating for loss of media with the maximum reached by three months; and (4) at one year, vessel-wall thickness approaching normal but with the intima becoming the thickest layer.

Mechanical factors influencing the development of vessel repair tissue strength are (1) the stress-strain relationship, the viscoelastic characteristics, and the ultimate strength of each component, e.g., collagen, elastin, smooth muscle cells, ground substance, and endothelial cells; (2) the structure of the composite and the geometric disposition of various basic components of the material as they deform under stress; (3) the ultrastructure of the basic components of the material such as crystalline material, vacancies, and nonhomogeneities; and (4) the type of loading (shear, tensile, or tearing), the history of loading, and the rate, including the hemodynamic forces.

It has been hypothesized that the difference in the collagen content of the tissue results in different rates of healing. Collagen is the component responsible for the extent of the tissue's mechanical strength.[120] Collagen-rich tissue such as skin and the aponeurosis do not reach the mechanical strength of intact tissue even after a year of healing,[121] while collagen-poor tissue attains this level after ten to 20 days.[120]

The tensile strength of the wound in the early stage of healing is related to the collagen content.[122] After two to three weeks, however, a further gain in strength is related to the qualitative change of the wound collagen,[123,124] which may be related to chemical and physical conditions.[125] For example, based on Wolff's law for soft tissue,[118] controlled tension across the wound may cause preferential alignment of the collagen molecules, which in turn may result in a more organized structure and an increase in strength.

One of the most important issues in the management of the patient who undergoes microvascular repair and replantation is when to begin mobilization of the replanted part. In one study, it was found that the tensile strength of the anastomosed vessel was 44% of the intact control vessel immediately after the operation and did not significantly increase for four to five months. Although early mobilization is not limited by the strength of the microarterial anastomoses, the in vivo determination of the stress on each of the vessels during passive and active motion must be considered.[96]

Suture Selection   A suture's holding capacity depends to a large extent on the integrity of the collagen network in tissue. In vein graft studies, the holding strength of suture decreased in the first week and then increased after two weeks.[116] Failure occurred when the intact suture loops pulled through the vessel wall. The cross-sectional area of the anastomosis sites increased significantly up to two weeks and then returned to control values by three weeks. The healing-strength curve obtained by Davis and associates[96] differs from the classic healing curve demonstrated in the skin wound of rats and in end-to-end anastomoses of the larger arteries repaired with absorbable sutures.[123,126,127] The suture-holding power of the tissue is the important factor. Equally spaced sutures or even running sutures should be placed to prevent strangulation of the tissue and excessive medial necrosis. The use of finer sutures and needles should further reduce the tissue disruption that weakens anastomoses initially.

Nonabsorbable sutures may cause foreign-body reaction and impair the ability of the vessel to complete long-term healing and further increase the strength. Thiede and associates[128] reported that anastomoses done with nonabsorbable sutures are stronger than those performed with absorbable sutures 30 to 70 days postoperatively, but the difference decreased at 100 to 150 days. The long-term results and clinical significance should be studied.

Stress on the Suture Line   Stress on the suture line may be induced by blood pressure, muscle attachments, and by resecting a segment of vessel and repairing the shortened vessel. One of the basic principles of microvascular surgery is that there should be no undue tension at the anastomosis site. However, what is undue tension? In Chow and associates'[129] study, different lengths of the femoral artery were resected (40%, 60%, 80% of the distance between the clamps with 250 g/mm² pressure) and the shortened vessel anastomosed. The patency rate was 100% for the 40% resected group but dropped to 75% for the 80% resected group. The average narrowing of the vessel after anastomosis was 16% and 52% for the 40% and 80% resected groups respectively. In future studies, the margin of safety should be defined, and, possibly, a gauge could be designed for tension measurement during surgery.

Complications of Vessel Repair   Aneurysms, thrombosis, leakage at the repair site, and rupture of the repair complicate vessel repairs. Too many sutures in the end-to-end anastomosis may lead to medial necrosis, causing aneurysms at the site of the end-to-end anastomoses.[130] Patency is the basic requirement for a successful repair. In Lauritzen and Bagge's[92] study, patency occurred in almost all end-to-end and end-in-end repairs. In Hyland and associates'[131] study, using the end-in-end technique, the patency was 90% within one week of the operation. The patency rate can be improved for vein repair by using a temporary stent technique.[132]

Platelet aggregation is necessary to prevent blood leakage by sealing the defects. Platelet accumulation can, however, exceed the desirable levels, and growing platelet aggregates then become a threat to unimpeded blood flow.[95] The fact that blood accumulation at the anastomosis site is higher in end-to-end than in end-in-end repairs may be attributed to the larger number of platelets required to seal the leak. When the end-in-end anastomosis maintains patency, the flow velocity increases at the stenotic segment, and platelet accumulation may be washed away rapidly.[133] On the other hand, the flow rate, which was decreased by 50%, may be caused by the stenosis.[94]

Small blood vessels have a naturally occurring contractility that resists dilatation. Caro and associates[99] have identified this as a circumferential compressive stress that increases with the decrease in the vessel's diameter. An external ring technique provides a direct method of overcoming the vessel's inherent circumferential compressive stress, provides maximal radial tethering force at the site of the anastomosis, and thus increases the patency.[134] The hemodynamic effect of the stenosis and circumferential compression should be studied further.

Methods of Improving the Repair Response

Although repair of even small vessels usually succeeds, improved surgical techniques could reduce complications, improve patency rates, and increase efficiency of vessel repair. Methods of controlling intimal hyperplasia and limiting tissue and vessel damage following injury have become important. However, the area of investigation that may have the greatest potential for improving the repair of musculoskeletal soft tissues following injury is the facilitation and control of angiogenesis. Angiogenesis is a fundamental component of diverse normal and pathologic processes. From the initiation of embryonic implantation, precisely controlled angiogenic signals accompany all phases of organogenesis and development and play a fundamental role in embryonic growth. In the adult, angiogenesis often accompanies injury, inflammation, surgery, and tumor growth.

Following an angiogenic stimulus[135]: (1) The vessels near the stimulus become dilated and permeable; endothelial junctions decrease and the cells appear "reactive." (2) The basement membrane dissolves, caused by activation of a variety of proteases, including plasminogen activator and type IV collagenase. (3) Endothelial cells migrate from the vascular lumen, through perivascular connective tissue and parenchyma toward the angiogenic stimulus. (4) Behind the leading front of migrating endothelial cells, replication begins in other endothelial cells. (5) Loops derived from anastomosis of proximate sprouts begin to form, creating capillary tubes. (6) Capillary loops and sprouts develop further; patency or canalization of loops occurs. These newly formed blood vessels then develop or differentiate further and lay down a basement membrane consisting of type IV collagen, laminin, and proteoglycans. Until a basement membrane develops fully with supporting cells, the newly formed blood vessels are permeable, as is seen in granulation tissue. (7) Pericytes and fibroblasts migrate to the sites of the capillary loops; these cells become associated with the "maturing" capillary. (8) Substantial remodeling, regression, and rearrangement of the newly formed capillaries occur, related to the continued presence or absence of angiogenic stimuli. The capillaries contract by means of fibroblasts and myofibroblasts.[135-143] Cells, matrix components, and soluble factors all help control this process.

Cells

A variety of normal tissues and tumors, both malignant and benign, are sources of angiogenic factors. The tissues containing these factors include lymphatic tissues,[136,144,145] thyroid tissue,[140] corpus luteum,[146] retina,[147] salivary glands,[148] and kidney.[140,141]

Cells and soluble factors generated by cells in inflammatory responses provide much of the stimulus for angiogenesis in normal and pathologic conditions. For example, lymphocytes and lymphokines induce an angiogenic response,[144,145] as do neutrophils[149,150] and macrophage-derived monokines.[151,152]

Macrophages, in particular, appear to be major contributors to angiogenesis signals. Angiogenesis is clearly related to inflammatory processes; however, there are conditions in which angiogenesis occurs without apparent inflammation. For example, Banda and associates[153] isolated an angiogenic factor from wound fluid that was nonmitogenic. Certain growth factors, such as transforming growth factor alpha (TGF-α) can promote angiogenesis without inflammation. It is known that serum can promote angiogenesis, and, in fact, platelets serve as an abundant source for a variety of growth factors. In addition to the role that inflammatory cells play in angiogenesis, there are data to support an autocrine mechanism of stimulating angiogenesis by endothelial cells. That is, during an angiogenic reaction, endothelial cells may be responsive to products made by the endothelial cells themselves.[154,155]

The mast cell's role in angiogenesis remains unknown, yet it may be among the most influential cells in angiogenesis. These cells have long been known to be associated with chronic inflammation and tumor growth. It has been suggested that neoplastic cells may themselves secrete angiogenesis factors.[156,157] Mast cells contain a variety of factors potentially important in angiogenesis, including heparin, histamine, and a variety of proteases.

What is clear from a number of studies is that the initial endothelial reaction in an angiogenic response is one of cellular migration and not necessarily replication. For example, Sholley and associates[158] showed that irradiated endothelial cells incapable of replication were able to show early responses to angiogenic stimulation. It also seems likely that endothelium derived from different sources might respond to different signals. For example, conditions that might promote neovascularization in bone might not work in the eye or tendon. Similarly, reagents evaluated and assayed in one particular model may not affect angiogenesis in related conditions. Therefore, it is highly desirable to develop in vitro and animal models to study endothelial cells and angiogenesis as they relate to important clinical conditions.

Neutral proteases derived from inflammatory cells—macrophages, neutrophils, and mast cells—may play a role in the degradation of basement membrane and the subsequent migration of endothelial or smooth muscle cells away from the site.[159] These neutral or other proteases may activate latent proenzymes, which then permit the degradation of collagens by specific type IV collagenases.[160-163]

Extracellular Matrix

Extracellular matrix macromolecules that may influence angiogenesis consist of (1) collagens, which are made up of genetically distinct types, i.e., type IV in basement membrane and types I and III in interstitial collagens; (2) noncollagenous adhesion molecules, including laminin, fibronectin, and others[164]; and (3) proteoglycans/glycosaminoglycans, such as heparan sulfate proteoglycan, hyaluronic acid, and chondroitin sulfate.

Collagens   Endothelial cells and smooth muscle cells are affected by the matrix or substratum to which they adhere.[146,165-169] Some of the earliest studies on the modulation of endothelial behavior by matrix components used two-dimensional matrices composed of collagen.[170-172] Perhaps more relevant are studies examining three-dimensional arrays of collagen or other molecules and the effects of this structure on cells.[167-169,173] Montesano and associates[167] observed in vitro the organization of endothelial cells into capillary tubes over time. In these experiments, collagen was coated onto a surface and endothelial cells were permitted to adhere; once a monolayer was formed, a second collagen layer was applied above the cells. The result was induced formation of branching and anastomosing capillary-like tubes from the preexisting monolayer of endothelial cells.

The studies of Schor and associates[143,168,169] showed that the extracellular matrix, principally collagen, dictated the endothelial cell's ability to respond to angiogenic stimuli. In these studies, angiogenic tissue extracts promoted endothelial cell proliferation on native type I collagen, but not on denatured collagen or uncoated tissue-culture dishes.[143,168,169,173]

The matrix on which cells lie also influences their differentiated state.[51,143] When endothelial cells were cultured on type IV collagen, cells organized into tube-like structures. However, on types I and III collagen, cells proliferated to form a confluent monolayer and only after prolonged culture were occasional tube-like capillary structures visualized.[51,166] Type IV collagen was capable of inducing an endothelial cell phenotype consistent with a more differentiated state whereby cells produced type IV collagen and laminin. This response is in contrast to the endothelial cell response when plated on types I and III collagen.[51,166] The ability of vascular endothelium to form capillary tube-like structures in vitro points to the key role that the endothelium plays in angiogenesis.

Noncollagenous Adhesion Molecules   With respect to affecting cell behavior in angiogenesis, fibronectin is the best characterized noncollagenous adhesion molecule. One of the early interactions involved in coagulation is the association of fibronectin with fibrin. This occurs by a direct binding and also by covalent binding via transglutaminase, coagulation factor XIII. This association appears to serve as a provisional matrix onto which epithelial cells migrate. Studies by Dvorak and associates[174] and Nicosia and associates[175] have implicated fibrin in tumor angiogenesis. McAuslan and associates[176] proposed that fibronectin serves as a signal for endothelial cell migration and subsequent capillary formation. The studies of Clark and associates[177] showed that blood vessel fibronectin increased in association with endothelial cell growth and capillary formation during wound healing. It is important to note that the earliest phase of an angiogenic response involves a nonmitogenic migration of capillary endothelial cells toward the angiogenic stimulus.[155,178,179] Also, Bowersox and Sorgente[180] showed that fibronectin could promote the chemotaxis of endothelial cells.

It seems probable that the stimulus for angiogenesis involving fibronectin does not necessarily involve a chemotactic response, but a haptotactic response. Haptotaxis is the stimulation of directed cell movement due to cellular adhesion gradients to substratum-attached components.[181] A substantial body of evidence has accrued on the role of fibronectin and laminin in haptotaxis of a variety of tumor cells and normal cells, such as neurons.[164,182–186] Most recently, it has been shown that there are specific domains of fibronectin that will promote the haptotactic migration of various cells.[185] Also, purified basement membrane components will promote the haptotactic migration of large-vessel endothelium (Thomas Herbst, personal communication).

As with other components of the extracellular matrix, fibronectin has the ability to bind to itself, to other matrix constituents, and to the surface of cells, via discrete domains on the molecule.[164,187–189] The primary structure of one adhesion sequence within fibronectin was deduced, by monoclonal antibody data and direct sequence analysis, to be a tetrapeptide sequence consisting of arginy-glycyl-aspartyl-serine (RGDS).[190] The RGDS peptide will directly promote the adhesion of certain cell types, and high levels of soluble RGDS will partially disrupt cell adhesion to intact fibronectin.[185,190,191] Cell adhesion to the RGDS sequence in fibronectin is thought to occur by the interaction of this sequence with a cell-surface glycoprotein complex termed integrin.[192,193]

Despite the importance of the RGDS/integrin complex in fibronectin-mediated cell adhesion, several lines of evidence point to the involvement of additional cellular receptors and different determinants in fibronectin. Many cell types form focal adhesions on intact fibronectin, which represent regions of close apposition between the plasma membrane and the substratum.[194,195] These sites also represent insertion points for actin-rich stress fibers and have been shown to contain several actin-associated cytoskeletal proteins. Focal adhesion sites also contain several classes of molecules implicated in cells adhesion, including integrin,[196] heparan sulfate and chondroitin sulfate proteoglycans,[194] and gangliosides.[187] The action of multiple cellular receptors has been implicated in adhesion plaque formation. Cells adherent to either RGDS-containing fragments or to heparin-binding, adhesive-promoting ligands, e.g., platelet factor IV or heparin-binding fragments of fibronectin, form only close contacts.[197,198] In contrast, cells adherent to both RGDS-containing fragments and heparin-binding ligands display fully developed focal adhesions.[198,199] Binding fragments of fibronectin inhibit focal adhesion formation, without drastically inhibiting the level of cell adhesion to intact fibronectin.[198] Collectively, these results argue for a role of heparin-binding domain(s) of fibronectin in promoting normal and malignant cell adhesion and in regulating phenotypic expression of cells. The nature of the cell adhesion and how the cellular cytoskeleton is triggered into action are keys in determining whether or not a cell will migrate.

Certain heparin-binding fragments of fibronectin promote the adhesion and spreading of metastatic melanoma cells by an RGDS-independent mechanism.[185] One fragment, which originates from the A-chain of the molecule,[188,189,200] also promotes the adhesion of neurons and the extension of neurites.[201] This fragment was shown to lack the RGDS sequence by direct sequencing, and was verified to contain part of the type IIIcs insert.[188,202] Monoclonal antibodies were generated against this fragment and to intact fibronectin. The results indicate that a cell-adhesion promoting activity of a 33-kD heparin-binding fragment of fibronectin can be distinguished from the RGDS sequence, and that this sequence and the RGDS sequence are used by cells to adhere to and spread on intact fibronectin.

In addition to fibronectin, proteins such as laminin, type IV collagen, and nidogen are also highly likely to modulate the behavior of endothelial and smooth muscle cells. Our understanding of the functional domains of laminin and type IV collagen is substantially less than that of fibronectin. In order to understand more fully the potential involvement of the proteins modulating angiogenesis in ligament, tendon, skeletal muscle, meniscus, and associated tissues additional work will be necessary.

Proteoglycans/Glycosaminoglycans   The role or potential role of proteoglycans or glycosaminoglycans in angiogenesis is somewhat of a paradox. Early studies isolated inhibitors of angiogenesis from cartilage and aorta.[203–206] Heparin derived from mast cell supernates could stimulate the migration of capillary endothelial cells.[207] Furthermore, it has been reported that heparin promotes angiogenesis in the chorioallantoic membrane, while protease inhibitors block angiogenesis.[208] The effect of heparin on angiogenesis is not related to its coagulation effect because non-anticoagulant heparin enhances angiogenesis. Perhaps the ability of heparin to promote angiogenesis stems from its ability to bind endothelial growth factors or angiogenic factors.[140,142,209-211] Heparin-affinity columns have recently been utilized by a number of investigators to purify a variety of growth factors. In contrast, Folkman and associates[139,212] and Fraser and Simpson[156] have observed that heparin and glucocorticoids inhibit angiogenesis. Other studies have shown that a synthetic heparin pentasaccharide inhibits angiogenesis.[212-214]

Another glycosaminoglycan, hyaluronate, has been studied less in relation to angiogenesis but may be of importance. Specifically, Feinberg and Beebe[215] examined the limb-bud development of chick embryos; in the peripheral mesoderm, a significant avascular zone was observed in areas that contained high levels of hyaluronic acid. Implants that were high in hyaluronic acid caused the formation of avascular zones, while similar implants containing other highly negatively charged molecules such as DNA had no such effect. Hyaluronic acid could serve either directly or indirectly as an inhibitor of angiogenesis in tumor growth, metastasis, excessive scarring, or chronic inflammation.

Soluble Factors

A variety of soluble polypeptides and other factors that promote angiogenesis have been described. Recently, significant strides have been made in defining, purifying, and sequencing certain of these components. For example, Thomas and associates[216] isolated a brain-derived acidic fibroblast growth factor (FGF) that was mitogenic for endothelial cells. Additionally, as with many of these angiogenic factors, it bound heparin, and, in the presence of heparin, promoted blood vessel growth in vivo, while heparin alone had no effect. Interestingly, the sequence of this acidic brain FGF was homologous with human interleukin-1 and the chondrosarcoma-derived endothelial cell growth factor described by Shing and associates.[211] Another polypeptide, hepatoma-derived growth factor (HDGF), was purified by Klagsbrun and associates[217] using heparin Sepharose chromatography and was found by the Western blotting technique to be structurally similar to basic FGF. Additionally, sequence analysis has found homologous peptides from tryptic fragments of HDGF and FGF.

Recently, Moscatelli and associates[218] isolated and purified a protein from human placenta that stimulated capillary endothelial cell migration, DNA synthesis, production of plasminogen activator, and activation of a procollagenase. Tumor promoters mimic this effect in endothelial cells.[219] Although this factor has a molecular weight comparable to other endothelial growth factors, the relationship of this protein to the brain FGF factor or HDGF is unknown.

Very recently, Schreiber and associates[220] showed that TGF-α and epidermal growth factor (EGF), which are structurally related proteins, promoted angiogenesis, and that the factors bound to endothelial cells and promoted their growth. TGF-α was more effective in promoting angiogenesis than EGF at equivalent concentrations. Schreiber and associates proposed that TGF-α, produced by tumor cells, contributed significantly, although not exclusively, to angiogenesis.

Of further interest is the observation that transforming growth factor beta (TGF-β) stimulated the production and accumulation of fibronectin and collagen in a variety of normal and tumor cells.[221] On the basis of the spectrum of cell responses to TGF-β, it is reasonable to believe that TGF-β stimulates fibronectin and collagen accumulation in endothelial cell cultures, although this was not specifically tested. Earlier studies by DiCorleto and Bowen-Pope[222] showed that cultured bovine aortic endothelial cells produced a polypeptide analogous to PDGF that could bind to PDGF receptors. In addition to the effects of factors on endothelial cells, there are significant data on the modulation of smooth-muscle cell activity by polypeptide growth factors. For example, not only does PDGF promote growth of smooth muscle cells, it also promotes their chemotactic migration.[223]

In addition to a host of soluble factors that may stimulate angiogenesis, the matrix or other components may modulate the cells' responsiveness to angiogenic signals. For example, specific matrix components may induce endothelial cells to regulate receptors up or down for a specific factor so that even though there may be high levels of a potent growth or angiogenic factor present, there may be a number of other conditions needed for a cell to respond.

Another area of consideration is the potential existence of inhibitors of angiogenesis. These might be useful to limit angiogenesis when it occurs in tumor growth or in neovascularization of the retina. However, normally occurring inhibitors of angiogenic stimuli might play a role in retarding a desired angiogenic response following muscloskeletal tissue injury. This inhibition could affect cartilage, meniscus, and tendon injuries. There is precedent for the existence of inhibitory factors in normal adult tissues in other systems. For example, in the central nervous system there is little to no neuron regeneration after injury. However, when peripheral nervous system implants are placed in the central nervous system, sprouting of central nervous system neurons occurs. It is not known at present whether the inhibitory substances were present in the microenvironment of the central nervous system or whether the peripheral nervous system transplants provided stimulating factors that counterbalanced a set of inhibitory factors.

We are on the threshold of exciting new ways to modify wound healing and angiogenesis. It is important to have appropriate in vitro and in vivo biologic models relevant to clinical problems. In the earliest stages of development, it may be sufficient to use microvascular endothelium to test the effects of factors that play a role in musculoskeletal angiogenesis and wound healing. The next step would then be to use in vivo animal models.

It seems highly probable that matrix proteins, growth factors, or other potentially oncogene-related gene products or receptors are important in wound healing. The challenge will be not only to discover these new factors but also to discern under what conditions these factors might work and their appropriate delivery systems.

Future Directions

Surgeons regularly repair vessel lacerations, replace injured segments of vessels, and cover large soft-tissue and bone defects by transferring tissue from one location to another and restoring its blood supply. However, advances in the following areas could significantly improve the treatment of injuries to the musculoskeletal soft tissues: (1) methods of limiting vessel and tissue damage following injury; (2) techniques of aligning and repairing vessels; (3) methods of determining the extent of vessel and tissue injury; (4) stimulation of collateral circulation; and (5) stimulation and control of angiogenesis

These advances will depend on gaining a clear understanding of the biology of peripheral vessels, including the origins and responses of the cells and the composition of the matrices they produce. This effort must include continued study of the cells, cell-cell interaction, and cell-matrix interactions and their modulation by physical and biochemical means. The in vitro study of isolated cells, including application of genetic techniques, can be expected to yield important insights.

Few mechanically controlled studies of the mechanism of peripheral vessel damage have been reported and such information, especially in avulsion injuries, is needed. Also needed is a rapid, clinically applicable method of evaluating areas of vessel and tissue damage following injury. Improvements in the techniques of repair may include tissue adhesives and alignment and fixation mechanisms that will make vessel repair more rapid and secure. The mechanisms that develop collateral circulation in musculoskeletal soft tissues should be explored since very little is known in this area.

One of the most direct ways to improve the results of vessel repair would be to minimize damage. There is a body of knowledge developing on the role of leukocyte-mediated endothelial damage. Current studies have suggested that oxygen radicals, peroxides, or other related intermediates are potent inflammatory mediators that damage vascular endothelium. A variety of enzymes are being used in this regard, including superoxide dismutase, catylase, peroxidase, and glutathione peroxidase. Some of these appear to offer theoretical advantages in minimizing oxygen radical effects, although most of the work performed in this field to date has dealt with the potential effects of superoxide dismutase.

Aside from advancing techniques of vessel repair, improving methods of assessing tissue and vessel damage, discovering new ways of limiting injury, and stimulating development of collateral circulation, there appear to be a number of exciting areas for investigation of angiogenesis. Rapid, directed growth of new vessels has the potential of improving repair of almost all injuries to the musculoskeletal soft tissues. A useful strategy might be to get more of the right materials to the right place at the right time to create the desired effect. Since endothelial migration and growth is essential for effective angiogenesis, one approach would be to use intact matrix molecules or basement membrane molecules that promote the adhesion and migration of endothelial cells. Suppression of smooth-muscle cell growth is also a key in maintaining patency of vessels. It is probable that peptide domains of matrix molecules that promote early aspects of an angiogenic response will be discovered. Future work could involve attempts to guide endothelial cells and potentially, therefore, to direct rapid vessel regrowth to specific sites using "molecular guidewires" formed from specific matrix molecules. Another approach might be to "seed" endothelium onto synthetic small-diameter vessels coated with matrix constituents or related chemically synthesized peptides that promote the continued attachment of endothelial cells. This effect would be highly desirable and might provide a small-diameter graft with a non-thrombogenic surface of either natural or synthetic origin that would remain patent.

An emerging area in the study of angiogenesis and soft-tissue repair, in general, is the role of specific growth factors. There are already a variety of growth factors known to promote angiogenesis. Current problems include how and when to deliver these molecules. Polymers of various materials could possibly be made which would contain growth factors that would be time released into a specific area. New factors may be discovered that selectively recruit specific cell types. It may ultimately be necessary to provide specific factors in a timed sequence to produce the desired effect.

References

1. Fawcett DW: Blood and lymph vascular systems, in *A Textbook of Histology*. Philadelphia, WB Saunders, 1986, pp 367–405.
2. Klitzman B, Johnson PC: Capillary network geometry and red cell distribution in hamster cremaster muscle. *Am J Physiol* 1982;242:H211-H219.
3. Granger JH, Meininger GA, Borders JL, et al: Microcirculation of skeletal muscle, in Mortillan NA (ed): *The Physiology and Pharmacology of the Micro Circulation*. New York, Academic Press, 1984, vol 2, pp 181–265.
4. Fagrell B: Microcirculation of the skin, in Mortillaro NA (ed): *The Physiology and Pharmacology of the Micro Circulation*. New York, Academic Press, 1984, vol 2, pp 133–180.
5. Mayne R: Normal biology and derangement in human diseases, in Vitto J, Perejda AJ (eds): *Connective Tissue Disease: Molecular Pathology of the Extracellular Matrix*. New York, Marcel Dekker, 1986, pp 163–183.
6. Davidson JM: Elastin: Structure and biology, in Vitto J, Perejda AJ (eds): *Connective Tissue Disease: Molecular Pathology of the Extracellular Matrix*. New York, Marcel Dekker, 1986, pp 29–54.
7. Mayne R, Zettergren JG, Mayne PM, et al: Isolation and partial characterization of basement membrane-like collagens from bovine thoracic aorta. *Artery* 1980;7:262–280.

8. Morton LF, Barnes MJ: Collagen polymorphism in the normal and diseased blood vessel wall: Investigation of collagens types I, III, and IV. *Atherosclerosis* 1982;42:41–54.

9. Chung E, Rhodes RK, Miller EJ: Isolation of three collagenous components of probable basement membrane origin from several tissues. *Biochem Biophys Res Commun* 1976;71:1167–1174.

10. Bartholomew JS, Anderson JC: Investigation of relationships between collagens, elastin and proteoglycans in bovine thoracic aorta by immunofluorescence techniques. *Histochem J* 1983;15:1177–1190.

11. Palotie A, Tryggvason K, Peltonen L, et al: Components of subendothelial aorta basement membrane: Immunohistochemical localization and role in cell attachment. *Lab Invest* 1983;49:362–370.

12. Madri JA, Dreyer B, Pitlick FA, et al: The collagenous components of the subendothelium: Correlation of structure and function. *Lab Invest* 1980; 43:303–315.

13. Oegema TR Jr, Hascall VC, Eisenstein R: Characterization of bovine aorta proteoglycan extracted with guanidine hydrochloride in the presence of protease inhibitors. *J Biol Chem* 1979;254:1312–1318.

14. Gardell S, Baker J, Caterson B, et al: Link protein and a hyaluronic acid-binding region as components of aorta proteoglycan. *Biochem Biophys Res Commun* 1980;95:1823–1831.

15. Mangkornkanok-Mark M, Eisenstein R, Bahu RM: Immunologic studies of bovine aortic and cartilage proteoglycans. *J Histochem Cytochem* 1981;29:547–552.

16. Kapoor R, Phelps CF, Wight TN: Physical properties of chondroitin sulphate/dermatan sulphate proteoglycans from bovine aorta. *Biochem J* 1986;240:575–583.

17. Aikawa J, Isemura M, Munakata H, et al: Isolation and characterization of chondroitin sulfate proteoglycans from porcine thoracic aorta. *Biochim Biophys Acta* 1986;883:83–90.

18. Wight TN, Ross R: Proteoglycans in primate arteries: I. Ultrastructural localization and distribution in the intima. *J Cell Biol* 1975;67:660–674.

19. Eisenstein R, Kuettner K: The ground substance of the arterial wall: Part II. Electron-microscopic studies. *Atherosclerosis* 1976;24:37–46.

20. Klein DJ, Oegema TR, Eisenstein R, et al: Renal localization of heparan sulfate proteoglycan by immunohistochemistry. *Am J Pathol* 1983;111:323–330.

21. Rohde H, Wick G, Timpl R: Immunochemical characterization of the basement membrane glycoprotein laminin. *Eur J Biochem* 1979;102:195–201.

22. Foidart JM, Bere EW Jr, Yaar M, et al: Distribution and immunoelectron microscopic localization of laminin, a noncollagenous basement membrane glycoprotein. *Lab Invest* 1980;42:336–342.

23. Woodley DT: The molecular organization of basement membrane, in Vitto J, Perejda AJ (eds): *Connective Tissue Disease: Molecular Pathology of the Extracellular Matrix.* New York, Marcel Dekker, 1986, pp 141–162.

24. Ruoslahti, Engvall E, Jalanko H, et al: Antigenic differences in nuclear proteins of normal liver and hepatoma: Identification of a nuclear protein present in hepatocytes but absent in hepatoma cells. *J Exp Med* 1977;146:1054–1067.

25. Schwartz SM: Dynamic maintenance of the endothelium, in Thilo-Körner DGS, Freshney RI (eds): *The Endothelial Cell: A Pluripotent Control Cell of the Vessel Wall.* Basel, S Karger, 1983, pp 113–125.

26. Chamley-Campbell J, Campbell GR, Ross R: The smooth muscle cell in culture. *Physiol Rev* 1979;59:1–61.

27. Thilo-Körner DGS, Heinrich D, Temme H: Endothelial cells in culture: A literature survey on isolation, harvesting, cultivation, medium and serum composition, cell counting, gas atmosphere, and confluency, in Thilo-Körner DGS, Freshney RI (eds): *The Endothelial Cell: A Pluripotent Control Cell of the Vessel Wall.* Basel, S Karger, 1983, pp 158–202.

28. Paaske WP, Sejrsen P: Transcapillary exchange of ^{14}C-inulin by free diffusion in channels of fused vesicles. *Acta Physiol Scand* 1977;100:437–445.

29. Simionescu N, Simionescu M, Palade GE: Structural basis of permeability in sequential segments of the microvasculature of the diaphragm. *Microvasc Res* 1978;15:17–36.

30. Simionescu M, Simionescu N: Functions of the endothelial cell surface. *Annu Rev Physiol* 1986;48:279–293.

31. Cryer A: Lipoprotein lipase: Molecular interactions of the enzyme. *Biochem Soc Trans* 1985;13:27–28.

32. Shimada K, Gill PJ, Silbert JE, et al: Involvement of cell surface heparin sulfate in the binding of lipoprotein lipase to cultured bovine endothelial cells. *J Clin Invest* 1981;68:995–1002.

33. Williams MP, Streeter HB, Wuseman FS, et al: Heparan sulphate and the binding of lipoprotein lipase to porcine thoracic aorta endothelium. *Biochim Biophys Acta* 1983;756:83–91.

34. Bengtsson G, Olivecrona T, Höök M, et al: Interaction of lipoprotein lipase with native and modified heparin-like polysaccharides. *Biochem J* 1980;189:625–633.

35. Marchum JA, Atha DH, Fritze LMS, et al: Cloned bovine aortic endothelial cells synthesize anticoagulantly active heparan sulfate proteoglycan. *J Biol Chem* 1986;261:7507–7517.

36. Björk I, Lindahl U: Mechanism of the anticoagulant action of heparin. *Mol Cell Biochem* 1982;48:161–182.

37. Lindahl U, Feingold DS, Rodén L: Biosynthesis of heparin. *Trends Biochem Sci* 1986;11:221–225.

38. Tollefsen DM, Peacock ME, Monafo WJ: Molecular size of dermatan sulfate oligosaccharides required to bind and activate heparin cofactor II. *J Biol Chem* 1986;261:8854–8858.

39. Guyton AC: *Textbook of Medical Physiology*, ed 7. Philadelphia, WB Saunders, 1986, pp 76–86.

40. Needleman P, Turk J, Jakschik BA, et al: Arachidonic acid metabolism *Annu Rev Biochem* 1986;55:69–102.

41. Castellot JJ Jr, Rosenberg RD, Karnovsky MJ: Endothelium, heparin, and the regulation of vascular smooth muscle cell growth, in Jaffe EA (ed): *Biology of Endothelial Cells.* Boston, Martinus Nijhoff, 1984, pp 118–128.

42. Yurchenco PD, Tsilibary EC, Charonis AS, et al: Models for the self-assembly of basement membrane. *J Histochem Cytochem* 1986;34:93–102.

43. Winterbourne DJ, Schor AM, Gallagher JT: Synthesis of glycosaminoglycans by cloned bovine endothelial cells cultured on collagen gels. *Eur J Biochem* 1983;135:271–277.

44. Kramer RH, Fuh G-M, Karasek MA: Type IV collagen synthesis by cultured human microvascular endothelial cells and its deposition into the subendothelial basement membrane. *Biochemistry* 1985;24:7423–7430.

45. Kinsella MG, Wight TN: Modulation of sulfated proteoglycan synthesis by bovine aortic endothelial cells during migration. *J Cell Biol* 1986;102:679–687.

46. Hiss D, Scott-Burden T, Gevers W: Disulfide-bonded heparan sulfate proteoglycans associated with the surfaces of cultured bovine vascular endothelial cells. *Eur J Biochem* 1987;162:89–94.

391

47. Wight TN, Hascall VC: Proteoglycans in primate arteries: III. Characterization of the proteoglycans synthesized by arterial smooth muscle cells in culture. *J Cell Biol* 1983;96:167–176.

48. Cheng C-F, Oosta GM, Bensadoun A, et al: Binding of lipoprotein lipase to endothelial cells in culture. *J Biol Chem* 1981;256:12893–12898.

49. Rauch U, Glössl J, Kresse H: Comparison of small proteoglycans from skin fibroblasts and vascular smooth-muscle cells. *Biochem J* 1986;238:465–474.

50. Leushner JRA, Haust MD: The effect of ascorbate on the synthesis of minor (non-interstitial) collagens by cultured bovine aortic smooth muscle cells. *Biochim Biophys Acta* 1986;883:284–292.

51. Madri JA, Pratt BM: Endothelial cell-matrix interactions: In vitro models of angiogenesis. *J Histochem Cytochem* 1986;34:85–91.

52. Gotlieb AI, Spector W, Wong MK, et al: In vitro reendothelialization: Microfilament bundle reorganization in migrating porcine endothelial cells. *Arteriosclerosis* 1984;4:91–96.

53. Merrilees MJ, Scott LJ: Effects of endothelial removal and regeneration on smooth muscle glycosaminoglycan synthesis and growth in rat carotid artery in organ culture. *Lab Invest* 1985;52:409–419.

54. Majack RA, Bornstein P: Heparin regulates the collagen phenotype of vascular smooth muscle cells: Induced synthesis of an M_r 60,000 collagen. *J Cell Biol* 1985;100:613–619.

55. Fung YC: *Biomechanics: Mechanical Properties of Living Tissue.* New York, Springer-Verlag, 1981.

56. Robinson DW, MacLeod A: Microvascular free jejunum transfer. *Br J Plast Surg* 1982;35:258–267.

57. Acland RD, Trachtenberg L: The histopathology of small arteries following experimental microvascular anastomosis. *Plast Reconstr Surg* 1977;60:868–875.

58. O'Brien BM: *Microvascular Reconstructive Surgery.* Edinburgh, Churchill Livingstone, 1977.

59. Tan E, O'Brien BM, Brennen M: Free flap transfer in rabbits using irradiated recipient vessels. *Br J Plast Surg* 1978;31:121–123.

60. Krag C, Holck S, DeRose G, et al: Healing of microvascular anastomoses: A comparative study using normal and irradiated recipient vessels for experimental free flaps in rabbits. *Scand J Plast Reconstr Surg* 1982;16:267–274.

61. Bush HL Jr, Jakubowski JA, Sentissi JM, et al: Neointimal hyperplasia occurring after carotid endarterectomy in a canine model: Effect of endothelial cell seeding vs perioperative aspirin. *J Vasc Surg* 1987;5:118–125.

62. Schwarcz TH, Yates GN, Ghobrial M, et al: Pathologic characteristics of recurrent carotid artery stenosis. *J Vasc Surg* 1987;5:280–288.

63. Clowes AW, Reidy MA, Clowes MM: Kinetics of cellular proliferation after arterial injury: I. Smooth muscle growth in the absence of endothelium. *Lab Invest* 1983;49:327–333.

64. Fuchs JCA, Mitchener JS III, Hagen P-O: Postoperative changes in autologous vein grafts. *Ann Surg* 1978;188:1–15.

65. Clowes AW, Kirkman TR, Clowes MM: Mechanisms of arterial graft failure: II. Chronic endothelial and smooth muscle cell proliferation in healing polytetrafluoroethylene prostheses. *J Vasc Surg* 1986;3:877–884.

66. Vlodaver Z, Edwards JE: Pathologic changes in aortic-coronary arterial saphenous vein grafts. *Circulation* 1971;44:719–728.

67. Schwartz SM, Ross R: Cellular proliferation in atherosclerosis and hypertension. *Prog Cardiovasc Dis* 1984;26:355–372.

68. Antoniades HN, Williams LT: Human platelet-derived growth factor: Structure and function. *Fed Proc* 1983;42:2630–2634.

69. Clowes AW: Arterial endothelial smooth muscle cell interactions, in Bergen JJ, Yao JST (eds): *Evaluation and Treatment of Upper and Lower Extremity Circulatory Disorders.* Orlando, Grune & Stratton, 1984, pp 25–38.

70. Lytle BW, Loop FD, Cosgrove DM, et al: Long-term (5 to 12 years) serial studies of internal mammary artery and saphenous vein coronary bypass grafts. *J Thorac Cardiovasc Surg* 1985;89:248–258.

71. Leather RP, Shah DM, Karmody AM: Infrapopliteal arterial bypass for limb salvage: Increased patency and utilization of the saphenous vein used "in situ." *Surgery* 1981;90:1000–1008.

72. Kinlough-Rathbone RL, Packham MA, Mustard JF: Vessel injury, platelet adherence, and platelet survival. *Arteriosclerosis* 1983;3:529–546.

73. Chesebro JH, Fuster V, Elveback LR, et al: Effect of dipyridamole and aspirin on late vein-graft patency after coronary bypass operations. *N Engl J Med* 1984;310:209–214.

74. McCann RL, Hagen P-O, Fuchs JCA: Aspirin and dipyridamole decrease intimal hyperplasia in experimental vein grafts. *Ann Surg* 1980;191:238–243.

75. Swedenborg J: Blood and surface factors of importance for vascular surgery. *Acta Chir Scand,* 1985, suppl 529, pp 3–6.

76. Hamilton RB, O'Brien BM, Morrison A, et al: Survival factors in replantation and revascularization of the amputated thumb—10 years experience. *Scand J Plast Reconstr Surg* 1984;18:163–173.

77. Mitchell GM, Morrison WA, Papadopoulos A, et al: A study of the extent and pathology of experimental avulsion injury in rabbit arteries and veins. *Br J Plast Surg* 1985;38:278–287.

78. Craig EV: A new high-pressure injection injury of the hand. *J Hand Surg* 1984;9A:240–242.

79. Ramos H, Posch JL, Lie KK: High-pressure injection injuries of the hand. *Plast Reconstr Surg* 1970;45:221–226.

80. Kon M, Sagi A: High-pressure water jet injury of the hand. *J Hand Surg* 1985;10A:412–414.

81. Dujovny M, Osgood CP, Barrionuevo PJ, et al: SEM evaluation of endothelial damage following temporary middle cerebral artery occlusion in dogs. *J Neurosurg* 1978;48:42–48.

82. Dujovny M, Wackenhut N, Kossovsky N, et al: Biomechanics of vascular occlusion in neurosurgery. *Acta Neurol Lat Am* 1980;26:123–127.

83. Dodson RF, Tagashira Y, Chu LW-F: Acute ultrastructural changes in the middle cerebral artery due to the injury and ischemia of surgical clamping. *Can J Neurol Sci* 1976;3:23–27.

84. Gregorius FK, Rand RW: Scanning electron microscopic observations of common carotid artery endothelium in the rat: II. Sutured arteries. *Surg Neurol* 1975;4:258–264.

85. Gertz SD, Rennels ML, Forbes MS, et al: Endothelial cell damage by temporary arterial occlusion with surgical clips: Study of the clip site by scanning and transmission electron microscopy. *J Neurosurg* 1976;45:514–519.

86. Gerstenkkorn GF, Kobayaski AS, Wiederhielm CA, et al: Structural analysis of an arteriole by direct stiffness method. *J Engr Ind* 1966;88:363–368.

87. Slayback JB, Bowen WW, Hinshaw DB: Intimal injury from arterial clamps. *Am J Surg* 1976;132:183–188.

88. Szilagyi DE, Whitcomb JG, Schenker W, et al: The laws of fluid flow and arterial grafting. *Surgery* 1960;47:55–73.

89. Rich NM, Spencer FC: *Vascular Trauma.* Philadelphia, WB Saunders, 1978.

90. Cook AF, Grossman JAI, Herpy ES, et al: The correlation of histology to needle type in microvascular anastomoses: A quantitative analysis. *Br J Plast Surg* 1982;35:156–157.

91. Pitt TTE, Humphries NM: Microarterial anastomoses in the rat: The influence of different suture materials on the patency, strength and the electron microscopic appearance of the vessels. *Br J Plast Surg* 1982;35:150–155.

92. Lauritzen C, Bagge U: A technical and biomechanical comparison between two types of microvascular anastomoses: An experimental study in rats. *Scand J Plast Reconstr Surg* 1979;13:417–421.

93. von Smitten K, Jiborn H, Ahonen J: Bursting strength of syngeneic aortic vein grafts in the rat. *Acta Chir Scand* 1981;147:115–119.

94. Wieslander JB, Åberg M: Blood flow in small arteries after end-to-end and end-in-end anastomoses: An experimental quantitative comparison. *J Microsurg* 1980;2:121–125.

95. Wieslander JB, Åberg M, Dougan P: Accumulation of isotope labelled platelets in small arteries after end-to-end and end-in-end anastomoses in the rabbit. *Br J Plast Surg* 1982;35:158–162.

96. Davis CB II, Smith BM, Hagen P-O, et al: The strength of microvascular anastomoses—An experimental evaluation in rats. *J Microsurg* 1982;3:156–161.

97. Sanders R, Green CJ, Tan WTL: The "wrap-around" end-to-side anastomosis for micro-vessels. *Br J Plast Surg* 1981;34:178–180.

98. Wieslander JB, Åberg M: Stenosis following end-in-end microarterial anastomosis: An angiographic comparison with the end-to-end technique. *J Microsurg* 1982;3:151–155.

99. Caro CG, Pedley TJ, et al: Solid mechanics and the properties of blood vessel walls, in *The Mechanics of Circulation*. Oxford, Oxford University Press, 1978, pp 85–105.

100. Paasche PE, Kinley CE, Dolan FG, et al: Consideration of suture line stresses in the selection of synthetic grafts for implantation. *J Biomech* 1973;6:253–259.

101. Marble AE, Sarwal SN, Watts KC, et al: A mathematical assessment of suture line stress in the end-to-side anastomosis: I. Steady flow. *J Biomech* 1979;12:941–944.

102. Sarwal SN, Marble AE, Kinley CE: A mathematical assessment of suture line stress in the end-to-side anastomosis: II. Pulsatile flow. *J Biomech* 1980;13:449–454.

103. Zoubos AB, Seaber AV, Urbaniak JR: The hemodynamic and histological differences in end-to-side anastomosis. *Trans Orthop Res Soc* 1985;10:314.

104. Östrup LT: Anastomosis of small veins with suture or Nakayama's apparatus: A comparative study. *Scand J Plast Reconstr Surg* 1976;10:9–17.

105. Thomsen MB, Östrup LT, Alm A: Arteriovenous fistulas constructed with Nakayama's ring-pin stapler: A comparison between stapler and suture anastomoses. *Acta Chir Scand* 1982;148:585–589.

106. Daniel RK, Olding M: An absorbable anastomotic device for microvascular surgery: Experimental studies. *Plast Reconstr Surg* 1984;74:329–336.

107. Greisler HP: Arterial regeneration over absorbable prostheses. *Arch Surg* 1982;117:1425–1431.

108. Borst HG, Haverich A, Walterbusch G, et al: Fibrin adhesive: An important hemostatic adjunct in cardiovascular operations. *J Thorac Cardiovasc Surg* 1982;84:548–553.

109. Gestring GF, Lerner R: Autologous fibrinogen for tissue-adhesion, hemostasis and embolization. *Vasc Surg* 1983;17:294–304.

110. Lupinetti FM, Stoney WS, Alford WC Jr, et al: Cryoprecipitate—topical thrombin glue: Initial experience in patients undergoing cardiac operation. *J Thorac Cardiovasc Surg* 1985;90:502–505.

111. Weber SC, Chapman MW: Adhesives in orthopaedic surgery: A review of the literature and *in vitro* bonding strengths of bone-bonding agents. *Clin Orthop* 194;191:249–261.

112. Dinges HP, Matras H, Kleitter G, et al: Histopathologische Untersuchungen zum Heilungsverlauf von Mikrogefässanastomosen bei Anwendung der kombinierten Naht- und Klebetechnik. *Vasa* 1978;7:161–166.

113. Karl P, Tilgner A, Heiner H: A new adhesive technique for microvascular anastomoses: A preliminary report. *Br J Plast Surg* 1981;34:61–63.

114. Davis RW, Noble PC, Weilbacher DG: CO_2 laser-assisted microvascular anastomosis: Histological and biomechanical features. *Trans Orthop Res Soc* 1987;12:334.

115. Lec S, Fung YC, Matsuda M, et al: The development of mechanical strength of surgically anastomosed arteries sutured with Dexon. *J Biomech* 1985;18:81–89.

116. Smitten K: Breaking strength and suture holding capacity of syngeneic aortic vein grafts in the rat. *Acta Chir Scand* 1981;147:545–549.

117. Smith M, Enquist IF: A quantitative study of impaired healing resulting from infection. *Surg Gynecol Obstet* 1967;125:965–973.

118. Forrester JG, Zederfeldt BH, Hayes TL, et al: Wolff's law in relation to the healing skin wound. *J Trauma* 1970;10:770–779.

119. Lidman D, Daniel RK: The normal healing process of microvascular anastomoses. *Scand J Plast Reconstr Surg* 1981;15:103–110.

120. Gottrup F: Healing of incisional wounds in stomach and duodenum: Collagen distribution and relation to mechanical strength. *Am J Surg* 1981;141:222–227.

121. Peacock EE Jr, Van Winkle W Jr: *Wound Repair*, ed 2. Philadelphia, WB Saunders, 1976, pp 145–203.

122. Dunphy JE, Udupa KN: Chemical and histochemical sequences in the normal healing of wounds. *N Engl J Med* 1955;253:847–851.

123. Levenson SM, Geever EF, Crowley LV, et al: The healing of rat skin wounds. *Ann Surg* 1965;161:293–308.

124. Madden JW, Peacock EE Jr: Studies on the biology of collagen during wound healing: Rate of collagen synthesis and deposition in cutaneous wounds of the rat. *Surgery* 1968;64:288–294.

125. Peacock EE Jr: Dynamic aspects of collagen biology: Part 1. Synthesis and assembly. *J Surg Res* 1967;7:433–445.

126. Lowenberg RI, Shumacker HB Jr: Experimental studies in vascular repair: II. Strength of arteries repaired by end to end suture, with some notes on growth of anastomosis in young animals. *Arch Surg* 1949;59:74–83.

127. Madden JW: Wound healing: Biologic and clinical features, in Sabiston DC Jr (ed): *Davis-Christopher Textbook on Surgery*. Philadelphia, WB Saunders, 1977, pp 271–294.

128. Thiede A, Lütjohann K, Beck C, et al: Absorbable and nonabsorbable sutures in microsurgery: Standardized comparable studies in rats. *J Microsurg* 1979;1:216–222.

129. Chow SP, Huang CD, Chan CW: Microvascular anastomosis of arteries under tension. *Br J Plast Surg* 1982;35:82–87.

130. Hayhurst JW, O'Brien BM: An experimental study of microvascular technique, patency rates and related factors. *Br J Plast Surg* 1975;28:128–132.

131. Hyland WT, Botens SR, Minasi JS: A re-appraisal and modification of the Lauritzen technique of microvascular anastomoses. *Br J Plast Surg* 1981;34:451–453.

132. Wei FC, Mancer K, Zuker RM: The temporary stent technique: An easier method of micro-venous anastomosis. *Br J Plast Surg* 1982;35:92–95.

133. Arfors K-E, Hint HC, Dhall DP, et al: Counteraction of platelet activity at sites of laser-induced endothelial trauma. *Br Med J* 1968;4:430–431.

134. Schenck RT, Weinrib HP, Labanauskas IG: The external ring technique for microvascular anastomosis. *J Hand Surg* 1983;8:105–109.

135. Cliff WJ: Observations on healing tissue: A combined light and electron microscopic investigation. *Philos Trans R Soc London [Biol Sci]* 1963;246:305–325.

136. Auerbach R: A review, in Pick E (ed): *Lymphokines*. New York, Academic Press, 1981, vol 4, p 69.

137. Folkman J: Tumor angiogenesis. *Adv Cancer Res* 1974;19:331–358.

138. Folkman J: What is the role of endothelial cells in angiogenesis?, editorial. *Lab Invest* 1984;51:601.

139. Folkman J: Angiogenesis: Initiation and modulation, in Nicolson GL, Milas L (eds): *Cancer Invasion and Metastasis: Biologic and Therapeutic Aspects*. New York, Raven Press, 1984, pp 201–207.

140. Folkman J: How is blood vessel growth regulated in normal and neoplastic tissue?—G.H.A. Clowes Memorial Award Lecture. *Cancer Res* 1986;46:467–473.

141. Folkman J, Cotran R: Relation of vascular proliferation to tumor growth. *Int Rev Exp Pathol* 1976;16:207–248.

142. Maciag T: Angiogenesis. *Prog Hemostasis Thromb* 1984;7:167–182.

143. Schor AM, Schor SL: Tumor angiogenesis. *J Pathol* 1983;141:385–413.

144. Auerbach R, Sidky YA: Nature of the stimulus leading to lymphocyte-induced angiogenesis. *J Immunol* 1979;123:751–754.

145. Auerbach R, Kubai L, Sidky Y: Angiogenesis induction by tumors, embryonic tissues, and lymphocytes. *Cancer Res* 1976;36:3435–3440.

146. Gospodarowicz D, Thakral KK: Production of a corpus luteum angiogenic factor responsible for proliferation of capillaries and neovascularization of the corpus luteum. *Proc Natl Acad Sci USA* 1978;75;847–851.

147. Federman JL, Brown GC, Felberg NT, et al: Experimental ocular angiogenesis. *Am J Ophthalmol* 1980;89:231–237.

148. Hoffman H, McAuslan B, Robertson D, et al: An endothelial growth-stimulating factor from salivary glands. *Exp Cell Res* 1976;102:269–275.

149. Fromer CH, Klintworth GK: An evaluation of the role of leukocytes in the pathogenesis of experimentally induced corneal vascularization: Comparison of experimental models of corneal vascularization. *Am J Pathol* 1975;79:537–554.

150. Fromer CH, Klintworth GK: An evaluation of the role of leukocytes in the pathogenesis of experimentally induced corneal vascularization: III. Studies related to the vasoproliferative capability of polymorphonuclear leukocytes and lymphocytes. *Am J Pathol* 1976;82:157–170.

151. Polverini PJ, Cotram RS, Gimbrone MA, et al: Activated macrophages induce vascular proliferation. *Nature* 1977;269:804–806.

152. Polverini PJ, Leibovich SJ: Induction of neovascularization *in vivo* and endothelial proliferation *in vitro* by tumor-associated macrophages. *Lab Invest* 1984;51:635–642.

153. Banda MJ, Knighton DR, Hunt TK, et al: Isolation of a nonmitogenic angiogenesis factor from wound fluid. *Proc Natl Acad Sci USA* 1982;79:7773–7777.

154. O'Donoghue MN, Zarem HA: Stimulation of neovascularization—comparative efficacy of fresh and preserved skin grafts. *Plast Reconstr Surg* 1971;48:474–477.

155. Sholley MM, Ferguson GP, Seibel HR, et al: Mechanisms of neovascularization: Vascular sprouting can occur without proliferation of endothelial cells. *Lab Invest* 1984;51:624–634.

156. Fraser RA, Simpson JG: Role of mast cells in experimental tumor angiogenesis, in *Development of the Vascular System*, Ciba Foundation Symposium 100. London, Pitman Books, 1983, pp 120–131.

157. Kessler DA, Langer RS, Pless NA, et al: Mast cells and tumor angiogenesis. *Int J Cancer* 1976;18:703–709.

158. Sholley MM, Gimbrone MA Jr, Cotran RS: Cellular migration and replication in endothelial regeneration: A study using irradiated endothelial cultures. *Lab Invest* 1977;36:18–25.

159. Schwartz LB, Austen KF: Acid hydrolases and other enzymes of rat and human mast cell secretory granules, in Becker EL, Simon AS, Austen KF (eds): *Biochemistry of the Acute Allergic Reactions*. New York, Alan R Liss, 1981, pp 103–121.

160. Glaser BM, Kalebic T, Garbisa S, et al: Degradation of basement membrane components by vascular endothelial cells: Role in neovascularization, in *Development of the Vascular System*, Ciba Foundation Symposium 100. London, Pitman Books, 1983, pp 150–162.

161. Gross JL, Moscatelli D, Rifkin DB: Increased capillary endothelial cell protease activity in response to angiogenic stimuli *in vitro*. *Proc Natl Acad Sci USA* 1983;80:2623–2627.

162. Loskutoff DJ, van Mourik JA, Erickson LA, et al: Detection of an unusually stable fibrinolytic inhibitor produced by bovine endothelial cells. *Proc Natl Acad Sci USA* 1983;80:2956–2960.

163. Rifkin DB, Moscatelli D, Gross J, et al: Proteases, angiogenesis, and invasion, in Nicolson GL, Milas L (eds): *Cancer Invasion and Metastasis: Biologic and Therapeutic Aspects*. New York, Raven Press, 1984, pp 187–200.

164. Furcht LT: Structure and function of the adhesive glycoprotein fibronectin. *Mod Cell Biol* 1983;1:53–117.

165. Kramer RH, Bensch KG, Davison PM, et al: Basal lamina formation by cultured microvascular endothelial cells. *J Cell Biol* 1984;99:692–698.

166. Madri JA, Williams SK: Capillary endothelial cell cultures: Phenotypic modulation by matrix components. *J Cell Biol* 1983;97:153–165.

167. Montesano R, Orci L, Vassalli P: In vitro rapid organization of endothelial cells into capillary-like networks is promoted by collagen matrices. *J Cell Biol* 1983;97:1648–1657.

168. Schor AM, Schor SL, Kumar S: Importance of a collagen substratum for stimulation of capillary endothelial cell proliferation by tumour angiogenesis factor. *Int J Cancer* 1979;24:225–234.

169. Schor SL, Schor AM, Bazill GW: The effects of fibronectin on the migration of human foreskin fibroblasts and Syrian hamster melanoma cells into three-dimensionals gels of native collagen fibres. *J Cell Sci* 1981;48:301–314.

170. Delvos U, Gajdusek C, Sage H, et al: Interactions of vascular wall cells with collagen gels. *Lab Invest* 1982;46:61–72.

171. Elsdale T, Bard J: Collagen substrata for studies on cell behavior. *J Cell Biol* 1972;54:626–637.

172. Folkman J, Haudenschild C: Angiogenesis *in vitro*. *Nature* 1980;288:551–556.

173. Schor AM, Schor SL, Weiss JB, et al: Stimulation by a low-molecular-weight angiogenic factor of capillary endothelial cells in culture. *Br J Cancer* 1980;41:790–799.

174. Dvorak HF, Dickersin GR, Dvorak AM, et al: Human breast carcinoma: Fibrin deposits and desmoplasia; Inflammatory cell type and distribution; Microvasculature and infarction. *J Natl Cancer Inst* 1981;67:335–345.

175. Nicosia RF, Tchao R, Leighton J: Angiogenesis-dependent tumor spread in reinforced fibrin clot culture. *Cancer Res* 1983;43:2159–2166.

176. McAuslan BR, Hannan GN, Reilly W, et al: Variant endothelial cells: Fibronectin as a transducer of signals for migration and neovascularisation. *J Cell Physiol* 1980;104:177–186.

177. Clark RAF, DellaPelle P, Manseau E, et al: Blood vessel fibronectin increases in conjunction with endothelial cell proliferation and capillary ingrowth during wound healing. *J Invest Dermatol* 1982;79:269–276.

178. Thorgeirsson G, Robertson AL Jr, Cowan DH: Migration of human vascular endothelial and smooth muscle cells. *Lab Invest* 1979;41:51–62.

179. Zetter BR: Migration of capillary endothelial cells is stimulated by tumour-derived factors. *Nature* 1980;285:41–43.

180. Bowersox JC, Sorgente N: Chemotaxis of aortic endothelial cells in response to fibronectin. *Cancer Res* 1982;42:2547–2551.

181. Carter SB: Principles of cell motility: The direction of cell movement and cancer invasion. *Nature* 1965;208:1183–1187.

182. Lacovara J, Cramer EB, Quigley JP: Fibronectin enhancement of directed migration of B16 melanoma cells. *Cancer Res* 1984;44:1657–1663.

183. McCarthy JB, Palm SL, Furcht LT: Migration by haptotaxis of a Schwann cell tumor line to the basement membrane glycoprotein laminin. *J Cell Biol* 1983;97:772–777.

184. McCarthy JB, Furcht LT: Laminin and fibronectin promote the haptotactic migration of B16 mouse melanoma cells in vitro. *J Cell Biol* 1984;98:1474–1480.

185. McCarthy JB, Hagen ST, Furcht LT: Human fibronectin contains distinct adhesion- and motility-promoting domains for metastatic melanoma cells. *J Cell Biol* 1986;102:179–188.

186. Rogers SL, Letourneu PC, Palm SL, et al: Neurite extension by peripheral and central nervous system neurons in response to substratum-bound fibronectin and laminin. *Dev Biol* 1983;98:212–220.

187. Yamada KM: Cell surface interactions with extracellular materials. *Annu Rev Biochem* 1983;52:761–799.

188. Hynes R: Molecular biology of fibronectin. *Annu Rev Cell Biol* 1985;1:67–90.

189. McCarthy JB, Basara ML, Palm SL, et al: The role of cell adhesion proteins—laminin and fibronectin—in the movement of malignant and metastatic cells. *Cancer Met Rev* 1985;4:124–152.

190. Pierschbacher MD, Ruoslahti E: Variants of the cell recognition site of fibronectin that retain attachment-promoting activity. *Proc Natl Acad Sci USA* 1984;81:5985–5988.

191. Yamada KM, Kennedy DW: Dualistic nature of adhesive protein function: Fibronectin and its biologically active peptide fragments can autoinhibit fibronectin function. *J Cell Biol* 1984;99:29–36.

192. Horwitz A, Duggan K, Greggs R, et al: The cell substrate attachment (CSAT) antigen has properties of a receptor for laminin and fibronectin. *J Cell Biol* 1985;101:2134–2144.

193. Tamkun JW, DeSimone DW, Fonda D, et al: Structure of integrin, a glycoprotein involved in the transmembrane linkage between fibronectin and actin. *Cell* 1986;46:271–282.

194. Lark MW, Laterra J, Culp LA: Close and focal contact adhesions of fibroblasts to a fibronectin-containing matrix. *Fed Proc* 1985;44:394–403.

195. Izzard CS, Izzard SL, DePasquale JA: Molecular basis of cell-substrate adhesions, in Haemmerli G, Strauli P (eds): *Motility of Vertebrate Cells in Culture and in the Organism*. Basel, S Karger, 1985, pp 1–22.

196. Damsky CH, Knudsen KA, Bradley D, et al: Distribution of the cell substratum attachment (CSAT) antigen on myogenic and fibroblastic cells in culture. *J Cell Biol* 1985;100:1528–1539.

197. Laterra J, Norton EK, Izzard CS, et al: Contact formation by fibroblasts adhering to heparan sulfate-binding substrata (fibronectin or platelet factor 4). *Exp Cell Res* 1983;146:15–27.

198. Woods A, Couchman JR, Johansson S, et al: Adhesion and cytoskeletal organisation of fibroblasts in response to fibronectin fragments. *EMBO J* 1986;5:665–670.

199. Beyth RJ, Culp LA: Complementary adhesive responses of human skin fibroblasts to the cell-binding domain of fibronectin and the heparan sulfate-binding protein, platelet factor-4. *Exp Cell Res* 1984;155:537–548.

200. Hayashi M, Yamada KM: Domain structure of the carboxy-terminal half of human plasma fibronectin. *J Biol Chem* 1983;258:3332–3340.

201. Rogers SL, McCarthy JB, Palm SL, et al: Neuron-specific interactions with two neurite-promoting fragments of fibronectin. *J Neurosci* 1985;5:369–378.

202. Kornblihtt AR, Umezawa K, Vibe-Pedersen K, et al: Primary structure of human fibronectin: Differential splicing may generate at least 10 polypeptides from a single gene. *EMBO J* 1985;4:1755–1759.

203. Eisenstein R, Sorgente N, Soble LW, et al: The resistance of certain tissues to invasion: Penetrability of explanted tissues by vascularized mesenchyme. *Am J Pathol* 1973;73:765–774.

204. Eisenstein R, Goren SB, Schumacher B, et al: The inhibition of corneal vascularization with aortic extracts in rabbits. *Am J Ophthalmol* 1979;88:1005–1012.

205. Langer GA, Frank JS, Nudd LM, et al: Sialic acid: Effect of removal on calcium exchangeability of cultured heart cells. *Science* 1976;193:1013–1015.

206. Langer R, Folkman J: Polymers for the sustained release of proteins and other macromolecules. *Nature* 1976;263:797–800.

207. Azizkhan RG, Azizkhan JC, Zetter BR, et al: Mast cell heparin stimulates migration of capillary endothelial cells in vitro. *J Exp Med* 1980;152:931–944.

208. Taylor S, Folkman J: Protamine is an inhibitor of angiogenesis. *Nature* 1982;297:307–312.

209. Maciag T, Mehlman T, Friesel R, et al: Heparin binds endothelial cell growth factor, the principal endothelial cell mitogen in bovine brain. *Science* 1984;225:932–935.

210. Shing Y, Folkman J, Sullivan R, et al: Heparin affinity: Purification of a tumor-derived capillary endothelial cell growth factor. *Science* 1984;223:1296–1299.

211. Shing Y, Folkman J, Haudenschild C, et al: Angiogenesis is stimulated by a tumor-derived endothelial cell growth factor. *J Cell Biochem* 1985;29:275–287.

212. Folkman J, Taylor S, Spillberg C: The role of heparin in angiogenesis, in *Development of the Vascular System*, Ciba Foundation Symposium 100. London, Pitman Books, 1983, pp 132–149.

213. Crum R, Folkman J: Anti-angiogenesis by steroids without glucocorticoid or mineralocorticoid activity in the presence of heparin, abstract. *J Cell Biol* 1984;99:158a.

214. Crum R, Szabo S, Folkman J: A new class of steroids inhibits angiogenesis in the presence of heparin or a heparin fragment. *Science* 1985;230:1375–1378.

215. Feinberg RN, Beebe DC: Hyaluronate in vasculogenesis. *Science* 1983;220:1177–1179.

216. Thomas KA, Rios-Candelore M, Gimenez-Gallego G, et al: Pure brain-derived acidic fibroblast growth factor is a potent angiogenic vascular endothelial cell mitogen with sequence hormology to interleukin 1. *Proc Natl Acad Sci USA* 1985;82:6409.

217. Klagsbrun M, Sasse J, Sullivan R, et al: Human tumor cells synthesize an endothelial cell growth factor that is structurally related to basic fibroblast growth factor. *Proc Natl Acad Sci USA* 1986;83:2448–2452.
218. Moscatelli D, Presta M, Rifkin DB: Purification of a factor from human placenta that stimulates capillary endothelial cell protease production, DNA synthesis, and migration. *Proc Natl Acad Sci USA* 1986;83:2091–2095.
219. Moscatelli D, Jaffe E, Rifkin DB: Tetradecanoyl phorbol acetate stimulates latent collagenase production by cultured human endothelial cells. *Cell* 1980;20:343–351.
220. Schreiber AB, Winkler ME, Derynck R: Transforming growth factor-α: A more potent angiogenic mediator than epidermal growth factor. *Science* 1986;232:1250–1253.
221. Ignotz RA, Massagué J: Transforming growth factor-β stimulates the expression of fibronectin and collagen and their incorporation into the extracellular matrix. *J Biol Chem* 1986;261:4337–4345.
222. DiCorleto PE, Bowen-Pope DF: Cultured endothelial cells produce a platelet-derived growth factor-like protein. *Proc Natl Acad Sci USA* 1983;80:1919–1923.
223. Grotendorst GR, Seppä HEJ, Kleinman HK, et al: Attachment of smooth muscle cells to collagen and their migration toward platelet-derived growth factor. *Proc Natl Acad Sci USA* 1981;78:3669–3672.

Section Eight
Articular Cartilage

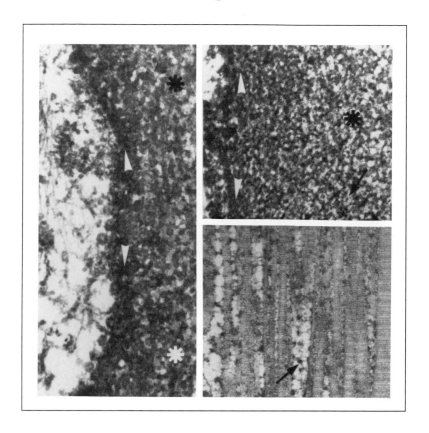

Group Leader

Lawrence Rosenberg, MD

Group Members

Joseph Buckwalter, MD
Richard Coutts, MD
Ernst Hunziker, MD
Van Mow, PhD

Group Participants

Mark Adams, MD, FRCPC
Mark Bolander, MD
Arnold Caplan, PhD
Bruce Caterson, PhD
Kenneth DeHaven, MD
David Eyre, PhD
Donald Fischman, MD
Helen Muir, PhD
Theodore Oegema, PhD
Eric Radin, MD
A. Hari Reddi, PhD
Melvin Rosenwasser, MD
Lawrence Shulman, MD

Synopsis

Articular cartilage consists of scattered chondrocytes surrounded by an extracellular matrix composed of a highly organized macromolecular framework filled with water. Chondrocytes create the molecular framework from three classes of molecules: collagens, proteoglycans, and noncollagenous proteins. A meshwork of type II collagen fibrils gives cartilage its form and tensile strength. A variety of quantitatively minor collagens may have important roles in organizing and maintaining the meshwork of type II collagen fibrils and possibly other functions. Cartilage proteoglycans and, in particular, the interaction of the proteoglycans with water, give cartilage its stiffness to compression, its resiliency, and contribute to its durability. Noncollagenous proteins appear to help organize and stabilize the articular cartilage matrix, attach chondrocytes to the extracellular matrix, and stabilize the chondrocyte phenotype. Although early studies suggested that articular cartilage was a homogeneous material, this is now known not to be the case. The cells and matrix vary considerably with depth from the articular surface, among regions of the same joint, and among joints. Even within a limited volume of cartilage at the same depth from the articular surface, three matrix compartments that differ in composition and organization can be identified: the pericellular matrix, the territorial matrix, and the interterritorial matrix. Studies of cartilage composition, organization, function, and repair must take this elaborate internal organization into consideration.

The capacity of normal cartilage to provide the tough, wear-resistant, nearly frictionless bearing material of diarthrodial joints depends on the material properties of the tissue. The concentration and organization of collagen, proteoglycan, and water influences the tensile, compression, shear, and permeability properties of articular cartilage. In addition, collagen cross-linking and proteoglycan molecular structure affect these material properties. Changes in the extracellular matrix, such as increased hydration, disruption of the collagen fibrillar network, and disaggregation and loss of proteoglycans, as in osteoarthritis, have profound effects on the material properties of the tissue, diminishing its ability to function as the bearing material of the joint.

The native cartilage repair response depends on the nature and extent of the injury, the location of the injury, and the state of the joint at the time of injury. Traumatic or surgical exposure of cartilage, immobilization, blunt trauma, infection, and other similar insults do not mechanically disrupt cartilage but may cause loss of matrix proteoglycans and a corresponding change in cartilage material properties. If the chondrocytes remain viable and the collagen meshwork remains intact, the cells replace the proteoglycans. If the injury progresses beyond the capacity of the cells to restore the matrix, the injury becomes irreversible; that is, injuries that cause massive loss of proteoglycans, disrupt the collagenous meshwork of the matrix, or cause cell death usually are not repaired with normal cartilage matrix. Mechanical disruption of cartilage may result from blunt or penetrating trauma. Blunt trauma that does not disrupt the collagen meshwork and kill chondrocytes probably is repaired satisfactorily. The chondrocytes may not be able to restore the normal cartilage after more severe blunt injuries. Trauma that penetrates cartilage but not subchondral bone leaves a defect that does not usually heal. Penetration of subchondral bone initiates the vascular response to injury and allows new cells to enter the defect. Some of these cells assume the morphologic features of chondrocytes and produce a cartilaginous matrix. Small defects can be repaired successfully, but larger defects usually fill with a fibrocartilaginous tissue that lacks the composition, structure, and durability of normal cartilage. The material proerties of this repair tissue have not been well defined. The ideal treatment of cartilage injury would restore the normal function and durability of the joint. No current method reliably accomplishes this in clinically significant defects, and it is not clear which current method or combination of current methods produces the best results in specific injuries.

Comparison of treatments requires evaluations of the quality of cartilage repair tissue. In general, this has been done by comparing the structure, composition, and material properties of the repair tissue with those of normal articular cartilage. This approach assumes that the more closely repair tissue resembles normal articular cartilage, the more likely it is that the repair tissue will function satisfactorily as a joint surface. Undoubtedly this assumption has value, but the best measure of cartilage repair tissue is its performance, that is, does it restore normal, pain-free joint function for a prolonged period.

Previous experimental work and clinical experience suggest that in most circumstances the best treatment of clinically significant cartilage injuries includes introducing a new population of cells by penetrating subchondral bone, protecting repair tissue from excessive loading, and encouraging controlled joint motion. For a new method to be considered an improvement, it must produce better long-term joint function than the current approach.

Chapter 9

Articular Cartilage: Composition and Structure

Joseph Buckwalter, MD
Ernst Hunziker, MD
Lawrence Rosenberg, MD
Richard Coutts, MD
Mark Adams, MD
David Eyre, PhD

Chapter Outline

Introduction

Synovial joint function depends on cartilage. It distributes load, minimizing peak stresses on subchondral bone; it can be deformed and regain its original shape; it has remarkable durability; and it provides an unequalled low-friction bearing surface. Despite these critical properties, the importance of in-depth study of cartilage may not be immediately apparent. The development of metal and plastic joint replacements has led many physicians to view cartilage as material that can be replaced easily when it is worn or damaged. If joint replacements could predictably restore the function of all painful, stiff joints, then further study of cartilage would be unlikely to lead to significantly improved medical care. However, critical review of the long-term results of joint replacements shows that they fall short of restoring normal joint function for many patients, that they may be associated with significant complications, and that they cannot be used for some types of joint problems or in some patients. When scientists compare cartilage with tissues such as muscle, nerve, or kidney, cartilage seems simple, homogeneous, and inert. Once formed, it appears to remain unchanged unless affected by disease or injury; it has a low level of metabolic activity, and it lacks blood vessels, lymphatic vessels, and nerves. Yet, the simple homogeneous appearance of cartilage at gross and light microscopic levels hides a highly ordered complex structure that gives cartilage properties unmatched by any available substitute. Understanding the properties of cartilage and articular cartilage injury and repair begins with detailed knowledge of the composition and structure of this complex tissue.

Articular Cartilage Composition

Cartilage consists of cells, that is, chondrocytes, embedded in an abundant extracellular matrix (Fig. 9–1). Unlike the situation in parenchymal tissue, cells contribute relatively little to the total volume of cartilage, usually 10% or less,[1-3] and the important functional properties of cartilage, including stiffness, durability, and distribution of load, depend on the extracellular matrix.

The gross and light microscopic appearances of cartilage (Fig. 9–1) led early investigators to believe that the matrix consisted of an inert material lacking internal structure. Virchow[4] noted that ". . . in hyaline cartilage, as in that lining the joint . . . the intercellular matter is perfectly homogeneous . . . as clear as water. . . ." This clear appearance of the matrix gave the tissue its name hyaline (Gr. *hyalos*, glass). Subsequent phase-contrast and electron microscopic studies suggested that the matrix consisted almost entirely of a collagen fibril network.[5] Both of these appearances are deceiving. Cartilage matrix is not homogeneous, nor does it consist primarily of collagen. As Virchow suspected, the largest component of the matrix is water, but his meticulous light microscopic studies could not disclose the elaborate framework of ma-

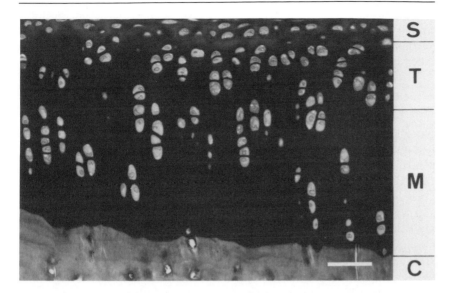

Fig. 9-1 *Articular cartilage from the medial femoral condyle of an 8-month-old rabbit. The cartilage matrix (extracellular substance) surrounding isolated chondrocytes is organized into four zones: the superficial zone (S), the transitional zone (T), the middle or deep zone (M), and calcified cartilage (C). In all zones the matrix appears to be homogeneous (glutaraldehyde-fixed in the presence of ruthenium hexamine trichloride, $\times$ 230; bar gauge = 50 μ).*

cromolecules—including collagens, proteoglycans, and noncollagenous proteins—that organizes and maintains water in the matrix.

Tissue Fluid

Tissue fluid contributes 60% to 80% of the wet weight of cartilage and consists of water with dissolved gases, small proteins, and metabolites.[6-13] The high water content distinguishes cartilage from most other connective tissues, makes essential contributions to the material properties of the tissue, and participates in joint lubrication.[14,15] Because the tissue fluid is not contained within cellular membranes, its volume, concentration, organization, and behavior depend on its interaction with the structural matrix molecules. If these macromolecules did not maintain and organize the tissue fluid and impede its flow through the matrix, cartilage would lose its ability to resist compression, its resilience, and its ability to contribute to joint lubrication. Because the cells rely on diffusion of nutrients and metabolites through the matrix rather than on a vascular supply, chondrocyte function also depends on the interaction of the tissue fluid and the matrix macromolecules. Details of the interaction between the tissue fluid and the matrix molecules remain uncertain,[6,8-11,16] but, since the tissue fluid can exchange with water outside the tissue and small solutes can move freely in the cartilage tissue fluid, the water must be loosely bound to the molecular framework in a way that maintains the hydration of the tissue but still allows exchange with fluid outside the tissue. Because matrix water concentration varies with the distance from the articular surface,[17] the relationship between water and the matrix macromolecules must also vary with the distance from the surface.

Structural Macromolecules

The structural macromolecules contribute 20% to 40% of the wet weight of cartilage and include collagens, proteoglycans, and noncollagenous proteins or glycoproteins. Chondrocytes synthesize these molecules from amino acids and sugars. The available evidence indicates that all chondrocytes are capable of synthesizing each of these molecules.[18] In most cartilages, collagen contributes about 50% of the tissue's dry weight, proteoglycans contribute 30% to 35%, and noncollagenous proteins and glycoproteins contribute about 15% to 20%.[12,13,19,20] Collagens form the fibrillar meshwork that gives cartilage its tensile strength and form.[21-23] Proteoglycans and the noncollagenous proteins complete the macromolecular framework by binding to the collagenous meshwork or becoming mechanically trapped within it. Water fills this framework, and chondrocytes attach themselves to the matrix macromolecules.

Collagens   Cartilage, like most other tissues, contains more than one genetically distinct type of collagen.[24-27] The principle collagen of articular cartilage, type II, forms the characteristic cross-banded fibrils seen by electron microscopy and accounts for 90% to 95% of the total cartilage collagen. Cartilage also contains at least two quantitatively minor collagens, types IX and XI (1α, 2α, 3α). In addition, Apone and associates[24] recently reported the presence of types V and VI. The functions and interrelationships of the quantitatively minor collagens remain unclear but they may contribute to the formation and stability of the type II collagen fibril meshwork.[27]

In addition to its presence in articular cartilage, type II collagen is found in other hyaline cartilages, notochord, nucleus pulposus, and vitreous.[26] In contrast, the other common fibrillar collagen, type I, is found in bone, cornea, dentin, fibrocartilage, skin, meniscus, anulus fibrosus, tendon, and heart valves.[26] Type I fibrils are larger, contain fewer hydroxylysine residues, and are less glycosylated than type II collagen fibrils. Comparison of the distributions of these two fibrillar collagens shows that type II is normally found in tissues with higher proteoglycan and water contents, suggesting that type II has specific properties that allow it to establish an ordered relationship with proteoglycans, thus helping to create and maintain a highly hydrated matrix.

Type IX collagen is assembled from three genetically distinct amino acid chains, and contains collagenous and noncollagenous domains.[27] One or possibly two chondroitin sulfate chains appear to be located in the noncollagenous domains. Type IX collagen may be involved in the organization and stabilization of the Type II collagen fibril meshwork, although this function has not been demonstrated.

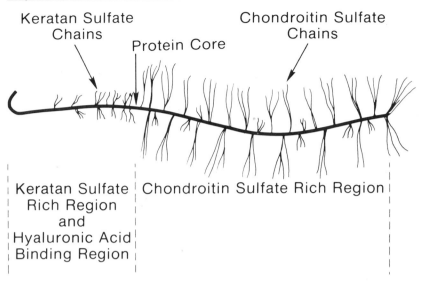

Keratan Sulfate
Chains

Chondroitin Sulfate
Chains

Protein Core

Keratan Sulfate
Rich Region
and
Hyaluronic Acid
Binding Region

Chondroitin Sulfate Rich Region

Fig. 9–2 *Schema of a proteoglycan monomer. The molecule consists of a central protein core with multiple covalently bound chondroitin sulfate and keratan sulfate chains. The oligosaccharides are not shown. (Reproduced with permission from Buckwalter JA: The fine structure of human intervertebral disc, in White AA III, Gordon SL (eds): American Academy of Orthopaedic Surgeons Symposium on Idiopathic Low Back Pain. St. Louis, CV Mosby Co, 1982, pp 108–143.)*

Type XI (1α, 2α, 3α) collagen consists of the native form or forms of three amino acid chains. It has been suggested that types 1α, 2α, and 3α collagen may be associated with the fibrils of type II collagen and involved in determining their diameter.[27]

The presence of types VI and VII collagen in articular cartilage has been reported only recently.[24] Thus far, no specific functions have been identified for them.

Proteoglycans Proteoglycans form the major macromolecule of cartilage ground substance.[28-30] They have multiple forms, including large aggregating proteoglycans that exist as individual monomers or as aggregates containing multiple monomers, large nonaggregating proteoglycans, and small nonaggregating proteoglycans.[19,30-34]

Aggregating monomers (Fig. 9–2) consist of protein core filaments with multiple covalently bound oligosaccharides and chondroitin and keratan sulfate chains. The glycosaminoglycans (Fig. 9–3), chondroitin sulfate and keratan sulfate, form about 95% of the molecule and protein contributes about 5%. Because each of the glycosaminoglycan chains contains a large number of negative charges, adjacent chains tend to repel each other, and, in solution, to maintain the molecule in an expanded form. The protein core filament consists of three regions (Fig. 9–2); the hyaluronic acid-binding region, the keratan sulfate-rich region, and the chondroitin sulfate-rich region.[35-38] The hyaluronic acid-

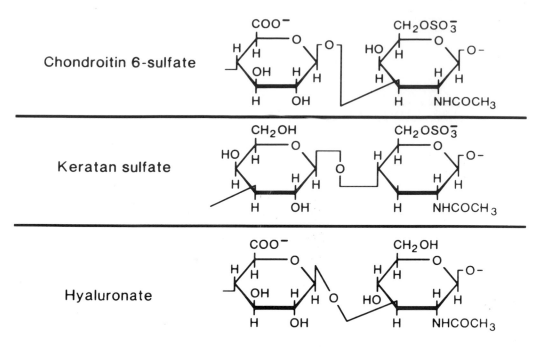

Fig. 9–3 *Schema of the disaccharides of chondroitin 6-sulfate (**top**), keratan sulfate (**center**), and hyaluronate (**bottom**). Glycosaminoglycans form from chains of disaccharides and each disaccharide has one or two negatively charged groups.*

binding region contains few, if any, glycosaminoglycan chains, and has the ability to associate reversibly with hyaluronic acid and link protein.[39–41] The keratan sulfate-rich region, located between the hyaluronic acid-binding region and the chondroitin sulfate-rich region, binds approximately 60% of the total keratan sulfate found in the molecule.[37,38] The chondroitin sulfate-rich region extends from the keratan sulfate-rich region, and binds 90% or more of the chondroitin sulfate.[38] At least two forms of aggregating monomers have been identified: chondroitin sulfate-rich proteoglycan and keratan sulfate-rich proteoglycan.

In the matrix, most of the aggregating proteoglycan monomers associate with hyaluronic acid filaments and link proteins to form aggregates (Fig. 9–4).[30,32,39,42,43] Hyaluronic acid forms the backbone of the aggregate, and may range in length from several hundred nanometers with a few associated monomers to more than 10,000 nm with more than 300 monomers.[19,31,32,42] Link proteins, 40-kD to 50-kD molecular weight proteins, stabilize the association between monomers and hyaluronic acid and play a role in directing the assembly of aggregates.[44] The importance of aggregate formation is not entirely clear. It may help anchor monomers within the matrix, preventing their displacement, it may organize and stabilize the relationship between

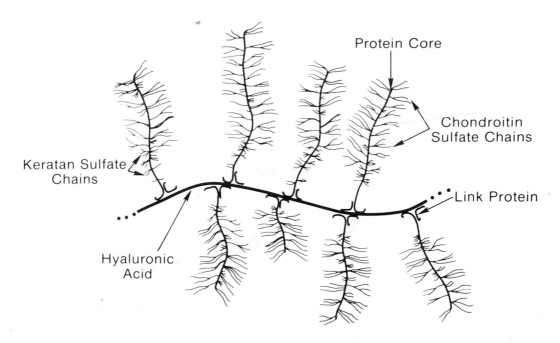

Fig. 9–4 *Schema of a proteoglycan aggregate. Multiple monomers noncovalently associate with link proteins and a central hyaluronic acid filament to form an aggregate. (Reproduced with permission from Buckwalter JA: The fine structure of human intervertebral disc, in White AA III, Gordon SL (eds): American Academy of Orthopaedic Surgeons Symposium on Idiopathic Low Back Pain. St. Louis, CV Mosby Co, 1982, pp 108–143.)*

proteoglycan monomers and type II collagen fibrils, and it probably helps to control the flow of water through the matrix.[45,46]

Because of their ability to interact with tissue fluid, proteoglycans help to give cartilage its stiffness to compression and resilience[21,22] and contribute to its durability as a joint surface. They have a structure that fills a large volume with negatively charged glycosaminoglycan chains that bind matrix water and cations. As shown in Figure 9–5, the anionically charged glycosaminoglycan chains tend to repel each other and hold themselves and the monomers stiffly extended. Compression may drive the glycosaminoglycan chains closer together, increasing their resistance to further compression and forcing the tissue fluid out of the molecular domain. Release of the pressure allows the molecules to expand again. Comparison of the maximal possible volume that proteoglycans can occupy in solution with their concentration in articular cartilage shows that they could, if fully extended, fill a volume many times larger than the tissue that contains them.[19] Therefore, proteoglycans in the matrix must be only partially hydrated and exert a constant pressure to expand, restrained only by the collagen fibril meshwork. If the collagen fibril meshwork is disrupted, the matrix

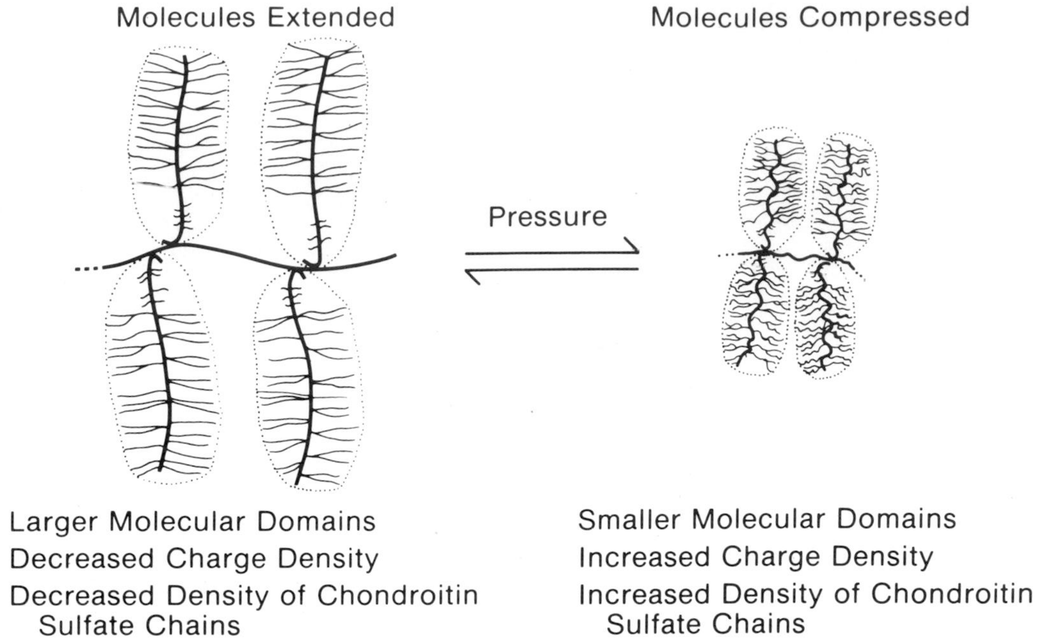

Molecules Extended   Molecules Compressed

Pressure

Larger Molecular Domains   Smaller Molecular Domains
Decreased Charge Density   Increased Charge Density
Decreased Density of Chondroitin   Increased Density of Chondroitin
  Sulfate Chains   Sulfate Chains

Fig. 9–5 *Schema of the reversible expansion of proteoglycan monomers in solution. The negative charges of the glycosaminoglycan chains repel each other and bind water, expanding the domain of the proteoglycans. Pressure may drive these negatively charged chains closer together. Release of the pressure allows the molecules to expand. In the tissue, the collagen meshwork presumably limits the expansion of proteoglycans. (Reproduced with permission from Buckwalter JA: Articular cartilage, in American Academy of Orthopaedic Surgeons* Instructional Course Lectures, XXXII. *St. Louis, CV Mosby Co, 1983, pp 349–370.)*

swells as proteoglycans expand, increasing the concentration of water and decreasing the proteoglycan concentration.

Currently available evidence indicates that the structure and composition of aggregating cartilage proteoglycans change with age.[31,47] With increasing age, the keratan sulfate and protein content of monomers increases, the chondroitin sulfate content decreases, chondroitin sulfate chains become shorter, average monomer size decreases, and the variability in monomer size increases. In addition, link protein may fragment,[48] the concentration of functional link protein decreases,[49] aggregates become smaller (Fig. 9–6),[31] the proportion of monomers that aggregate decreases,[50] hyaluronate filament length decreases,[31] and the proportion of hyaluronic acid-binding region fragments of the protein core may increase.[51] These changes may be caused by age-related alterations in chondrocyte synthetic function, by degradation of proteoglycans in the matrix, or by a combination of the two. Although the functional significance of the age changes in matrix proteoglycans

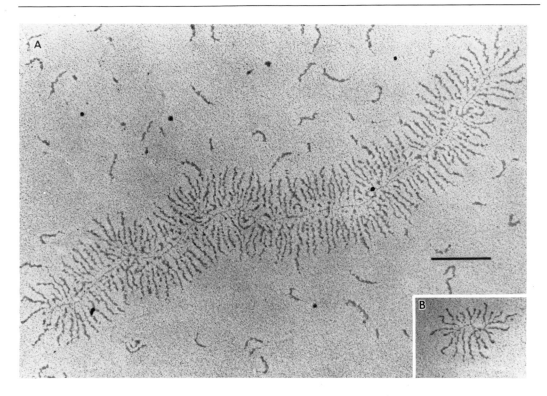

Fig. 9–6 *Electron micrographs of bovine articular cartilage proteoglycan aggregates. Proteoglycan aggregate from a skeletally immature calf consists of a central hyaluronic acid filament and multiple attached monomers (bar gauge = 500 μ).* **Inset,** *A proteoglycan aggregate from a skeletally mature steer. (Reproduced with permission from Buckwalter JA, Kuettner KE, Thonar EJ: Age-related changes in articular cartilage proteoglycans: Electron microscopic studies. J Orthop Res 1985;3:251–257.)*

remains uncertain, they suggest that the ability of the cells to restore a normal matrix may also change with age.

The source of the large nonaggregating proteoglycans remains uncertain. Some of the large nonaggregating chondroitin sulfate-containing proteoglycans may form from the breakdown of the aggregating proteoglycan monomers, while others probably represent a distinct population of proteoglycans. In addition to the large aggregating and nonaggregating proteoglycans, small nonaggregating proteoglycans have been identified in cartilage. These molecules may contain chondroitin sulfate and another glycosaminoglycan, dermatan sulfate.[33] At least some of the small proteoglycans are preferentially located in the more superficial cartilage layers of adult cartilage and may have strong specific interactions with collagen fibrils. Their function is unknown.

Noncollagenous Proteins Although less is known about noncollagenous proteins than about either collagens or proteoglycans, they form a significant component of the macromolecular framework. They consist primarily of protein and may have a small number of attached monosaccharides or oligosaccharides. It appears that at least some of these molecules help organize and maintain the macromolecular structure of the matrix and the relationship between the chondrocytes and the matrix. Link protein, chondronectin, and anchorin CII may be considered noncollagenous proteins. Unlike link protein, chondronectin and anchorin CII help chondrocytes adhere to the molecular framework of the matrix. Undoubtedly, other noncollagenous proteins exist.

Hewitt and associates[52] described chondronectin as a 150-kD molecular weight cartilage matrix molecule that mediates the adhesion of chondrocytes to type II collagen and may help stabilize the chondrocyte phenotype. It does not bind to collagen by itself but will do so in the presence of chondroitin sulfate. It may help stabilize the chondrocyte phenotype.

Anchorin CII is a 34-kD molecular weight protein located in the pericellular space of cartilage matrix directly adjacent to chondrocytes but not in the intercellular matrix. Studies by von der Mark and associates[53] indicate that anchorin CII mediates binding of type II collagen to chondrocytes. The relationship between chondronectin-mediated chondrocyte adhesion to type II collagen and anchorin CII-mediated adhesion remains unclear.

Chondrocytes

The material properties of cartilage depend on its extracellular matrix but the existence and maintenance of the matrix depends on chondrocytes. Like most other mesenchymal cells, chondrocytes surround themselves with their extracellular matrix and do not form cell-to-cell contacts.[1,2,54,55] In addition to the organelles responsible for matrix synthesis, such as endoplasmic reticulum and Golgi membranes (Fig. 9–7), chondrocytes may contain intracytoplasmic filaments and glycogen, and most articular chondrocytes have a cilium that may be involved in the regulation of matrix turnover.

Although the mechanisms have not been well defined, it is clear that the matrix influences chondrocyte function. Changes in matrix composition—such as a decrease in proteoglycan content, the loss of type II collagen, an increase or decrease in hyaluronic acid concentration, and probably others—alter chondrocyte synthetic function. The matrix may also transmit mechanical signals to chondrocytes through changes in tension on the cell membrane or it may act as an electromechanical transducer. Understanding how chondrocytes respond to mechanical signals may help explain what types of loads are necessary for maintenance of normal cartilage, and lead to the use of controlled loading to improve the quality of cartilage repair.

During formation of articular cartilage, chondrocytes proliferate rapidly and synthesize large volumes of matrix. With maturation, these processes slow and the cell numerical density decreases. Under normal circumstances, chondrocytes in mature articular cartilage rarely, if ever, divide, and the cell numerical density of cartilage, particularly in the superficial layer, declines with age.[3,56] These age-related decreases in

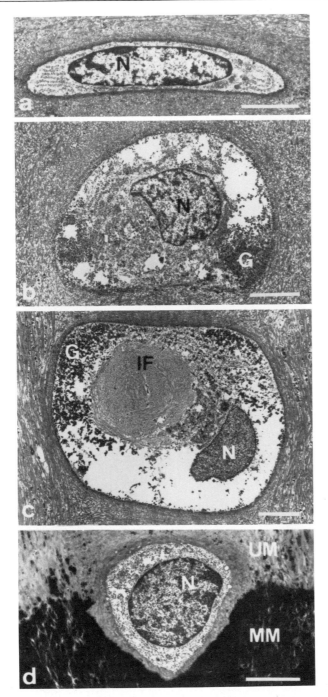

Fig. 9–7 *Electron micrographs of mature articular cartilage chondrocytes from the femoral medial condyle of a rabbit.* **Top:** *Superficial zone (× 5,400).* **Top center:** *Transitional zone (× 4,650).* **Bottom center:** *Middle zone (× 3,830).* **Bottom:** *Calcifying zone (× 5,000). N, nucleus; G, glycogen granules; IF, intermediate filaments; MM, mineralized matrix; UN, unmineralized matrix (bar gauge = 3 μ).*

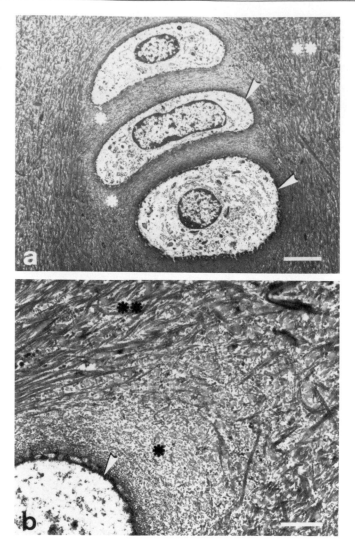

Fig. 9–8 *Two electron micrographs of mature articular cartilage chondrocytes from the femoral medial condyle of a rabbit and their surrounding matrix, which is divided into various compartments. Protruding from the surface of each chondrocyte are many short, slender cell processes that extend through the pericellular matrix (arrowheads) to the border of the territorial matrix (*). Note that in the presence of ruthenium hexamine trichloride, chondrocytes are preserved in an unshrunken state and that there are no artificial lacunae. (**) interterritorial matrix (top, × 3,250, bar gauge = 3 μ; bottom, × 10,500, bar gauge = 1 μ).*

cell density and cell synthetic function may limit the ability of the cells to restore a damaged matrix.

Articular Cartilage Structure

Although articular cartilage has the same basic composition as other hyaline cartilages, it develops a unique internal organization to per-

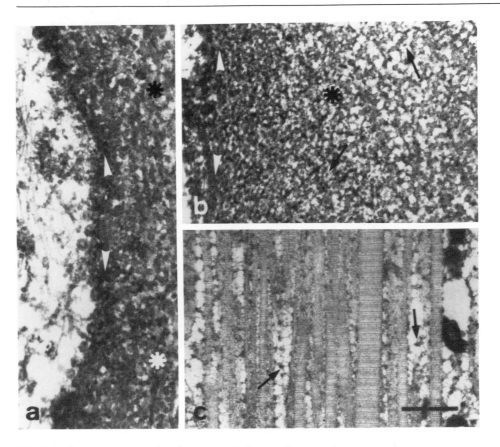

Fig. 9–9 *Electron micrographs of mature articular cartilage matrix compartments from the medial femoral condyle of a rabbit.* **Left:** *The pericellular matrix (white arrowheads) consists of a narrow, dense coat of proteoglycans and possibly glycoproteins as well as other nonfibrillar molecules. Note the intimate contact between matrix components and the chondrocyte plasma membrane.* **Top right:** *The territorial matrix (* at left and top right) consists of a dense network of fine collagen that may extend around one cell or a group of cells (a territorium or chondron). Note the absence of a clearly delineated border between this compartment and the pericellular matrix.* **Bottom right:** *The interterritorial matrix typically consists of parallel-oriented collagen fibrils and fibers with matrix granules (arrows at top right and bottom right) interspersed between them. The matrix granules are proteoglycans precipitated with ruthenium hexamine trichloride (× 48,000; bar gauge = 0.3 μ).*

form its specialized role as a joint surface. The structure, composition, and presumably the function of articular cartilage varies between the joint surface and subchondral bone.[57–60] The differences in articular cartilage with depth have been traditionally described by dividing articular cartilage into layers or zones referred to as the superficial zone, the middle or transition zone, the deep or radial zone, and the zone of calcified cartilage (Figs. 9–1 and 9–7).[19,55,60] Within these zones or layers, distinct matrix regions or compartments can be identified, including the pericellular matrix, the territorial matrix, and the interterritorial matrix (Figs. 9–8 and 9–9).[55,61]

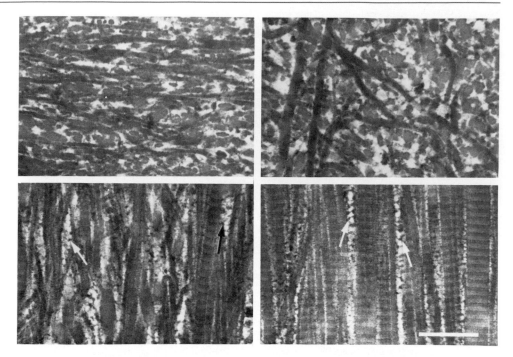

Fig. 9–10 *Electron micrographs of interterritorial matrix at different levels through mature articular cartilage from the medial femoral condyle of an 8-month-old rabbit. In the superficial zone* **(top left)***, the slender collagen fibrils lie parallel to the surface plane of the articular cartilage and interfibrillar matrix granules are almost completely absent. In the transitional zone* **(top right)***, the only zone in which the collagen fibrils or fibers are not highly organized in a parallel arrangement, reorientation takes place. In the middle zone* **(bottom left)***, and the deep zone* **(bottom right)***, the collagen fibers are vertical to the joint surface. Fibril and fiber diameters gradually increase as they approach the calcifying cartilage, as does the matrix granule (arrows) content (proteoglycans precipitated with ruthenium hexamine trichloride,* $\times$ *35,000; bar gauge = 0.5 μ).*

Zones of Articular Cartilage

Cell shape and size, collagen fibril diameter, collagen fibril orientation relative to the articular surface, matrix water content, and proteoglycan content all change with depth from the articular surface (Fig. 9–10).[9,62] There may also be alterations in the relationships between collagens and proteoglycans, in the degree of proteoglycan aggregation, and in the concentration of link protein.[60] It also appears that chondrocytes from different zones are phenotypically different, that is, the chondrocytes of articular cartilage may consist of several populations of cells that have different functions.

Superficial Zone   The thinnest of the zones, the superficial zone, forms the gliding surface of the joint (Figs. 9–1, 9–7 and 9–10). Directly adjacent to the joint lies a thin cell-free layer of matrix. It consists primarily of fine fibrils with little polysaccharide. This most superficial layer presumably corresponds to the thin, clear film that can be mechanically stripped from articular cartilage, and it also probably represents the "lamina splendens," so named because phase-contrast study of articular cartilage sections shows a conspicuous bright layer at the articular surface.[5] Deep to this most superficial layer lie elongated chondrocytes that arrange themselves so that their long axis is parallel to the articular surface (Fig. 9–7, *top*). These relatively inactive cells contain endoplasmic reticulum, Golgi membranes, and mitochondria.

Little or no hyaluronic acid is present in this region, but immunoelectron microscopy has identified proteoglycan monomers and link protein on or in association with collagen fibrils.[60] Proteoglycans associated with collagen in this zone resist extraction more than proteoglycans associated with collagen in the other zones, suggesting a stronger association between collagen and proteoglycan. Possibly this strong association is an important feature of the superficial zone, which is subjected most directly to the shearing stresses of joint motion.

Transitional Zone   The transitional zone occupies several times the volume of the superficial zone (Fig. 9–1). The cells become more spherical and fill their cytoplasm with endoplasmic reticulum, Golgi membranes, mitochondria, glycogen, and, occasionally, intracytoplasmic filaments (Fig. 9–7, *top center*). The collagen fibrils of this zone are larger than those of the superficial zone (Fig. 9–10).

Deep Zone   This zone usually forms the largest part of articular cartilage (Fig. 9–1). The cells resemble the spheroidal cells in the transitional zone, but tend to arrange themselves in an almost columnar pattern perpendicular to the joint surface. They typically contain large amounts of intermediate filaments and glycogen granules (Fig. 9–7, *bottom center*). This zone contains the largest collagen fibrils of articular cartilage and has the highest proteoglycan content as well as the lowest water content (Fig. 9–10). Thus, from the superficial zone to the deepest portion of the deep or radial zone, water content decreases and proteoglycan content increases.

Zone of Calcified Cartilage   This zone separates the softer hyaline cartilage from the stiffer subchondral bone (Fig. 9–1). Collagen fibrils of the radial zone penetrate directly into the calcified cartilage (Fig. 9–7, *bottom*). The cells of this zone are usually smaller, with relatively little cytoplasm and almost no endoplasmic reticulum.

It is tempting to believe that the differences in matrix composition and organization among zones reflect differences in mechanical function. That is, the superficial zone may primarily resist shearing forces, the transitional zone may allow the change in orientation of collagen fibrils from the superficial zone to the radial zone, and the radial zone may primarily help to resist and distribute compressive loads. The zone of calcified cartilage would then provide a transition in material properties between hyaline cartilage and bone, as well as anchor the hyaline cartilage to the bone.

Matrix Regions of Articular Cartilage

The matrix regions (Figs. 9–8 and 9–9) differ in their proximity to chondrocytes, their collagen content, their collagen fibril diameter, the orientation of their collagen fibrils, and their proteoglycan and non-collagenous protein content and organization.[55,60,61]

Pericellular Matrix Chondrocyte cell membranes appear to attach directly to a thin layer of pericellular matrix that completely surrounds the cell (Figs. 9–8 and 9–9, *left*). The pericellular matrix is the smallest of the matrix regions and appears to contain little or no fibrillar collagen. It may contain some of the minor collagen types, but the predominant macromolecules of the pericellular matrix appear to be proteoglycans, noncollagenous proteins, and glycoproteins, such as anchorin CII and chondronectin.

Territorial Matrix An envelope of territorial matrix surrounds the pericellular matrix of individual chondrocytes and in some cases pairs and clusters of chondrocytes and their pericellular matrices (Figs. 9–8 and 9–9, *top right*). In the radial zone, a territorial matrix surrounds each chondrocyte column. The thin collagen fibrils of the territorial matrix nearest to the cells appear to adhere to the pericellular matrix. At a distance from the cell they decussate and intersect at various angles, forming a fibrillar basket around the cells that may provide mechanical protection of chondrocytes when the cartilage is deformed.[61] An abrupt increase in collagen fibril diameter and a transition from the basket-like orientation of fibrils to a more parallel arrangement marks the boundary between the territorial and interterritorial matrices (Fig. 9–8). However, many fibrils connect the territorial and interterritorial matrices, making it difficult to separate them mechanically.

Interterritorial Matrix The interterritorial matrix forms the largest matrix compartment of articular cartilage (Figs. 9–1 and 9–8). It is responsible for the mechanical properties of articular cartilage, and studies of the composition and material properties of articular cartilage necessarily describe the interterritorial matrix. The organization and orientation of the large collagen fibrils of this region as they pass from the articular surface to the deeper zones have been subjects of considerable interest (Figs. 9–9, *bottom right*, and 9–10). Benninghoff[63] described a predominantly tangential orientation of the collagen fibrils in the superficial zone, arching decussating arcades in the middle zone, and a radial orientation in the deep zone. Although some have challenged Benninghoff's description of collagen fibril arcades in the middle or transitional zone, most morphologic studies (Fig. 9–10) generally confirm that the principal interterritorial matrix collagen fibrils in the superficial zone lie parallel to the joint surface and that those of the radial or deep zone lie perpendicular to the joint surface.[62] The fibrils of the transitional zone exhibit less anisotropic ordering.

Future Directions

Recent work has significantly advanced our understanding of cartilage; nonetheless, our knowledge of its structure, composition, formation, and function remains incomplete. To understand how an articular surface that provides the normal function of a synovial joint can be restored after injury, it is important to learn more about the cells and matrices of normal articular cartilage.

The collagen meshwork gives cartilage its form and tensile strength, yet little is known about how this highly organized molecular framework is established and maintained. It seems likely that mechanical loads and joint motion have a role in establishing the orientation and organization of the matrix collagen fibrils, but this relationship has not been clearly demonstrated. Type II collagen forms most of the meshwork, but the quantitatively minor collagen types may also be important. Answers to the following questions would help explain the role of these molecules: What are the functions of the quantitatively minor collagen types? How do these minor collagen types interact with type II collagen and, possibly, with chondrocytes, proteoglycans, and noncollagenous proteins? Do they form part of the type II collagen fiber meshwork? Do they help organize and stabilize this meshwork?

A better understanding of the basic structure of proteoglycans has made it possible to refine our questions concerning their role in cartilage. Some of the questions raised by recent work are: What is the source of large nonaggregating proteoglycans? What is the function of the large and small nonaggregating proteoglycans? How are proteoglycan aggregates assembled in the matrix? What form do proteoglycans assume in the matrix?

Systematic study of noncollagenous proteins has just begun, but it is clear that these molecules may help organize and stabilize the matrix as well as influence chondrocyte function. Only a few noncollagenous proteins have been identified. A complete description of these molecules is essential. This may be accomplished by a combination of traditional biochemical techniques and isolation and characterization of cDNA and genomic clones of cartilage proteins. Analysis at the nucleotide sequence level should provide information on the sequence and structural characteristics of these proteins as well as on the regulation of their genetic expression. Once a complete list of these molecules has been established, it will be necessary to define their functions and interactions.

Cartilage repair resembles the embryonic formation and development of the tissue. A better understanding of how articular cartilage forms and develops may suggest ways to improve the repair response.

It is clear that the chondrocytes and cartilage matrix change with age. The full extent and significance of these age-related changes have not been defined and it is not known whether these changes affect the tissue's response to injury. It is important to know if chondrocytes—or cells that differentiate into chondrocytes following injury—lose the ability to synthesize and assemble the molecular components of the matrix with increasing age.

Although the basic details of protein synthesis are relatively well defined, a number of significant questions concerning the synthesis of cartilage macromolecules remain unanswered. Among these are the following: Exactly where within the cell are the glycosaminoglycan chains synthesized? How are these chains modified after synthesis? Certain proteins, including factors acting by autocrine and paracrine mechanisms, heat-shock proteins, and cell cycle-dependent regulatory proteins (proto-ontogenes), influence synthesis of matrix macromolecules, but the details are not well understood.

Recombinant techniques may allow us to identify the proteins that regulate matrix synthesis in normal and repair tissue. These techniques have the advantage of permitting quantification of genetic messages for the structural components of cartilage.

Cell shape also may affect synthesis of matrix molecules. Chondrocytes and chondrocyte-like cells with relatively spherical shapes synthesize hyaline cartilage macromolecules including proteoglycans and type II collagen. When these same cells are flattened, they synthesize the molecules found in more fibrous tissues. It is not known how cell shape regulates synthesis or if it is possible to devise methods of influencing cell shape in vivo.

It is not known how proteoglycans, collagen, and noncollagenous proteins interact to form the macromolecular frame work of the matrix or how this framework is stabilized. Furthermore, it is not clear how chondrocytes, loads applied to the tissue, and joint motion influence matrix composition and structure.

Maintenance of the normal cartilage matrix requires turnover of the matrix macromolecules. Understanding how this is accomplished may lead to methods of treating cartilage injuries.

The formation, development, and maintenance of cartilage require continuing, complex interactions between chondrocytes and the matrix. These interactions must be equally important in repair. It is essential, therefore, to determine how the extracellular matrix influences chondrocyte function. In particular, what specific changes in matrix composition and organization influence chondrocytes? Does the matrix transmit mechanical or electrical signals to chondrocytes that influence cell function? If so, how does this occur?

References

1. Hamerman D, Rosenberg LC, Schubert M: Diarthrodial joints revisited. *J Bone Joint Surg* 1970;52A:725–774.
2. Stockwell RA: Chondrocytes. *J Clin Pathol* 1978;12(suppl):7–13.
3. Stockwell RA: The cell density of human articular and costal cartilage. *J Anat* 1967;101:753–763.
4. Virchow R: *Cellular Pathology*, Chance F (trans). London, Churchill, 1860, p 46.
5. MacConaill MA: The movements of bones and joints: 4. The mechanical structure of articulating cartilage. *J Bone Joint Surg* 1951;33B:251–257.
6. Jaffe FF, Mankin HJ, Weiss C, et al: Water binding in the articular cartilage of rabbits. *J Bone Joint Surg* 1974;56A:1031–1039.
7. Linn FC, Sokoloff L: Movement and composition of interstitial fluid of cartilage. *Arthritis Rheum* 1965;8:481–494.
8. Mankin HJ: The water of articular cartilage, in Simon WH (ed): *The Human Joint in Health and Disease*. Philadelphia, University of Pennsylvania Press, 1978, pp 37–42.

9. Maroudas A: Physiochemical properties of articular cartilage, in Freeman MAR (ed): *Adult Articular Cartilage*, ed 2. Tunbridge Wells, Pitman Medical, 1979, pp 215–290.

10. Maroudas A: Physical chemistry of articular cartilage and the intervertebral disc, in Sokoloff L (ed): *The Joints and Synovial Fluid*. New York, Academic Press, 1980, vol 2, pp 239–291.

11. Maroudas A, Schneiderman R: "Free" and "exchangeable" or "trapped" and "non-exchangeable" water in cartilage. *J Orthop Res* 1987;5:133–138.

12. Muir IHM: The chemistry of the ground substance of joint cartilage, in Sokoloff L (ed): *The Joints and Synovial Fluid*. New York, Academic Press, 1980, vol 2, pp 27–94.

13. Muir IHM: Biochemistry, in Freeman MAR (ed): *Adult Articular Cartilage*, ed 2. Tunbridge Wells, Pitman Medical, 1979, pp 145–214.

14. Brand RA: Joint lubrication, in Albright JA, Brand RA (eds): *The Scientific Basis of Orthopaedics*, ed 2. Norwalk, Appleton & Lange, 1987, pp 373–386.

15. McCutchen CW: Lubrication of joints, in Sokoloff L (ed): *The Joints and Synovial Fluid*. New York, Academic Press, 1978, pp 437–483.

16. Torzilli PA, Rose DE, Dethmers DA: Equilibrium water partition in articular cartilage. *Biorheology* 1982;19:519–537.

17. Brocklehurst R, Bayliss MT, Maroudas A, et al: The composition of normal and osteoarthritic articular cartilage from human knee joints: With special reference to unicompartmental replacement and osteotomy of the knee. *J Bone Joint Surg* 1984;66A:95–106.

18. Smith PH: Autoradiographic evidence for the concurrent synthesis of collagen and chondroitin sulfates by chick sternal chondrocytes. *Connect Tissue Res* 1972;1:181–188.

19. Buckwalter JA: Articular cartilage, in American Academy of Orthopaedic Surgeons *Instructional Course Lectures, XXXII*. St. Louis, CV Mosby Co, 1983, pp 349–370.

20. Muir H, Bullough P, Maroudas A: The distribution of collagen in human articular cartilage with some of its physiological implications. *J Bone Joint Surg* 1970;52B:554–563.

21. Harris ED Jr, Parker HG, Radin EL, et al: Effects of proteolytic enzymes on structural and mechanical properties of cartilage. *Arthritis Rheum* 1972;15:497–503.

22. Kempson GE: The mechanical properties of articular cartilage, in Sokoloff L (ed): *The Joints and Synovial Fluid*. New York, Academic Press, 1980, vol 2, pp 177–238.

23. Kempson GE, Muir H, Pollard C, et al: The tensile properties of the cartilage of human femoral condyles related to the content of collagen and glycosaminoglycans. *Biochim Biophys Acta* 1973;297:456–472.

24. Apone S, Wu JJ, Eyre DR: Collagen heterogeneity in articular cartilage: Identification of five genetically distinct molecular species. *Trans Orthop Res Soc* 1987;12:109.

25. Burgeson RE, Hollister DW: Collagen heterogeneity in human cartilage: Identification of several new collagen chains. *Biochem Biophys Res Commun* 1979;87:1124–1131.

26. Eyre DR: Collagen: Molecular diversity in the body's protein scaffold. *Science* 1980;207:1315–1322.

27. Mayne R, Irwin MH: Collagen types in cartilage, in Kuettner KE, Schleyerbach R, Hascall VC (eds): *Articular Cartilage Biochemistry*. New York, Raven Press, 1986, pp 23–38.

28. Comper WD, Laurent TC: Physiological function of connective tissue polysaccharides. *Physiol Rev* 1978;58:255–315.

29. Greenwald RA, Moy WW, Seibold J: Functional properties of cartilage proteoglycans. *Semin Arthritis Rheum* 1978;8:53–67.

30. Hascall VC: Interaction of cartilage proteoglycans with hyaluronic acid. *J Supramol Struct* 1977;7:101–120.
31. Buckwalter JA, Kuettner KE, Thonar EJM: Age-related changes in articular cartilage proteoglycans: Electron microscopic studies. *J Orthop Res* 1985;3:251–257.
32. Buckwalter JA, Rosenberg LC: Electron microscopic studies of cartilage proteoglycans: Direct evidence for the variable length of the chondroitin sulfate-rich region of proteoglycan subunit core protein. *J Biol Chem* 1982;257:9830–9839.
33. Rosenberg LC, Choi HU, Tang L-H, et al: Isolation of dermatan sulfate proteoglycans from mature bovine articular cartilages. *J Biol Chem* 1985;260:6304–6313.
34. Heinegård D, Paulson M: Structure and metabolism of proteoglycans, in Piez KA, Reddi AH (eds): *Extracellular Matrix Biochemistry*. New York, Elsevier, 1984, pp 277–328.
35. Hascall VC, Riolo RL: Characteristics of the protein-keratan sulfate core and of keratan sulfate prepared from bovine nasal cartilage proteoglycan. *J Biol Chem* 1972;247:4529–4538.
36. Hascall VC, Sajdera SW: Physical properties and polydispersity of proteoglycan from bovine nasal cartilage. *J Biol Chem* 1970;245:4920–4930.
37. Heinegård D: Polydispersity of cartilage proteoglycans: Structural variations with size and buoyant density of the molecules. *J Biol Chem* 1977;252:1980–1989.
38. Heinegård D, Axelsson I: Distribution of keratan sulfate in cartilage proteoglycans. *J Biol Chem* 1977;252:1971–1979.
39. Hardingham TE, Muir H: Binding of oligosaccharides of hyaluronic acid to proteoglycans. *Biochem J* 1973;135:905–908.
40. Hardingham TE, Muir H: The specific interaction of hyaluronic acid with cartilage proteoglycans. *Biochim Biophys Acta* 1972;279:401–405.
41. Hascall VC, Heinegård D: Aggregation of cartilage proteoglycans: I. The role of hyaluronic acid. *J Biol Chem* 1974;249:4232–4241.
42. Buckwalter JA, Rosenberg L: Structural changes during development in bovine fetal epiphyseal cartilage. *Coll Rel Res* 1983;3:489–504.
43. Kimura JH, Hardingham TE, Hascall VC: Assembly of newly synthesized proteoglycan and link protein into aggregates in cultures of chondrosarcoma chondrocytes. *J Biol Chem* 1980;255:7134–7143.
44. Buckwalter JA, Rosenberg LC, Tang L-H: The effect of link protein on proteoglycan aggregate structure: An electron microscopic study of the molecular architecture and dimensions of proteoglycan aggregates reassembled from the proteoglycan monomers and link proteins of bovine fetal epiphyseal cartilage. *J Biol Chem* 1984;259:5361–5363.
45. Hardingham TE, Muir H, Kwan MK, et al: Viscoelastic properties of proteoglycan solutions with varying proportions present as aggregates. *J Orthop Res* 1987;5:36–46.
46. Mow VC, Mak AF, Lai WM, et al: Viscoelastic properties of proteoglycan subunits and aggregates in varying solution concentrations. *J Biomech* 1984;17:325–338.
47. Thonar EJ, Buckwalter JA, Kuettner KE: Maturation-related differences in the structure and composition of proteoglycans synthesized by chondrocytes from bovine articular cartilage. *J Biol Chem* 1986;261:2467–2474.
48. Mort JS, Poole AR, Roughley PJ: Age-related changes in the structure of proteoglycan link proteins present in normal human articular cartilage. *Biochem J* 1983;214:269–272.
49. Plaas AHK, Sandy JD: Age-related decrease in the link-stability of proteoglycan aggregates formed by articular chondrocytes. *Biochem J* 1984;220:337–340.

50. Buckwalter JA, Roughley PJ: Age-related changes in human articular cartilage proteoglycans. *Trans Orthop Res Soc* 1987;12:152.

51. Roughley PJ, White AR, Poole AR: Identification of hyaluronic acid-binding protein that interferes with the properties of high-buoyant density proteoglycan aggregates from adult articular cartilage. *Biochem J* 1985;231:129.

52. Hewitt AT, Varner HH, Silver MH, et al: The role of chondronectin and cartilage proteoglycan in the attachment of chondrocytes to collagen, in Kelley RO, Goetinck PF, MacCabe JA (eds): *Limb Development and Regeneration: Part B*. New York, Alan R Liss, 1982, pp 25–33.

53. von der Mark K, Mollenhauer J, Pfäffle M, et al: Role of anchorin CII in the interaction of chondrocytes with extracellular collagen, in Kuettner KE, Schleyerbach R, Hascall VC (eds): *Articular Cartilage Biochemistry*. New York, Raven Press, 1986, pp 125–141.

54. Ghadially FN: Fine structure of joints, in Sokoloff L (ed): *The Joints and Synovial Fluid*. New York, Academic Press, 1978, vol 1, pp 105–176.

55. Schenk RK, Eggli PS, Hunziker EB: Articular cartilage morphology, in Kuettner KE, Schleyerbach R, Hascall VC (eds): *Articular Cartilage Biochemistry*. New York, Raven Press, 1986, pp 3–22.

56. Stockwell RA, Meachim G: The chondrocytes, in Freeman MAR (ed): *Adult Articular Cartilage*, ed 2. Tunbridge Wells, Pitman Medical, 1979, pp 69–144.

57. Franzén A, Inerot S, Hejderup S-O, et al: Variations in the composition of bovine hip articular cartilage with distance from the articular surface. *Biochem J* 1981;195:535–543.

58. Jones IL, Larsson S-E, Lemperg R: The glycosaminoglycans of human articular cartilage: Concentration and distribution in different layers in the adult individual. *Clin Orthop* 1977;127:257–264.

59. Jones IL, Lemperg R: The glycosaminoglycans of human articular cartilage: Molecular weight distribution of chondroitin sulphate in different layers in the adult individual. *Clin Orthop* 1978;134:364–370.

60. Poole AR, Pidoux I, Reiner A, et al: An immunoelectron microscope study of the organization of proteoglycan monomer, link protein, and collagen in the matrix of articular cartilage. *J Cell Biol* 1982;93:921–937.

61. Poole CA, Flint MH, Beaumont BW: Morphological and functional interrelationships of articular cartilage matrices. *J Anat* 1984;138:113–138.

62. Clark JM: The organization of collagen in cryofractured rabbit articular cartilage: A scanning electron microscopic study. *J Orthop Res* 1985;3:17–29.

63. Benninghoff A: Form und Bau der Gelenkknorpel in ihren Beziehungen zur Funktion: Zweiter Teil. Der aufbau des Gelenkknorpels in Seinen Beziehungen zur Funktion. *Z Zellforsch Mikrosk Anat* 1925;2:783–862.

Chapter 10

Articular Cartilage: Biomechanics

Van Mow, PhD
Melvin Rosenwasser, MD

Chapter Outline

Introduction

Understanding the mechanical behavior of articular cartilage requires consideration of its solid and fluid phases. This biphasic (porous solid and imcompressible fluid) description emphasizes the role interstitial water plays in determining the biomechanical properties of cartilage and helps explain how nutrients and the synthetic products of the cell pass through the matrix.[1-7] Such a description permits the determination of the intrinsic biomechanical properties of the porous solid matrix, leading in turn to an understanding of the roles played by collagen and proteoglycans.[7-12]

In diarthrodial joints, articular cartilage serves as a wear-resistant,[13] smooth,[14-16] nearly frictionless,[17-28] load-bearing surface.[6,29-36] The composition and physicochemical properties of articular cartilage,[37-39] the ultrastructural organization of the collagen network,[14,15,40-46] and the molecular organization of collagen and proteoglycans[47-56] all have profound influences on the intrinsic mechanical properties of the extracellular matrix as well as on the fluid transport and diffusional properties of the tissue.[4,5,7,12,37,39,57-62] These properties, in turn, provide this tissue with its normal friction, lubrication, wear, and load-bearing characteristics.[22,23,26-28,32,63-67]

Changes associated with osteoarthritis, injuries, and other degenerative processes alter the normal structure-function relationships existing within the tissue. For example, specific compositional, molecular, and structural changes observed in degenerated or fibrillated tissues include decreased proteoglycan and increased water contents, collagen fibril network disorganization, and proteoglycan disaggregation.[68-77] Some of these compositional and molecular structural changes have been shown to alter the swelling and intrinsic mechanical properties of the tissue.[1,6,7,12,39,63,78-85] Similar structural relationships need to be elucidated for injured and repair cartilage.[69]

Nutrient Transport Through the Matrix

Chondrocytes lack a blood supply, yet they must synthesize and secrete large volumes of matrix molecules to compensate for the turnover of proteoglycans[72,86,87] and the attrition of collagen caused by normal joint wear and tear.[13,88,89] To understand how chondrocytes can maintain their metabolic activity, it is necessary to understand the processes by which electrolytes and nutrients are transported through the extracellular matrix.[37,57,61,62,90,91] In normal adult articular cartilage, the substances required by the chondrocytes are derived from the synovial fluid. This means that the required nutrients must traverse a significant distance, efficiently and rapidly, through a porous matrix to reach the cells. The pore size of the matrix has been estimated to range from 2.5 to 6.5 nm.[4,23,37,39] The transport of these substances through cartilage can occur by diffusion, convection, or a combination of these two mechanisms.[4,6,37,61,62,92]

Convection is caused by interstitial fluid flow associated with cartilage deformation in response to joint loading. In this manner, joint motion may be important in maintaining articular cartilage homeostasis. There is ample evidence of this.[93-103] For example, it has been shown that the biochemical composition and the tensile and swelling properties of normal, fibrillated, and osteoarthritic human knee joint cartilages depend strongly on whether or not the tissue is normally subjected to load-bearing.[78,79]

Material Properties of Normal Articular Cartilage

Like many other musculoskeletal soft tissues, articular cartilage exhibits a viscoelastic response when subjected to loads and deformations, that is, it creeps under a constant applied load and stress-relaxes under a constant applied deformation.[4,5,7,9,59,104-115] However, unlike tendons and ligaments, the viscoelastic response of articular cartilage depends on two fundamentally different physical mechanisms: (1) the intrinsic viscoelastic properties of the macromolecules that form the organic solid matrix[4,9,116]; and (2) the frictional drag arising from the flow of the interstitial fluid through the porous-permeable solid matrix.[4,5,59] Each mechanism contributes to the overall viscoelastic properties of cartilage under tension, compression, and shear.[2,4,12]

Although the molecular structures of collagen and proteoglycans, their roles in organizing the solid matrix of normal articular cartilage,[53,54,117,118] and their roles, in conjunction with water, in determining the material properties of normal articular cartilage are well understood,[4,5,7,10-12,30,37-39,60,78,116,119] comparable understanding does not exist for atrophic, osteoarthritic, or healing cartilage. Our knowledge of normal cartilage can provide the basis for research necessary to elucidate similar structure-function relationships for osteoarthritic, atrophic, and healing articular cartilage.

Structure and Properties of Cartilage Collagen

In cartilage, a strong cohesive network is achieved by intramolecular and intermolecular cross-links formed in collagen monomers and collagen fibrils.[49,120-126] Although there have been many studies of the chemistry of collagen cross-links (see the excellent review by Nimni[54]), no studies have correlated the nature of collagen cross-links with the material properties of articular cartilage. The density and biochemical characteristics of these cross-links must in some way strengthen the collagen network of articular cartilage and other collagenous tissues.[49,54,105,107-111,120,122,124,125] Aside from a recent abstract by Schmidt and associates,[127] little or no information exists regarding collagen cross-link density or strength relative to the overall material properties of articular cartilage. This avenue of investigation should be pursued.

The most important mechanical properties of collagen fibers are their tensile stiffness and strength. Although a single collagen fiber has not been tested under tension, much is known abut the tensile properties of structures with a high collagen content such as tendons and

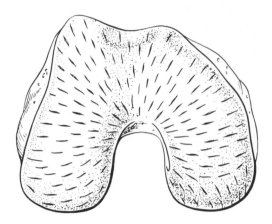

Fig. 10–1 *A split-line pattern on the surface of human femoral condyles. (Reproduced from Hultkrantz.[135])*

ligaments (see Sections I and II).[107-113,128] The tensile properties of articular cartilage have been determined in a number of studies.[78,79,83,84,112,114,115,129-131] These studies have shown that adult articular cartilage is highly nonhomogeneous, that is, there is a pronounced variation of tensile stiffness and strength with depth. (In skeletally immature animals, stiffness and strength does not decrease with depth.[131] No explanation is currently available as to why such a dramatic change occurs in these tensile properties with the closing of the physis at the onset of skeletal maturity.) Studies have shown that specimens from the superficial zone have much higher stiffness and failure stresses than those of the deeper zones. These tensile characteristics are consistent with collagen composition data and our understanding of the ultrastructural organization of cartilage.[14,15,41-46,78,112,115,129,131-135] In all reported cases, the tensile modulus and strength of normal articular cartilage are far less than those of tendons and ligaments. This is most probably because the collagen content is significantly lower in articular cartilage than in tendon or ligament and because the collagen organization and collagen ultrastructure of articular cartilage are more complex.

Articular cartilage is anisotropic, that is, its material properties are different for tissue specimens obtained from different orientations relative to the joint surface.[78,83,112,114,115,131] It is thought that this anisotropy is related to the varying collagen fibril arrangements within the planes parallel to the articular surface. In tension, this anisotropy is usually described with respect to the direction of the articular surface "split-lines" pattern.[43,132-135] These split lines are elongated fissures produced by piercing the articular surface with a small round awl (Fig. 10–1); the split-line pattern is generally believed to delineate the collagen fibril direction over the articular surface.[132-135] Although the origin of the split-line pattern is unknown, it has been repeatedly shown that the pattern is related to the directional variation of the tensile stiffness and strength of articular cartilage.[82,83,112,114,115,131] These studies showed that cartilage specimens are stiffer and stronger in the direction parallel to the split-line pattern than in the direction perpendicular to the split-line pattern. However, scanning electron microscopic examinations of collagen ultrastructure have not confirmed the existence of preferred fibril orientations in planes parallel to the articular surface.[14-16,41-43] To date, no explanation exists regarding the cause of this pronounced anisotropic tensile behavior nor do we understand the possible functional significance of this material anisotropy.

Viscoelastic Properties of Cartilage Proteoglycans

The physical and mechanical properties of musculoskeletal tissues like tendon, meniscus, and cartilage depend not only on differences in the proportions of collagen and proteoglycans but also on the organization of these molecules.[78,79,129,136]

Proteoglycan aggregates help immobilize proteoglycans within the collagen meshwork, adding structural rigidity to the extracellular matrix.[8,50,53,117,137,138] Recent studies of the flow properties of concentrated proteoglycan solutions under shear have demonstrated that proteoglycans can form elastic networks in solution.[8,136,139] These studies have also described the influence of aggregation on the strength and energy-storing capacity of molecular networks formed by the proteoglycans in solution. Solutions of aggregates store more energy elastically than solutions of monomers; this energy-storage capacity increases as the proportion of aggregates relative to the monomers increases. Networks formed by proteoglycan aggregates also have greater rupture strength than those formed by monomers.[139] Finally, all viscoelastic properties depend on the concentration of proteoglycans in solution. At the highest concentrations reported, 50 mg/ml, similar to those existing in cartilage, the magnitude of the shear modulus ranges from 1 to 10 Pa. Thus, proteoglycan aggregation and link protein promote the formation of a stronger molecular network in the interfibrillar space in cartilage. This ability to form networks in solution enhances the immobilization of proteoglycans within the fine collagen meshwork of cartilage.

At present, little information exists regarding the way in which changes in proteoglycan structure during aging and osteoarthritis affect the properties of the extracellular matrix. Further, no information exists regarding changes in proteoglycan structure during cartilage healing and repair.

The Influence of Water on Cartilage Properties

Water is the most abundant component of articular cartilage. Water concentration is highest near the articular surface (about 80%) and decreases in a nearly linear fashion with increasing depth to approximately 65% in the deep zone.[39,61,62,89] Average water content in cartilage has been reported to range from 60% to 80% depending on the condition of the tissue and the method of measurement.[1,4,39,61,62,77,140] Because of the sulfate and carboxyl charge groups fixed along the glycosaminoglycan chains along the proteoglycans, this fluid contains many free mobile cations (for example, Na^+ and Ca^{2+}) that greatly influence the mechanical behavior of cartilage.[4,6,10,37,39,57,60,141-143] The fluid component of articular cartilage is also essential to the health of this avascular tissue because it permits gas, nutrients, and waste products to move back and forth between chondrocytes and the synovial fluid.[37,39,61,62,73,96]

Only a small percentage of the water in cartilage is intracellular and about 30% may be strongly associated with the collagen fibrils; the latter figure, though, is highly debated.[39,57,61,62,144] Most of the water occupies the intermolecular space and is free to move when a load or pressure gradient is applied to the tissue.[1-7,23,37,39,145,146] Edwards[145] estimated that, when loaded, about 70% of the water within cartilage may be moved. This movement is important in controlling cartilage surface deformation, mechanical behavior, and joint lubrication.[1-7,23,37,38,57-59,145-149]

Aside from providing fundamental information on how the movement of interstitial fluid contributes to the viscoelasticity of cartilage, studies have also provided valuable insights into the "pore" structure of the extracellular matrix.[4,23,37] According to measured permeability coefficients and uniform tube Poiseuille flow models, the average equivalent "pore" diameter in cartilage ranges from 3 to 6 nm. Thus, while the average porosity (defined as total weight − dry weight/total weight) of cartilage ranges from 60% to 85%, the average equivalent "pore" is of molecular size. This means that the collagen and proteoglycan making up the solid matrix are highly dispersed in the interstitial fluid, resulting in a microporous material with extremely low permeability. The flow of the interstitial fluid and the transport of solutes through this microporous solid matrix are resisted by a high frictional drag[4-6,59,147-149] and transport of larger molecules (<20,000 daltons) is likely to be blocked by steric exclusion.[90] Recent studies have shown that certain high polymers can be transported rapidly through concentrated polymeric solutions by the formation of "ordered structures."[150,151] This phenomenon may explain the rapid transport of large collagen and proteoglycans fragments through the microporous matrix of cartilage and may be very important in the understanding of extracellular matrix genesis.[53] No studies have been reported on the mechanism of transport of macromolecules such as collagen and proteoglycans through the microporous matrix of cartilage. Studies along this line can provide important information on the kinetics of healing and the repair of injured cartilage as well.

Structural Interactions Among Cartilage Components

Normal Cartilage Composition The chemical structure of glycosaminoglycans influences both the conformation of the proteoglycan aggregates and the properties of normal articular cartilage. The closely spaced sulfate and carboxylate groups on the glycosaminoglycan disaccharide repeating units create strong intramolecular and intermolecular repulsive forces. This charge repulsion extends and stiffens proteoglycan aggregates in the interfibrillar space formed by the collagen network. Mobile cations such as Na^+ and Ca^{2+} in the solution are attracted to the fixed negatively charged groups on the glycosaminoglycans,[37,39,57,152] creating a substantial osmotic swelling pressure (Donnan osmotic effect) approaching 0.35 MPa.[39] The magnitude of this swelling pressure is related to the charge density along the glycosaminoglycan chains.[37,39,57,152] This swelling pressure is, in turn, resisted and balanced by tension developed in the collagen network,[10,39,85] confining the proteoglycans to about 20% of their free solution domain.[53,117] Consequently, this swelling pressure subjects the surrounding fine collagen meshwork to a state of "pre-stress" even in the absence of external loads.[10,85] Furthermore, cartilage proteoglycans are nonhomogeneously distributed throughout the matrix with their concentration generally being highest in the middle zone and lowest in the superficial and deep zones.[39,53,74,77,140,153] This lack of homogeneity in cartilage produces a nonhomogeneous swelling phenomenon.[60,154] The functional significance of the heterogeneous distributions and organization of collagen and proteoglycan, and the lack of homogeneity of stresses in collagen and proteoglycan swelling pressure throughout the depth of the tissue, is as yet unknown.

When an external load is applied to the cartilage surface, there is an instantaneous deformation caused primarily by a change in the proteoglycan molecular domain (Fig. 10–2). This external load can also cause the interstitial fluid pressure in the porous solid matrix to exceed the Donnan osmotic swelling pressure; consequently, the interstitial fluid begins to flow and exudation occurs. With increasing depletion of the interstitial fluid, the proteoglycan concentration within the solid matrix increases, which, in turn, increases the Donnan osmotic swelling pressure, charge-charge repulsive force, and bulk compressive stress until they are in equilibrium with the applied external load. In this manner, the physicochemical properties of the proteoglycan gel trapped within the collagen meshwork enable cartilage to resist compression. This mechanism complements the role played by collagen fibers, which are strong in tension but can easily buckle under compression.

The ability of proteoglycans to resist compression thus arises from two sources: (1) the Donnan osmotic swelling pressure associated with the tightly packed fixed anionic groups on the glycosaminoglycans[39,57]; and (2) the bulk compressive stiffness of the proteoglycan aggregates entangled in the collagen network.[4,10,60,141,142,155] It has been shown that Donnan osmotic pressure and bulk compressive stiffness of the collagen-proteoglycan solid matrix contribute, in nearly equal proportions, to the overall compressive stiffness of the tissue.[141,142,155] However, just how the composition and molecular structure of the proteoglycans and collagen of articular cartilage change during osteoarthritis, healing, and repair are not known. For example, loss of

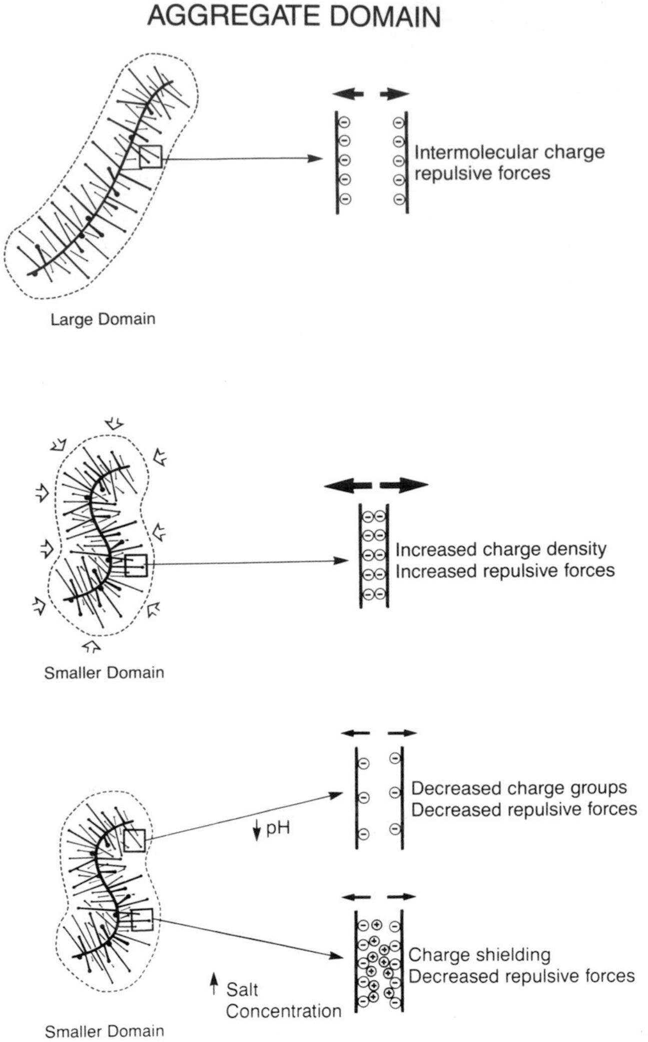

AGGREGATE DOMAIN

Fig. 10–2 *Aggregate domain.* **Top:** *Schema of proteoglycan aggregate in solution with the sulfate and carboxyl groups being ionized; repelling forces are associated with the fixed negative charge groups on the glycosaminoglycan chains (at right).* **Center:** *Compressive stresses decreasing the aggregate solution domain, which in turn increases the charge density and thus the Donnan osmotic pressure and intermolecular charge repulsive forces.* **Bottom:** *Decreasing solution pH and/or increasing the solution ionic concentration decreases the Donnan osmotic pressure and intermolecular charge repulsive forces.*

proteoglycan in disease states or changes in collagen ultrastructure during healing or repair alter cartilage biomechanics.

Collagen and proteoglycans also interact and these interactions are of great functional importance.[44,53,69,118,156–160] These collagen-proteoglycan interactions probably involve proteoglycan monomers, hyaluronate, type II collagen, and an unknown bonding agent, possibly a minor type IX collagen.[53,157,161] A small number of the proteoglycans are closely associated with collagen and may serve as a bonding agent between the collagen fibrils, spanning distances too great for collagen cross-links to develop.[53,117,118,157,160] Proteoglycans are also thought to play an important role in maintaining the ordered structure and mechanical properties of the collagen fibrils.[44,53,69,118,157,158,160,162-164] Recent studies of the chemical composition of type IX collagen suggest that it may mediate the interaction of type II collagen in articular cartilage and proteoglycan.[161]

Proteoglycan-proteoglycan interactions may also occur within cartilage. This possibility is suggested by the recent rheologic investigations of concentrated solutions of proteoglycan monomers and aggregates.[8,136,139,164] Solutions of proteoglycans can store energy elastically; this can only occur if the proteoglycan molecules are forming networks in solution. These studies have also shown that the strengths of these interactions differ in networks formed by monomers and networks formed by aggregates.

Alterations in Cartilage Composition   Akizuki and associates[78,79] determined that the tensile modulus and swelling properties of adult human knee joint cartilage depend not only on the collagen content but, surprisingly, more strongly on the collagen-proteoglycan ratio. This finding indicates that both collagen and proteoglycan, as well as the organization of the collagen and proteoglycan matrix, contribute to the tensile properties of the matrix. A number of studies have shown that there is a gradual decrease in tensile fatigue properties with age[165,166] and in tensile stiffness and tensile strength with mild fibrillation.[78,79,167] The shear stiffness of articular cartilage must also derive from its collagen content or from the collagen-proteoglycan interaction.[4,8,9,11] A change in the collagen content relative to the proteoglycan content or a change in collagen ultrastructure would, therefore, alter the shear properties of cartilage. In addition, Akizuki and associates[78] showed that there are chaotic changes in these tensile properties in osteoarthritic cartilage. This is probably a reflection of the well-documented chaotic changes in the collagen ultrastructure[14,42,168,169] and proteoglycan organization[47,138,170-172] in osteoarthritic tissues. These studies suggest that changes in the collagen ultrastructural arrangement and molecular conformation of proteoglycans that occur during fibrillation and osteoarthritis result in a loss of the tensile strength of the tissue.

The mechanical properties of the cartilage treated with proteolytic enzymes to degrade collagen and proteoglycan have been assessed.[173-175] All these studies provide support for Akizuki and associates' finding that normal collagen ultrastructure and proteoglycan molecular conformation are required to maintain the normal range of cartilage mechanical properties. Most recently, Schmidt and associates[127] showed that both tensile stiffness and tensile strength of normal bovine articular cartilage depend strongly on the density of hydroxyproline cross-links in cartilage, and that elastase treatment dramatically decreases these properties.

These results suggest that disruption of the collagen network may be a key factor leading to the development of osteoarthritis. Also, loosening of the collagen network is generally believed to be responsible for the increased swelling, and thus water content, and loss of proteoglycans in osteoarthritic cartilage.[68,73,85] These changes have profound effects on the material properties of articular cartilage.[1,78–81] Thus, the integrity of the collagen fibril and the intermolecular interaction between collagen and proteoglycan are believed to be two key factors controlling tissue hydration and material properties.

In conclusion, collagen and proteoglycan macromolecules interact to form a porous, composite, fiber-reinforced matrix possessing all the essential mechanical characteristics of a solid swollen with water and able to resist the stresses and strains of joint articulation. Studies have demonstrated a strong correlation between cartilage component interactions and the material and mechanical properties of normal articular cartilage. In repair cartilage, in which collagen content and ultrastructure are expected to be different, the viscoelastic spectrum in shear and the reduced stress-relaxation function are expected to change correspondingly.

Mechanical Behavior of Normal Articular Cartilage

During walking or running, forces on the joint surface may vary from almost zero to several times body weight over a period of one to two seconds.[176,177] The contact areas also vary in a complex manner; typically they measure only several square centimeters and they migrate considerably over the joint surface.[29,30,33–35] Thus, under physiologic loading conditions, articular cartilage is a highly stressed material. Therefore, it is important to know the response of the tissue under rapidly fluctuating and high loading conditions, especially how the solid phase and the fluid phase (the interstitial water with the inorganic salts dissolved in it) determine the time- and history-dependent viscoelastic response of the tissue.[1,4–7,59,113,114,141,142,148,178,179]

Viscoelastic Response

If a material is subjected to the action of a constant (time-independent) load or a constant deformation, and its response is varying (time-dependent), then the mechanical behavior of the material is said to be viscoelastic. In general, the response of such a material can be theoretically modeled as a combination of the responses of a viscous fluid and an elastic solid.

The two fundamental responses of a viscoelastic material are creep and stress relaxation. Creep occurs when a viscoelastic solid is subjected to the action of a constant load. Typically, a viscoelastic solid responds with a rapid initial deformation followed by a slow (time-dependent), progressively increasing deformation known as creep, until an equilibrium state is reached. Stress relaxation occurs when a viscoelastic solid is subjected to the action of a constant deformation. Typically, a viscoelastic solid responds with a high initial stress followed by a slow (time-dependent), progressively decreasing stress required to maintain the deformation.

Viscoelastic phenomena such as creep and stress relaxation may be caused by different mechanisms. For solid polymeric materials, these phenomena result from internal friction caused by the motion of the long polymeric chains within the stressed material.[180] The viscoelastic behavior of tendons and ligaments results primarily from this mechanism.[7,113] For articular cartilage, however, the compressive viscoelastic behavior is primarily caused by the flow of the interstitial fluid.[4,5,59,147,148] The component of articular cartilage viscoelasticity resulting from interstitial fluid flow is called biphasic viscoelastic behavior[4,5,7,59,147,148] and the component of viscoelasticity resulting from macromolecular motion is called flow-independent[116] or intrinsic viscoelastic behavior.[9,181,182]

Biphasic Creep Response in Compression

The biphasic creep response of articular cartilage in a one-dimensional confined compression experiment is shown in Figure 10–3. In this case, a constant compressive stress σ_o is applied onto the tissue at point A and the tissue is allowed to creep to its final equilibrium value ϵ_∞. For articular cartilage, creep is caused primarily by the exudation of interstitial fluid. Initially, exudation occurs rapidly, as evidenced by the early rapid rate of increased deformation, and diminishes gradually until flow cessation occurs. During creep, the load applied at the surface is balanced by the compressive stress developed within the collagen-proteoglycan solid matrix and by frictional drag generated by the flow of the interstitial fluid during exudation. Creep ceases when the compressive stress developed within the solid matrix balances the applied stress; at this point, fluid flow ceases and the equilibrium deformation ϵ_∞ is reached.

Typically, for relatively thick human and bovine articular cartilages, which measure up to 4 mm, it takes four to 16 hours to reach creep equilibrium. For rabbit cartilage, which is on the order of 1 mm thick, it takes approximately one hour to reach creep equilibrium. Theoretically, it can be shown that the time it takes to reach creep equilibrium varies inversely with the square of the thickness of the tissue.[59,178] Under relatively high loading conditions, >1.0 MPa, about 70% of the total fluid content may be squeezed from the tissue.[145] Furthermore, in vitro studies demonstrate that if the tissue is immersed in physiologic saline, this exuded fluid is fully recoverable when the load is removed.[183,184]

Since the rate of creep is predominantly governed by the rate of fluid exudation, it can be used to determine an "apparent" permeability coefficient of the tissue.[4,59] This provides an indirect measurement of tissue permeability, k. At equilibrium, since no fluid flow occurs, the equilibrium deformation can be used to measure the intrinsic com-

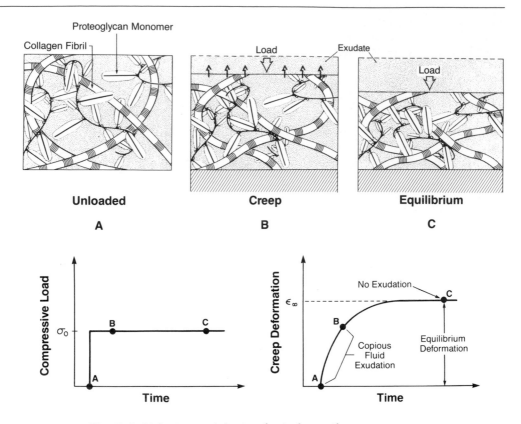

Fig. 10–3 *Biphasic creep behavior of articular cartilage.*

pressive modulus (H_A) of the porous collagen-proteoglycan solid matrix.[1,4,59,185] The intrinsic compressive modulus, H_A, is approximately 0.8 MPa for both human and bovine cartilage.[1,7] Since these coefficients are intrinsic material properties of the porous collagen-proteoglycan solid matrix, it is, therefore, meaningful to determine how they vary with matrix composition. Thus, it was determined that k varies directly whereas H_A varies inversely with water content[1,6,7,186]; H_A also varies directly with the tissue's uronic acid content, that is, with proteoglycan content.[7,185,186]

Biphasic Stress-Relaxation Response in Compression

The biphasic viscoelastic stress-relaxation response of articular cartilage in a one-dimensional compression experiment is shown in Figure 10–4. In this case, a constant compression rate is applied to the tissue until u_O is reached; beyond point B, the deformation u_O is maintained.[4,59,187,188] During the compression phase, the stress rises continuously until σ_O is reached, corresponding to u_O; during the stress-relaxation phase, the stress continuously decays along the curve B-C-D-E until the equilibrium stress σ_∞ is reached.

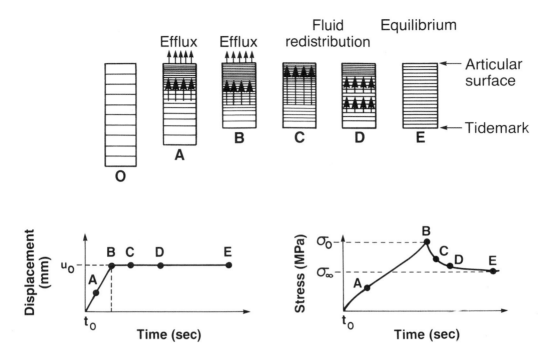

Fig. 10–4 *Stress-relaxation of cartilage under controlled displacement.*

The mechanism responsible for stress rise and stress relaxation is also shown in Figure 10–4. The stress rise in the compression phase is associated with fluid exudation, whereas stress relaxation is associated with fluid redistribution within the porous collagen-proteoglycan solid matrix. During the compressive phase, the high stress is generated by the forced exudation of the interstitial fluid and the compaction of the solid matrix near the surface. Stress relaxation is in turn caused by the relief or rebound of the high-compaction region near the surface of the solid matrix. This stress-relaxation process ceases when the compressive stress developed within the solid matrix reaches the stress generated by the intrinsic compressive modulus of the solid matrix corresponding to u_0.[4,59,187] Analysis of this stress-relaxation process leads to the conclusion that, under physiologic loading conditions, excessive stress levels are difficult to maintain in cartilage since stress relaxation rapidly attenuates the stress. This must necessarily lead to the rapid spreading of the contact area in the joint during articulation.

The compressive stress-relaxation phenomenon in cartilage may also be used to determine the intrinsic material properties of the tissue. Because the kinetics of stress relaxation are largely governed by the flow of the interstitial fluid, they may be used to determine the intrinsic permeability of cartilage. Similarly, the equilibrium stress is determined by the intrinsic compressive modulus of the solid matrix.[4,187] The values of k and H_A measured by the biphasic stress-relaxation method are consistent with those measured by the creep method.[1,59,175] It is important to determine these values for healing and repair cartilage.

Permeability

Fluid-filled porous materials may or may not be permeable depending on whether or not the "pores" of the material are connected. Cartilage behaves as a porous structure with connected "pores." Porosity of the tissue is a geometric concept defined by the ratio (β) of the pore volume within the tissue to the total volume of the tissue. For articular cartilage, porosity can only be measured as a weight ratio (wet weight $-$ dry weight/wet weight); this provides only an approximation of the true porosity of the tissue. By this measure, the porosity or water content of cartilage has been determined to range between 60% and 80%.[1,4,13,37,39,68,73]

Permeability is a measure of the ease with which fluid can flow through a porous material. Permeability is inversely proportional to the frictional drag exerted by the fluid flowing through the material. Thus, permeability is a measure of the resistive force required to cause the fluid to flow at a given speed through the porous-permeable material. This resistive force is generated by the interaction of the viscous interstitial fluid and the pore walls of the porous-permeable material. The permeability coefficient k is related to the frictional drag coefficient K by the relationship $k = \beta^2/k$.[4,7,149] Articular cartilage has a very low permeability; thus, high frictional resistive forces are generated when fluid flows through the porous solid matrix of the tissue.

An experimental method to measure permeability directly is shown in Figure 10–5. Here, a specimen of the tissue is held fixed in a chamber subjected to the action of a pressure gradient, the imposed upstream pressure P_1 being greater than the downstream ambient pressure P_2. The thickness of the specimen is denoted by h and the cross-sectional area of permeation is defined by A. Darcy's law, used to determine k in this simple experimental setup, yields $k = Qh/A(P_1 - P_2)$ where Q is the volumetric discharge per unit time through the specimen whose area of permeation is A. Using a pressure gradient of approximately 0.1 MPa, the permeability of articular cartilage was first determined to range from 0.11×10^{-15} m^4/(N.s) to 0.76×10^{-15} m^4/N.s by McCutchen,[23] Edwards,[145] and Maroudas.[58] The permeability of cartilage under high pressures, up to 3 MPa, and compressive strains was later obtained by Mansour and Mow[146] and Mow and associates.[4,6,59]

The latter experiments provide valuable insights into cartilage function under conditions that resemble those found in physiologic situations. In diarthrodial joints, pressures may range up to several MPa[29,30,33–35] and compressive strains may range up to 20%.[31] At these high pressures and compressive strains, the permeability of cartilage decreases exponentially with both compression and applied fluid pressure[4,146] (Fig. 10–5). In terms of pore structure, compression de-

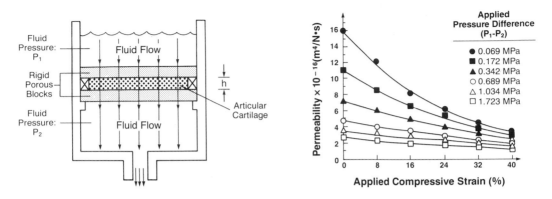

Fig. 10–5 *Left: Experimental configuration used in measuring the permeability of articular cartilage, involving the application of a pressure gradient $(P_1 - P_2)/h$ across a sample of the tissue (h = tissue thickness). **Right:** Experimental curves for articular cartilage permeability show strong dependence on compressive strain and applied pressure.*

creases the average "pore diameter" within the solid matrix, increasing frictional resistance or decreasing permeability. The decrease in tissue permeability with applied pressure has been shown to derive from compaction of the solid matrix resulting from the frictional drag caused by the permeating fluid.[149] This nonlinear permeability suggests that the tissue has a mechanical feedback system that limits the rate of fluid exudation at high pressures and loads. Whether this mechanism exists in repair or healing cartilage is unknown.

Behavior Under Tension

Articular cartilage also exhibits viscoelastic behavior under tension.[7,112,114,130,189] This viscoelastic behavior is attributable to both the internal friction associated with the molecular motion of collagen and proteoglycan[189,] and the flow of the interstitial fluid.[7,119] To examine the intrinsic mechanical response of the collagen-proteoglycan solid matrix under tension, it is necessary to negate the fluid flow effects. To do this, one must perform reasonably slow, strain rate experiments[7,115,131] or an incremental strain experiment in which stress relaxation is allowed to progress toward equilibration at each increment of strain.[60,78,79] Typically, in a low strain rate tensile experiment, a rate on the order of 1%/sec is used and the specimens are usually pulled to failure. Unfortunately, in such procedures, the time required to negate the effect of interstitial fluid flow also negates any manifestation of the intrinsic viscoelastic behavior of the solid matrix. Thus, only equilibrium intrinsic mechanical properties of the solid matrix can be determined from these tensile tests.

A typical stress-strain curve for a specimen of articular cartilage obtained during a low strain rate test is shown in Figure 10–6. Failure

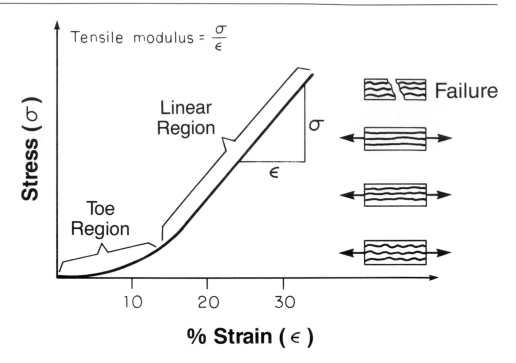

Fig. 10-6 *Typical tensile stress-strain curve for soft connective tissues and articular cartilage. The drawings at the right of the curve show the configuration of the collagen fibrils at various stages of loading and stretch.*

usually occurs at strains greater than 30%; in some cases, the strain may be as large as 120%.[83,112,114,131] The strain at failure has been observed to increase with increasing depth from the surface.[83,112,131] From these tensile studies, it is seen that the tissue has a nonlinear tensile stress-strain behavior, that is, it tends to stiffen with increasing strain when the strain becomes large. There is a nonlinear toe region where the collagen fibers are reorganized to the direction of load followed by a linear region where, it is believed, the collagen fibers are stretched[7,107,108,112,129,131] (Fig. 10-6). Figure 10-7 shows collagen alignment as a function of tensile strain.[190] Thus, over the entire range of strain in tension, articular cartilage cannot be described by a single Young's modulus. Rather, a tangent modulus, defined by the tangent to the stress-strain curve, at a prescribed strain (or stress) level, can be used to describe the tensile stiffness of the tissue. Faced with this, many investigators have utilized an exponential relationship to describe the tensile stress-strain behavior of biologic tissues.[113,191-194] For articular cartilage, this approach was adopted by Woo and associates[7,112,114] and Roth and Mow.[131] From these studies, the tangent modulus of cartilage strips was determined to range from 5 to 50 MPa depending on the levels of stress (or strain).

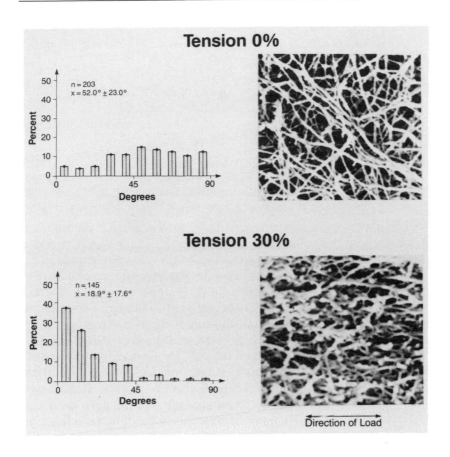

Fig. 10-7 *Collagen fibril alignment is clearly demonstrated by the scanning electron micrographs (× 10,000) of cartilage strips under 0% and 30% stretch. The histograms at left, calculated from the micrographs, represent the percent of collagen fibers oriented in the direction of the applied tension.*

The intrinsic tensile stress-strain behavior of the fiber-reinforced solid matrix remains to be determined. To do this, it is assumed that at low strains the collagen-proteoglycan matrix behaves, even under tension, like a cohesive fiber-reinforced solid material. In physiologic conditions, the compressive strain, and hence the tensile strain, in normal hip joint cartilage is less than 15%.[31] Utilizing these data, Myers and associates[60,154] and Akizuki and associates[78,79] determined the equilibrium tangent modulus of the collagen-proteoglycan solid matrix. Indeed, these studies have shown that the equilibrium tensile stress-strain behavior is linear for strains less than 15%. This linearity provides strong indirect evidence that, at low strain levels, the collagen and proteoglycan of articular cartilage form a cohesive solid material. Thus, even though cartilage is viscoelastic, the equilibrium stress-strain behavior of the tissue has been shown in a repeatable manner to be elastic. In this condition, the collagen-proteoglycan solid matrix possesses a pseudo strain energy potential.[195,196]

In normal adult articular cartilage, the superficial zone is much stiffer and stronger than the middle and deep zones.[7,78,82,83,112,114,129,131] The tensile behaviors of these zones are modulated by the swelling behavior of the tissue and thus depend on the nature of the electrolyte within the interstitial space.[10,60,79,143,197] The specific anisotropic and nonhomogeneous tensile and swelling properties of healing and repair cartilage and their dependence on the fixed-charge density and interstitial ionic conditions are not known.

Behavior Under Pure Shear

In tension and compression, only the equilibrium intrinsic properties of the collagen-proteoglycan solid matrix can be determined. This is because the volumetric change, which always occurs within a material when it is subjected to tension or compression, causes interstitial fluid flow and thus a biphasic viscoelastic response. If, however, articular cartilage is tested under pure shear under infinitesimal strain conditions, no pressure gradients or volumetric changes occur within the material; thus, no interstitial fluid flow effects are induced.[9,116,181] These experiments can be used to determine the flow-independent or intrinsic viscoelastic properties of the collagen-proteoglycan solid matrix. Pure shear can be achieved by subjecting thin circular cylindrical specimens to axial torsion.[9,181,182] Simple shear tests, such as those performed by Hayes and Bodine,[116] can approximate the pure shear test for thin specimens. The dynamic complex shear modulus (as a function of frequency) and the reduced stress-relaxation function in shear have been determined by these investigators.[9,12,116,181,198]

The dynamic complex modulus is defined by the storage modulus G_1 (strain energy stored per unit volume) and the loss modulus G_2 (energy loss by viscous dissipation per unit volume). The viscoelastic spectrum is defined by G_1 and G_2 determined over a range of frequencies. It is often convenient to use the magnitude of the dynamic shear modulus, $|G^*|$ given by:

$$|G^*| = [(G_1)^2 + (G_2)^2]^{1/2}$$

(1)

and the phase shift angle δ given by:

$$\delta = \tan^{-1}(G_2/G_1).$$

(2)

The magnitude of the dynamic shear modulus is a measure of the total resistance offered by the viscoelastic material in shear. The phase shift angle δ, the angle between the applied sinusoidal strain and the sinusoidal torque response, measures frictional energy dissipated relative to the energy stored within the material. For a pure elastic material

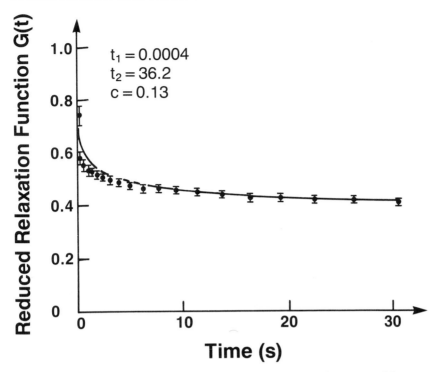

$t_1 = 0.0004$
$t_2 = 36.2$
$c = 0.13$

Fig. 10–8 *Typical reduced shear stress-relaxation curve expressed in terms of the mean of ten cycles of stress relaxation. The solid line represents the theoretical prediction of the quasilinear viscoelastic theory of Fung.*[195,196]

with no internal frictional dissipation, the phase shift angle δ is zero; for a pure viscous fluid, the phase shift angle δ is 90 degrees.

The measured magnitude of the dynamic shear modulus for normal bovine articular cartilage ranges from 1 MPa to 3 MPa, and the measured phase shift angle ranges from 9 to 20 degrees.[9,116,182] The transient shear stress-relaxation function of the collagen-proteoglycan solid matrix has also been determined.[182] It has been shown that both the steady and the transient results can be described by the quasilinear viscoelastic theory.[182,195,196] Figure 10–8 compares a theoretical prediction of the quasilinear viscoelastic theory with the transient stress-relaxation data of articular cartilage under shear. Indeed, the theory can provide a good description of the flow-independent viscoelastic data.

Lubrication of Articular Cartilage and Diarthrodial Joints

From an engineering perspective, there are two types of lubrication. One is boundary lubrication involving a monolayer of lubricant molecules adsorbed on each bearing surface. The other is fluid-film lu-

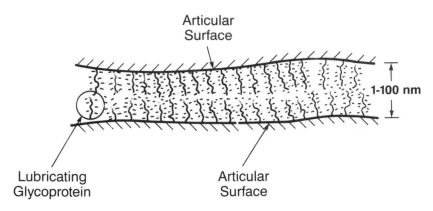

Articular
Surface

1-100 nm

Lubricating
Glycoprotein

Articular
Surface

Fig. 10–9 *Boundary lubrication of the articular cartilage. The load is carried by a monolayer of the lubricating glycoprotein adsorbed onto the articular surfaces.*

brication in which a thin fluid film provides greater surface separation.[63,64,67,199] Both types of lubrication appear to occur in diarthrodial joints.

Boundary Lubrication

During diarthrodial joint function, relative motion of the articulating surfaces occurs. The surface of cartilage is protected by the adsorbed layer of boundary lubricant, preventing direct surface-to-surface contact, thus eliminating most of the surface wear. Boundary lubrication is essentially independent of the physical properties of either the lubricant (that is, its viscosity) or the bearing material (that is, its stiffness), but instead depends almost entirely on the chemical properties of the lubricant.[17,24,64,65] In synovial joints, a specific glycoprotein, "lubricin," appears to be the synovial fluid constituent responsible for boundary lubrication.[200,201] Lubricin is adsorbed to each articulating surface (Fig. 10–9). These two apposing layers, ranging in combined thickness from 1 to 100 nm, are able to carry loads and appear to be effective in reducing friction.

Fluid-Film Lubrication

In contrast to boundary lubrication, fluid-film lubrication occurs when there is a thin film of lubricant between the sliding bearing surfaces. This generally results in greater separation between the two surfaces and no surface-to-surface contact. The load on the bearing is thus supported by a hydrodynamic pressure in this fluid film. Usually, fluid-film thickness associated with engineering bearings is less than 25 μ. Two classic modes of fluid-film lubrication defined in engineering are hydrodynamic and squeeze film lubrication (Fig. 10–10). These two modes apply to bearings composed of relatively undeformable material

Rigid Bearings

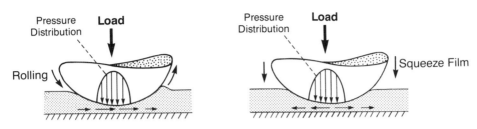

Thin Fluid Film and High Pressures

Deformable Bearings

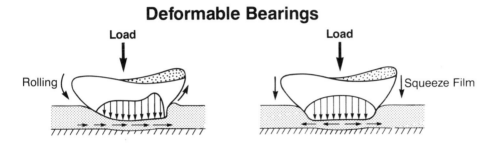

Thick Fluid Film and Low Pressures

*Fig. 10–10 Comparison of rolling (hydrodynamic) **(top left)** and squeeze film lubrication **(top right)** of rigid surfaces, and elastohydrodynamic lubrication of deformable bearing surfaces under a rolling (hydrodynamic) **(bottom left)** and a squeeze film action **(bottom right)**. Surface deformation of elastohydrodynamically lubricated bearings increases the contact area, thus increasing the load-carrying capacity of these bearings.*

such as stainless steel. Hydrodynamic lubrication occurs when non-parallel rigid bearing surfaces lubricated by a fluid film move tangentially on each other, forming a converging wedge of fluid. A lifting pressure is generated by fluid viscosity as the bearing motion drags the fluid into the gap between the surfaces (Fig. 10–10, *top left*). Squeeze film lubrication occurs when the rigid bearing surfaces move perpendicularly toward each other (Fig. 10–10, *top right*). In the gap between the two surfaces, the fluid viscosity generates the pressure required to force the fluid lubricant out. The squeeze film mechanism is sufficient to carry high loads for short durations. Eventually, however, the fluid film becomes so thin that contacts between the asperities (peaks) on the two bearing surfaces occurs.

A variation of these modes of fluid-film lubrication occurs when the bearing material is not rigid but relatively soft, such as in the articular

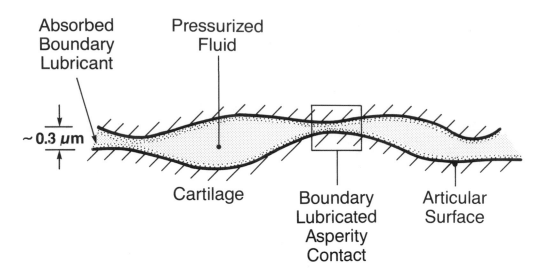

Absorbed Boundary Lubricant

Pressurized Fluid

~ 0.3 μm

Cartilage

Boundary Lubricated Asperity Contact

Articular Surface

Fig. 10–11 *Mixed-mode lubrication operating in the joint; boundary lubrication occurs at the asperities of the surfaces and fluid-film lubrication takes place in areas with more widely separated surfaces.*

cartilage covering the joint surface. This type of lubrication, termed elastohydrodynamic, operates when the relatively soft bearing surfaces undergo either a sliding (hydrodynamic) or squeeze film action, in which the pressure generated in the fluid film substantially deforms the surfaces (Fig. 10–10, *bottom*). The deformations of the soft material tend to increase the surface area, thus beneficially altering film geometry. By increasing the bearing contact area, the lubricant is less able to escape from between the bearing surfaces, a longer-lasting lubricant film is generated, and the stress of articulation is lower and more sustainable.[63,64,67,202,203] Elastohydrodynamic lubrication enables bearings to increase their load-carrying capacity greatly.[64,65]

Loading Requirements and Lubrication

In any bearing, the effective mode of lubrication depends on the applied loads and on the velocity of the bearing surfaces. Adsorption of the synovial fluid glycoprotein, lubricin, to articular surfaces seems to be most important under severe loading conditions, that is, contact surfaces with high loads, low relative speeds, and long duration. Under these conditions, as the surfaces are pressed together, the boundary lubricant monolayers interact, preventing direct contact between the articular surfaces. Conversely, fluid-film lubrication operates under less severe conditions, when loads are low and/or oscillate in magnitude, and when the contact surfaces are moving at high relative speeds.

Boundary-lubricated surfaces typically have a coefficient of friction one or two orders of magnitude higher than surfaces lubricated by a

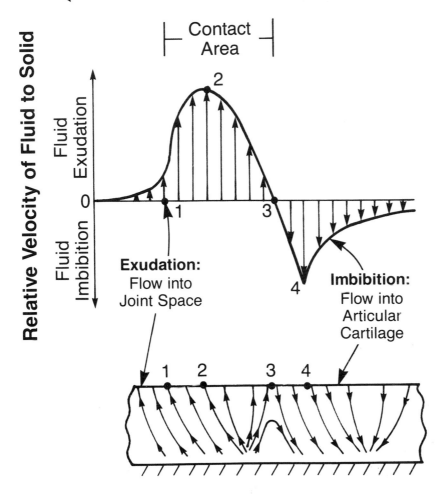

Fig. 10–12 *Calculated fluid exudation and imbibition as articular cartilage is subjected to the action of a sliding load. The direction of fluid flow is indicated graphically (at top) and within the cartilage (at bottom).*

fluid film. Since intact synovial joints have an extremely low coefficient of friction (0.001 to 0.05),[19,20,22–28,64] synovial joints must be lubricated, at least in part, by the fluid-film mechanism. In the joint, situations may also occur in which the fluid-film thickness is of the same order of magnitude as the mean articular surface asperity. Under such conditions, boundary lubrication between the asperities may occur while fluid film lubrication occurs at other regions over the joint surface. This is called a mixed mode of lubrication (Fig. 10–11). In mixed lubrication, it is probable that most of the friction is generated in

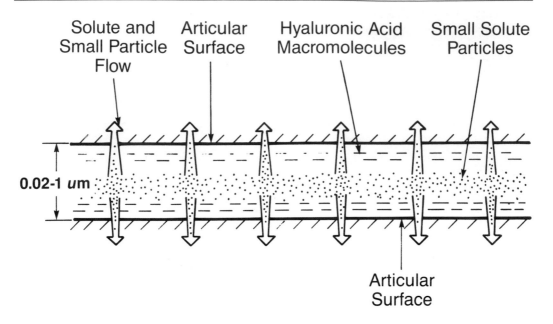

Fig. 10–13 *Ultrafiltration of the synovial fluid at the articular surfaces. The small solute molecules and solvent escape into the articular cartilage and into the joint space laterally, leaving the large hyaluronic acid macromolecules in the joint space.*

boundary-lubricated areas, whereas most of the load is carried by the fluid film.

Synovial joint surfaces are different from the typical engineering bearings described above in that they are made of a porous-permeable material that can exude and imbibe fluid. As a joint moves, the loaded articular surfaces slide over each other, and cartilage exudes fluid in front of and beneath the leading half of the moving load[6,67,204,205] (Fig. 10–12). Once the area of peak stress has passed a given point, the cartilage then starts to resorb the fluid as it returns to its original dimensions.

Another mechanism by which the articular cartilage surfaces may be protected during joint articulation has been proposed by Walker and associates[26–28] and Maroudas.[66] This mechanism, termed "boosted lubrication" by Walker and associates, depends on the ability of the solvent component of the synovial fluid to pass into the articular cartilage during squeeze film action, leaving a concentrated pool of hyaluronic acid-protein complex to lubricate the surfaces. In other words, an ultrafiltration process occurs at the articular surface as a result of squeeze film lubrication. According to this theory, this mechanism develops because it becomes progressively more difficult, as the two articular surfaces approach each other, for the hyaluronic acid macromolecules in the synovial fluid to escape from the gap between the surfaces as the gap narrows (Fig. 10–13). The water and small solute molecules can still escape the high-pressure region and pass into the articular cartilage and/or flow laterally into the joint space at the periphery of the joint.

An understanding of joint lubrication requires fundamental knowledge of the material properties of cartilage and synovial fluid and of the loads acting on the joint. Under conditions of healing and repair, it is probable that the loads and motion acting on the joint, and the material properties of the cartilage and synovial fluid, change. To determine the stresses and strains to which these tissues are subjected and the likelihood of their long-term viability, it is first necessary to determine the material properties of healing and repair cartilage.

Wear of Articular Cartilage

Wear is the removal of material from solid surfaces by either chemical or mechanical action. For mechanical action, there are two components of wear: (1) interfacial wear caused by the interaction of bearing surfaces and (2) fatigue wear caused by bearing-surface deformation under load. For articular cartilage, chemical wear may be considered the loss resulting from biochemical degradation.

Interfacial Wear

Interfacial wear occurs when bearing surfaces come into direct contact with no fluid lubricant and adsorbed molecules separating them. This type of wear can take place in two ways: adhesion or abrasion. Adhesive wear occurs when, as the bearing surfaces come into contact at focal points, they adhere to each other and are torn off from the surface during sliding. Abrasive wear, conversely, occurs when a soft material is scraped by a harder one; the harder material can be either an opposing bearing or loose particles between the bearing surfaces. The low rates of interfacial wear of articular cartilage determined under in vitro condition[13] suggest that direct surface-to-surface contact between the asperities of the two cartilage surfaces rarely occurs, although this mechanism is not ruled out under in vivo conditions. The different modes of effective lubrication working in concert make interfacial wear of articular cartilage seem unlikely. Nevertheless, adhesive and abrasive wear may take place in an impaired or degenerated synovial joint. Once the cartilage surface sustains ultrastructural defects and/or decreases in mass, it becomes softer and more permeable.[1,6,78,79] Thus, fluid from the lubricant film separating the bearing surfaces may leak away more easily through a more permeable cartilage surface. This loss of lubricating fluid from between the surfaces increases the probability of direct contact between the asperities and undoubtedly exacerbates the abrasion process.[63–65,67]

Fatigue Wear

Fatigue wear of bearing surfaces does not result from surface-to-surface contact, but rather from the accumulation of microscopic damage within the bearing material under repetitive stressing. Bearing surface failure may occur with the repeated application of high loads over a relatively short period, or with the repetition of low loads over an extended period (note that the magnitude of the low loads may be much lower than the ultimate strength of the material). This fatigue wear, caused by cyclically repeated deformation of the bearing materials, can take place even in well-lubricated bearings.

In synovial joints, the cyclical variation in total joint load during most physiologic activities causes repetitive articular cartilage stressing or deformation. In addition, during rotation and sliding, a specific region of the articular surface "moves in and out" of the loaded contact area, repetitively stressing that articular region. Loads imposed on articular cartilage are supported by the collagen-proteoglycan matrix and by the resistance generated by fluid movement throughout the matrix. Thus, repetitive joint movement and loading cause repetitive stressing of the solid matrix and repeated exudation and imbibition of the tissue's interstitial fluid.[6,36,204,205] These processes give rise to two possible mechanisms by which fatigue damage may accumulate in articular cartilage: (1) disruption of the collagen-proteoglycan solid matrix by accumulation damages and (2) proteoglycan "washout" caused by repetitive and massive exudation and imbibition of the interstitial fluid.

Impact Loading

A third mechanism of damage and resultant articular wear is associated with joint impact loading. With normal compressive physiologic loading, articular cartilage undergoes surface compaction with the lubricating fluid being exuded through this compacted region near the surface (Fig. 10–13). As described above, however, fluid redistribution within the articular cartilage occurs with time to relieve the stress in this compacted region. This process of stress relaxation takes place quite quickly; the stress may decrease by 63% within two to five seconds.[4,59,147,148] If, however, loads are supplied so quickly that there is insufficient time for internal fluid redistribution to relieve the compacted region, the high stresses produced in the collagen-proteoglycan matrix may produce damage. This phenomenon could well explain why Radin and associates[206-209] found dramatic articular cartilage damage with repeated impact loads.

Once the collagen-proteoglycan solid matrix of cartilage is disrupted, the rate of damage accumulation resulting from any of the wear mechanisms mentioned above is likely to increase. With a continuing high rate of attrition from biomechanical effects, such as increased velocity of interstitial fluid flow and abnormally high stresses, and from biochemical degradative effects, such as increased proteolytic effects, further disruption of the collagen-proteoglycan solid matrix may begin to occur at a rate that exceeds the rate at which chondrocytes can replenish the extracellular matrix.[87,210,211] Thus, failure of cartilage occurs as an imbalance between biomechanical and biochemical wear rates and the ability of the chondrocytes to maintain the extracellular matrix.

Future Directions

Many avenues are open for exploration of the biomechanics of normal articular cartilage and repair cartilage. They range from studies of the relationships between the mechanical properties of cartilage and the molecular organization of the matrix macromolecules to studies of the functional behavior of the repair tissues in situ in diarthrodial joints. At each level of structural organization (molecular, ultrastructural, light microscopic, and macroscopic) the structure-function relationships must be defined. These relationships must be further developed for normal articular cartilage and must be extended to repair cartilage. Important topics for studies of repair cartilage include the following:

(1) The relationship between the mechanical properties and the composition and organization of repair cartilage, including the degree of variability within the repair tissue, the morphologic features of the tissue, the structural characteristics of the molecular framework of the tissue, including collagen cross-linking, the molecular architecture of the proteoglycans of repair cartilage, and the organization of these macromolecules.

(2) The Donnan osmotic pressure of repair cartilage and its variations within the tissue and within "normal" cartilage over adjacent regions on the joint surface.

(3) The strength of the bonding between the repair tissue and the adjacent "normal" cartilage.

(4) The influence of joint motion and loading on the biomechanical properties of repair cartilage.

(5) The friction, lubrication, and wear characteristics of repair cartilage; especially important are the friction and wear characteristics in situ in a diarthrodial joint.

(6) Stresses and strains in repair cartilage and in its adjacent surrounding "normal" tissue in situ within a diarthrodial joint.

Acknowledgments

This work was sponsored, in part, by grants from the National Science Foundation and the National Institutes of Health. The authors wish to thank Ms. Lori Setton for her assistance in preparing and editing this manuscript.

References

1. Armstrong CG, Mow VC: Variations in the intrinsic mechanical properties of human articular cartilage with age, degeneration, and water content. *J Bone Joint Surg* 1982;64A:88–94.
2. Mak AF: The apparent viscoelastic behavior of articular cartilage: The contributions from the intrinsic matrix viscoelasticity and interstitial fluid flows. *J Biomech Eng* 1986;108:123–130.

3. Mak AF, Lai WM, Mow VC: Biphasic indentation of articular cartilage: Part I. Theoretical analysis. *J Biomech* 1987;20:703–714.

4. Mow VC, Holmes MH, Lai WM: Fluid transport and mechanical properties of articular cartilage: A review. *J Biomech* 1984;17:377–394.

5. Mow VC, Kwan MK, Lai WM, et al: A finite deformation theory for nonlinearly permeable soft hydrated biological tissues, in Schmid-Schönbein GW, Woo SL-Y, Zweifach BW (eds): *Frontiers in Biomechanics*. New York, Springer-Verlag, 1986, pp 153–179.

6. Mow VC, Lai WM: Recent developments in synovial joint biomechanics. *SIAM Rev* 1980;22:275–317.

7. Woo SL-Y, Mow VC, Lai WM: Biomechanical properties of articular cartilage, in Skalak R, Chien S (eds): *Handbook of Bioengineering*. New York, McGraw-Hill, 1987, pp 4.1–4.44.

8. Hardingham TE, Muir H, Kwan MK, et al: Viscoelastic properties of proteoglycan solutions with varying proportions present as aggregates. *J Orthop Res* 1987;5:36–46.

9. Mow VC, Lai WM, Holmes MH: Advanced theoretical and experimental techniques in cartilage research, in Huiskes R, van Campen DH, de Wijn JR (eds): *Biomechanics: Principles and Applications*. The Hague, Martinus Nijhoff, 1982, pp 47–74.

10. Myers ER, Armstrong CG, Mow VC: Swelling pressure and collagen tension, in Hukins DWL (ed): *Connective Tissue Matrix*. London, MacMillan Press, 1984, pp 161–186.

11. Myers ER, Mow VC: Biomechanics of cartilage and its response to biomechanical stimuli, in Hall BK (ed): *Cartilage: Vol 1. Structure, Function, and Biochemistry*. New York, Academic Press, 1983, pp 313–341.

12. Myers ER, Zhu W, Mow VC: Viscoelastic properties of articular cartilage and meniscus, in Nimni ME (ed): *Collagen: Chemistry, Biology and Biotechnology*. Boca Raton, CRC Press, in press.

13. Lipshitz H, Glimcher MJ: *In vitro* studies of the wear of articular cartilage: Part II. *Wear* 1979;52:297.

14. Clarke IC: Articular cartilage: A review and scanning electron microscope study: I. The interterritorial fibrillar architecture. *J Bone Joint Surg* 1971;53B:732–750.

15. Clarke IC: Surface characteristics of human articular cartilage—A scanning electron microscope study. *J Anat* 1971;108:23–30.

16. Gardner DL, McGillivray DC: Living articular cartilage is not smooth: The structure of mammalian and avian joint surfaces demonstrated *in vivo* by immersion incident light microscopy. *Ann Rheum Dis* 1971;30:3–9.

17. Charnley J: The lubrication of animal joints in relation to surgical reconstruction by arthroplasty. *Ann Rheum Dis* 1960;19:10–19.

18. Jones ES: Joint lubrication. *Lancet* 1936;1:1043–1044.

19. Linn FC: Lubrication of animal joints: I. The arthrotripsometer. *J Bone Joint Surg* 1967;49A:1079–1098.

20. Linn FC: Lubrication of animal joints: II. The mechanism. *J Biomech* 1968;1:193–205.

21. MacConaill MA: The function of intra-articular fibrocartilages, with special references to the knee and inferior radio-ulnar joints. *J Anat* 1932;66:210–227.

22. Malcom LL: *An Experimental Investigation of the Frictional and Deformational Responses of Articular Cartilage Interfaces to Static and Dynamic Loading*, thesis. University of California, San Diego, 1976.

23. McCutchen CW: The frictional properties of animal joints. *Wear* 1962;5:1.

24. McCutchen CW: Boundary lubrication by synovial fluid: Demonstration and possible osmotic explanation. *Fed Proc* 1966;25:1061–1068.

25. Unsworth A, Dowson D, Wright V: Some new evidence on human joint lubrication. *Ann Rheum Dis* 1975;34:277–285.

26. Walker PS, Dowson D, Longfield MD, et al: "Boosted lubrication" in synovial joints by fluid entrapment and enrichment. *Ann Rheum Dis* 1968;27:512–520.

27. Walker PS, Sikorski J, Dowson D, et al: Behaviour of synovial fluid on surfaces of articular cartilage: A scanning electron microscope study. *Ann Rheum Dis* 1969;28:1–14.

28. Walker PS, Unsworth A, Dowson D, et al: Mode of aggregation of hyaluronic acid protein complex on the surface of articular cartilage. *Ann Rheum Dis* 1970;29:591–602.

29. Ahmed AM, Burke DL: In-vitro measurement of static pressure distribution in synovial joints: Part I. Tibial surface of the knee. *J Biomech Eng* 1983;105:216–225.

30. Ahmed AM, Burke DL, Yu A: In-vitro measurement of static pressure distribution in synovial joints: Part II. Retropatellar surface. *J Biomech Eng* 1983;105:226–236.

31. Armstrong CG, Bahrani AS, Gardner DL: *In vitro* measurement of articular cartilage deformations in the intact human hip joint under load. *J Bone Joint Surg* 1979;61A:744–755.

32. Askew MJ, Mow VC: The biomechanical function of the collagen ultrastructure of articular cartilage. *J Biomech Eng* 1978;100:105–115.

33. Brown TD, Shaw DT: *In vitro* contact stress distributions in the natural human hip. *J Biomech* 1983;16:373–384.

34. Brown TD, Shaw DT: *In vitro* contact stress distribution on the femoral condyles. *J Orthop Res* 1984;2:190–199.

35. Hodge WA, Fijan RS, Carlson KL, et al: Contact pressures in the human hip joint measured *in vivo*. *Proc Natl Acad Sci USA* 1986;83:2879–2883.

36. Kwan MK, Lai WM, Mow VC: Fundamentals of fluid transport through cartilage in compression. *Ann Biomed Eng* 1984;12:537–558.

37. Maroudas A: Biophysical chemistry of cartilaginous tissues with special reference to solute and fluid transport. *Biorheology* 1975;12:233–248.

38. Maroudas A: Fluid transport in cartilage. *Ann Rheum Dis* 1975;34(suppl 2):77–81.

39. Maroudas A: Physicochemical properties of articular cartilage, in Freeman MAR (ed): *Adult Articular Cartilage*, ed 2. Tunbridge Wells, Pitman Medical, 1979, pp 215–290.

40. Ghadially FN: Fine structure of joints, in Sokoloff L (ed): *The Joints and Synovial Fluid*. New York, Academic Press, 1978, vol 1, pp 105–176.

41. Lane JM, Weiss C: Review of articular cartilage collagen research. *Arthritis Rheum* 1975;18:553–562.

42. Minns RJ, Steven FS: The collagen fibril organization in human articular cartilage. *J Anat* 1977;123:437–457.

43. Mow VC, Lai WM, Redler I: Some surface characteristics of articular cartilage: I. A scanning electron microscopy study and a theoretical model for the dynamic interaction of synovial fluid and articular cartilage. *J Biomech* 1974;7:449–456.

44. Poole CA, Flint MH, Beaumont BW: Morphological and functional interrelationships of articular cartilage matrices. *J Anat* 1984;138:113–138.

45. Redler I, Mow VC, Zimny ML, et al: The ultrastructure and biomechanical significance of the tidemark of articular cartilage. *Clin Orthop* 1975;112:357–362.

46. Weiss C, Rosenberg L, Helfet AJ: An ultrastructural study of normal young adult human articular cartilage. *J Bone Joint Surg* 1968;50A:663–674.

47. Buckwalter JA, Kuettner KE, Thonar EJM: Age-related changes in articular cartilage proteoglycans: Electron microscopic studies. *J Orthop Res* 1985;3:251–257.

48. Buckwalter JA, Rosenberg LC: Electron microscopic studies of cartilage proteoglycans: Direct evidence for the variable length of the chondroitin sulfate-rich region of proteoglycan subunit core protein. *J Biol Chem* 1982;257:9830–9839.

49. Eyre DR: Collagen: Molecular diversity in the body's protein scaffold. *Science* 1980;207:1315–1322.

50. Hardingham TE: The role of link-protein in the structure of cartilage proteoglycan aggregates. *Biochem J* 1979;177:237–247.

51. Hardingham TE, Ewins RJF, Muir H: Cartilage proteoglycans: Structure and heterogeneity of the protein core and the effects of specific protein modifications on the binding to hyaluronate. *Biochem J* 1976;157:127–143.

52. Muir IHM: Biochemistry, in Freeman MAR (ed): *Adult Articular Cartilage*, ed 2. Tunbridge Wells, Pitman Medical, 1979, pp 145–214.

53. Muir H: Proteoglycans as organizers of the intercellular matrix. *Biochem Soc Trans* 1983;11:613–622.

54. Nimni ME: Collagen: Structure, function, and metabolism in normal and fibrotic tissues. *Semin Arthritis Rheum* 1983;13:1–86.

55. Rosenberg L: Structure of cartilage proteoglycans, in Burleigh PMC, Poole AR (eds): *Dynamics of Connective Tissue Macromolecules*. New York, American Elsevier, 1975, pp 105–128.

56. Rosenberg L, Hellman W, Kleinschmidt AK: Electron microscopic studies of proteoglycan aggregates from bovine articular cartilage. *J Biol Chem* 1975;250:1877–1883.

57. Maroudas A: Physicochemical properties of cartilage in the light of ion exchange theory. *Biophys J* 1968;8:575–595.

58. Maroudas A: Distribution and diffusion of solutes in articular cartilage. *Biophys J* 1970;10:365–379.

59. Mow VC, Kuei SC, Lai WM, et al: Biphasic creep and stress relaxation of articular cartilage in compression: Theory and experiments. *J Biomech Eng* 1980;102:73–84.

60. Myers ER, Lai WM, Mow VC: A continuum theory and an experiment for the ion-induced swelling behavior of articular cartilage. *J Biomech Eng* 1984;106:151–158.

61. Torzilli PA, Rose DE, Dethmers DA: Equilibrium water partition in articular cartilage. *Biorheology* 1982;19:519–537.

62. Torzilli PA: Influence of cartilage conformation on its equilibrium water partition. *J Orthop Res* 1985;3:473–483.

63. Armstrong CG, Mow VC: Friction, lubrication and wear of synovial joints, in Owen R, Goodfellow J, Bullough P (eds): *Scientific Foundations of Orthopaedics and Traumatology*. London, William Heinemann, 1980, pp 223–232.

64. Dowson D: Modes of lubrication in human joints. *Proc Inst Mech Eng* 1967;1813J:45.

65. Dowson D, Unsworth A, Cooke AF, et al: Lubrication of joints, in Dowson D, Wright V (eds): *An Introduction to the Biomechanics of Joints and Joint Replacement*. London, Institute of Mechanical Engineering, 1981, pp 120–145.

66. Maroudas A: Hyaluronic acid film. *Proc Inst Mech Eng* 1967;181:122.

67. Mow VC, Mak AF: Lubrication of diarthrodial joints, in Skalak R, Chien S (eds): *Handbook of Bioengineering*. New York, McGraw-Hill, 1987, pp 5.1–5.34.

68. Bollet AJ, Nance JL: Biochemical findings in normal and osteoarthritic articular cartilage: II. Chondroitin sulfate concentration and chain length, water, and ash content. *J Clin Invest* 1966;45:1170–1177.

69. Donohue JM, Buss D, Oegema TR Jr, et al: The effects of indirect blunt trauma on adult canine articular cartilage. *J Bone Joint Surg* 1983;65A:948–957.

70. Floman Y, Eyre DR, Glimcher MJ: Induction of osteoarthrosis in the rabbit knee joint: Biochemical studies on the articular cartilage. *Clin Orthop* 1980;147:278–286.

71. Mankin HJ, Dorfman H, Lippiello L, et al: Biochemical and metabolic abnormalities in articular cartilage from osteo-arthritic human hips: II. Correlation of morphology with biochemical and metabolic data. *J Bone Joint Surg* 1971;53A:523–537.

72. Mankin HJ, Lippiello L: The turnover of adult rabbit articular cartilage. *J Bone Joint Surg* 1969;51A:1591–1600.

73. Mankin HJ, Thrasher AZ: Water content and binding in normal and osteoarthritic human cartilage. *J Bone Joint Surg* 1975;57A:76–80.

74. Maroudas A, Venn M: Chemical composition and swelling of normal and osteoarthrotic femoral head cartilage: II. Swelling. *Ann Rheum Dis* 1977;36:399–406.

75. McDevitt C, Gilbertson E, Muir H: An experimental model of osteoarthritis: Early morphological and biochemical changes. *J Bone Joint Surg* 1977;59B:24–35.

76. McDevitt CA, Muir H: Biochemical changes in the cartilage of the knee in experimental and natural osteoarthritis in the dog. *J Bone Joint Surg* 1976;58B:94–101.

77. Venn MF, Maroudas A: Chemical composition and swelling of normal and osteoarthrotic femoral head cartilage: I. Chemical composition. *Ann Rheum Dis* 1977;36:121–129.

78. Akizuki S, Mow VC, Muller F, et al: Tensile properties of human knee joint cartilage: I. Influence of ionic conditions, weight bearing, and fibrillation on the tensile modulus. *J Orthop Res* 1986;4:379–392.

79. Akizuki S, Mow VC, Müller F, et al: Tensile properties of human knee joint cartilage: II. Correlations between weight bearing and tissue pathology and the kinetics of swelling. *J Orthop Res* 1987;5:173–186.

80. Altman RD, Tenenbaum J, Latta L, et al: Biomechanical and biochemical properties of dog cartilage in experimentally induced osteoarthritis. *Ann Rheum Dis* 1984;43:83–90.

81. Hoch DH, Grodzinsky AJ, Koob TJ, et al: Early changes in material properties of rabbit articular cartilage after meniscectomy. *J Orthop Res* 1983;1:4–12.

82. Kempson GE: Mechanical properties of articular cartilage and their relationship to matrix degradation and age. *Ann Rheum Dis* 1975;34(suppl 2):111–113.

83. Kempson GE: Mechanical properties of articular cartilage, in Freeman MAR (ed): *Adult Articular Cartilage*, ed 2. Tunbridge Wells, Pitman Medical, 1979, pp 333–414.

84. Kempson GE, Freeman MAR, Swanson SAV: Tensile properties of articular cartilage. *Nature* 1968;220:1127–1128.

85. Maroudas A: Balance between swelling pressure and collagen tension in normal and degenerate cartilage. *Nature* 1976;260:808–809.

86. Mankin HJ, Johnson ME, Lippiello L: Biochemical and metabolic abnormalities in articular cartilage from osteoarthritic human hips: III. Distribution and metabolism of amino sugar-containing macromolecules. *J Bone Joint Surg* 1981;63A:131–139.

87. Maroudas A: Glycosaminoglycan turn-over in articular cartilage. *Philos Trans R Soc Lond* 1975;271B:292–313.

88. Lipshitz H, Etheredge R III, Glimcher MJ: *In vitro* wear of articular cartilage: I. Hydroxyproline, hexosamine, and amino acid composition of bovine articular cartilage as a function of depth from the surface; hydroxyproline content of the lubricant and the wear debris as a measure of wear. *J Bone Joint Surg* 1975;57A:527–534.

89. Lipshitz H, Etheredge R III, Glimcher MJ: Changes in the hexosamine content and swelling ratio of articular cartilage as functions of depth from the surface. *J Bone Joint Surg* 1976;58A:1149–1153.

90. Maroudas A: Transport of solutes through cartilage: Permeability to large molecules. *J Anat* 1976;22:335–347.

91. Winlove CP, Parker KH: Diffusion of macromolecules in hyaluronate gels: I. Development of methods and preliminary results. *Biorheology* 1984;21:347–362.

92. McKibbin B, Maroudas A: Nutrition and metabolism, in Freeman MAR (ed): *Adult Articular Cartilage*, ed 2. Tunbridge Wells, Pitman Medical, 1979, pp 461–486.

93. Arokoski J, Jurvelin J, Kiviranta I, et al: Combined biomechanical and histochemical analysis of articular cartilage in the canine knee. Presented at the 15th Symposium of the European Society of Osteoarthrology, Kuopio, Finland, 1986.

94. Caterson B, Lowther DA: Changes in the metabolism of the proteoglycans from sheep articular cartilage in response to mechanical stress. *Biochim Biophys Acta* 1978;540:412–422.

95. Hillberry BM, Van Sickle DC, Maturo CJ, et al: A model for studying the influence of mechanical usage on articular cartilage. Presented at the 15th Symposium of the European Society of Osteoarthrology, Kuopio, Finland, 1986.

96. Honner R, Thompson RC: The nutritional pathways of articular cartilage: An autoradiographic study in rabbits using ^{35}S injected intravenously. *J Bone Joint Surg* 1971;53A:742–748.

97. Palmoski MJ, Brandt KD: Running inhibits the reversal of atrophic changes in canine knee cartilage after removal of a leg cast. *Arthritis Rheum* 1981;24:1329–1337.

98. Palmoski MJ, Colyer RA, Brandt KD: Joint motion in the absence of normal loading does not maintain normal articular cartilage. *Arthritis Rheum* 1980;23:325–334.

99. Palmoski M, Perricone E, Brandt KD: Development and reversal of a proteoglycan aggregation defect in normal canine knee cartilage after immobilization. *Arthritis Rheum* 1979;22:508–517.

100. Salter RB, Field P: The effects of continuous compression on living articular cartilage: An experimental investigation. *J Bone Joint Surg* 1960;42A:31–49.

101. Thaxter TH, Mann RA, Anderson CE: Degeneration of immobilized knee joints in rats: Histological and autoradiographic study. *J Bone Joint Surg* 1965;47A:567–585.

102. Troyer H: The effect of short-term immobilization on the rabbit knee joint cartilage: A histochemical study. *Clin Orthop* 1975;107:249–257.

103. Van Sickle DC, Hillberry BM: *In vivo* biomechanical modulation of adult articular cartilage. Presented at the 15th Symposium of the European Society of Osteoarthrology, Kuopio, Finland, 1986.

104. Elliott DH, Crawford GNC: The thickness and collagen content of tendon relative to the strength and cross-sectional area of muscle. *Proc R Soc Lond* 1965;162B:137–146.

105. Harkness RD: Mechanical properties of collagenous tissues, in Gould BS (ed): *Treatise on Collagen: Vol 2. Biology of Collagen, Part A*. London, Academic Press, 1968, pp 247–310.

106. Hayes WC, Mockros LF: Viscoelastic properties of human articular cartilage. *J Appl Physiol* 1971;31:562–568.

107. Viidik A: Elasticity and tensile strength of the anterior cruciate ligament in rabbits as influenced by training. *Acta Physiol Scand* 1968;74:372–380.

108. Viidik A: Tensile strength properties of Achilles tendon systems in trained and untrained rabbits. *Acta Orthop Scand* 1969;40:261–272.

109. Viidik A: Interdependence between structure and function in collagenous tissues, in Viidik A, Vuust J (eds): *Biology of Collagen*. London, Academic Press, 1980, pp 257–280.

110. Viidik A: Properties of tendons and ligaments, in Skalak R, Chien S (eds): *Handbook of Bioengineering*. New York, McGraw-Hill, 1987, pp 6.1–6.19.

111. Viidik A, Ekholm R: Light and electron microscopic studies of collagen fibers under strain. *Z Anat Entwicklungsgesch* 1968;127:154–164.

112. Woo SL-Y, Akeson WH, Jemmott GF: Measurements of nonhomogeneous, directional mechanical properties of articular cartilage in tension. *J Biomech* 1976;9:785–791.

113. Woo SL-Y, Gomez MA, Akeson WH: The time and history-dependent viscoelastic properties of the canine medial collateral ligament. *J Biomech Eng* 1981;103:293–298.

114. Woo SL-Y, Lubock P, Gomez MA, et al: Large deformation nonhomogeneous and directional properties of articular cartilage in uniaxial tension. *J Biomech* 1979;12:437–446.

115. Woo SL-Y, Simon BR, Kuei SC, et al: Quasi-linear viscoelastic properties of normal articular cartilage. *J Biomech Eng* 1980;102:85–90.

116. Hayes WC, Bodine AJ: Flow-independent viscoelastic properties of articular cartilage matrix. *J Biomech* 1978;11:407–419.

117. Hascall VC, Hascall GK: Proteoglycans, in Hay ED (ed): *Cell Biology of Extracellular Matrix*. New York, Plenum Press, 1981, pp 39–63.

118. Junqueira LC, Montes GS: Biology of collagen-proteoglycan interaction. *Arch Histol Jpn* 1983;46:589–629.

119. Li JT, Armstrong CG, Mow VC: Effect of strain rate on mechanical properties of articular cartilage in tension. *Am Soc Mech Eng* 1983;56:117–120.

120. Eyre DR, Oguchi H: The hydroxypyridinium crosslinks of skeletal collagens: Their measurement, properties and a proposed pathway of formation. *Biochem Biophys Res Commun* 1980;92:403–410.

121. Fujimoto D: Evidence for natural existence of pyridinoline crosslink in collagen. *Biochem Biophys Res Commun* 1980;93:948–953.

122. Harkness RD, Nimni ME: Chemical and mechanical changes in the collagenous framework of skin induced by thiol compounds. *Acta Physiol Hung* 1968;33:325–343.

123. Nimni ME: Collagen: Its structure and function in normal and pathological connective tissues. *Semin Arthritis Rheum* 1974;4:95–150.

124. Eyre DR, Grynpas MD, Shapiro FD, et al: Mature crosslink formation and molecular packing in articular cartilage collagen. *Semin Arthritis Rheum* 1981;11(suppl 1):46–47.

125. Eyre DR, Koob TJ, Van Ness KP: Quantitation of hydroxypyridinium crosslinks in collagen by high-performance liquid chromatography. *Anal Biochem* 1984;137:380–388.

126. Torchia DA, Batchelder LS, Fleming WW, et al: Mobility and function in elastin and collagen, in *Mobility and Function in Proteins and Nucleic Acids*, Ciba Foundation Symposium 93. London, Pitman, 1983, pp 98–115.

127. Schmidt MB, Schoonbeck JM, Mow VC, et al: The relationship between collagen crosslinking and the tensile properties of articular cartilage. *Trans Orthop Res Soc* 1987;12:134.

128. Haut RC, Little RW: Rheological properties of canine anterior cruciate ligaments. *J Biomech* 1969;2:289–298.

129. Kempson GE, Muir H, Pollard C, et al: The tensile properties of the cartilage of human femoral condyles related to the content of collagen and glycosaminoglycans. *Biochim Biophys Acta* 1973;297:456–472.

130. Ranu HS: *Rheological Behavior of Articular Cartilage Under Tensile Loads*, thesis. University of Surrey, 1967.

131. Roth V, Mow VC: The intrinsic tensile behavior of the matrix of bovine articular cartilage and its variation with age. *J Bone Joint Surg* 1980;62A:1102–1117.

132. Roth V, Mow VC, Grodzinsky AJ: Biophysical and electromechanical properties of articular cartilage, in Simmons DJ, Kunin AS (eds): *Skeletal Research: An Experimental Approach.* New York, Academic Press, 1979, pp 301–341.

133. Benninghoff A: Form und Bau der Gelenkknorpel in ihren Beziehungen zur Funktion: Erste Mitteilung: Die modellierenden und formerhaltenden Faktoren des Knorpelreliefs. *Z Anat Entwicklungsgesch* 1925;76:43–63.

134. Benninghoff A: Form und Bau der Gelenkknorpel in ihren Beziehungen zur Funktion: Zweiter Teil. II. Der aufbau des Gelenkknorpel in seinen Beziehunen zu Funktion. *Z Zellforsch Mikrosk Anat* 1925;2:783.

135. Hultkrantz W: Ueber die Spaltrichtungen der Gelenkknorpel. *Verh Anat Ges* 1898;12:248–256.

136. Mow VC, Mak AF, Lai WM, et al: Viscoelastic properties of proteoglycan subunits and aggregates in varying solution concentrations. *J Biomech* 1984;17:325–338.

137. Muir IHM: The chemistry of the ground substance of joint cartilage, in Sokoloff L (ed): *The Joints and Synovial Fluid.* New York, Academic Press, 1980, vol 2, pp 27–94.

138. Muir H: Molecular approach to the understanding of osteoarthrosis. *Ann Rheum Dis* 1977;326:199–208.

139. Mak AF, Mow VC, Lai WM: Predictions of the number and strength of proteoglycan-proteoglycan interactions from viscometric data. *Trans Orthop Res Soc* 1983;8:3.

140. Venn MF: Variation of chemical composition with age in human femoral head cartilage. *Ann Rheum Dis* 1978;37:168–174.

141. Eisenberg SR, Grodzinsky AJ: Swelling of articular cartilage and other connective tissues: Electromechanochemical forces. *J Orthop Res* 1985;3:148–159.

142. Eisenberg SR, Grodzinsky AJ: The kinetics of chemically induced nonequilibrium swelling of articular cartilage and corneal stroma. *J Biomech Eng* 1987;109:79–89.

143. Mow VC, Myers ER, Roth V, et al: Implications for collagen-proteoglycan interactions from cartilage stress relaxation behavior in isometric tension. *Semin Arthritis Rheum* 1981;11(suppl 1):41–43.

144. Maroudas A, Schneiderman R: "Free" and "exchangeable" or "trapped" and "non-exchangeable" water in cartilage. *J Orthop Res* 1987;5:133–138.

145. Edwards J: Physical characteristics of articular cartilage. *Proc Inst Mech Eng* 1967;181:16.

146. Mansour JM, Mow VC: The permeability of articular cartilage under compressive strain and at high pressures. *J Bone Joint Surg* 1976;58A:509–516.

147. Holmes MH: A theoretical analysis for determining the nonlinear hydraulic permeability of a soft tissue from a permeation experiment. *Bull Math Biol* 1985;47:669–683.

148. Holmes MH: Finite deformation of soft tissue: Analysis of a mixture model in uni-axial compression. *J Biomech Eng* 1986;108:372–381.

149. Lai WM, Mow VC: Drag-induced compression of articular cartilage during a permeation experiment. *Biorheology* 1980;17:111–123.

150. Laurent TC, Preston BN, Sundelöf L-O: Transport of molecules in concentrated systems. *Nature* 1979;279:60–62.

151. Preston BN, Laurent TC, Comper WD, et al: Rapid polymer transport in concentrated solutions through the formation of ordered structures. *Nature* 1980;287:499–503.

152. Schubert M, Hamerman D: *A Primer on Connective Tissue Biochemistry.* Philadelphia, Lea & Febiger, 1968.

153. Muir H, Bullough P, Maroudas A: The distribution of collagen in human articular cartilage with some of its physiological implications. *J Bone Joint Surg* 1970;52B:554–563.

154. Myers ER: *Kinetic and Equilibrium Swelling Studies of Connective Tissues,* thesis. Troy, New York, Rensselaer Polytechnic Institute, 1984.

155. Mow VC, Schoonbeck JM: Contribution of Donnan osmotic pressure towards the biphasic compressive modulus of articular cartilage. *Trans Orthop Res Soc* 1984;9:262.

156. Hascall VC: Interaction of cartilage proteoglycans with hyaluronic acid. *J Supramol Struct* 1977;7:101–120.

157. Poole AR, Pidoux I, Reiner A, et al: An immunoelectron microscope study of the organization of proteoglycan monomer, link protein, and collagen in the matrix of articular cartilage. *J Cell Biol* 1982;93:921–937.

158. Broom ND, Poole CA: Articular cartilage collagen and proteoglycans. *Arthritis Rheum* 1983;26:1111–1119.

159. Dunham J, Shackleton DR, Nahir AM, et al: Altered orientation of glycosaminoglycans and cellular changes in the tibial cartilage in the first two weeks of experimental canine osteoarthritis. *J Orthop Res* 1985;3:258–268.

160. Shepard N, Mitchell N: The localization of articular cartilage proteoglycan by electron microscopy. *Anat Rec* 1977;187:463–476.

161. Bruckner P, Vaughan L, Winterhalter KH: Type IX collagen from sternal cartilage of chicken embryo contains covalently bound glycosaminoglycans. *Proc Natl Acad Sci USA* 1985;82:2608–2612.

162. Scott JE, Orford CR, Hughes EW: Proteoglycan-collagen arrangements in developing rat tail tendon: An electron-microscopical and biochemical investigation. *Biochem J* 1981;195:573–581.

163. Serafini-Fracassini A, Smith JW: *The Structure and Biochemistry of Cartilage.* Edinburgh, Churchill Livingstone, 1974.

164. Mak AF, Mow VC, Lai WM, et al: Assessment of proteoglycan-proteoglycan interactions from solution biorheological behaviors. *Trans Orthop Res Soc* 1982;7:169.

165. Weightman B: *In vitro* fatigue testing of articular cartilage. *Ann Rheum Dis* 1975;34(suppl 2):108–110.

166. Weightman B: Tensile fatigue of human articular cartilage. *J Biomech* 1976;9:193–200.

167. Takei T, Myers ER, Mow VC: Quantitation of tensile and swelling behavior of mildly fibrillated human articular cartilage. *Trans Orthop Res Soc* 1984;9:365.

168. Redler I, Zimny ML: Scanning electron microscopy of normal and abnormal articular cartilage and synovium. *J Bone Joint Surg* 1970;52A:1395–1404.

169. Redler I: A scanning electron microscopic study of human normal and osteoarthritic articular cartilage. *Clin Orthop* 1974;103:262–268.

170. Sweet MBE, Thonar EJ, Marsh J: Age-related changes in proteoglycan structure. *Arch Biochem Biophys* 1979;198:439–448.

171. Bayliss MT, Ali SY: Age-related changes in the composition and structure of human articular-cartilage proteoglycans. *Biochem J* 1978;176:683–693.

172. Axelsson I, Bjelle A: Proteoglycan structure of bovine articular cartilage: Variation with age and in osteoarthrosis. *Scand J Rheumatol* 1979;8:217–221.

173. Kempson GE, Tuke MA, Dingle JT, et al: The effects of proteolytic enzymes on the mechanical properties of adult human articular cartilage. *Biochim Biophys Acta* 1976;428:741–760.

174. Li JT, Mow VC, Koob TJ, et al: Effect of chondroitinase ABC treatment on the tensile behavior of bovine articular cartilage. *Trans Orthop Res Soc* 1984;9:35.

175. Stahurski TM, Armstrong CG, Mow VC: Variation of the intrinsic aggregate modulus and permeability of articular cartilage with trypsin digestion. *Am Soc Mech Eng* 1981;43:137–140.

176. Paul JP: Joint kinetics, in Sokoloff L (ed): *The Joints and Synovial Fluid.* New York, Academic Press, 1980, vol 2, pp 139–176.

177. Seireg A, Arvikar RJ: The prediction of muscular load sharing and joint forces in the lower extremities during walking. *J Biomech* 1975;8:89–102.

178. Armstrong CG, Lai WM, Mow VC: An analysis of the unconfined compression of articular cartilage. *J Biomech Eng* 1984;106:165–173.

179. Simon SR, Radin EL, Paul IL, et al: The response of joints to impact loading: II. *In vivo* behavior of subchondral bone. *J Biomech* 1972;5:267–272.

180. Ferry JD: Illustrations of viscoelastic behavior of polymeric systems, in *Viscoelastic Properties of Polymers*, ed 2. New York, John Wiley & Sons, 1970.

181. Roth V, Schoonbeck JM, Mow VC: Low frequency dynamic behavior of articular cartilage under torsional shear. *Trans Orthop Res Soc* 1982;7:150.

182. Zhu WB, Lai WM, Mow VC: Intrinsic quasi-linear viscoelastic behavior of the extracellular matrix of cartilage. *Trans Orthop Res Soc* 1986;11:407.

183. Elmore SM, Sokoloff L, Norris G, et al: Nature of "imperfect" elasticity of articular cartilage. *J Appl Physiol* 1963;18:393–396.

184. Sokoloff L: Elasticity of articular cartilage: Effect of ions and viscous solutions. *Science* 1963;141:1055–1057.

185. Armstrong CG, Mow VC: Biomechanics of normal and osteoarthrotic articular cartilage, in Straub LR, Wilson PD Jr (eds): *Clinical Trends in Orthopaedics.* New York, Thieme-Stratton, 1982, pp 189–197.

186. Roth V, Mow VC, Lai WM, et al: Correlation of intrinsic compressive properties of bovine articular cartilage with its uronic acid and water content. *Trans Orthop Res Soc* 1981;6:49.

187. Holmes MH, Lai WM, Mow VC: Singular perturbation analysis of the nonlinear, flow-dependent compressive stress relaxation behavior of articular cartilage. *J Biomech Eng* 1985;107:206–218.

188. Holmes MH, Lai WM, Mow VC: Compression effects on cartilage permeability, in Hargens AR (ed): *Tissue Nutrition and Viability.* New York, Springer-Verlag, 1986, pp 73–100.

189. Simon BR, Coats RS, Woo SL-Y: Relaxation and creep quasilinear viscoelastic models for normal articular cartilage. *J Biomech Eng* 1984;106:159–164.

190. Wada T, Akizuki S: An ultrastructural study of solid matrix in articular cartilage under uniaxial tensile stress. *J Jpn Orthop Assoc*, in press.

191. Fung YCB: Elasticity of soft tissues in simple elongation. *Am J Physiol* 1967;213:1532–1544.

192. Kenedi RM, Gibson T, Daly CH: Bio-engineering studies of the human skin: The effects of unidirectional tension, in Jackson SF, Harkness RD, Tristram GR (eds): *Structure and Function of Connective and Skeletal Tissue.* London, Butterworths, 1965, pp 388–395.

193. Morgan FR: The mechanical properties of collagen fibers: Stress-strain curves. *J Soc Leather Trades Chem* 1960;44:171–182.

194. Ridge MD, Wright V: The description of skin stiffness. *Biorheology* 1964;2:67–74.

195. Fung Y-CB: Stress-strain-history relations of soft tissues in simple elongation, in Fung YC, Perrone N, Anliker M (eds): *Biomechanics: Its Foundations and Objectives.* Englewood Cliffs, Prentice-Hall, 1972, pp 181–208.

196. Fung Y-CB: Quasi-linear viscoelasticity of soft tissues, in *Biomechanics: Mechanical Properties of Living Tissues.* New York, Springer-Verlag, 1981, pp 226–238.

197. Grodzinsky AJ, Roth V, Myers E, et al: The significance of electromechanical and osmotic forces in the nonequilibrium swelling behavior of articular cartilage in tension. *J Biomech Eng* 1981;103:221–231.

198. Kwan MK, Woo SL-Y, Amiel D, et al: Neocartilage generated from rib perichondrium: A long term multidisciplinary evaluation. *Trans Orthop Res Soc* 1987;12:277.

199. Bowden FP, Tabor D: *Friction and Lubrication.* London, Methuen, 1967.

200. Swann DA, Radin EL, Hendren RB: The lubrication of articular cartilage by synovial fluid glycoproteins, abstract. *Arthritis Rheum* 1979;22:665–666.

201. Swann DA, Silver FH, Slayter HS, et al: The molecular structure and lubricating activity of lubricin isolated from bovine and human synovial fluids. *Biochem J* 1985;225:195–201.

202. Higginson GR, Litchfield MR, Snaith J: Load-displacement-time characteristics of articular cartilage. *Int J Mech Sci* 1976;18:481.

203. Higginson GR, Unsworth T: The lubrication of natural joints, in Dumbleton JH (ed): *Tribology of Natural and Artificial Joints*, Tribology Series, 3. Amsterdam, Elsevier, 1981, pp 47–73.

204. Lai WM, Mow VC: Stress and flow fields in articular cartilage, in *Advances in Civil Engineering Through Engineering Mechanics.* American Society of Civil Engineering, 1979, pp 202–205.

205. Mow VC, Lai WM: The optical sliding contact analytical rheometer (OSCAR) for flow visualization at the articular surface, in Wells MK (ed): *Advances in Bioengineering.* New York, American Society of Mechanical Engineers, 1979, pp 97–99.

206. Radin EL, Martin RB, Burr DB, et al: Effects of mechanical loading on the tissues of the rabbit knee. *J Orthop Res* 1984;2:221–234.

207. Radin EL, Parker HG, Pugh JW, et al: Response of joints to impact loading: III. Relationship between trabecular microfractures and cartilage degeneration. *J Biomech* 1973;6:51–57.

208. Radin EL, Paul IL, Lowy M: A comparison of the dynamic force transmitting properties of subchondral bone and articular cartilage. *J Bone Joint Surg* 1970;52A:444–456.

209. Radin EL, Paul IL: Response of joints to impact loading: I. In vitro wear. *Arthritis Rheum* 1971;14:356–362.

210. Howell DS: Etiopathogenesis of osteoarthritis, in Moskowitz RW, Howell DS, Goldberg VM, et al (eds): *Osteoarthritis: Diagnosis and Management.* Philadelphia, WB Saunders, 1984, pp 129–146.

211. Mankin HJ, Brandt KD: Biochemistry and metabolism of cartilage in osteoarthritis, in Moskowitz RW, Howell DS, Goldberg VM, et al (eds): *Osteoarthritis: Diagnosis and Management.* Philadelphia, WB Saunders, 1984, pp 43–79.

Chapter 11

Articular Cartilage: Injury and Repair

Joseph Buckwalter, MD
Lawrence Rosenberg, MD
Richard Coutts, MD
Ernst Hunziker, MD
A. Hari Reddi, PhD
Van Mow, PhD

Chapter Outline

Introduction

The cartilage repair response has been the focus of investigations for more than 250 years. In 1743 Hunter[1] noted that "ulcerated cartilage is a troublesome thing . . . once destroyed it is not repaired." A little more than 100 years later, Paget[2] reported that there are ". . . no instances in which a lost portion of cartilage has been restored, or a wounded portion repaired with new and well formed cartilage. . . ." Studies during the last 150 years generally confirmed the work of Hunter, Paget, and others by showing that cartilage is repaired under certain conditions, but, in most circumstances, the repair tissue lacks the molecular composition and organization, the material properties, and the durability of normal articular cartilage.[3-21]

Despite the many studies of the cartilage repair response, the quality and durability of the repair following a specific cartilage injury cannot be predicted with certainty. Some of the uncertainty stems from inherent differences in the repair response among different individuals and animals, but an equally important source of confusion is the failure to define precisely the injury, the age of the individual, the condition of the joint before injury, the quality, extent, and durability of the repair, and the long-term function of the joint. It is not sufficient to report whether cartilage repair occurs or not, or whether repair tissue does or does not grossly resemble articular cartilage. Improved understanding of cartilage repair depends on experimental and clinical observations that define the nature and extent of the injury, the state of the cartilage and the joint at the time of injury, the age of the individual or animal, and the structure, composition, function, and durability of the repair tissue resulting from that injury.

Acute Cartilage Injury

Acute injuries to cartilage be classified into two general groups: loss of matrix macromolecules without mechanical damage to cells or the collagen fibril meshwork and mechanical disruption of cells and matrix. Progressive loss of matrix macromolecules leads to mechanical disruption of the articular surface, and mechanical disruption may release factors that may stimulate matrix degradation; thus, the two types of injury may overlap.

Loss of Matrix Macromolecules

Cartilage exposure by traumatic or surgical disruption of the synovial membrane, infection and other inflammatory diseases, prolonged joint immobilization, some types of anti-inflammatory agents, and possibly joint irrigation can stimulate degradation of proteoglycans or suppress proteoglycan synthesis.[22-26] These insults may also have other effects on the matrix and the cells, but the loss of matrix proteoglycans is the most obvious initial change.

Prompt termination of the process responsible for the loss of matrix proteoglycans allows the chondrocytes to replace the lost matrix components and the cartilage may regain its normal composition and function. In contrast, if the process continues, the articular damage becomes irreversible. It is not clear at what point this occurs. Presumably, the cartilage can be restored to its normal condition only if the loss of matrix proteoglycans does not exceed what the cells can rapidly produce, if the chondrocytes are viable and capable of synthesizing the appropriate proteoglycans, and if the collagen meshwork is intact.

Mechanical Injury to Cartilage

Blunt trauma, penetrating injuries, frictional abrasion, or sharp concentration of weightbearing forces kill chondrocytes and disrupt the matrix. The response of articular cartilage to penetrating injury depends on the depth of injury; that is, injuries limited to cartilage elicit a different repair response than injuries involving cartilage and subchondral bone.[3-5,7,11,17,18] The volume and surface area of cartilage injury may also influence the extent and quality of the repair.[8] Although the study of blunt trauma has been less extensive, such injuries also stimulate a repair response that depends on the extent and severity of the injury.[23,26-28] The responses of cartilage to abrasion and sharp concentration of weightbearing loads have not been well described.

Blunt Trauma   Blunt trauma to cartilage occurs frequently, even in the absence of fractures, and may be the source of significant long-term joint dysfunction. The degree of cartilage injury and the eventual result of the repair response varies with the intensity of the blunt trauma. On the basis of available information, the intensity of acute blunt trauma can be classified as (1) greater than normal loading, but less than that required to fracture bone or cartilage and (2) sufficient to fracture bone and cartilage.[23,26,28,29]

Physiologic levels of impact loading do not appear to produce cartilage injury. However, Donohue and associates[23] found that impact above this level but less than that necessary to produce fracture of cartilage or bone caused cartilage swelling, increased collagen fibril diameter, and alteration of the relationship between collagen and proteoglycans.

Blunt trauma that produces fractures or fissuring of the cartilage matrix, fractures of the subchondral bone, disruption of the collagen fibril network, or cell injury and death stimulates the same type of repair response seen with penetrating injuries. In these types of injuries, the response of the cartilage depends primarily on the extent of injury to the cartilage matrix and cells and whether or not the subchondral bone is fractured.

Penetrating Injury   The zone of calcified cartilage separates the articular cartilage from the marrow cells and blood vessels that participate in the inflammatory response to injury. The characteristics of the cartilage response to injury, therefore, depend on whether a defect is restricted to the substance of the articular cartilage or extends into subchondral bone.[5] The success of the repair may also depend on the extent of the injury as measured by the volume and surface area of injured cartilage, and on the location of the injury in the joint. Available evidence indicates that repair of larger defects is less satisfactory than repair of smaller defects[8] and that osteochondral defects in locations that articulate with another cartilage surface heal less well than defects that do not articulate with cartilage.[30] However, these variables have not been studied extensively.

Most of the information concerning the responses of articular cartilage to penetrating injury, including the following description (L. Rosenberg, unpublished data), is based on experimental studies of rabbit knee joints. Although the initial response of cartilage to injury probably is similar in different species, the long-term results of cartilage repair may differ. Compared with human cartilage, rabbit knee cartilage is thin, usually less than 1 mm thick. The molecular composition and organization of rabbit cartilage are slightly different, and rabbit knees have different mechanical properties.

Laceration of articular cartilage perpendicular to the surface kills chondrocytes at the site of the injury and creates a wedge-shaped matrix defect. Because the cells in the marrow and the endothelial cells of subchondral blood vessels do not have access to lesions limited to articular cartilage, these lesions do not elicit an inflammatory response. Fibrin, neutrophils, and macrophages do not adhere to the surfaces of the articular cartilage defect, fibroblasts do not migrate into the defect, and reparative granulation tissue does not appear. Chondrocytes near the lesion make the only apparent response to the injury. Some of them proliferate, forming small clusters of new cells and synthesize new matrix that remains in the region of the cells. The chondrocytes do not migrate to the lesion and the new matrix they synthesize does not fill the defect; thus their limited response does not repair the injury. Although lesions restricted to articular cartilage do not heal, they seldom progress. A wedge-shaped defect perpendicular to the articular surface examined one year after injury is almost identical in appearance to a defect examined 24 hours after injury.

Studying lesions restricted to articular cartilage is important because they are analogous in some ways to the early lesions found in osteoarthritis. In the earliest phase of osteoarthritis, articular cartilage degeneration begins at the cartilage surface with fraying of the bundles of collagen fibers in the superficial zone. The cartilage then begins to tear along roughly vertical lines, creating vertical clefts and fissures. With time, the clefts and fissures become deeper. In the more superficial regions of the fibrillated cartilage, proteoglycan is lost from the extracellular matrix, particularly in the interterritorial regions. The collagen content of the extracellular matrix does not change appreciably. Microscopically, the areas in which proteoglycan is lost show decreased staining with cationic dyes such as safranin O. In response to the loss of proteoglycan and the injury to cartilage, chondrocytes proliferate, forming clusters or brood capsules. The matrix between the chondrocyte clusters and immediately surrounding them stains heavily with safranin O as the chondrocytes synthesize increased amounts of proteoglycan, in an attempt to replace proteoglycan lost from the interterritorial matrix. The formation of chondrocyte clusters involves the proliferation of chondrocytes and synthesis of proteoglycans but, as in acute cartilage injuries, the cells and the newly synthesized proteoglycans remain constrained by extracellular matrix. The chondrocytes do not migrate into the areas of the defects or fill the defects with newly synthesized matrix.

As the clefts and fissures grow deeper, the chondrocytes at the surfaces of the fibrillated cartilage degenerate and die, and the tips of the fibrillated cartilage tear loose and are released into the joint. Like experimental lacerations of cartilage, these osteoarthritic lesions do not heal, if we define healing as the replacement of lost tissue. Unlike the experimental lacerations, the osteoarthritis lesions progress. The articular cartilage becomes progressively thinner, and is finally worn down to subchondral bone. Throughout this process one of the most striking characteristics is the lack of repair within the clefts and fissures of the fibrillated cartilage.

Superficial lacerations made tangential or parallel to the articular surface follow a course similar to that described for lacerations perpendicular to the articular surface.[11] Cells directly adjacent to the injury site may die, whereas others show signs of increased proliferative and synthetic activity. A layer of new matrix may form over the injured surface, but its durability is unpredictable. Some clinical experience suggests that fibrillated human cartilage may respond to superficial abrasion or shaving by forming a smoother surface that improves joint function and decreases symptoms. This clinical observation has not yet been verified by controlled experiments, and one report[31] indicated that shaving damaged human medial femoral cartilage did not stimulate repair. Shaving normal rabbit patellar cartilage neither stimulated significant repair nor caused degeneration.[32]

Thus, most currently available evidence indicates that chondrocytes respond to penetrating injuries limited to cartilage by proliferating and increasing their synthetic activity, but that, in general, they fail to restore a normal matrix even when the defect is small. The repair response may decrease with age as the cell numerical density and the capacity of the cells to respond to injury decreases. Because of the low cell density and the inability of chondrocytes to migrate into a defect, repair of defects limited to cartilage at any age would require a considerable proliferative and synthetic effort from relatively few cells.

Accidents and athletic activities commonly cause injuries that extend through cartilage into subchondral bone. Osteochondritis dissecans, a form of ischemic necrosis of subchondral bone, can also lead to loss of a portion of an articular surface. Many individuals suffering pain and disability from traumatic osteochondral injuries or osteochondritis dissecans are not candidates for joint replacement because of their age, weight, activity level, or other factors. For this reason, attempts to increase our understanding of full-thickness cartilage injury and the search for methods of improving repair of these defects must continue.

When a full-thickness articular cartilage defect is created, marrow cells capable of participating in an inflammatory response have access to the defect. Even without treatment, some full-thickness defects undergo repair during the first two months after injury. Although the repair histologically appears to be satisfactory, the chemical composition of the reparative tissue is not the same as that of normal articular cartilage, the structure of the reparative tissue is imperfect, and the integrity of the reparative tissue is not maintained. By six months after injury, fibrillation, fissuring, and extensive degenerative changes occur in the reparative tissues of approximately one half of full-thickness defects.

The following account briefly describes the histologic evolution of the repair process in full-thickness defects 2 mm in diameter created in the distal femoral articular cartilage and subchondral bone of mature white New Zealand rabbits. Other work indicates that defects 1 mm in diameter heal more completely, whereas those 3 mm in diameter frequently heal less well (J.A. Buckwalter and M.V. Olmstead, unpublished data).

Forty-eight hours after injury, a fibrin clot containing mononuclear cells filled the defect. Five days later, fibroblasts and collagen fibers, oriented parallel to the articular surface, had replaced the fibrin clot. After two weeks, metaplasia to cartilage began in the fibroblastic tissue, and islands of chondrocytes separated by broad areas of extracellular matrix stained with safranin O had appeared. One month after injury, most of the "fibroblast-like" cells had differentiated into chondrocytes separated by extracellular matrix that stained heavily with safranin O. At two months, the repair seemed to be satisfactory. The microscopic appearance of the repair tissue resembled that of cartilage, a proteoglycan-rich extracellular matrix had been formed, and the thickness of the reparative tissue was approximately the same as that of the adjacent normal articular cartilage. By six months, a normal-appearing zone of calcified cartilage and layer of subchondral bone had formed at the level of the zone of calcified cartilage and subchondral bone of the adjacent normal cartilage.

Six months after injury, however, striking abnormalities were present in the reparative tissue. None of the full-thickness defects had appreciable staining of the extracellular matrix with safranin O (Fig. 11–1), indicating that proteoglycan, which was present throughout the extracellular matrix at two months, had decreased significantly six months after injury. Moreover, in 50% of the full-thickness defects, the reparative tissue had undergone fibrillation and erosive changes, and was greatly decreased in thickness. Similar changes were noted in full-thickness defects examined one year after injury (Fig. 11–2).

A study of healing of full-thickness defects in pig joints demonstrated that the intrinsic biomechanical properties of repair cartilage were inferior to those of normal articular cartilage.[33] This study, the only one of biomechanical changes in repair tissue, supported the general impression that healed cartilage differs biomechanically and histologically from normal cartilage.

What is the biochemical basis for the imperfect repair of full-thick-

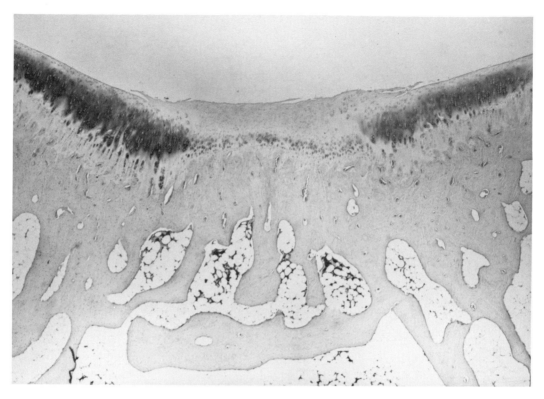

Fig. 11–1 *Light micrograph of a full-thickness articular cartilage lesion 2 mm in diameter six months after injury. The repair tissue is in the center. The subchondral bone and the articular cartilage have been restored. The dark stain within the articular cartilage is safranin O, indicating the presence of proteoglycans. Note that the repair tissue has almost no safranin O staining in comparison with the uninjured cartilage on either side.*

ness articular cartilage defects? Our knowledge of the chemical composition and structure of the reparative tissue formed in full-thickness defects is limited, and the abnormalities and deficiencies in the reparative tissue must be more adequately described so that conditions can be defined to elicit more satisfactory repair. However, one definite abnormality in the reparative tissue is that it contains appreciable amounts of type I collagen. Furukawa and associates[10] determined the amounts of type I and type II collagen present in healing cartilage. They created defects 3 mm in diameter that extended into subchondral bone, in rabbit patellar grooves. Biochemical studies on the reparative tissue were carried out three, four, six, 12, 24, and 48 weeks after creation of the defects.

At three to four weeks, when the reparative tissue was differentiating into hyaline cartilage, less than 40% of the collagen was type II. At eight weeks, more than 40% of the collagen in the reparative tissue

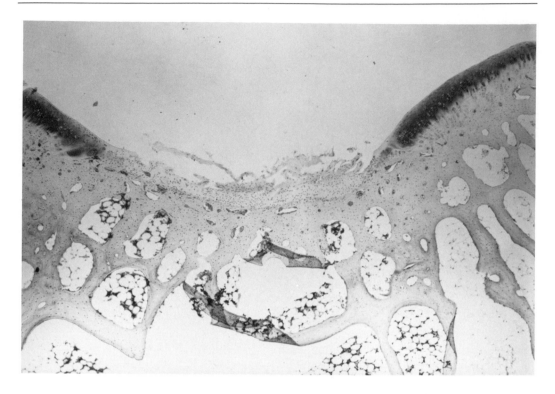

Fig. 11–2 *One year after injury, a full-thickness articular cartilage lesion 2 mm in diameter has undergone extensive fibrillation.*

was type I. It should be emphasized that at eight weeks the reparative tissue appeared to have differentiated into a satisfactory hyaline cartilage, based on its histologic appearance and the high concentration of proteoglycans in its extracellular matrix as indicated by safranin O staining. Between two months and six months, the reparative tissue still contained 25% to 33% type I collagen. This is the period during which proteoglycans seem to disappear from the reparative tissue, and during which it undergoes fibrillation and degeneration. Even at one year after injury, approximately 20% of the collagen in the reparative tissue was type I.

These observations, viewed in relation to the biologic processes involved in repair and regeneration in other tissues, make it possible to describe the requirements for satisfactory repair of full-thickness articular cartilage defects. Stem cells in a location remote from the articular cartilage defect must be stimulated to proliferate and differentiate into a population of progenitor cells capable of differentiating into chondrocytes. These progenitor cells must migrate into the cartilage defect. These events appear to take place in the initial phase of repair after a full-thickness articular cartilage defect is created, as the defect fills with fibroblast-like cells that assume the rounded shape of chondrocytes and begin to synthesize an apparently cartilaginous matrix. However, to restore the matrix, these cells must then differentiate into chondrocytes whose phenotypic expression is identical or very similar to that of chondrocytes in normal articular cartilage. They then must form an extracellular matrix composed of the same macromolecules, with the same elaborate structure as those formed in normal cartilage, in the same proportions, and organized in the same way.

On the basis of the available information,[10] it appears that this does not occur. Following cartilage injury, differentiation of fibroblast-like progenitor cells into chondrocytes is woefully imperfect, and it does not produce a population of chondrocytes whose phenotypic expression is similar to that of chondrocytes in normal mature cartilage.

Examination of the repair matrix demonstrates that the cells do not differentiate into normal chondrocytes. The extracellular matrix of normal articular cartilage consists primarily of type II collagen, cartilage-type proteoglycans, a variety of matrix proteins, and water. The cartilage-type proteoglycan monomers consist of chondroitin sulfate and keratan sulfate chains covalently bound to a protein core. In contrast, the reparative tissue formed in full-thickness defects contains large amounts of type I collagen, even months after a tissue is formed that resembles cartilage histologically.

Connective tissues such as tendon and ligament that synthesize and secrete type I collagen usually simultaneously synthesize and secrete relatively large amounts of dermatan sulfate-containing proteoglycans and relatively small amounts of large cartilage-like aggregating proteoglycans. Compared with the large cartilage-specific proteoglycan, the dermatan sulfate-containing proteoglycan is much smaller, with feeble elastic properties. Substitution of small dermatan sulfate proteoglycans for large cartilage-type proteoglycans may contribute to the decreased capacity of the reparative tissue to resist wear, and to its degeneration.

There have been no studies to show that, in the repair of full-thickness articular cartilage defects, huge cartilage-specific aggregating proteoglycans with potent elastic properties are partially replaced by small dermatan-sulfate proteoglycans. However, developments in the last few years now make it feasible to examine this possibility. Cartilage-specific proteoglycan monomers and link proteins have been isolated to homogeneity. Antibodies raised against these molecules can be used to demonstrate by immunohistochemical methods the distribution of these molecules in normal articular cartilage and alterations in the distribution of these molecules in the reparative tissue formed by full-thickness defects. Dermatan sulfate proteoglycans have also been isolated to homogeneity and characterized. Antibodies raised against the dermatan sulfate proteoglycans can be used in immunohistochemical studies to determine whether disproportionate amounts of the dermatan sulfate proteoglycans are present in reparative tissue.

The differences in the repair of defects restricted to articular cartilage and injuries extending into subchondral bone raise the possibility that bone matrix plays a critical role in cartilage repair.[34,35] The possibility that bone matrix contains growth and differentiation factors governing cartilage and bone formation suggests that studies of bone-matrix-induced cartilage and bone formation have implications for cartilage repair.

Subcutaneous implantation of demineralized bone matrix results in local differentiation of mesenchymal stem cells into cartilage and bone.[36,37] Plasma fibronectin binds to the implanted bone matrix and facilitates attachment and migration of cells. Cells in the vicinity of the matrix proliferate, indicating the mitogenic potential of the matrix components and perhaps associated growth factors.[38] After proliferation, the cells differentiate into chondrocytes on days 6 to 8 and produce a cartilaginous matrix. Thus, bone matrix contains factors that can stimulate the formation of hyaline cartilage. An injury that penetrates subchondral bone may release these factors whereas an injury limited to cartilage may not. Conceivably, differences in the concentration of these factors may help explain some of the variability in cartilage repair. For example, small defects, 1 mm in diameter, may contain a sufficient concentration of the bone matrix factor to stimulate successful cartilage repair. Repair of larger defects may require a higher concentration of these bone matrix factors than normally occurs after injury.

Methods of Facilitating Cartilage Repair

The disability and pain that result from failure of synovial joints have stimulated the search for ways of facilitating cartilage repair. For repair tissue to perform satisfactorily as a joint surface, it must restore normal pain-free motion of the synovial joint for a prolonged period and it must prevent deterioration of the joint. It has been assumed that for this to occur, the repair tissue must have a structure and composition, material properties, and durability similar to those of the natural articular surface. A number of methods of promoting cartilage repair, or chondrogenesis, have been explored. These include cartilage shaving, perforation or abrasion of subchondral bone, changing the loading of the injured articular surface, motion, resurfacing with periosteum or perichondrium, digestion or extraction of matrix proteoglycans at the site of superficial defects, laser stimulation of chondrocytes, implantation of immature chondrocytes, implantation of gels, pulsed electromagnetic fields, and chondrogenesis-stimulating factors. The current clinically established method of promoting cartilage repair consists of perforating or abrading subchondral bone, initial protection of the repair tissue from excessive load, and, frequently, a controlled program of joint motion.

Shaving or Abrasion of Fibrillated Cartilage

Clinical experience indicates that arthroscopic shaving or abrasion of fibrillated and irregular cartilage, particularly patellar cartilage,[39,40] may relieve symptoms. Although this procedure is performed frequently, its efficacy has not been documented by well-controlled long-term clinical studies. Schmid and Schmid[31] reported that shaving the traumatically damaged medial femoral articular cartilage in humans did not restore a smooth congruent articular surface and may have caused increased fibrillation and cell necrosis in and adjacent to the original defect. Investigations of the effect of shaving fibrillated articular cartilage in animals have not been reported, but shaving normal rabbit patellar cartilage does not stimulate significant repair.[32]

Perforation or Abrasion of Subchondral Bone

Perforation or abrasion of subchondral bone elicits the native repair response. This includes formation of a fibrin clot and vascular invasion of the repair tissue. The new cell population that enters the defect produces a much more extensive repair response than that found in injuries limited to cartilage alone. Clinical experience[41-43] and a study of cartilage repair in rabbits[16] show that large areas of the articular surface can be repaired by the technique of making multiple small drill holes into subchondral bone. The repair tissue grows from the drill holes and spreads over the exposed bone. In the experimental study, the repair tissue initially had a hyaline appearance, but with time it became more fibrous and frequently began to fibrillate and break down. Clinical experience also shows that the repair tissue produced by this method occasionally functions well as a joint surface for relatively prolonged periods but the results are not predictable. Abrasion of subchondral bone may produce better results than making multiple drill holes but this has not been demonstrated.

Changing the Loads Applied to Injured Cartilage

Patients with osteoarthritis may gain symptomatic relief after an osteotomy that decreases the loads on severely involved areas of a joint. In some individuals, the change in load also appears to stimulate repair of the damaged cartilage.[44] Decreasing load by preventing weightbearing, limiting the use of a joint, or osteotomy may also facilitate repair of acute cartilage injuries or at least prevent damage to the initial repair tissue, although these possibilities have not been extensively investigated.

Motion and Loading

DePalma and associates[9] noted that early motion and weightbearing had a beneficial effect of the repair of full-thickness cartilage defects. Salter and associates[45,46] carried out extensive investigations of the effects of continuous passive motion on healing of rabbit cartilage. They found no beneficial effect on the repair of injuries limited to cartilage. However, in defects 1 mm in diameter that penetrated the subchondral bone, continuous passive motion accelerated repair and produced a tissue that closely resembled hyaline cartilage morphologically and histochemically. This tissue appeared to have greater durability than the repair tissue found in animals treated with immobilization or intermittent active motion. Although passive motion appears to have potential as a method of facilitating cartilage repair, the available evidence suggests that passive motion may not be as effective in promoting repair of larger defects (J.A. Buckwalter, M.V. Olmstead, and A. Pedrini, unpublished data)

Periosteal and Perichondrial Grafts

Clinical experience with resurfacing of upper extremity and knee joints[20,47,48] and experimental studies[49-53] with resurfacing rabbit knees have shown that perichondrium and periosteum have the potential to facilitate repair of relatively large cartilage defects. Like penetrating or abrading subchondral bone, this approach introduces a new cell population that participates in the repair of the articular surface. There is evidence that passive motion further improves the quality of periosteal graft repair[51,52] In contrast, another report indicated that passive motion treatment of perichondrial grafts did not improve the quality of the repair tissue evaluated one year after injury.[50]

Although periosteal and perichondrial grafts can produce cartilage repair tissue, questions concerning this method of facilitating repair remain. Do the periosteum and perichondrium from skeletally mature individuals and animals have the same potential for facilitating repair as these tissues from skeletally immature animals? Does the tissue produced by perichondrium and periosteum have long-term durability similar to that of normal cartilage?

Extraction of Proteoglycans or Suppression of Proteoglycan Synthesis

Several investigators have hypothesized that one of the difficulties in healing of defects limited to cartilage has been the lack of a fibrin clot to serve as a temporary matrix that would allow cell migration into the defect (L. Rosenberg and L.L. Johnson, unpublished data). If a fibrin clot could be induced, the quality of the repair tissue might be improved. Since matrix proteoglycans inhibit platelet aggregation and clot formation in superficial cartilage defects, mild extraction of proteoglycans or suppression of proteoglycan synthesis might improve repair.

There is limited experimental and clinical evidence (L. Rosenberg and L.L. Johnson, unpublished data) to suggest that saline irrigation of defects to extract proteoglycans or use of enzyme solutions that degrade proteoglycans allows fibrin clot formation and improves the repair response. The long-term results of these approaches to improving cartilage repair have not been reported.

Laser Stimulation of Cartilage Repair

Schultz and associates[54] found that superficial cartilage defects exposed to low-dose laser energy demonstrated a reparative process superior to that found in untreated injuries and in injuries treated with higher doses of laser energy. The mechanism of this response to low-dose laser energy is unclear, as are the long-term results of the laser-stimulated repair.

Transplantation of Chondrocytes

Another method of introducing a new cell population to repair a cartilage defect is the implantation of chondrocytes grown and maintained in culture or chondrocytes harvested from another site.[55] Chondrocytes could be maintained in a defect by implanting them in a gel, and then might synthesize a new cartilage matrix. Itay and associates[55] reported improved repair of chick articular cartilage defects following implantation of homologous embryonic chick chondrocytes embedded in a resorbable gel.

Gels or Fibrous Implants

Filling a cartilage defect with a collagen or fibrin gel,[56] a carbon fiber implant,[57] or other synthetic gels or implants could potentially facilitate repair. Mesenchymal cells might invade the gel, adhere to it, proliferate, and synthesize a new matrix. Experimental evidence suggests that using such implants may enhance the quality of the repair tissue.[56,57] Gels or implants containing chondrocytes[55] and/or chondrogenesis-stimulating factors may also facilitate cartilage repair.

Electromagnetic Fields

Chondrocytes may increase synthesis of cartilage-like proteoglycans when stimulated by pulsed electromagnetic fields.[58,59] Although this effect may be useful, thus far pulsed electromagnetic fields have not been shown to facilitate repair of actual cartilage defects.

Chondrogenesis-Stimulating Factors (Including Growth Factors)

A number of substances are known to stimulate chondrogenesis, including growth hormone and proteins extractable from bone matrix. In one experiment, treatment of surgically debrided patellofemoral surfaces in adult rabbit knees with purified bovine growth hormone appeared to stimulate cartilage repair.[60] In another set of experiments, a 31-kD bone matrix protein was found to stimulate chondrogenesis in limb-bud cell cultures.[61] The potential of this protein and other chondrogenesis-stimulating factors, including growth factors, appears to be significant, but further work is required to identify the most useful factors, find ways of delivering them to the site of injury, determining the ideal dose-response relationship, and assessing the quality of the repair tissue.

Future Directions

Recent investigations suggest that successful repair of clinically significant articular cartilage defects may be possible. The requirements for regeneration of articular cartilage include (1) a population of cells that will migrate into or can be placed in a cartilage defect where they will proliferate and differentiate into chondrocytes, (2) a matrix for the cells to invade and replace or remodel into cartilage or cells capable of creating such a matrix, (3) mechanical stimuli to enhance the formation and development of articular cartilage, (4) protection of the repair cartilage from excessive loads, and (5) maintenance or restoration of the normal shape and conformation of the joint.

Specific investigations that may help meet these requirements include the following:

Several investigators have reported that partial-thickness cartilage defects do not progress and do not appear to affect joint function adversely. This may not be true for all regions of the joint. For example, partial-thickness defects in articular cartilage regions subjected to high loads may progress while similar defects in regions subjected to smaller loads may remain unchanged. If some partial-thickness defects do progress, it would be appropriate to investigate ways of repairing them. Possible methods of stimulating repair of partial-thickness defects include shaving, extraction of proteoglycans to stimulate fibrin clot formation, use of gels and implants, chondrocyte transplantation, and use of growth factors.

Some clinical reports suggest that altering the loads applied to cartilage repair tissue may stimulate repair and improve the function and durability of the repair tissue. This possibility should be investigated experimentally to define the relationship between joint loading and cartilage repair.

Animal experiments and clinical experience indicate that, at least for certain types of cartilage injuries, joint motion improves cartilage repair. Further study is needed to determine exactly what combinations of motion and loading produce the best results for specific injuries.

Clinical observations suggest that abrasion of damaged cartilage to bleeding subchondral bone stimulates successful repair in humans. This possibility should be investigated further.

Sources of cells that may be capable of repairing cartilage defects include cartilage adjacent to the injury, periosteum and perichondrium, bone marrow, other cartilages, perivascular cells, and transplanted embryonic chondrocytes. Despite a number of studies, it is not clear which of the sources provides cells with the greatest potential for satisfactory repair. Studies should be designed to test the potential of different cell types to repair cartilage defects.

The subchondral bone matrix may provide a factor or factors that stimulate cartilage repair. One potential approach is to characterize bioactive factors in the cartilage and bone matrix that have the potential to initiate cartilage differentiation, perhaps by stimulating mesenchymal stem cell differentiation. Syftestad and Caplan report promising results with a 31-kD bone matrix protein.[61] Another such agent may be the newly isolated osteogenin. The protein, which has an apparent molecular weight of 22 kD, in conjunction with insoluble collagenous matrix initiates cartilage formation.[62]

Additional approaches include the assessment of the role of transforming growth factor beta,[63,64] a 25-kD multifunctional growth regulator that induces cartilage proteoglycan synthesis[65] in the mesenchymal cells of rat muscle. Platelet-derived growth factor,[66] fibroblast growth factor,[67] and insulin-like growth factor[68] are potentially important substances that may play a role in articular cartilage repair. Although the use of growth factors in tissue culture is relatively straightforward, assessing their actions in vivo must take into account such things as the role of degradation, the choice of drug-delivery system, and the extracellular matrix milieu. This is a formidable challenge for students of articular cartilage repair and regeneration.

Acknowledgment

Supported in part by National Institutes of Health grant RO1 AM34758–02.

References

1. Hunter W: On the structure and diseases of articulating cartilage. *Philos Trans R Soc London* 1743;9:267.
2. Paget J: Healing of injuries in various tissues. *Lect Surg Pathol* 1853;T:262.
3. Bennett GA, Bauer W: Further studies concerning the repair of articular cartilage in dog joints. *J Bone Joint Surg* 1935;17:141–150.
4. Bennett GA, Bauer W, Maddock SJ: A study of the repair of articular cartilage and the reaction of normal joints of adult dogs to surgically created defects of articular cartilage, "joint mice" and patellar displacement. *Am J Pathol* 1932;8:499–524.
5. Buckwalter JA: Articular cartilage, in American Academy of Orthopaedic Surgeons *Instructional Course Lectures, XXXII.* St. Louis, CV Mosby Co, 1983, pp 349–370.
6. Calandruccio RA, Gilmer WS Jr: Proliferation, regeneration, and repair of articular cartilage of immature animals. *J Bone Joint Surg* 1962;44A:431–455.
7. Campbell CJ: The healing of cartilage defects. *Clin Orthop* 1969;64:45–63.
8. Convery FR, Akeson WH, Keown GH: The repair of large osteochondral defects: An experimental study in horses. *Clin Orthop* 1972;82:253–262.
9. DePalma AF, McKeever CD, Subin DK: Process of repair of articular cartilage demonstrated by histology and autoradiography with tritiated thymidine. *Clin Orthop* 1966;48:229–242.
10. Furukawa T, Eyre DR, Koide S, et al: Biochemical studies on repair cartilage resurfacing experimental defects in the rabbit knee. *J Bone Joint Surg* 1980;62A:79–89.
11. Ghadially FN, Thomas I, Oryschak AF, et al: Long-term results of superficial defects in articular cartilage: A scanning electron-microscope study. *J Pathol* 1977;121:213–217.
12. Mankin HJ: Localization of tritiated thymidine in articular cartilage of rabbits: II. Repair in immature cartilage. *J Bone Joint Surg* 1962;44A:688–698.
13. Mankin HJ: The response of articular cartilage to mechanical injury. *J Bone Joint Surg* 1982;64A:460–466.
14. Mankin HJ: The reaction of articular cartilage to injury and osteoarthritis: Part I. *N Engl J Med* 1974;291:1285–1292.

15. Mankin HJ: The reaction of articular cartilage to injury and osteoarthritis: Part II. *N Engl J Med* 1974;291:1335–1340.

16. Mitchell N, Shepard N: The resurfacing of adult rabbit articular cartilage by multiple perforations through the subchondral bone. *J Bone Joint Surg* 1976;58A:230–233.

17. Meachim G: The effect of scarification on articular cartilage in the rabbit. *J Bone Joint Surg* 1963;45B:150–161.

18. Meachim G, Roberts C: Repair of the joint surface from subarticular tissue in the rabbit knee. *J Anat* 1971;109:317–327.

19. Redfern P: On the healing of wounds in articular cartilage. *Monthly J Med Sci* 1851;13:210.

20. Rubak JM: Reconstruction of articular cartilage defects with free periosteal grafts: An experimental study. *Acta Orthop Scand* 1982;53:175–180.

21. Thompson RC Jr: An experimental study of surface injury to articular cartilage and enzyme responses within the joint. *Clin Orthop* 1975;107:239–248.

22. Curtiss PH Jr: Cartilage damage in septic arthritis. *Clin Orthop* 1969;64:87–90.

23. Donohue JM, Buss D, Oegema TR Jr, et al: The effects of indirect blunt trauma on adult canine articular cartilage. *J Bone Joint Surg* 1983:65A:948–957.

24. Parsons JR, McManus E, Johnson E: Time dependent histologic and mechanical alteration of articular cartilage with joint sepsis. *Trans Orthop Res Soc* 1982;7:217.

25. Reagan BF, McInerny VK, Treadwell BV, et al: Irrigating solutions for arthroscopy: A metabolic study. *J Bone Joint Surg* 1983;65A:629–631.

26. Reimann I, Christensen SB, Diemer NH: Observations of reversibility of glycosaminoglycan depletion in articular cartilage. *Clin Orthop* 1982;168:258–264.

27. Radin EL, Ehrlich MG, Chernack R, et al: Effect of repetitive impulsive loading on the knee joints of rabbits. *Clin Orthop* 1978;131:288–293.

28. Repo RU, Finlay JB: Survival of articular cartilage after controlled impact. *J Bone Joint Surg* 1977;59A:1068–1076.

29. Dekel S, Weissman SL: Joint changes after overuse and peak overloading of rabbit knees *in vivo*. *Acta Orthop Scand* 1978;49:519–528.

30. Stover SS, Pool RR, Fischer AT: Healing in osteochondrial defects: A comparison of articulating and non-articulating locations. *Trans Orthop Res Soc* 1987;12:275.

31. Schmid A, Schmid F: Results after cartilage shaving studied by electron microscopy. *Am J Sports Med* 1987;15:386–387.

32. Mitchell N, Shepard N: Effect of patellar shaving in the rabbit. *J Orthop Res* 1987;5:388–392.

33. Whipple RR, Gibbs MC, Lasi WM, et al: Biphasic properties of repaired cartilage at the articular surface. *Trans Orthop Res Soc* 1985;10:340.

34. Reddi AH: Role of subchondral bone matrix factors in the repair of articular cartilage, in Verbruggen G., Veys EM (eds): *Degenerative Joints*. Amsterdam, Excerpta Medica, 1985, vol 2, pp 271–274.

35. Reddi AH: Extracellular bone matrix dependent local induction of cartilage and bone. *J Rheumatol* 1983;10(suppl 11):67–69.

36. Reddi AH, Wientroub S, Muthukumaran N: Biologic principles of bone induction. *Orthop Clin North Am* 1987;18:207–212.

37. Reddi AH, Huggins C: Biochemical sequences in the transformation of normal fibroblasts in adolescent rats. *Proc Natl Acad Sci USA* 1972;69:1601–1605.

38. Sampath TK, DeSimone DP, Reddi AH: Extracellular bone matrix-derived growth factor. *Exp Cell Res* 1982;142;460–464.

39. Johnson LL: *Diagnostic and Surgical Arthroscopy.* St. Louis, CV Mosby Co, 1980.

40. O'Donoghue DH: Treatment of chondral damage to the patella. *Am J Sports Med* 1981;9:12–10.

41. Haggart GE: The surgical treatment of degenerative arthritis of the knee joint. *J Bone Joint Surg* 1940;22:717–729.

42. Insall J: The Pridie debridement operation for osteoarthritis of the knee. *Clin Orthop* 1974;101:61–67.

43. Magnuson PB: Joint debridement: Surgical treatment of degenerative arthritis. *Surg Gynecol Obstet* 1941;73:1–9.

44. Radin EL, Burr DB: Hypothesis: Joints can heal. *Semin Arthritis Rheum* 1984;13:293–302.

45. Salter RB, Minster RR, Bell RS, et al: Continuous passive motion and the repair of full-thickness articular cartilage defects: A one-year follow-up. *Trans Orthop Res Soc* 1982;7:167.

46. Salter RB, Simmonds DF, Malcolm BW, et al: The biological effect of continuous passive motion on healing of full-thickness defects in articular cartilage: An experimental study in the rabbit. *J Bone Joint Surg* 1980;62A:1232–1251.

47. Engkvist O, Johansson SH: Perichondrial arthroplasty: A clinical study in twenty-six patients. *Scand J Plast Reconstr Surg* 1980;14:71–87.

48. Pastacaldi P, Engkvist O: Perichondrial wrist arthroplasty in rheumatoid patients. *Hand* 1979;11:184–190.

49. Kleiner JB, Coutts RD, Woo SL-Y, et al: The short term evaluation of different treatment modalities upon full thickness articular cartilage defects: A study of rib perichondrial chondrogenesis. *Trans Orthop Res Soc* 1986;11:282.

50. Kwan MK, Woo SL-Y, Amiel D, et al: Neocartilage generated from rib perichondrium: A long-term multidisciplinary evaluation. *Trans Orthop Res Soc* 1987;12:277.

51. O'Driscoll SW, Keeley FW, Salter RB: The chondrogenic potential of free autogenous periosteal grafts for biological resurfacing of major full-thickness defects in joint surfaces under the influence of continuous passive motion: An experimental study in the rabbit. *J Bone Joint Surg* 1986;68A:1017–1035.

52. O'Driscoll SW, Salter RB: The induction of neochondrogenesis in free intra-articular periosteal antografts under the influence of continuous passive motion: An experimental study in the rabbit. *J Bone Joint Surg* 1984;66A:1248–1257.

53. Zarnett R, Delaney JP, O'Driscoll SW, et al: Cellular origin and evolution of neochondrogenesis in major full-thickness defects of a joint surface treated by free autogenous periosteal grafts and subjected to continuous passive motion in rabbits. *Clin Orthop* 1987;222:267–274.

54. Schultz RJ, Krishnamurthy S, Thelmo W, et al: Effects of varying intensities of laser energy on articular cartilage: A preliminary study. *Lasers Surg Med* 1985;5:577–588.

55. Itay S, Abramovici A, Nevo Z: Use of cultured embryonal chick epiphyseal chondrocytes as grafts for defects in chick articular cartilage. *Clin Orthop* 1987;220:284–303.

56. Speer DP, Chvapil M, Volz RG, et al: Enhancement of healing in osteochondral defects by collagen sponge implants. *Clin Orthop* 1979;144:326–335.

57. Hart JAL: The use of carbon fibre implants for articular cartilage defects. Presented at the 47th Annual Meeting of the Australian Orthopaedic Association, Melbourne, 1987.

58. Aaron RK, Ciomber DM, Jolly G: Modulation of chondrogenesis and chondrocyte differentiation by pulsed electromagnetic fields. *Trans Orthop Res Soc* 1987;12:272.

481

59. Aaron RK, Plaas AAK: Stimulation of proteoglycan synthesis in articular chondrocyte cultures by a pulsed electromagnetic field. *Trans Orthop Res Soc* 1987;12:273.

60. Dunn AR, Sampsell R: Regrowth of articular cartilage by direct hormonal induction with growth hormone following full-thickness surgical debridement. Presented at the 53rd Annual Meeting of the American Academy of Orthopaedic Surgeons, New Orleans, Louisiana, Feb 25, 1986.

61. Syftestad G, Caplan A: A 31,000 dalton bone matrix protein stimulates chondrogenesis in chick limb and bud cell cultures. *Trans Orthop Res Soc* 1986;11:278.

62. Sampath TK, Muthukumaran N, Reddi AH: Isolation of osteogenin, an extracellular matrix-associated, bone-inductive protein, by heparin affinity chromatography. *Proc Natl Acad Sci USA* 1987;84:7109–7113.

63. Sporn MB, Roberts A, Wakefield LM, et al: Transforming growth factor-β: Biological function and chemical structure. *Science* 1986;233:532–534.

64. Seyedin SM, Thompson AY, Bentz H, et al: Cartilage-inducing factor-A: Apparent identity to transforming growth factor-β. *J Biol Chem* 1986;261:5693–5695.

65. Cheung HS, Cottrell WH, Stephenson K, et al: In vitro collagen biosynthesis in healing and normal rabbit articular cartilage. *J Bone Joint Surg* 1978;60A:1076–1081.

66. Huang JS, Proffit RT, Baenziger J-U, et al: Human platelet derived growth factor, in *Differentiation and Function of Hematopoietic Cell Surfaces*. New York, Alan R Liss, 1982.

67. Zapf J, Froesch ER: Insulin-like growth factors/somatomedins: Structure, secretion, biological actions and physiological role. *Horm Res* 1986;24:121–130.

68. Hauschka PV, Mavrakos AE, Iafrati MD, et al: Growth factors in bone matrix: Isolation of multiple types by affinity chromatography on heparin-sepharose. *J Biol Chem* 1986;261:12665–12674.

Section Nine
Meniscus

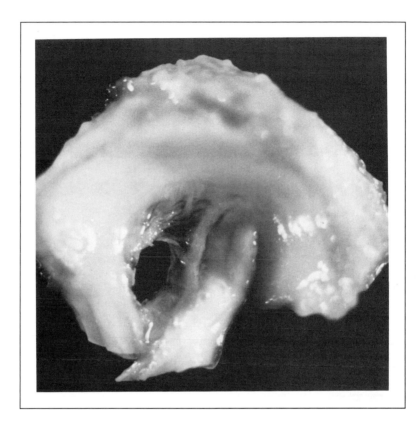

Group Leader	Group Members	Group Participants
Steven Arnoczky, DVM	Mark Adams, MD, FRCPC	Wayne Akeson, MD
	Kenneth DeHaven, MD	Mark Bolander, MD
	David Eyre, PhD	Thomas Dameron, Jr., MD
	Van Mow, PhD	Cyril Frank, MD
		Edward Grood, PhD
		Ernst Hunziker, MD
		Jack Lewis, PhD
		Helen Muir, PhD
		Barry Oakes, MD
		Lawrence Rosenberg, MD
		Melvin Rosenwasser, MD
		Ileen Stewart, MS
		Peter Torzilli, PhD

Synopsis

The meniscus plays an essential role in the normal function and health of the knee. However, studies of meniscal tissue properties and meniscal function are in their infancy. Many investigations have lacked scientific rigor, including some reports that are of dubious validity. This is surprising in view of the frequency of meniscal injuries and the number of clinical procedures used to treat these injuries. The implication that meniscectomy and meniscal injury lead to degenerative joint disease provides an additional reason to improve our understanding of the structure, function, and capacity for repair of the meniscus.

The menisci are C-shaped disks of fibrocartilage interposed between the condyles of the femur and tibia. The microarchitecture of the meniscus is dominated by the distinctive orientation of the collagen fiber network that, together with the proteoglycan-interstitial water component of the extracellular matrix, gives the meniscus its mechanical properties. Although the meniscus is known to be viscoelastic and anisotropic, the exact responses of the tissue to compression, tension, and shear have yet to be elucidated. Similarly, the exact mechanisms by which the menisci perform their functions within the joint (load-bearing, shock absorption, increasing joint stability, and joint lubrication) and the mechanisms by which the meniscal cells maintain the matrix remain unknown.

Meniscal injury can occur as an isolated lesion or in association with other joint injuries (most commonly a rupture of the anterior cruciate ligament. The meniscal lesions usually result from compression and/or shear forces. The location of the lesion within the meniscus dictates the tissue's response to injury. If the lesion occurs in the periphery of the meniscus, that is, in the area of the blood supply (peripheral 25% of the lateral meniscus; peripheral 30% of the medial meniscus), the meniscus is capable of mounting a reparative response similar to those of other vascular tissues (exudation, organization, vascularization, proliferation, and remodeling). The response to injury originates within the perimeniscal capillary plexus as well as in the meniscal synovial fringe over the area of the injury. This response can repair meniscal lesions by means of fibrovascular scar tissue that eventually changes into a tissue resembling fibrocartilage. The biochemical and biomechanical character of this tissue has yet to be determined. If the injury occurs in the avascular portion of the meniscus, little, if any, response ensues and the lesion remains relatively unchanged. This inability of meniscal lesions in the avascular portion of the tissue to heal has provided the rationale for partial meniscectomy.

Experimental studies have investigated potential ways of stimulating repair of lesions in the avascular region of the meniscus by extending the vascular response to injury through access channels and synovial abrasion. Although these techniques have been successful in the laboratory and in limited clinical trials, the long-term results are unknown. Recently, laboratory studies have shown that an exogenous fibrin clot and its associated chemotactic and mitogenic factors can induce and support an avascular reparative response within the meniscus. Although these results are preliminary, they represent exciting new concepts in our understanding of meniscal repair.

Future research must be directed at further characterizing the composition, structure, and function of the meniscus in health and disease. In particular, these studies should be directed towards establishing models that relate the mechanical functions of the meniscus to its composition and structure. Prospective clinical studies must be developed to define the natural course of various types of meniscal injuries, the possible relationships between these injuries and degenerative joint disease, the most effective techniques of meniscal repair, and the long-term results of meniscal repair. The techniques and information developed from these basic and clinical studies will allow us to understand the response of the meniscus to injury and help us to evaluate new methods of diagnosing and treating meniscal injuries.

Chapter 12
Meniscus

Steven Arnoczky, DVM
Mark Adams, MD, FRCPC
Kenneth DeHaven, MD
David Eyre, PhD
Van Mow, PhD
Michael A. Kelly, MD
Christopher S. Proctor, MD

Chapter Outline

Chapter Outline   *continued*

Introduction

Of all the structures within or about the knee, the meniscus is perhaps the least understood. Once described as the functionless remains of a leg muscle,[1] the menisci were relegated to an obscure role within the knee and were often completely removed after injury. However, basic science and clinical investigations into the function of the menisci have identified these structures as integral components in the complex biomechanics of the knee and have underscored the importance of preserving them whenever possible.

Although meniscal repair and preservation have thus become routine in orthopaedic surgery, relatively little is known about the overall biomechanical and biologic character of the meniscus. Obviously such information is needed for a complete understanding of this enigmatic tissue.

Meniscal Structure

Evolution

Millions of years ago when animals began to walk on land, the structure of their femoral-tibial joints changed to suit new mechanical demands. Part of this adaptation was the evolution of menisci, which extend the tibial condyles, mating the tibia to the femur.[2,3]

The first creatures with meniscus-like structures were amphibians. In some primitive amphibians with tails the space between the femur and tibia is entirely filled with a pliant fibrovascular tissue. In the salamander, which possess a true joint cavity, this tissue is limited to the medial portion of the joint and has been referred to as a medial meniscus. The femoral-tibial joint of the bullfrog contains two distinct elliptical structures between the tibial and femoral condyles.

In the femoral-tibial joint of the crocodile, the menisci appear as two large, imperforate masses of fibrocartilage. The lizard possesses smaller menisci; the lateral meniscus is discoid and the medial meniscus is circular, with a central perforation through which intra-articular ligaments pass.

Birds possess similar meniscal structures. Their medial meniscus is C-shaped, while their lateral meniscus remains discoid. Tetrapedal and bipedal mammals have the most developed fibrocartilaginous menisci. Although the attachments of the menisci to the tibia vary among the different orders, the crescent shape of the menisci prevails in most cases. Discoid lateral menisci do occur at times in dogs and humans, but these appear to be anomalies within the expected range of morphologic variation.

Development

In the human embryo, the menisci first exist as a condensation of intermediate-layer mesenchymal tissue and are clearly defined by the

489

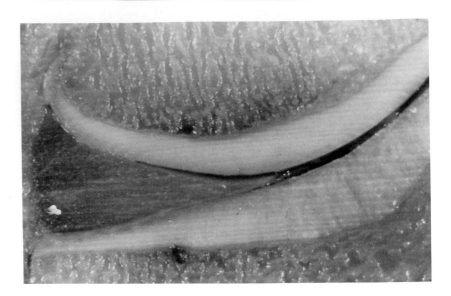

Fig. 12-1 *Frontal section of the medial compartment of a human knee illustrates the articulation of the menisci with the condyles of the femur and tibia. (Reproduced with permission from Warren RF, Arnoczky SP, Wickiewicz TL: Anatomy of the knee, in Nicholas JA, Hershman EB (eds):* The Lower Extremity and Spine in Sports Medicine. *St. Louis, CV Mosby Co, 1986, pp 657–694.)*

eighth week. They grossly resemble the semilunar shape of the adult menisci.[4,5] At this stage, the meniscus consists largely of fibroblasts without much extracellular matrix, and vascular channels permeate the entire structure. As the fetus develops, the matrix of the meniscus becomes more collagenous, with a gradual orientation of the collagen bundles into a circumferential arrangement. Joint motion is thought to be an important factor in the orientation of collagen fibers at this time. The increases in collagen content and fibrous organization continue into the postnatal period. Vessels are present throughout much of the meniscus at birth; however, by mid to late adolescence the internal and intermediate regions of the menisci appear to be avascular.

Gross Anatomy

Properly regarded as extensions of the tibia, the menisci serve to deepen the surfaces of the articular fossae of the head of the tibia for reception of the condyles of the femur.[2,6] The peripheral border of each meniscus is thick, convex, and attached to the inside capsule of the joint; the opposite border tapers to a thin, free edge. The proximal surfaces of the menisci are concave and in contact with the condyles of the femur; their distal surfaces are flat and rest on the head of the tibia (Fig. 12–1).

The semicircular medial meniscus is approximately 3.5 cm in length and considerably wider posteriorly than it is anteriorly (Fig. 12–2). The anterior horn of the medial meniscus is attached to the tibial plateau in the area of the anterior intercondylar fossa in front of the anterior cruciate ligament (ACL). The posterior fibers of the anterior horn attachment merge with the transverse ligament, which connects the anterior horns of the medial and lateral menisci. The posterior horn of the medial meniscus is firmly attached to the posterior intercondylar fossa of the tibia between the attachments of the lateral meniscus and the posterior cruciate ligament (PCL). The periphery of the medial meniscus is attached to the joint capsule throughout its length. The tibial portion of this capsular attachment is often referred to as the coronary ligament. At its midpoint, the medial meniscus is more firmly attached to the femur and tibia through a condensation in the joint capsule known as the deep medial ligament.

The nearly circular lateral meniscus covers a larger portion of the tibial articular surface than does the medial meniscus; it is approximately the same width from front to back (Fig. 12–2). The anterior horn of the lateral meniscus is attached to the tibia in front of the intercondylar eminence and behind the attachment of the ACL, with which it partially blends. The posterior horn of the lateral meniscus is attached behind the intercondylar eminence of the tibia in front of the posterior end of the medial meniscus (Fig. 12–2, *bottom*). Although the lateral meniscus is not attached to the lateral collateral ligament, there is a loose peripheral attachment to the joint capsule. The tendon of the popliteus passes immediately adjacent to the posterolateral aspect of the lateral meniscus. A groove in the peripheral surface of the meniscus accommodates the passage of the tendon. The ligaments of Humphrey (the anterior meniscofemoral ligament) and Wrisberg (the posterior meniscofemoral ligament) run from the posterior horn of the lateral meniscus to the medial femoral condyle, either just in front of or behind the origin of the PCL.

Microanatomy

Light microscopy shows that the meniscus is a fibrocartilaginous tissue made up of an interlaced network of collagen fibers interspersed with cells[7] (Fig. 12–3).

Cells   As in other connective tissues, the cells of the meniscus are responsible for synthesizing and maintaining the extracellular matrix of the tissue. It has long been debated whether the cells of the meniscus are fibroblasts, chondrocytes, or a mixture of both and whether the tissue should be classified as fibrous tissue or fibrocartilage.[8] Although these issues have yet to be resolved, two basic cell types have been described within the meniscus: a fusiform cell found in the superficial zone of the meniscus and an ovoid or polygonal cell found throughout the remainder of the tissue.[9] Although the fusiform cells resemble fibroblasts, they also resemble the chondrocytes found in the superficial (tangential) zone of articular cartilage.[8,9]

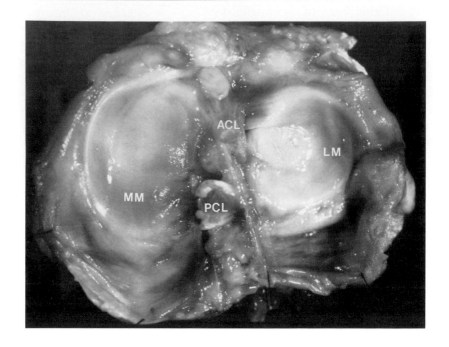

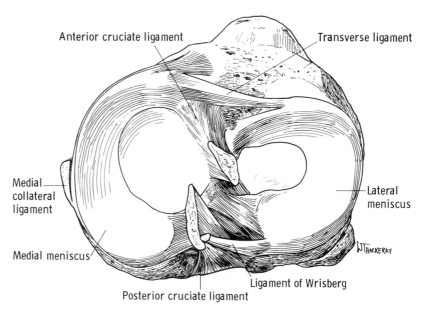

Fig. 12–2 *Photograph (top) and drawing (bottom) of a human tibial plateau show the position and attachments of the medial meniscus (MM) and lateral meniscus (LM). ACL, anterior cruciate ligament; PCL, posterior cruciate ligament. (Reproduced with permission from Warren RF, Arnoczky SP, Wickiewicz TL: Anatomy of the knee, in Nicholas JA, Hershman EB (eds): The Lower Extremity and Spine in Sports Medicine. St. Louis, CV Mosby Co, 1986, pp 657–694.)*

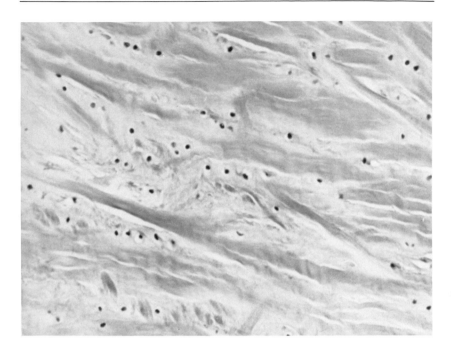

Fig. 12–3 *Human meniscus shows the histologic appearance of meniscal fibrocartilage (hematoxylin and eosin, × 100).*

Collagen Fiber Orientation   The extracellular matrix of the meniscus is primarily composed of collagen fibers. The orientation of these fibers appears to be related directly to the function of the meniscus. In a classic study Bullough and associates[10] described the orientation of the collagen fibers within the menisci (Fig. 12–4). They noted that, although the principal orientation of the collagen fibers within the meniscus is circumferential, a few small, radially disposed fibers appear on both the tibial and femoral surfaces of the menisci as well as within the middle zone (Fig. 12–5). It is thought that these radial fibers act as "ties" to provide the structural rigidity required of this extremely fibrous material and may help resist any longitudinal splitting of the menisci resulting from undue compression (Fig. 12–6). Subsequent light and electron microscopic examinations of the meniscus revealed three different collagen framework layers: a superficial layer, a surface layer just beneath the superficial layer, and a middle layer.[11,12] The superficial layer is composed of a network of fine fibrils woven into a mesh-like matrix. The surface layer is composed in part of irregularly aligned collagen bundles. In the middle layer the collagen fibers are larger and coarser and are oriented in a parallel, circumferential direction (Fig. 12–7).

Vascular Anatomy   The vascular supply to the medial and lateral menisci of the human knee originates predominantly from the lateral

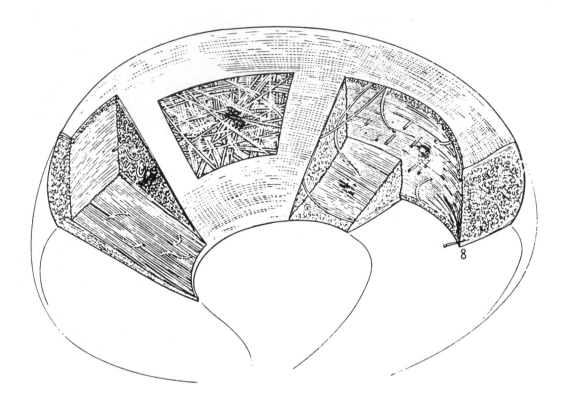

Fig. 12-4 *Schema of the meniscus shows collagen fiber orientation. (Modified with permission from Bullough PG, Munuera L, Murphy J, et al: The strength of the menisci of the knee as it relates to their fine structure.* J Bone Joint Surg *1970;52B:564-570.)*

and medial geniculate arteries (both inferior and superior).[13-15] Branches from these vessels give rise to a perimeniscal capillary plexus within the synovial and capsular tissues of the knee joint. This branching network of vessels supplies the peripheral border of the meniscus throughout its attachment to the joint capsule (Fig. 12-8). These perimeniscal vessels are oriented in a predominantly circumferential pattern, with radial branches directed toward the center of the joint (Fig. 12-9). Anatomic studies have shown that the vessels penetrate 10% to 30% of the width of the medial meniscus and 10% to 25% of the width of the lateral meniscus.

The middle geniculate artery, along with a few terminal branches of the medial and lateral geniculate arteries, also supplies vessels to the menisci through the vascular synovial covering of the anterior and posterior horn attachments. These synovial vessels penetrate the horn attachments and give rise to endoligamentous vessels that enter the meniscal horns for a short distance and end in terminal capillary loops.

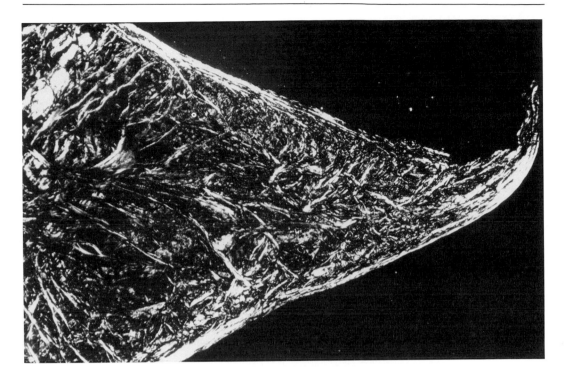

Fig. 12-5 *Cross section of a medial meniscus viewed under polarized light demonstrates orientation of collagen fibrils. Note radial orientation of fibers on articular surfaces of the meniscus (× 40). (Reproduced with permission from Arnoczky SP, Torzilli PA: The biology of cartilage, in Hunter LY, Funk FJ Jr (eds): Rehabilitation of the Injured Knee. St. Louis, CV Mosby Co, 1984, pp 148–209. Original illustration courtesy of Peter Bullough, Hospital for Special Surgery, New York.)*

A small proliferation of vascular synovial tissue is also present throughout the peripheral attachments of the medial and lateral menisci on both the femoral and tibial articular surfaces. (An exception is the posterolateral portion of the lateral meniscus adjacent to the area of the popliteal tendon.) This "meniscal synovial fringe" extends for a short distance (1 to 3 mm) over the articular surfaces of the menisci and contains small, terminally looped vessels (Fig. 12–10). Although this vascular synovial tissue adheres intimately to the articular surfaces of the menisci, it does not normally contribute vessels to the meniscal tissue.

Neuroanatomy Although the vascular anatomy of the menisci has been well described, the nerve supply to the tissue has yet to be determined in detail.[15-17] Studies in animal menisci demonstrated the presence of type I and type II nerve endings in the ligament-like meniscal horns and posterior meniscofemoral ligaments. Similarly, in human menisci the meniscal horns are much more intensely innervated

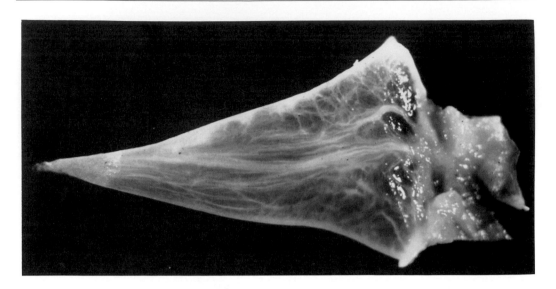

Fig. 12–6 *Cross section of a lateral meniscus shows radial orientation of fibrous "ties" within the substance of the meniscus. (Reproduced with permission from Arnoczky SP, Torzilli PA: The biology of cartilage, in Hunter LY, Funk FJ Jr (eds):* **Rehabilitation of the Injured Knee.** *St. Louis, CV Mosby Co, 1984, pp 148–209.)*

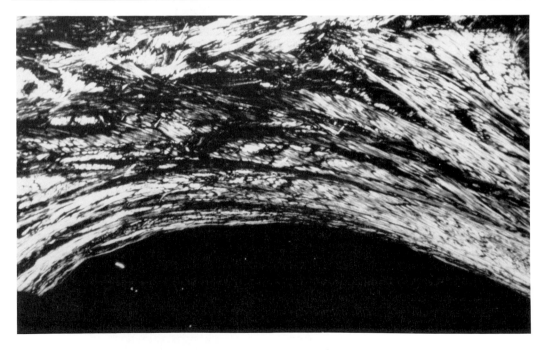

Fig. 12–7 *Longitudinal section of a medial meniscus viewed under polarized light demonstrates the orientation of the coarse, deep, circumferentially oriented collagen fibers.*

Fig. 12–8 *Superior aspect of a medial meniscus after vascular perfusion with india ink and tissue clearing with a modified Spaltholz technique. Note the vascularity at the periphery of the meniscus as well as at the anterior and posterior horn attachments. (Reproduced with permission from Arnoczky SP, Warren RF: Microvasculature of the human meniscus.* Am J Sports Med *1982;10:90–95.)*

than the meniscal bodies, the central thirds of which are totally devoid of innervation. Because the nerve endings found within the meniscus are associated with sensory characteristics, it has been postulated that this afferent arc of the nervous system may provide proprioceptive information relating to joint position.

Extracellular Matrix Composition

The dominant morphologic characteristic of the meniscus is the distinctive weave of the collagen fiber network. The two predominant fiber orientations—circumferentially along the long axis in the interior of the meniscus, and at right angles to this wrapped around in the plane of a distinct 100-μ thick surface zone—have been established by polarized light microscopy,[10] scanning electron microscopy,[11] and X-ray diffraction.[12] Fewer studies have dealt with the biochemistry of

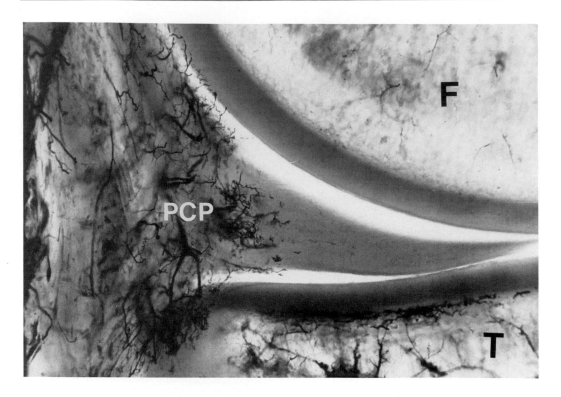

Fig. 12–9 *Frontal section 5 mm thick of the medial compartment of the knee (Spaltholz, × 3). Branching radial vessels from the perimeniscal capillary plexus (PCP) penetrate the peripheral border of the medial meniscus. F, femur; T, tibia. (Reproduced with permission from Arnoczky SP, Warren RF: Microvasculature of the human meniscus.* Am J Sports Med *1982;10:90–95.)*

meniscal fibrocartilage than with that of hyaline articular cartilage, but some statements can be made about its overall composition. Peters and Smillie[18] provided the first comprehensive description of the composition of meniscal extracellular matrix. Ingman and associates[19] confirmed that the mean composition of adult human meniscal tissue was about 75% collagen, 8% to 13% noncollagenous protein, and 1% hexosamine. They also noted some minor age-related changes.

Collagens

Meniscus contains at least four types of collagen. Type I collagen is the most abundant, making up about 90% of the total collagen. Type II collagen accounts for 1% to 2%, type V from 1% to 2%, and type VI for about 1% (J. J. Wu and D. R. Eyre, unpublished data). Although type III collagen was tentatively identified by electrophoresis, the molecule has not been identified with certainty.

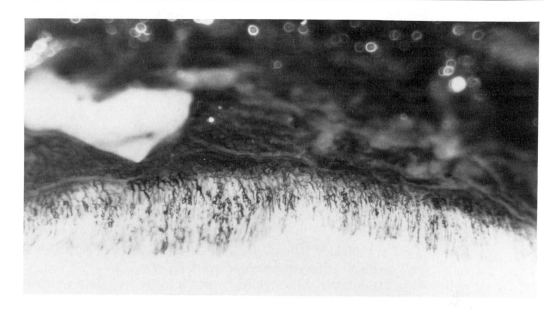

Fig. 12–10 *Peripheral border of a medial meniscus after vascular perfusion with india ink demonstrates the terminally looped capillary vessels of the synovial fringe extending over the femoral surface of the meniscus.*

Type I Collagen   The fibrous framework of the meniscus is formed primarily by coarse fibrils of type I collagen. The presence of type I collagen is the key biochemical feature that distinguishes fibrocartilage from hyaline cartilage, since the latter contains type II collagen but no type I collagen. Meniscal type I collagen must be the product of the same genes that code for type I collagen in all other connective tissues, that is, the $\alpha1(I)$ and $\alpha2(I)$ chains must have the same primary amino acid sequence as those in bone, tendon, and skin. However, meniscal type I collagen shows post-translationally regulated chemical differences in having higher hydroxylysine and hydroxylysine glycoside contents than type I collagen found in other connective tissues.[20] Meniscal collagen is also heavily cross-linked by hydroxylysine-based aldehydes, with one of the highest levels of mature, hydroxypyridinium cross-linking residues that can be found in a connective tissue rich in type I collagen.[21,22] Human meniscus contains 1 to 1.5 moles of hydroxypyridinium per mole of collagen.

Type II Collagen   There seems to be a small but consistent proportion of type II collagen in meniscal fibrocartilage.[20] It is recovered from a pepsin digest of the tissue in a 1.2M NaCl acid precipitate together with type V collagen, and can be fully resolved from type I collagen, which precipitates at 0.9M NaCl. This behavior differs from that of pepsin-solubilized type II collagen of hyaline cartilage, most of which precipitates at 0.9M NaCl. Differences in post-translational modifications (more hydroxylysine and hydroxylysine glycosides in meniscal type II collagen) seems to be responsible for this difference in solubility. Analysis after microdissection of the articulating surface zone of the meniscus from the interior tissue showed no major differences in the relative amount of type II collagen, although two to three times more type V collagen seems to be concentrated in the surface zone than in the interior.[20]

Type III Collagen   In pepsin digests of 18- to 24-month-old bovine menisci, type III collagen accounted for about 10% of the collagen that precipitated at 0.7M NaCl, the rest being type I.[20] Since the outer one third and ligamentous ends of the meniscus were removed, an origin in the fibrocartilage matrix rather than the vasculature origin seemed likely. However, meniscal type III collagen was not rigorously characterized nor was its distribution in the tissue examined.

Type V Collagen   About 1% to 2% of the collagen in meniscus is type V, with an apparent composition of $[\alpha 1(V)]_2\ \alpha 2(V)$.[20] In fact, all tissues that contain type I collagen also seem to include a similarly small proportion of type V collagen. The function of type V collagen remains unknown, although there is a growing suspicion that it may be copolymerized within the same fibrils as type I collagen.

Type VI Collagen   Denaturant (4M guanidine hydrochloride) or glycosidase (streptococcal hyaluronidase) extracts of meniscal fibrocartilage contain the intact form of type VI collagen as a prominent component on electrophoresis (equivalent to about 1% of the tissue's dry weight).[23] Type VI collagen is widely distributed in soft connective tissues as a disulfide-bonded polymer. Its precise ultrastructural distribution and function in the extracellular matrix are unclear, although it seems to be the structural protein of the interfibrillar microfilaments evident in many soft connective tissues. However, many of its properties, notably its ease of extraction, seem to make type VI collagen fit better in the class of interfibrillar glycoproteins than in the class of fibrillar collagens.[23]

Elastin

Peters and Smillie[18] reported an elastin content for normal human meniscus of 0.6% of the dry weight. There appear to be no further biochemical studies of meniscal elastin, nor have ultrastructural studies provided insights on the distribution of this minor constituent within the tissue.

Proteoglycans

The mean amount of glycosaminoglycan in human meniscus seems to be about 10% of that in hyaline articular cartilage. However, considerable variation in content with age and site in the tissue is evident. Several kinds of proteoglycan molecules are represented in meniscus, and the overall composition differs from that of hyaline cartilage proteoglycans. Habuchi and associates[24] first demonstrated that a significant fraction of the hexuronic acid of human meniscus was iduronic acid and, hence, that dermatan sulfate was present. Most of the meniscal glycosaminoglycans appeared to be copolymeric dermatan sulfate-chondroitin sulfate chains.

McNicol and Roughley[25] extracted proteoglycans from adult human menisci in 4M guanidine hydrochloride, fractionated them by density-gradient centrifugation and agarose column chromatography, and described their composition and molecular properties. Two basic types of proteoglycans, a large aggregating species and a smaller nonaggregating molecular species, were identified. A density-gradient fraction of high-buoyant density contained the large proteoglycan molecules that were capable of forming aggregates with hyaluronic acid.[26] Electron microscopy showed that these aggregates had the same structure as aggregates from fetal articular cartilage.[26] Like the aggregating proteoglycans of hyaline cartilage, which are rich in chondroitin sulfate, the molecules contained chondroitin sulfate, keratan sulfate, and sialic acid-containing oligosaccharides. In size and composition they more closely resembled proteoglycans from the articular cartilage of younger individuals rather than articular cartilage proteoglycans from individuals of the same age as the menisci. The molecules did not contain dermatan sulfate. A proteoglycan fraction of lower density from meniscus rich in dermatan sulfate did not aggregate with hyaluronic acid and belonged to the "small proteoglycan" molecular category.[25,26]

Adams and associates[27,28] reported on the proteoglycans of adult canine menisci. The glycosaminoglycan content was one eighth that of articular cartilage from the same animals, although the actual concentration varied within different regions of the meniscus. The meniscal glycosaminoglycan composition (based on chondroitinase digestion products) was constant at 60% chondroitin 6-sulfate, 25% chondroitin 4-sulfate, 10% chondroitin, and 5% dermatan sulfate. Of the total uronic acid in the tissue, 6% was in the form of hyaluronic acid. A subsequent study showed that the lower dermatan sulfate content resulted from a species difference between humans and dogs, or from exclusion of the ligamentous horns of the canine menisci.[29]

Webber and associates[30] studied the proteoglycans synthesized by rabbit meniscal fibrocartilage in explant culture in vitro. These newly synthesized proteoglycan molecules were incapable of forming aggregates with hyaluronic acid, and the authors emphasized the chemical and physical differences between meniscal proteoglycans and hyaline cartilage proteoglycans. Similar conclusions were drawn on the basis of glycosaminoglycan composition about the majority of proteoglycans in porcine meniscus.[31] However, proteoglycans synthesized in vivo and extracted from both human and canine menisci have shown the ability to form aggregates with hyaluronate.[25,26,28]

Site-dependent variations in hexosamine and uronic acid content were quantified in porcine meniscus.[22] The inner third (the thin edge) had two to four times more than outer two thirds (the thick edge) of the wedge-shaped tissue. There also was a trend to higher concentrations in the anterior horn compared with the posterior horn in both the medial meniscus and the lateral meniscus.

Noncollagenous Matrix Proteins

Denaturant extracts of adult canine meniscus were fractionated by density-gradient centrifugation and then analyzed electrophoretically for noncollagenous proteins.[32] Proteoglycan link proteins were detected immunochemically and by electrophoresis. Another protein, called 116-kD protein (its molecular size after disulfide reduction), was also present and accounted for about 4% of the protein in the tissue extract. Its immunofluorescence localization suggests that it is concentrated on collagen bundles in the tissue.[33] This protein has also been found in articular cartilage. Its function is unknown.

Metabolism and Turnover

Little is known about the rates of turnover of the extracellular matrix of meniscus. Most studies of meniscal metabolism have stemmed from comparative analyses of experimental animal models for inducing osteoarthrosis.[34,35]

Meniscal Function

The functional behavior of the meniscus is best understood when it is viewed as having two distinct phases: a fluid phase (the interstitial water with the inorganic salts dissolved in it) and a solid phase (the collagen-proteoglycan organic solid matrix composing the fluid-filled porous-permeable medium). When the meniscus is viewed as a biphasic medium, one can understand the fundamental mechanism by which the interstitial fluid flow and the porous solid matrix deformation contribute to the overall mechanical behavior of the meniscus as a tissue, and how the meniscus functions in the knee joint. Like articular cartilage, the mechanical behavior of meniscal tissue is viscoelastic when subjected to loading. It is probable that this viscoelastic response depends on two fundamental physical mechanisms: (1) the intrinsic viscoelastic properties of the macromolecules (collagen, proteoglycans, and other "minor" structural macromolecules) composing the organic solid matrix and (2) the frictional drag exerted by the interstitial fluid as it flows through the porous-permeable solid matrix. At the molecular level, little is known about how the structure-function relationship and interstitial fluid flow contribute to the viscoelastic response of meniscal tissue.

Recent investigations have dispelled the earlier belief that the menisci are functionless.[1] These investigations have demonstrated that the menisci serve a number of functions in the knee, including load-bearing, shock absorption, increasing joint stability, and joint lubrication. Indeed, one animal model for osteoarthritis uses meniscectomy to alter the mechanics of the knee joint.[36] Despite this, little is known about how meniscal tissue behaves under load and, therefore, little is known of how menisci, as discrete anatomic structures, function in the knee joint. In order to understand how the meniscus as a structure functions under physiologic loading conditions in the knee, it is necessary to know precisely the intrinsic mechanical properties of meniscal tissue under compression, tension, and shear. Once the intrinsic properties of meniscal tissue are known, precise and accurate models of the meniscus as a structure can be developed. These structural models may then be used to simulate and to predict meniscal function in the knee.

Behavior Under Confined Compression

Favenesi and associates[37] were the first to determine the intrinsic compressive modulus and the permeability of the meniscus. From the normal menisci of young, skeletally mature steers, they cored circular plugs (6.35 mm in diameter) of tissue perpendicular to the femoral surface. Each core was then split into surface and deep disks and each disk was then studied by means of the biphasic confined-compression test protocol developed by Armstrong and Mow[38] for articular cartilage. The geometric configuration of the confined-compression experiment permitted the use of the simple one-dimensional mathematical analysis developed by Mow and associates.[39] The linear biphasic theory describes the compressive creep behavior of soft, hydrated connective tissues such as articular cartilage and can be used to describe this creep behavior to the first approximation.

From this study, the creep responses of the meniscus and articular cartilage can be compared. This comparison revealed that although the creep behaviors of the two tissues are similar in character, these tissues have significantly different material coefficients. The average intrinsic elastic modulus for the meniscus is 0.411 MPa, a value less than one half that of articular cartilage, and the average permeability of the meniscus, determined from the transient phase of the compression creep test, is 8.13×10^{-16} m^4/N.s, six times less than articular cartilage.[40] In compression then, the meniscus is softer and much less permeable than articular cartilage.

From this initial study, it appears that the compressive viscoelastic creep behavior of the meniscus is governed, to a large degree, by the interstitial fluid flowing through the tissue during compression. Unfortunately, the exact details of the biphasic deformational behavior of meniscal tissue and the structure-function relationship of the macromolecules and water composing the meniscal tissue remain unknown. For example, it is not known if the permeability of the meniscus depends on compression, as is the case with articular cartilage, nor is it known whether the tissue exhibits compressive stiffening effects under finite deformation. In addition, it has not been determined whether the predominantly circumferential arrangement of collagen fiber architecture provides the tissue with an anisotropic compressive behavior, nor whether the nonhomogeneous distribution of proteoglycan throughout the tissue provides variations in the tissue's compressive properties. This type of knowledge is required in order to know how the meniscus functions in the knee under severe physiologic loading conditions. All these fundamental aspects of the mechanical behavior of meniscal tissue under compression must be determined before we can construct a reliable model to simulate and predict meniscal function in the knee.

Behavior Under Uniaxial Tension

The first attempt to determine the tensile behavior of the meniscus dates back to 1949 when Mathur and associates[41] noted medial and lateral differences in the failure strength of menisci. This observation provided an indication of the lack of homogeneity in the mechanical response of meniscal tissue under tension. Bullough and associates[10] were the first to demonstrate definitively the anisotropy of meniscal tissue under tension. These investigators reported that human meniscal tissue samples were stiff in the circumferential direction, with an approximate Young's modulus of 7.55 MPa, and soft in the radial direction, with a modulus of 0.78 MPa. More recently, Uezaki and associates[42] reported a circumferential tensile modulus on the order of 17.5 MPa for whole meniscus, thus averaging the effects of the tissue's lack of structural homogeneity. The test protocol used by Uezaki and associates, however, called for storing and testing specimens in silicone oil; thus, the physiologic relevance of these measurements is in doubt.

Modifying techniques developed for testing articular cartilage, several investigators[43-45] determined the strength and stiffness of normal bovine meniscal tissue and related these intrinsic material properties to the orientation, location, and depth of the specimens within the meniscus. Using a special freezing and microtomy process, they removed precisely shaped thin strips of meniscus parallel to the femoral surface of the meniscus. The first slice removed contained the surface layer, which was 150 to 200 μ thick; subsequent slices were 400 μ thick. Dumbbell-shaped specimens were then cut from each slice, either in a radial or a circumferential orientation, and tested under tension at constant low strain rates (0.05 cm/min). The strain rates were kept low to ensure that the biphasic fluid flow effects were kept to a minimum. In this manner, the intrinsic tensile properties of the meniscal tissue were determined.

The tensile stress-strain behavior, the tensile failure characteristics, and the morphologic characteristics of the failed specimens (obtained photographically during the tests), led to a number of interesting observations regarding the structure-function relationships of meniscal tissue. Surface specimens (1) showed no directional (circumferential and radial) variation in their tensile stress-strain behavior, (2) contracted significantly with elongation, and (3) failed transversely across the cross-sectional area with severe narrowing. From these observations it was concluded that the surface of the meniscus behaves in an isotropic manner. This conclusion is consistent with the perception that the surface layer is composed of fine collagen fibers woven in a random fashion parallel to the articulating surface.

It was also observed that deeper specimens (1) showed significant directional variations (circumferential vs radial orientation) in their tensile stress-strain behavior, (2) did not exhibit a pronounced lateral contraction with elongation, and (3) failed by accumulation of successive fiber bundle failures at different locations across the cross-sectional area. From these observations it was concluded that the deep meniscal tissue behaves in an anisotropic manner. This conclusion is consistent with the evidence that the deeper zone of meniscal tissue is composed of large collagen-fiber bundles, oriented predominantly in the circumferential direction, with each bundle appearing to act relatively independent of the others.

These conclusions were further supported by the stress-strain behavior of the deep meniscal specimens after the initiation of failure. The behavior is characterized by a gradually decreased load with successive formation of oblique cracks across the major fiber bundles composing the specimen. During this stage of specimen failure, the larger fiber bundles appeared to "pull out" of the remaining intact material, causing shear between the material and the fiber being pulled out. During this process, it is possible that the small radial fibers spanning the large circumferential fiber bundles are recruited to share some of the tensile loads. This action may break radial fibers. Thus, in the circumferentially oriented specimens, rupture and "pull out" of large fiber bundles may cause radial fiber damage and, hence, planes of weakness between the large fiber bundles. It can be hypothesized, therefore, that clinically observed bucket-handle tears of the meniscus occur along these planes of weakness. This type of tissue failure may be investigated by testing meniscal tissue specimens (prepared with the large fiber bundles aligned parallel to the axis of the cylindrical test specimen) in the shear mode where the fine radial collagen fibers joining the large fiber bundles are subject to tension.

Because the tensile stress-strain behavior of meniscal tissue is nonlinear, a single Young's modulus cannot be used to describe the stiffness characteristics of the specimens. A tangent modulus, or a modulus defined within a certain range of strain, must be used (Fig. 12–11). Average values for the tensile modulus (in the linear region of the stress-strain curve) of the surface meniscal tissue is 60 MPa. Meniscal tissue from the deeper zones is highly anisotropic and nonhomogeneous; the average tensile modulus of circumferentially oriented specimens subjacent to the surface is 200 MPa whereas those from deeper layers averaged 140 MPa and the average tensile modulus of radially oriented specimens from the subjacent and deep zones is only 10 MPa. Samples oriented circumferentially had very high failure stresses and low failure strains whereas those oriented radially had very low failure stresses and high failure strains.

From these studies, it appears that the tensile behavior of meniscal tissue is nonlinear, anisotropic, and nonhomogeneous, and that the tensile behavior depends on the structural organization of the collagen (large circumferentially oriented fiber bundles and small radially oriented fibers) and proteoglycans composing the tissue. Little is known about the details of the tensile deformational behavior and how this

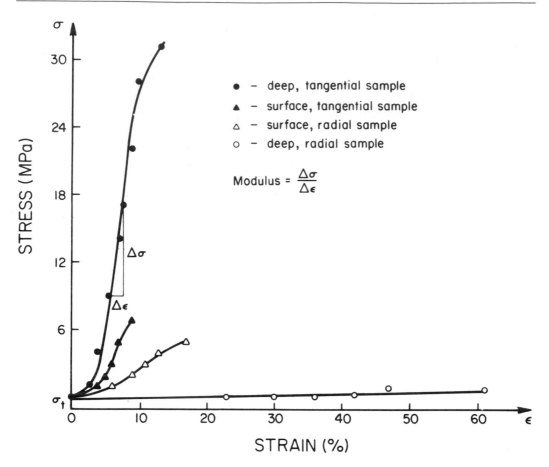

Fig. 12–11 *Typical stress-strain curve for the four anatomic variations of the deep and surface fibers of the meniscus.*

behavior is related to the molecular and structural organization of the tissue. Further, little is known about how these properties and the structure-function relationships influence the function of the meniscus as a load-bearing structure in the knee. In addition, nothing is known about the viscoelastic tensile behaviors of the meniscus, such as creep and stress-relaxation, or how the tissue behaves when subjected to various strain-rates under tensions. We need such knowledge to understand how the meniscus responds under the prolonged loading or rapid loading conditions that can occur in physiologic conditions. Furthermore, the finite deformational efforts under large tensile strains, the anisotropic effects associated with the arrangement of collagen fiber architecture, and the lack of homogeneity in material properties associated with the nonuniform distribution of collagen and proteoglycans must be better characterized. In order to know how the meniscus functions as a load-bearing structure in the knee, we must determine all these tensile characteristics of meniscal tissue. No reliable model can be constructed to simulate or predict meniscal function in the knee without knowledge of these structure-function relationships that determine meniscal tissue behavior.

Behavior Under Pure Shear

Because it is certain that the meniscus is subjected to significant levels of stress and strain in shear and that shear failure may be the mechanism causing bucket-handle tears, it is surprising that no rigorously controlled studies have ever been performed to determine the intrinsic shear properties of meniscal tissue. This may be the result of the inherent technical difficulties associated with performing a pure shear experiment on soft, hydrated connective tissues such as the meniscus. To create pure shear, a precise amount of torsional shear stress or strain must be imposed onto precisely prepared thin wafers of meniscal tissue. Such techniques are presently being developed in some laboratories; as yet, however, no results have been reported. Obviously, knowledge of the shear properties of meniscal tissue, including the effect of structural anisotropy and lack of homogeneity, must be determined before a complete biomechanical description of the meniscus can be developed.

Structure-Function Relationships

The material properties and functional abilities of the meniscus depend on the properties and interactions of its individual components. For example, the transient compressive behavior (creep and stress-relaxation) of a biphasic material is governed both by the polyanionic proteoglycans and by the frictional resistance to fluid flow through the solid matrix. Although the compressive stiffness of articular cartilage is significantly influenced by proteoglycan molecules, the proteoglycan content of the meniscus is one tenth that of articular cartilage. Therefore, polyanionic proteoglycans may contribute little to the transient or dynamic compressive stiffness of the meniscus. It has been shown, however, that the permeability of the meniscus is extremely low and may thus be the dominant factor determining its dynamic compressive behavior. Furthermore, the tensile behavior of the meniscus has been shown to be anisotropic and nonhomogeneous, and these characteristics appear to be strongly dependent on collagen fiber arrangement. The surface layer, with its randomly oriented fiber pattern, exhibits an isotropic behavior under tension. This fine structural organization of the surface layer appears to be necessary in that it provides the meniscus with a fine membrane enclosing the coarse fibrous material of the interior as well as a smooth and nearly frictionless articulating surface. The deeper layers, however, with their large, circumferentially oriented fiber bundles, are strong in the circumferential direction and weak in the radial direction.

Thus, it appears that there is a strong structure-function relationship between the organization of the extracellular matrix of the meniscus and its material properties. Aside from the observations presented above, however, little is known about the molecular details of these relationships. For example, it is not clear how the proteoglycan component of the tissue influences its mechanical properties or how the meniscus, with such a small proteoglycan content, can have a relatively high water content.[18,27] Indeed, nothing is known of how the porous-permeable, solid, collagen-proteoglycan extracellular matrix of the meniscus controls the transport of water through the tissue; is this mechanism similar to that known for articular cartilage, in which the proteoglycan content is ten times greater? In view of this lower concentration, does the proteoglycan component play an important role in meniscal function? What other components or mechanisms are important in regulating the water content of the meniscus and what limits the permeability of the meniscus to just a fraction (one sixth) of that of articular cartilage? Are these differences in material properties functionally important?

Functional Roles of the Meniscus

The importance of the menisci in normal knee function is well documented. Although this considerable body of knowledge has established the biomechanical roles of the meniscus, little detail is known regarding the mechanisms by which the menisci actually carry out these functions. Data on the intrinsic material properties of meniscal tissue are sparse.

Load-Bearing   In 1936 King[46] first suggested that the meniscus is directly involved in force transmission across the knee. It has since been demonstrated by many investigators that this is one of its primary roles. Kettelkamp and Jacobs,[47] using radiographic techniques, were among the first to identify the contact areas within a knee. Walker and Erkman,[48] using cast models, also identified the contact areas. Using miniature pressure transducers, these investigators also measured the pressure profiles over the meniscal-tibial surfaces. From these studies, it was determined that the menisci carry large loads and that the contact areas within the knee change with flexion and rotation. Fukubayashi and Kurosawa[49] obtained similar results using pressure-sensitive film to measure the pressure profile. More recently, Ahmed and Burke[50] measured the static pressure distribution of the tibial plateau using a thin film "micro-indentation transducer." The latter investigation demonstrated that at least 50% of the compressive load of the knee joint is transmitted through the meniscus in extension with 85% being transmitted in 90 degrees of flexion.

In an attempt to develop an understanding of how the meniscus transmits loads in the knee, Shrive and associates[51] analyzed a section of the meniscus and proposed a model of load distribution within the tissue. If it is assumed that the meniscal surfaces are frictionless, then only normal stresses exist; thus, the direction of the stress vectors acting on these surfaces is simply defined by the normal vector of the meniscus surface. Because of the wedge shape of the meniscus, the

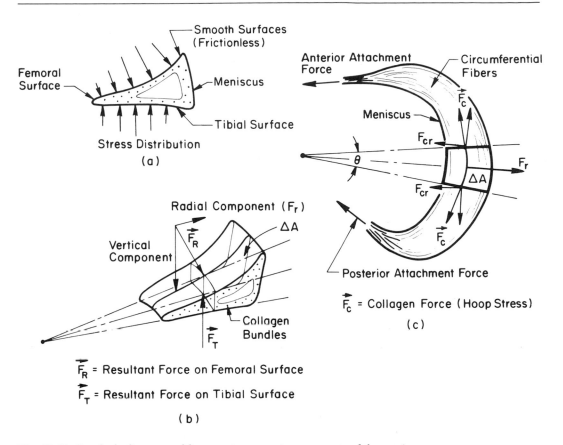

Smooth Surfaces
(Frictionless)

Femoral
Surface

Meniscus

Tibial Surface

Stress Distribution

(a)

Anterior Attachment
Force

Circumferential
Fibers

Meniscus

$\vec{F}_c$

F_{cr}

θ

F_r

ΔA

F_{cr}

$\vec{F}_c$

Posterior Attachment Force

$\vec{F}_c$ = Collagen Force (Hoop Stress)

(c)

Radial Component (F_r)

ΔA

Vertical
Component

$\vec{F}_R$

Collagen
Bundles

$\vec{F}_T$

$\vec{F}_R$ = Resultant Force on Femoral Surface

$\vec{F}_T$ = Resultant Force on Tibial Surface

(b)

Fig. 12–12 *Free-body diagrams of forces acting on various segments of the meniscus.*

resultant compressive force acting on the femoral surface does not parallel the resultant compressive force acting on the tibial surface. Thus, there must be a net radial component that tends to push the meniscus radially outward (Fig. 12–12). This radial expansion stretches the large collagen bundles in the circumferential direction and is resisted by the tibial attachments at the anterior and posterior horns of the meniscus. Thus, the radial component of the loading force is balanced by tensile stresses developed in the circumferentially oriented collagen fiber bundles.

From the available experimental data on knee joint mechanics, it is clear that the meniscus must resist large compressive loads. In view of the low intrinsic compressive modulus and low permeability, interstitial fluid flow through the porous-permeable, collagen-proteoglycan solid matrix appears to be the mechanism that stiffens the meniscus in compression. The large tensile hoop stress required to balance the radial component of applied force appears to be provided by the high

509

tensile modulus of the circumferentially oriented collagen bundles and their strong insertion onto the intercondylar region of the tibia.

Although these heuristic arguments are attractive and plausible, there is no concrete scientific evidence that the menisci actually function in the manner proposed by Shrive and associates.[51] We need to develop a three-dimensional computer model of the meniscus, incorporating all the known material behaviors of the tissue, in order to predict the stresses and strains within the meniscus when it is subjected to the action of loading similar to that found in the knee. For example, the simple predictions from the axisymmetric material and loading model used by Shrive and associates cannot predict the incidence of posterior horn tears in the meniscus. Is this the result of local variations in the intrinsic material properties of the meniscus or of differences in the loads acting over the tissue, or both? No information is available on the level of shear stress and strain within the tissue; thus, it is not possible to ascertain if the hypothesis related to bucket-handle tears is correct.

Shock Absorption   Another presumed function of the meniscus is shock absorption. During normal gait, impulse loading generates intermittent shock waves that propagate from the heel up to and through the knee. Several investigators[52-54] have determined the compressive load-deformation response of the knee with and without the meniscus and suggested that the soft, viscoelastic meniscus may function to attenuate these shock waves. More recently, Voloshin and Wosk[55] used an accelerometer to determine that the normal knee has a shock-absorbing capacity about 20% higher than knees that have undergone meniscectomy. The musculoskeletal system's inability to absorb shock has been strongly implicated in the development of osteoarthritis.[56] Thus, menisci appear to play an important role in shock absorption in the knee. This function is in turn provided by the intrinsic material properties of the meniscal tissue; its low intrinsic stiffness and permeability appear to be ideal for this purpose. It is clear from the previous discussion on material properties that the solid organic matrix is softer and less permeable than that of articular cartilage. With loading and deformation, the interstitial fluid is forced to flow through a porous matrix of extremely low permeability; thus, the meniscus acts as an efficient medium to dissipate and absorb energy and, therefore, to attenuate the shocks within the knee joint during daily activities.

It is important to realize that this attractive scenario is based solely on circumstantial evidence. For example, there are no studies characterizing the nature of energy dissipation in meniscal tissue nor how this behavior may affect shock absorption within the knee. In addition, no studies have been undertaken to determine the structural-functional behavior of the meniscus; thus, it is not known which element of the tissue is specifically involved in energy storage or energy dissipation.

Joint Stability   In addition to their involvement in load transmission and shock absorption, the menisci are also thought to increase knee stability by maintaining proper relative positioning of the femur with respect to the tibia. Some investigators believed that the menisci provide this stabilizing function by reducing the incongruency between the convex femoral condyles and the relatively flat tibial plateaus.[57] Others attempted to evaluate the stabilizing effects of the menisci in the knee experimentally by load-deformational studies in the anterior-posterior and medial-lateral directions and by torque-rotational studies in the varus-valgus and screw axis directions.[58-66] The results of these studies suggest that the extent of knee stability provided by the menisci is influenced by a number of factors, including articular surface geometry, knee flexion angle, compressive joint load, and ligamentous and capsular integrity. Although increased knee joint laxity after meniscectomy has been reported,[61-63] a consensus of opinion concerning the stabilizing role of the meniscus in the otherwise intact knee has not been achieved. Many investigators believe that meniscectomy alone does not significantly increase joint instability[59,60,65,67] (defined as an increase of deformation or rotation at given amounts of force and torque). However, when meniscectomy is associated with ligament insufficiencies, especially ACL deficiency, joint instability is significantly increased.[59-61,65-67] Shoemaker and Markolf[66] investigated the stabilizing actions of the meniscus in the ACL-deficient knee and concluded that, when a contact force was applied, "compression and buttressing" of the menisci resisted anterior tibial subluxation. Furthermore, as previously suggested by Levy and associates,[65] the posterior horn of the medial meniscus is most important in providing this stability. Thus, it seems that as the primary stabilizing structures of the knee are damaged, the menisci play an increasingly important role in overall knee joint stability.

Diarthrodial joint articulation, specifically knee joint articulation, depends on the passive constraints provided by the ligaments and intra-articular structures such as the menisci, as well as by actively involved related muscle groups. Precise gliding motion of the articulating surfaces is required to generate the thin (25-μ) fluid film necessary for joint lubrication.[68] Because the knee is an extremely complex anatomic structure, each of the structures providing support and constraint is required for proper knee function. To understand knee joint mechanics, it is necessary to know the precise manner in which the menisci function under physiologic conditions. At present no information is available, in terms of the three-body problem, describing the articulation of the femoral condyle, meniscus, and tibial plateau. In light of the nonlinear and biphasic nature of the load-bearing tissues of the knee, the three-body problem offers an extraordinary challenge that is yet to be resolved. Little understanding of meniscal function is possible without an understanding of this problem.

Joint Lubrication   MacConaill[69] suggested that the meniscus may play an important role in the lubrication of the knee joint. Using engineering analysis, he postulated that the joint surfaces were separated by a thin layer of fluid. This fluid was thought to be continuously dragged into the gap between the wedge-shaped meniscus and the femoral and tibial articulating surfaces. By providing joint conformity, MacConaill hypothesized, the meniscus promotes the viscous hydrodynamic action required for full fluid-film lubrication. Classic hydrodynamic lubrication, however, requires continuous, high-speed relative motion between two bearing surfaces. Because joint motions in vivo are, at best, intermittent, the classic hydrodynamic lubrication mechanism of fluid-film generation is highly unlikely in diarthrodial joints. However, the low coefficient of friction exhibited by diarthrodial joints argues strongly that fluid-film lubrication operates during some portion of normal physiologic range of motion of the joint, but the exact manner in which this occurs is unknown.[68]

McCutchen[70] proposed that fluid exudation from articular cartilage may be the mechanism of lubricant film generation in diarthrodial joints. Mow and associates[71] provided mathematical analyses to describe the exact mode of fluid transport through articular cartilage during function. Normal articular cartilage contains, on average, 80% water; 70% of this water is freely exchangeable with the surrounding bathing medium (for example, synovial fluid).[40,72] Thus, sufficient fluid volume is available in articular cartilage to provide the 10 to 25 μ of fluid film required for joint lubrication in a diarthrodial joint (providing that the hydrodynamics of fluid flow permits the formation of this lubricant film). It is not clear, however, whether the interstitial fluid of the menisci, in which the average water content is 74%,[18,27] serves as a source of lubricant since meniscus permeability is only one sixth that of articular cartilage. That is to say, it is not known if fluid exudation occurs across the meniscal surface in the same manner as it occurs across the articular cartilage surface. Thus, no information exists as to whether the meniscus actually serves a function in knee joint lubrication. Indeed, fundamental knowledge in this area is needed for the understanding of the biomechanics of knee joints. If the meniscus does play a role in knee joint lubrication, then changing meniscal geometry will affect the pressure profile over the femoral and tibial surfaces, thus altering the stresses transmitted to the articular cartilages. These changes will affect the long-term viability of cartilage and hence the well-being of the knee.

Mechanical Effects of Meniscectomy

It is apparent from the above discussion that the meniscus does serve a load-bearing function in the knee joint; hence, investigators have evaluated the changes in load distribution at the knee joint after meniscectomy. Kurosawa and associates[54] demonstrated a 30% to 50% reduction in the contact area after meniscectomy in otherwise intact knees. More recently, Ahmed and Burke[50] determined that medial meniscectomy reduces the contact area by 50% to 70%, with the greater reductions occurring at greater loads. It was also concluded that this reduced contact area alters the pressure distribution within the knee so that: (1) the peak stress is increased; (2) the high-pressure areas are focused over small regions of the tibial plateau; and (3) there is a marked increase in the pressure gradients. Bourne and associates[73] concluded that partial and total medial meniscectomy significantly alters the strain distribution in the proximal tibia. These investigations demonstrated that meniscectomy profoundly alters the manner in which load-bearing occurs at the knee joint.

Fairbank[74] was the first to postulate that loss of the load-bearing role of the meniscus after total meniscectomy could lead to degenerative changes in the knee joint. These changes include narrowing of the involved joint space, the formation of an osteophytic ridge on the femoral condyle over the side of the removed meniscus, and flattening of the marginal half of the femoral articular surface. These degenerative changes were later well documented and it is now thought that meniscectomy increases the risk of degenerative joint disease.[75–77]

Moskowitz and associates[36] developed an experimental model of osteoarthritis in rabbits by performing a meniscectomy. In this, and similar osteoarthritis models, the histologic,[32,78,79] biochemical,[80,81] and biomechanical[82] changes in articular cartilage were determined. These changes have been shown to resemble closely changes in articular cartilage in the naturally occurring human osteoarthritic process. Furthermore, Cox and associates[78] reported that the amount of degenerative change that occurs after meniscectomy is directly proportional to the amount of meniscus removed.

In humans, post-meniscectomy degenerative changes appear first in the area of the tibial plateau not normally covered by the meniscus, and then progress into the previously covered areas.[83–85] Bullough and associates[86] showed that there are biochemical differences between these two regions of normal human tibial plateau. They found that the water content and proteoglycan extractability were increased in the uncovered areas. In addition to these biochemical changes, the material properties of the uncovered articular cartilage are biomechanically inferior to those of the covered articular cartilage.[87] These findings may well explain why articular cartilage degeneration after meniscectomy occurs first in regions not normally covered by the meniscus. Although these important changes in articular cartilage appear to be related to meniscal function, no study assessing the exact effects of meniscectomy has been performed in humans. Obviously, this is the result of the difficulty of obtaining the required specimens from human knee joints that have undergone meniscectomy. This difficulty could be addressed by a small, easily inserted, arthroscopically directed probe that measures the material properties of the knee joint's articular cartilage in vivo. The technical difficulties of developing such a probe have contributed to our lack of knowledge in this important area of investigation.

Meniscal Injury and Repair

Traumatic Tears

Traumatic tears of the menisci usually occur in young (13- to 40-year-old), active people, especially those who engage in running, cutting, contact, or agility sports. The injuries may or may not be contact-related. Noncontact mechanisms include acceleration or deceleration stresses combined with a sudden change in direction, or squatting and twisting. These types of movements subject the menisci to a combination of compression and shear forces that can tear the tissue. External contact forces (valgus, varus, or hyperextension) applied to the knee in various degrees of flexion and rotation can also result in compression and shear forces on the menisci. Contact forces applied to the knee are more likely than noncontact forces to cause associated injury to knee ligaments (collateral ligaments, cruciate ligaments, or posterior capsule), although the ACL is frequently injured by noncontact mechanisms.

Traumatic tears of the menisci are commonly associated with ACL tears caused by either contact or noncontact mechanisms as long as the injured extremity is weightbearing at the time of injury. Under these circumstances, approximately one half of the acute ACL tears will also have significant associated meniscal lesions. However, if the ACL tear is caused by a nonweightbearing mechanism (such as can occur in skiing), the menisci are rarely torn.

Traumatic injuries usually produce longitudinal or transverse vertical tears.

Vertical Longitudinal Tears These tears are by far the most frequent traumatic tears associated with the medial and lateral menisci, with medial tears outnumbering lateral tears approximately 2.5 to 1. Peripheral or almost peripheral tears of sufficient length (more than 1 cm) are unstable with stress and usually cause symptoms. These injuries frequently require surgical treatment. Short (less than 5 mm), full-thickness tears and partial-thickness tears are usually stable, rarely cause symptoms, and can usually be left alone.

Vertical Transverse Tears These tears are much less common than vertical longitudinal tears, and almost always involve the middle third of the lateral meniscus. Occasionally they occur in the medial meniscus. Stable lesions less than 5 mm long may be left alone, but those longer than 5 mm are unstable, usually cause clinical problems, and frequently warrant surgical treatment. Unstable radial tears can extend anteriorly or posteriorly, becoming parrot-beak or flap tears.

Degenerative Tears

Degenerative meniscal tears (horizontal cleavage/flap or degenerative/complex) occur with increasing frequency with advancing age, particularly in individuals over the age of 40 years. Although people more than 40 years of age do sustain traumatic tears, degenerative lesions typically have no history of specific injury or the degree of stress is considered minor (for example, squatting to pick up something, getting up from a chair, or out of a car). These degenerative lesions can involve either or both of the menisci and are frequently associated with degenerative articular cartilage.

When a degenerative meniscal tear is present in a joint with other degenerative changes, it is frequently difficult, if not impossible, to determine whether the clinical symptoms are related to the meniscus, the degenerative articular cartilage changes, or both. These degenerative articular cartilage and meniscal changes are believed to be independent, concurrent processes rather than a causally related sequence of events.

Although these categories of meniscal lesions provide some idea as to the possible mechanism of meniscal injury, they do not give any indication of the ability of the meniscus to mount a reparative response. This ability (or inability) has been thought to depend on the location of the lesion with respect to the blood supply of the meniscus. Thus, we must examine the response of the meniscus to injuries within the vascular and avascular portions of the tissue.

Injuries in the Vascular Portions of the Meniscus

Although Annandale[88] was credited with the first surgical repair of a torn meniscus in the 1880s, it was not until 1936, when King[89] published his classic experiment on meniscal healing in dogs, that the biologic limitations of meniscal healing were set forth. King demonstrated that for meniscal lesions to heal they must communicate with the meniscal blood supply in the periphery of the meniscus. Once the lesion is in contact with the vascular bed, the inflammatory phase of wound repair is possible and the lesion heals in a manner similar to other vascular tissue: exudation of blood and the formation of a wound hematoma and fibrin scaffold, migration of vessels from the adjacent vascular areas, migration and proliferation of cells within the wound, and remodeling. Indeed, experimental studies[90–94] have shown that after injury within the vascular zone of the meniscus a fibrin clot forms and acts as a scaffolding for the ingrowth of vessels from the perimeniscal capillary plexus (Fig. 12–13, *top*). This vascular ingrowth is accompanied by migration and proliferation of undifferentiated mesenchymal cells. Eventually, the lesion is filled with a cellular fibrovascular scar tissue that "glues" the wound's edges together (Fig. 12–13, *center*). Although experimental studies[90,91] have shown that spontaneous repair of complete radial lesions occurs by ten weeks, the repair tissue resembles fibrovascular scar tissue (Fig. 12–13, *bottom*). Modification of this scar tissue into normal-appearing meniscal fibrocartilage, however, requires several months. A similar repair scenario was observed in complete radial meniscal lesions that were sutured.[91] The healing response within the vascular area of the meniscus appears to be extrinsic in origin. The exact contribution of the meniscal cells to the repair response (intrinsic response) is unknown. Although this repair tissue appears to provide a functional union of the meniscal segments, the material properties of the tissue have yet to be determined.

Injuries in the Avascular Portions of the Meniscus

Although King's study provided an encouraging explanation of meniscal repair, it also provided the more sobering conclusion that the

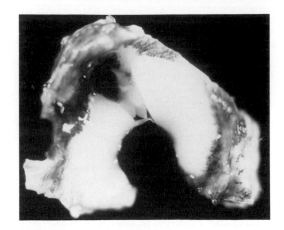

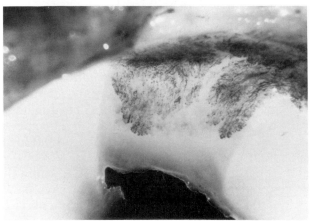

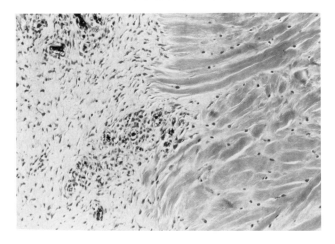

Fig. 12–13 *Canine medial meniscus perfused with india ink after complete transection.* **Top:** *At two weeks the vessels have proliferated from the perimeniscal capillary plexus into the organized clot.* **Center:** *At six weeks a fibrovascular scar has united the two segments of the menisci and there is proliferation of vascular synovial tissue over the scar.* **Bottom:** *The healing meniscus at the junction of the fibrovascular scar and the adjacent normal meniscal tissue (hematoxylin and eosin, × 100). (Reproduced with permission from Arnoczky SP, Warren RF: The microvasculature of the meniscus and its response to injury: An experimental study of the dog. Am J Sports Med 1983;11:131–141.)*

avascular regions of the meniscus, like articular cartilages, are incapable of repair.[89] He concluded that the meniscal cells are incapable of mounting a reparative response after injury and thus depend on an extrinsic source of repair, a vascular supply. To determine if this conclusion is correct, several studies have examined the response of the meniscus to injury in its avascular zone.

Sharp Incision Experimental studies in several species have shown that after sharp incision in the avascular zone of the meniscus there is no gross or microscopic evidence of repair or reaction.[89,90,92–97] Electron microscopic examination of these lesions after six months demonstrated that, although the wall of the incision developed a spongy appearance and became encrusted with electron-dense material, the fibrocartilage adjacent to the incision showed no noteworthy changes.[95] Cellular proliferation, such as the chondrocyte clusters seen in injured or osteoarthritic cartilage, was not evident, nor was there any sign of overt cell death. Thus, the sharp incision (excision) of avascular meniscal tissue apparently has neither a stimulatory nor degenerative effect upon the meniscus.

Electrosurgery Electrocautery for meniscectomy has been tried in both clinical and experimental settings.[98,99] Using various electrocautery tips and electrocautery generators with power outputs from 127 W (at a 500-ohm load) to 375 W (at a 500-ohm load), Miller and associates[98] and Schosheim and Caspari[99] evaluated the effect of electrocautery on the adjacent meniscal tissue by routine histologic techniques. Schosheim and Caspari noted that approximately 0.05 to 0.15 mm of tissue necrosis occurred after electrosurgical resection of rabbit menisci. Six weeks after surgery, they observed a hypercellular response adjacent to the resection margin. It should be noted, however, that since the rabbit meniscus is more vascular than the menisci of other species, this response may not reflect a purely meniscal fibrochondrocyte response and may, in fact, reflect a vascular response.

Laser The use of an infrared carbon dioxide laser as tool for meniscectomy has been evaluated experimentally in rabbits.[100,101] An infrared light with a wavelength of 10.6 μ was absorbed by meniscal fibrocartilage with by-products of heat, water vapor, and a small residue of carbon ash. The adjacent meniscal tissue contained viable chondrocytes within 50 μ of the margin of resection and the collagen fiber architecture was not altered. As with the sharp dissection, laser resection of the meniscus resulted in no apparent stimulation of the meniscal cells.

Joint Instability Models

Morphologic Changes ACL transection profoundly affects canine menisci. One week after transection a mild streakiness appears in the posterior medial meniscus. Small vertical and horizontal tears usually appear by four weeks. By 12 to 16 weeks the posterior medial meniscus invariably tears. By 32 weeks the posterior portion of the medial me-

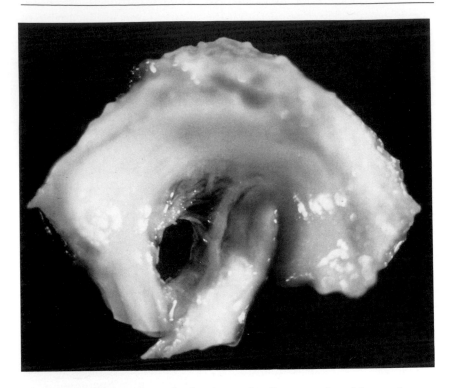

Fig. 12–14 *Medial meniscus of a dog six months after transection of the anterior cruciate ligament. Note the degenerative tearing of the posterior portion of the meniscus.*

niscus is shattered and the anterior portion shows changes, but usually remains intact even as late as 64 weeks after surgery (Fig. 12–14).[102] The lateral meniscus shows only a slight streaking by 12 weeks and a few progressive changes.[103]

Biochemical Changes The water content of the menisci increases as soon as one week after ACL transection and remains increased. The medial meniscus changes more than the lateral meniscus.[104]

Incorporation of ³H-proline into hydroxyproline is 11.2-fold higher in menisci from knees that have been operated on than in menisci from control knees.[34] This increase is of the same order as that in articular cartilage.

After ACL transection in early experimental osteoarthritis, the canine meniscus shows a more than 30% decrease in chondroitin sulfate and a more than 50% decrease in keratan sulfate compared with control values. In late experimental and natural spontaneous osteoarthritis, there was a more than 70% increase in both substances compared with control values.[104] Similar responses have been reported in human menisci. Whether the early changes in the composition of the tissue result from preferential loss of proteoglycans rich in keratan sulfate or from faster synthesis of proteoglycans rich in chondroitin sulfate is not known. Structural changes in the meniscal proteoglycans have not been defined to the same degree as those in articular cartilage proteoglycans.

After ACL transection, proteoglycan synthesis increases to a much greater extent in menisci than it does in articular cartilage both in vivo and in vitro.[35]

Degenerative areas of menisci have a higher content of noncollagenous protein than do normal menisci.[19] It is not known whether this represents imbibition of exogenous proteins by fibrillated tissue or neosynthesis and accumulation of endogenous proteins. The 550-kD matrix protein increases in the articular cartilage in this model of osteoarthritis,[105] but such studies have not been reported for menisci.

Effects on Articular Cartilage   It has been suggested that any increased incidence of osteoarthritis in individuals who have undergone meniscectomy might be related to the original trauma or subsequent meniscal degeneration. However, the increased incidence of articular cartilage degeneration in the knees of rabbits,[106–108] dogs,[109,110] and monkeys[111] after total or partial meniscectomy alone argues against this. The lesions in rabbits seldom, if ever, progress to full-thickness cartilage loss. The damage appears greater in the rabbit after partial meniscectomy. However, in dogs, the articular cartilage lesions were worse after total meniscectomy and were worse if there was no meniscal regeneration.[109] In monkeys, the articular cartilage lesions progressed to a focal, full-thickness cartilage loss.

Changes With Spontaneous Joint Degeneration

Incidence   Meniscal degeneration occurs with the development of osteoarthritis. In 100 random autopsy examinations, meniscal horizontal cleavage lesions were found in 18.4% of knees with grade 0 osteoarthritis and 61.5% of knees with grade 3 osteoarthritis.[112]

Biochemical Changes   Peters and Smillie[18] noted an increase in the glycosaminoglycan concentration in menisci with horizontal cleavage lesions. Ghosh and associates[113] showed the hexosamine concentrations in degenerate human menisci to be increased by almost 30% over those in normal human menisci. Herwig and associates[114] noted that with increasing grades of degeneration there was an increase in the glycosaminoglycan content per dry weight (but not wet weight in their study). Thus, menisci in spontaneous and experimental canine osteoarthritis and in spontaneous human osteoarthritis show evidence of increased proteoglycans. The magnitude and direction of these changes make it unlikely that this represents a dissection artifact.[29]

Crystals   The incidence of calcium pyrophosphate dihydrate (CPPD) crystals in menisci increases with increasing age,[115] as does the incidence of hydroxyapatite calcification in the vessels of the outer one third of the meniscus. Hydroxyapatite can form as dystrophic calcification in areas of menisci that have been subjected to trauma. Bjelle[116] suggested that changes in the proteoglycan matrix may predispose to CPPD deposition. Thus, it is conceivable that the proteoglycan changes in menisci in which osteoarthritis is developing may lead to CPPD deposition. These crystals can evoke an inflammatory response, which could in turn evoke or activate proteases and catabolic factors, thus further aggravating the degenerative process.

Age-Related Changes

Morphologic Changes   Meachim[117] studied the morphologic changes in the mensici of 94 subjects, aged 21 to 94 years, concentrating on the lateral meniscus and excluding subjects with evidence of joint disease. Fraying of the middle portion of the free meniscal edge was minimal in the younger subjects, but became progressively more common and severe with age. In the medial menisci studied, the changes were more common in the anterior portion. Histologic foci of "eosinophilic necrosis of collagen" within the substance of the menisci unrelated to surface changes were uncommon in younger menisci, but became common by 80 years of age.

Biochemical Changes   McNichol and Roughley[25] found that extractable proteoglycans increased in menisci from older humans, a finding opposite to that of articular cartilage. As in articular cartilage, the proportion of chondroitin 6-sulfate to chondroitin 4-sulfate increased with age.

Crystals   The incidence of CPPD deposition in the substance of the meniscus and hydroxyapatite deposition in the vessels of the outer third increases with age.[115]

Meniscal Regeneration

The rationale for total meniscectomy was often based on the ability of a meniscus or a meniscus-like structure to regenerate after total removal of the meniscus. However, there exists some controversy about whether this is possible.[118-121] This controversy may have resulted from confusion about the extent of meniscectomy (whether it was partial or total) and the fact that much of the data regarding meniscal regeneration is derived from investigations in animals.

Experiments in rabbits, sheep, and dogs after total meniscectomy have demonstrated that there is a regrowth of a structure similar in shape and texture to the removed meniscus (Fig. 12–15)[78,122,123] This regenerated tissue initially has the histologic appearance of fibrous connective tissue without cartilage cells. Long-term (seven-month) evaluation of meniscal regeneration in dogs, however, has shown that the regenerated tissue eventually resembles fibrocartilage and contains cells that appear to be chondrocytes.[122]

For this meniscus-like tissue to regenerate, however, the entire meniscus must be resected to expose the vascular synovial tissue. In partial meniscectomy, the excision must extend into the peripheral vasculature of the meniscus. Smillie[121] observed that in humans the most perfect replica of a regenerated meniscus follows total meniscectomy.

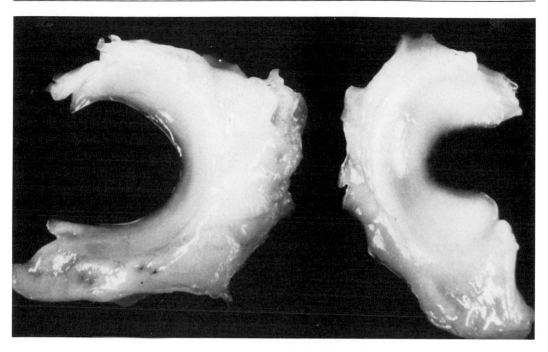

Fig. 12–15 *Medial meniscus of a dog.* **Left:** *Regeneration seven months after total meniscectomy.* **Right:** *Normal medial meniscus of the opposite limb. (Reproduced with permission from Arnoczky SP, Torzilli PA: The biology of cartilage, in Hunter LY, Funk FJ Jr (eds):* Rehabilitation of the Injured Knee. *St. Louis, CV Mosby Co, 1984, pp 148–209. Original illustration courtesy of Gretchen Flo, Michigan State University.)*

Experimental studies in rabbits have shown the importance of the peripheral synovial tissue in meniscal regeneration. None of the animals had evidence of tissue regrowth at 12 weeks following total meniscectomy and synovectomy. However, regrowth of a meniscus-like structure followed total meniscectomy alone in 83% (15 of 18) of the animals.[123]

Although the frequency and degree of meniscal regeneration in humans has not been established, the presence of regenerated meniscus-like tissue has been reported after total meniscectomy and in cases of knee arthroplasty.[120,121,124,125] These regenerated tissues grossly resemble normal menisci, and consist of a fibrocartilage-like tissue formed from chondrocytes and a dense connective tissue matrix.[125]

Thus, it appears that the synovial and peripheral meniscal vasculatures are capable of generating a connective tissue replacement for the removed meniscus. It should be noted, however, that this regeneration is not always complete and does not occur in all cases.

Meniscal Remodeling

Although meniscal regeneration occurs only after total or subtotal (into the peripheral vascular zone) meniscectomy, a remodeling response has been observed in the avascular zone of the meniscus after partial meniscectomy (Fig. 12–16, *top*). Because previous studies have shown the meniscus to be incapable of mounting a reparative response in the absence of a blood supply, the origin of this remodeling response was quite puzzling.

In an experimental study Arnoczky and associates[126] demonstrated that this reparative response probably results from an extrameniscal accretion of new tissue adjacent to the meniscectomy site. They theorized that a hemarthrosis fills the dead space adjacent to the meniscectomy site after partial meniscectomy, creating a scaffolding for the formation of new tissue. Indeed, it was observed that a fibrin clot, populated by mononuclear cells (presumably free-floating synovial cells), formed adjacent to the partial meniscectomy surface. These cells assumed the form of fibroblasts and eventually fibrochondrocytes and synthesized a homogeneous matrix (Fig. 12–16, *center*). Although this tissue was grossly and histologically different from the normal meniscus (Fig. 12–16, *bottom*), it appeared to remodel the inner rim of the meniscus functionally. In the experimental study only 67% of the menisci showed evidence of the remodeling response. Although the exact percentage of menisci that remodel after partial meniscectomy in humans is not known, arthroscopic "second looks" confirm that it does not always occur. This lends support to the theory that remodeling depends on an extrinsic source (the presence of an organized hematoma adjacent to the meniscectomy surface) and not on an intrinsic response from the meniscus itself.

Meniscal Repair

The importance of the meniscus in normal knee function has been established through an understanding of its function within the joint and by observing the degeneration of the joint in its absence. This understanding has led to efforts to repair and preserve as much of the meniscus as possible.

The basic indication for meniscal repair has been a vertical tear of significant length (more than 10 mm) within the vascular zone of the meniscus. These repairs have been carried out through "open" approaches, arthroscopic techniques, or a combination of both. Although the results, to date, have been encouraging, only 15% to 20% of meniscal tears actually occur within the vascular zone and are thus amenable to repair by the traditional techniques. In addition, the degenerative changes within the menisci that occur with age may further limit the ability of the tissue to be repaired.[112]

New areas of research must look toward methods of enhancing the reparative response of the meniscus and extending the potential areas of repair within the meniscus. In addition, the biomechanical and physiologic character of the repair tissue must undergo closer scrutiny to determine the extent of its physical and chemical maturation.

Improving the Repair Response

Classically, meniscal healing has been determined by the relationship of the lesion to the peripheral synovial blood supply of the me-

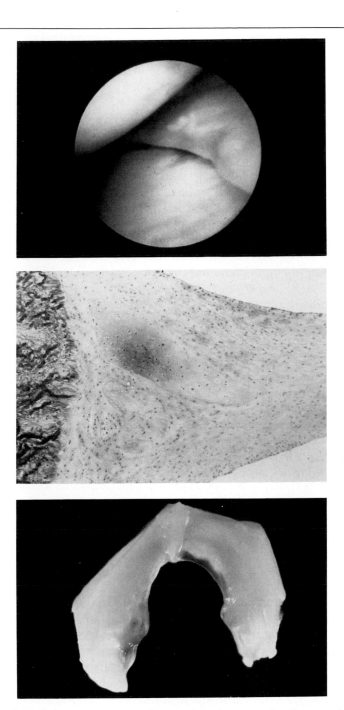

Fig. 12–16 *Medial meniscus.* **Top:** *Arthroscopic view nine months after partial arthroscopic meniscectomy in the avascular zone of a human meniscus. Note the remodeling of the meniscectomy surface with a translucent tissue.* **Center:** *The tissue covering the meniscectomy surface at 26 weeks. The tissue appears histologically to be fibrocartilage but is still markedly different from the normal adjacent meniscus (hematoxylin and eosin, × 100).* **Bottom:** *A canine meniscus 26 weeks after partial meniscectomy. Note that the meniscectomy surface has been remodeled by a translucent tissue that has reestablished a uniform inner border to the meniscus and has conformed to normal meniscal contours. (Reproduced with permission from Arnoczky SP, Warren RF, Kaplan N: Meniscal remodeling following partial meniscectomy: An experimental study in the dog.* Arthroscopy *1985;1:247–252.)*

niscus.[89–92] Thus, in evaluating methods of improving the reparative response of menisci, we must identify the extent of the blood supply in vivo and find ways to extend this vascularity into the heretofore avascular areas of the menisci. In addition, the concept of avascular repair must be explored.

Evaluating the Extent of Meniscal Blood Supply

To date, clinical methods of evaluating the extent of the blood supply within the meniscus have been limited to the visualization of a bleeding peripheral surface. The use of newer techniques may allow for the in vivo identification and quantification of blood flow within the meniscus. This, in turn, may prove valuable in determining if the blood supply is sufficient to support the reparative response.

The currently available techniques for evaluating the blood flow of structures within the knee include laser Doppler, tissue fluorometry, microsphere injections, and hydrogen ion washout techniques. In the clinical situation, however, only the laser Doppler and tissue fluorometry techniques are practical.

Laser Doppler Flowmetry   Laser Doppler flowmetry has been used to measure blood flow within capillary beds.[127–129] A 2-mW helium-neon red laser light is directed through an optical fiber to a probe. When the tip of the probe is placed in direct contact with tissue, the light enters the tissue and is repeatedly reflected, refracted, and gradually absorbed. The laser Doppler flowmeter has a measuring volume that can be regarded as a hemisphere that has a radius or depth of penetration of 1 to 1.5 mm, with its center under the endpoint of the optical fiber on the probe tip at the tissue surface. All blood cells traversing this volume are struck by the light, partly reflecting it, so the light undergoes a Doppler shift. In the measuring volume, there is a mixture of unshifted and Doppler-shifted light. The frequency of distribution and magnitude of the Doppler-shifted light are related to the number of blood cells moving through the measured volume and to their velocity, yet are independent of the movement direction of the individual blood cells. The light is picked up by two efferent optical fibers arranged in parallel, with the afferent fiber carrying the light to the probe head. The efferent fibers conduct the original light frequencies and the Doppler-shifted light frequencies to the photodetectors, where the light frequencies are converted to electrical signals. The output signal, blood cell flux, is expressed in terms of volts and is proportional to blood blow. Because it is independent of direction, blood cell flux is not an absolute indicator of blood flow, and since the laser Doppler sensor beam therefore ignores directional changes, end capillary loops will artificially increase flow measurements. Another disadvantage is that the laser sensor beam penetrates only 1.5 mm into the tissue surface. Thus, it is very sensitive to surface capillary geometry and it ignores deep flow.

This technique has been used experimentally to evaluate blood flow in sheep menisci.[129] The highest blood cell flux values were found at the periphery of the meniscus and at the anterior and posterior horn attachments. This is in agreement with other in vivo studies.[97]

Fluorometry   Fluorescence has long been used in ophthalmic and plastic surgery to evaluate tissue viability. Newer tissue fluorometers can be used in daylight and are sensitive enough to detect microscopic doses of fluorescein in the tissues. Although these instruments have digital displays to quantify the fluorescence, they do not actually measure blood flow or volume.

Fanton and Andrish,[97] in an experimental study in dogs, demonstrated the ability of fluorescein dye to delineate the blood supply of the meniscus and proposed using this technique in a clinical situation. It should be noted that although fluorescein may be useful in demonstrating the presence of a blood supply, it is still complicated by the potential problems of nausea, anaphylaxis, hypotension, and acute pulmonary edema.[130]

It should be remembered that, although quantification of blood supply may be important to determining the potential for tissue repair, the effective level of microcirculation necessary to heal a meniscal tear remains unknown.

Extending the Blood Supply Into the Avascular Areas of the Meniscus

Because the role of the peripheral vasculature in the repair of meniscal lesions has been so well established, it seems that extension of blood vessels into the avascular portions of the meniscus would be the most direct way to improve meniscal healing. This manipulation of the peripheral vasculature has been attempted in several ways.

Vascular Access Channels   Initial attempts to extend the peripheral vascularity of the meniscus into the avascular zone used the creation of vascular access channels. In one experimental study,[90] a longitudinal lesion in the avascular portion of the medial meniscus of dogs was connected, at its midportion, to the peripheral vasculature of the meniscus by a full-thickness vascular access channel (Fig. 12–17, *top*). Vessels from the peripheral meniscal tissues later migrated into the channel and healed the meniscal lesion by the proliferation of fibrovascular scar tissue (Fig. 12–17, *bottom*). This same mechanism was successful in healing lesions in the avascular portion of sheep[95] and rabbit[96] menisci. Nonetheless, the function of the meniscus may be compromised through the destruction of the integrity of the peripheral meniscal rim. It should be noted, however, that because the vasculature of the meniscus can penetrate up to 33% of the meniscal width, a vascular access channel can be created without completely disrupting the peripheral rim of the meniscus.

A modification of this technique uses the creation of a conduit or tunnel to facilitate ingrowth of vessels to an avascular lesion. In one such experimental study, the investigators utilized a 1.85-mm trephine to create a vascular conduit that connected a lesion in the avascular portion of the meniscus with the peripheral vasculature.[131] The results

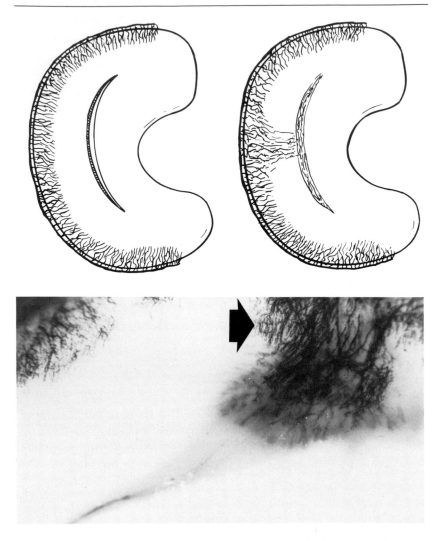

Fig. 12–17 *Canine meniscus.* ***Top:*** *Schema shows location of a longitudinal lesion in the avascular portion of the meniscus (at left) and creation of a vascular access channel at its midportion to permit vascular ingrowth from peripheral tissues at right).* ***Bottom:*** *At four weeks, perfusion with india ink shows vessels entering from the vascular access channel (arrow) and progressing within a fibrin scaffold into the anterior limb of the lesion. (Reproduced with permission from Arnoczky SP, Warren RF: The microvasculature of the meniscus and its response to injury: An experimental study in the dog.* Am J Sports Med *1983;11:131–141.)*

indicated that such a conduit works only if the knee is immobilized and weightbearing is prevented. Conduits of different sizes were not evaluated.

Synovial Pedicle Flaps   In an effort to bring vessels to an avascular meniscal lesion without incising meniscal tissues, the use of synovial pedicle flaps has been explored. In these experimental studies, a flap of vascular synovial tissue adjacent to the meniscus was sutured into a longitudinal lesion made in the avascular portion of menisci of dogs, sheep, and rabbits.[95,96,131] The results indicted that in some instances the synovial flap effected a reparative response in these avascular lesions. However, as was seen in the study of vascular conduits, immobilization and the prevention of weightbearing were often critical in determining the success of this technique. Perhaps the attritional wear of the synovial flap secondary to joint motion and/or weight-bearing may limit its efficacy in all but immobilized joints.

Synovial Abrasion   Another technique of "manipulating" the vascular supply of the meniscus uses synovial abrasion to incite a proliferative response in the synovial fringe of the meniscus.[132] As noted previously, the meniscal synovial fringe is a vascular synovial tissue that extends over the femoral and tibial articular surfaces of the meniscus. Although it does not contribute vessels to the meniscus under normal circumstances, it plays a major role in the healing of meniscal lesions in contact with the peripheral vasculature of the meniscus. It was thought that by stimulating (through rasping or abrading) the synovial fringe a proliferative vascular response could be extended over the meniscal surface to previously avascular areas of the meniscus. Although clinical results are encouraging, the exact extent and quality of the repair have yet to be determined.

Avascular Meniscal Repair

For more than 50 years it has been axiomatic that meniscal lesions outside the peripheral vasculature of the meniscus could not be repaired. This principle is based on the belief that meniscal cells are incapable of mounting a reparative response and that a blood supply is a prerequisite for wound repair.

In a recent study, however, Webber and associates[133] showed that meniscal fibrochondrocytes are indeed capable of proliferation and matrix synthesis when exposed to chemotactic and mitogenic factors normally present in the wound hematoma. Using cell cultures, they demonstrated that meniscal cells exposed to platelet-derived growth factor were able to proliferate and synthesize an extracellular matrix.

In normal wound repair, hemorrhage from vascular injury gives rise to a fibrin clot that provides a scaffolding on which to support a reparative response. In addition, the clot produces substances, such as platelet-derived growth factor and fibronectin, that act as chemotactic and mitogenic stimuli of reparative cells.

In an experimental study, Arnoczky and associates[134] evaluated the ability of an exogenous fibrin clot to stimulate and support a reparative response in the avascular portion of the meniscus. In this study full-thickness defects 2 mm in diameter were made in the avascular portion of the medial menisci of dogs. The defects were filled with an exogenous fibrin clot, and the healing response evaluated at intervals from one week to six months postoperatively.

The defects filled with a fibrin clot healed through a proliferation of fibrous connective tissue that eventually became a fibrocartilaginous tissue. The fibrin clot appeared to act as a chemotactic and mitogenic stimulus for reparative cells as well as providing a scaffolding for the repair process (Fig. 12–18, *top*). The origin of these repair cells was not determined but they were thought to arise from the synovium as well as the adjacent meniscal tissue.[135] Although the repair tissue was histologically (Fig. 12–18, *center*) and grossly (Fig. 12–18, *bottom*) different from the normal adjacent meniscal tissue, it may represent a functional scar that can effectively repair lesions in the avascular portion of the meniscus.

Future studies must be directed at identifying the growth factors essential in stimulating the meniscal fibrochondrocytes to proliferate and synthesize an extracellular matrix. In addition, the capabilities and limitations of the meniscal cellular response as well as the contribution of extrameniscal cells such as synovial cells must be studied in detail.

Future Directions

The menisci are integrally involved in normal knee joint function, and injury to these structures may lead to progressive degenerative disease of the joint. Unfortunately, few details are known about the way in which the intrinsic properties of meniscal tissue provide the tissue's mechanical functions in the knee joint and critical examination of the potential for successful repair of meniscal injuries has just begun. Investigations in the following areas will lead to improved treatment of meniscal injuries and possibly to methods of restoring meniscal function when the meniscus cannot be repaired successfully.

Meniscus Composition, Structure, and Nutrition

Microassays, immunohistochemical techniques, and modern ultrastructural analyses should be used to better define the composition and fine structure of the meniscus. Special attention should be given to collagen types, locations, and interactions with proteoglycans and other matrix components. Better localization and quantification of neural elements within the menisci and their associated structures should also be included. In addition, studies are needed to characterize the metabolism of the matrix components and to identify the relative contributions of synovial fluid components to meniscus nutrition.

Meniscal Cells

The meniscus has traditionally been characterized as fibrocartilage like the anulus fibrosis, pseudarthroses, and ligament attachments. It is important to determine whether meniscal cells are identical to those found in other fibrocartilaginous tissues or whether they are phenotypically different.

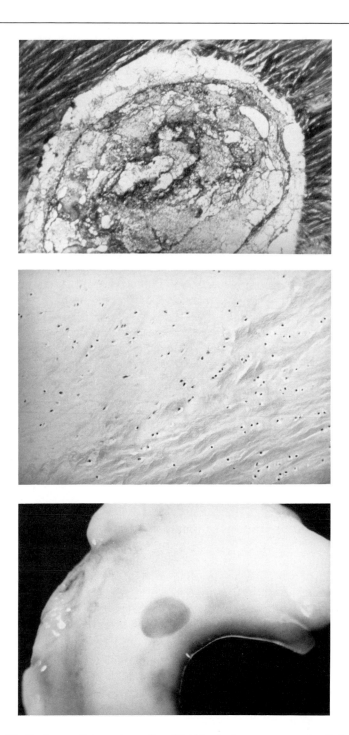

Fig. 12–18 *Canine medial meniscus.* ***Top:*** *The histologic appearance of a fibrin clot within the defect at one week (PTAH, × 50).* ***Center:*** *The meniscus-defect junction at six months. The tissue within the defect is fibrocartilage although it is histologically different from meniscal fibrocartilage (hematoxylin and eosin, × 100).* ***Bottom:*** *Gross appearance at six months. (Reproduced with permission from Arnoczky SP, Warren RF, Spivak J: Meniscal repair using an exogenous fibrin clot: An experimental study in dogs. J Bone Joint Surg, in press.)*

Meniscus Function

Although important functions have been ascribed to the meniscus, little detailed mechanical information is available on how the meniscus carries out these functions or how the tissue responds to various stresses. Studies characterizing the mechanical functions of the meniscus should be directed at establishing models to relate mechanical functions to the composition and structure of the meniscus and the interactions among the meniscus, the synovial fluid, and the adjacent articular cartilage. This approach could also help define the mechanisms of meniscal injury.

Meniscus Models

Precise biomechanical models of meniscal tissue, similar to that developed for articular cartilage, will help define the functional and deformational behavior of the meniscus under physiologically relevant loads. These models of the meniscus must include the nonlinearity, anisotropy, lack of homogeneity, and biphasic characteristics exhibited by the meniscal tissue. In addition, accurate geometric forms should be developed so that differences between the medial and lateral meniscus and differences between the anterior and posterior aspects of the meniscus can be included. To develop such models, we need more detailed descriptions of the material behavior of meniscal tissue under tensile, compressive, and shear conditions. The development of these models to describe meniscal function will lead to an understanding of the mechanical response of menisci in normal and abnormal environments. These models will provide the means to evaluate plausible causes of meniscal injury and degeneration, and the effects of different surgical procedures on meniscal repair, including partial and total meniscectomy.

Meniscal Injury and Degenerative Joint Disease

The clinical implications of meniscal injury have been documented by several investigators. However, the exact progression of the degenerative sequelae of the various types of meniscal injuries has yet to be defined. This could be accomplished by a prospective clinical study documenting the natural course of various types of meniscal injury and correlating magnetic resonance imaging and arthroscopic findings with the clinical symptoms and changes in articular cartilage and subchondral bone.

Meniscal Repair

Although the ability of the vascularized region of the meniscus to heal has been established, we need further studies of the exact mechanisms by which this occurs and biochemical, cellular, and mechanical evaluations of the repaired tissue. Although meniscal repairs are frequently performed, it is not clear which methods of surgical repair produce the best results, nor have the long-term results of surgical repair been well documented. Some reports have challenged the belief that the meniscus is incapable of mounting a reparative response in its avascular region. The possibility of successfully repairing an avascular, dense fibrous structure needs additional study.

References

1. Sutton JB: *Ligaments: Their Nature and Morphology.* London, MK Lewis & Co., 1897.
2. Bullough PG, Vosburgh F, Arnoczky SP, et al: The menisci of the knee, in Insall JN (ed): *Surgery of the Knee.* New York, Churchill Livingstone, 1984, pp 135–146.
3. Van Sickle DC, Kincaid SA: Comparative arthrology, in Sokoloff L (ed): *The Joints and Synovial Fluid.* New York, Academic Press, 1978, vol 1, pp 1–47.
4. Clark CR, Ogden JA: Development of the menisci of the human knee joint: Morphological changes and their potential role in childhood meniscal injury. *J Bone Joint Surg* 1983;65A:538–547.
5. Kaplan EB: The embryology of the menisci of the knee joint. *Bull Hosp Joint Dis* 1955;16:111–124.
6. Warren R, Arnoczky SP, Wickiewicz TL: Anatomy of the knee, in Nicholas JA, Hershman EB (eds): *The Lower Extremity and Spine in Sports Medicine.* St. Louis, CV Mosby Co, 1986, pp 657–694.
7. Wheater PR, Burkitt HG, Daniels VG: *Functional Histology.* New York, Churchill Livingstone, 1979.
8. Ghadially FN: *Fine Structure of Synovial Joints.* London, Butterworths, 1983.
9. Ghadially FN, Thomas I, Yong N, et al: Ultrastructure of rabbit semilunar cartilages. *J Anat* 1978;125:499–517.
10. Bullough PG, Munuera L, Murphy J, et al: The strength of the menisci of the knee as it relates to their fine structure. *J Bone Joint Surg* 1970;52B:564–570.
11. Yasui K: Three dimensional architecture of human normal menisci. *J Jpn Orthop Assoc* 1978;52:391–399.
12. Aspden RM, Yarker YE, Hukins DWL: Collagen orientations in the meniscus of the knee joint. *J Anat* 1985;140:371–380.
13. Arnoczky SP, Warren RF: Microvasculature of the human meniscus. *Am J Sports Med* 1982;10:90–95.
14. Danzig L, Resnick D, Gonsalves M, et al: Blood supply to the normal and abnormal menisci of the human knee. *Clin Orthop* 1983;172:271–276.
15. Day B, MacKenzie WG, Shim SS, et al: The vascular and nerve supply of the human meniscus. *Arthroscopy* 1985;1:58–62.
16. O'Connor BL, McConnaughey JS: The structure and innervation of cat knee menisci, and their relation to a "sensory hypothesis" of meniscal function. *Am J Anat* 1978;153:431–442.
17. Wilson AS, Legg PG, McNeur JC: Studies on the innervation of the medial meniscus in the human knee joint. *Anat Rec* 1969;165:485–492.
18. Peters TJ, Smillie IS: Studies on the chemical composition of the menisci of the knee joint with special reference to the horizontal cleavage lesion. *Clin Orthop* 1972;86:245–252.
19. Ingman AM, Ghosh P, Taylor TKF: Variation of collagenous and non-collagenous proteins of human knee joint menisci with age and degeneration. *Gerontologia* 1974;20:212–223.
20. Eyre DR, Wu JJ: Collagen of fibrocartilage: A distinctive molecular phenotype in bovine meniscus. *FEBS Lett* 1983;158:265–270.
21. Eyre DR, Oguchi H: The hydroxypyridinium crosslinks of skeletal collagens: Their measurement, properties and a proposed pathway of formation. *Biochem Biophys Res Commun* 1980;92:403–410.

22. Nakano T, Thompson JR, Aherne FX: Distribution of glycosaminoglycans and the nonreducible collagen crosslink, pyridinoline in porcine menisci. *Can J Vet Res* 1986;50:532–536.

23. Wu JJ, Eyre DR, Slatter HS: Type VI collagen of the intervertebral disc: Biochemical and electronmicroscopic characterization of the native protein. *Biochem J*, to be published.

24. Habuchi H, Yamagata T, Iwata H, et al: The occurrence of a wide variety of dermatan sulfate-chondroitin sulfate copolymers in fibrous cartilage. *J Biol Chem* 1973;248:6019–6028.

25. McNicol D, Roughley PJ: Extraction and characterization of proteoglycan from human meniscus. *Biochem J* 1980;185:705–713.

26. Roughley PJ, McNicol D, Santer V, et al: The presence of a cartilage-like proteoglycan in the adult human meniscus. *Biochem J* 1981;197:77–83.

27. Adams ME, Muir H: The glycosaminoglycans of canine menisci. *Biochem J* 1981;197:385–389.

28. Adams ME, McDevitt CA, Ho A, et al: Isolation and characterization of high-buoyant-density proteoglycans from semilunar menisci. *J Bone Joint Surg* 1986;68A:55–64.

29. Adams ME, Ho YA: Localization of glycosaminoglycans in human and canine menisci and their attachments. *Connect Tissue Res* 1987;16:269–279.

30. Webber RJ, Norby DP, Malemud CJ, et al: Characterization of newly synthesized proteoglycans from rabbit menisci in organ culture. *Biochem J* 1984;221:875–884.

31. Norby DP, Goldberg VM, Malemud CJ, et al: Proteoglycans from pig menisci—major differences from proteoglycans of articular cartilage. *Trans Orthop Res Soc* 1982;7:116.

32. Fife RS: Identification of link proteins and a 116,000-dalton matrix protein in canine meniscus. *Arch Biochem Biophys* 1985;240:682–688.

33. Fife RS, Hook GL, Brandt KD: Topographic localization of a 116,000-dalton protein in cartilage. *J Histochem Cytochem* 1985;33:127–133.

34. Eyre DR, McDevitt CA, Billingham, MEJ, et al: Biosynthesis of collagen and other matrix proteins by articular cartilage in experimental osteoarthrosis. *Biochem J* 1980;188:823–837.

35. Sandy JD, Adams ME, Billingham MEJ, et al: In vivo and in vitro stimulation of chondrocyte biosynthetic activity in early experimental osteoarthritis. *Arthritis Rheum* 1984;27:388–397.

36. Moskowitz RW, Davis W, Sammarco J, et al: Experimentally induced degenerative joint lesions following partial meniscectomy in the rabbit. *Arthritis Rheum* 1973;16:397–405.

37. Favenesi JA, Shaffer JC, Mow VC: Biphasic mechanical properties of knee meniscus. *Trans Orthop Res Soc* 1983;8:57.

38. Armstrong CG, Mow VC: Variations in the intrinsic mechanical properties of human articular cartilage with age, degeneration, and water content. *J Bone Joint Surg* 1982;64A:88–94.

39. Mow VC, Kuei SC, Lai WM, et al: Biphasic creep and stress relaxation of articular cartilage in compression?: Theory and experiments. *J Biomech Eng* 1980;102:73–84.

40. Woo SL-Y, Mow VC, Lai WM: Biomechanical properties of articular cartilage, in Skalak R, Chien S (eds): *Handbook of Bioengineering*. New York, McGraw-Hill, 1987, pp4.1–4.41.

41. Mathur PD, McDonald JR, Ghormley RK: A study of the tensile strength of the menisci of the knee. *J Bone Joint Surg* 1949;31A:650–654.

42. Uezaki N, Kobayashi A, Matsushige K: The viscoelastic properties of the human semilunar cartilage. *J Biomech* 1979;12:65–73.

43. Roth V, Mow VC: The intrinsic tensile behavior of the matrix of bovine articular cartilage and its variation with age. *J Bone Joint Surg* 1980;62A:1102–1117.

44. Mow VC, Whipple RR: Biology and mechanical properties of cartilage and menisci, in AAOS Committee on Basic Sciences: *Williamsburg Seminar, 1984: Resource for Basic Science Educators.* Park Ridge, American Academy of Orthopaedic Surgeons, 1984, pp 165–197.

45. Whipple RR, Wirth CR, Mow VC: Anisotropic and zonal variations in the tensile properties of the meniscus. *Trans Orthop Res Soc* 1985;10:367.

46. King D: The function of semilunar cartilages. *J Bone Joint Surg* 1936;18:1069–1076.

47. Kettelkamp DB, Jacobs AW: Tibiofemoral contact area—determination and implications. *J Bone Joint Surg* 1972;54A:349–356.

48. Walker PS, Erkman MJ: The role of the menisci in force transmission across the knee. *Clin Orthop* 1975;109:184–192.

49. Fukubayashi T, Kurosawa H: The contact area and pressure distribution pattern of the knee: A study of normal and osteoarthrotic knee joints. *Acta Orthop Scand* 1980;51:871–879.

50. Ahmed AM, Burke DL: In-vitro measurement of static pressure distribution in synovial joints: Part I. Tibial surface of the knee. *J Biomech Eng* 1983;105:216–225.

51. Shrive NG, O'Connor JJ, Goodfellow JW: Load-bearing in the knee joint. *Clin Orthop* 1978;131:279–287.

52. Seedholm BB, Hargreaves DJ: Transmission of the load in the knee joint with special reference to the role of the menisci. *Eng Med* 1979;8:220–228.

53. Krause WR, Pope MH, Johnson RJ, et al: Mechanical changes in the knee after meniscectomy. *J Bone Joint Surg* 1976;58A:599–604.

54. Kurosawa H, Fukubayashi T, Nakajima H: Load-bearing mode of the knee joint: Physical behavior of the knee joint with or without menisci. *Clin Orthop* 1980;149:283–290.

55. Voloshin AS, Wosk J: Shock absorption of meniscectomized and painful knees: A comparative in vivo study. *J Biomed Eng* 1983;5:157–161.

56. Radin EL, Rose RM: Role of subchondral bone in the initiation and progression of cartilage damage. *Clin Orthop* 1986;213:34–40.

57. Smillie JS: *Injuries of the Knee Joint.* London, Churchill Livingstone, 1970, p 31.

58. Brantigan OC, Voshell AF: The mechanics of the ligaments and menisci of the knee joint. *J Bone Joint Surg* 1941;23:44–66.

59. Wang C-J, Walker PS: Rotary laxity of the human knee joint. *J Bone Joint Surg* 1974;56A:161–170.

60. Hsieh H-H, Walker PS: Stabilizing mechanisms of the loaded and unloaded knee joint. *J Bone Joint Surg* 1976;58A:87–93.

61. Markolf KL, Bargar WL, Shoemaker SC, et al: The role of joint load in knee stability. *J Bone Joint Surg* 1981;63A:570–585.

62. Markolf KL, Mensch JS, Amstutz HC: Stiffness and laxity of the knee—The contributions of the supporting structures. *J Bone Joint Surg* 1976;58A:583–593.

63. Oretorp N, Gillquist J, Liljedahl S-O: Long term results of surgery for non-acute anteromedial rotatory instability of the knee. *Acta Orthop Scand* 1979;50:329–336.

64. Fukubayashi T, Torzilli PA, Sherman MF, et al: An *in vitro* biomechanical evaluation of anterior-posterior motion of the knee: Tibial displacement, rotation, and torque. *J Bone Joint Surg* 1982;64A:258–264.

65. Levy IM, Torzilli PA, Warren RF: The effect of medial meniscectomy on anterior-posterior motion of the knee. *J Bone Joint Surg* 1982;64A:883–888.

66. Shoemaker SC, Markolf KL: The role of the meniscus in the anterior-posterior stability of the loaded anterior cruciate-deficient knee. *J Bone Joint Surg* 1986;68A:71–79.

67. Bargar WL, Moreland JR, Markolf KL, et al: *In vivo* stability testing of post-meniscectomy knees. *Clin Orthop* 1980;150:247–252.

68. Mow VC, Mack AF: Lubrication of diarthrodial joints, in Skalak R, Chien S (eds): *Handbook of Bioengineering*. New York, McGraw-Hill, 1987, pp5.1–5.34.

69. MacConaill MA: The function of intra-articular fibrocartilages, with special reference to the knee and inferior radio-ulnar joints. *J Anat* 1932;66:210–227.

70. McCutchen CW: The frictional properties of animal joints. *Wear* 1962;5:1.

71. Mow VC, Holmes MH, Lai WM: Fluid transport and mechanical properties of articular cartilage: A review. *J Biomech* 1984;17:377–394.

72. Torzilli PA, Rose DE, Dethmers DA: Equilibrium water partition in articular cartilage. *Biorheology* 1982;19:519–537.

73. Bourne RB, Finlay JB, Papadopoulos P, et al: The effect of medial meniscectomy on strain distribution in the proximal part of the tibia. *J Bone Joint Surg* 1984;66A:1431–1437.

74. Fairbank TJ: Knee joint changes after meniscectomy. *J Bone Joint Surg* 1948;30B:664–670.

75. Jackson JP: Degenerative changes in the knee after meniscectomy. *Br Med J* 1968;2:525–527.

76. Dandy DJ, Jackson RW: The diagnosis of problems after meniscectomy. *J Bone Joint Surg* 1975;57B:349–352.

77. Minns RJ, Muckle DS: The role of the meniscus in an instability model for osteoarthritis in the rabbit knee. *Br J Exp Pathol* 1982;63:18–24.

78. Cox JS, Nye CE, Schaefer WW, et al: The degenerative effects of partial and total resection of the medial meniscus in dogs' knees. *Clin Orthop* 1975;109:178–183.

79. Korkala O, Karaharju E, Grönblad M, et al: Articular cartilage after meniscectomy: Rabbit knees studied with the scanning electron microscope. *Acta Orthop Scand* 1984;55:273–277.

80. Moskowitz RW: Osteoarthritis: Studies with experimental models. *Arthritis Rheum* 1977;20:S104–S108.

81. Moskowitz RW, Howell DS, Goldberg VM, et al: Cartilage proteoglycan alterations in an experimentally induced model of rabbit osteoarthritis. *Arthritis Rheum* 1979;22:155–163.

82. Hoch DH, Grodzinsky AJ, Koob TJ, et al: Early changes in material properties of rabbit articular cartilage after meniscectomy. *J Orthop Res* 1983;1:4–12.

83. Keyes EL: Erosions of the articular surfaces of the knee joint. *J Bone Joint Surg* 1933;15:369–371.

84. Bennett GA, Waine H, Bauer W: *Changes in the Knee Joint at Various Ages: With Particular Reference to the Nature and Development of Degenerative Joint Disease*. New York, The Commonwealth Fund, 1942.

85. Bullough P, Goodfellow J, O'Connor J: The relationship between degenerative changes and load-bearing in the human hip. *J Bone Joint Surg* 1973;55B:746–758.

86. Bullough PG, Yawitz PS, Tafra L, et al: Topographical variations in the morphology and biochemistry of adult canine tibial plateau articular cartilage. *J Orthop Res* 1985;3:1–6.

87. Akizuki S, Mow VC, Müller F, et al: Tensile properties of human knee joint cartilage: I. Influence of ionic conditions, weight bearing, and fibrillation on the tensile modulus. *J Orthop Res* 1986;4:379–392.

88. Annandale T: An operation for displaced semilunar cartilage. *Br Med J* 1885;1:779.

89. King D: The healing of semilunar cartilages. *J Bone Joint Surg* 1936;18:333–342.

90. Arnoczky SP, Warren RF: The microvasculature of the meniscus and its response to injury: An experimental study in the dog. *Am J Sports Med* 1983;11:131–141.

91. Cabaud HE, Rodkey WG, Fitzwater JE: Medial meniscus repairs: An experimental and morphologic study. *Am J Sports Med* 1981;9:129–134.

92. Heatley FW: The meniscus—can it be repaired?: An experimental investigation in rabbits. *J Bone Joint Surg* 1980;62B:397–402.

93. Pontarelli WR, Albright JP, Martin RK, et al: Meniscus suture repair. *Trans Orthop Res Soc* 1983;8:58.

94. Redman DS, Haynes DW: Meniscal repair: An experimental study in the rabbit. *Trans Orthop Res Soc* 1983;8:59.

95. Ghadially FN, Wedge JH, Lalonde JMA: Experimental methods of repairing injured menisci. *J Bone Joint Surg* 1986;68B:106–110.

96. Veth RPH, den Heeten GJ, Jansen HWB, et al: Repair of the meniscus: An experimental investigation in rabbits. *Clin Orthop* 1983;175:258–262.

97. Fanton GS, Andrish JT: Meniscofluoresis: An aid in determining prognosis of meniscal tears. *Cleve Clin Q* 1983;50:379–383.

98. Miller GK, Drennan DB, Maylahn DJ: The effect of technique on histology of arthroscopic partial meniscectomy with electrosurgery. *Arthroscopy* 1987;3:36–44.

99. Schosheim PM, Caspari RB: Evaluation of electrosurgical meniscectomy in rabbits. *Arthroscopy* 1986;2:71–76.

100. Whipple TL, Caspari RB, Meyers JF: Laser subtotal meniscectomy in rabbits. *Lasers Surg Med* 1984;3:297–304.

101. Whipple TL, Caspari RB, Meyers JF: Arthroscopic laser meniscectomy in a gas medium. *Arthroscopy* 1985;1:2–7.

102. Arnoczky SP, Lane JM, Marshall JL, et al: Meniscal degeneration due to knee instability: An experimental study in the dog. *Trans Orthop Res Soc* 1979;4:79.

103. Adams ME, Pelletier JP: The anterior cruciate ligament transection model of osteoarthritis, in Greenwald, Diamond (eds): *Animal Models of Rheumatic Disease*. Boca Raton, CRC Press, in press.

104. Adams ME, Billingham MEJ, Muir H: The glycosaminoglycans in menisci in experimental and natural osteoarthritis. *Arthritis Rheum* 1983;26:69–76.

105. Fife RS: Alterations in a cartilage matrix glycoprotein in canine osteoarthritis. *Arthritis Rheum* 1986;29:1493–1500.

106. Shapiro F, Glimcher MJ: Induction of osteoarthrosis in the rabbit knee joint: Histologic changes following meniscectomy and meniscal lesions. *Clin Orthop* 1980;147:287–295.

107. Moskowitz RW: Experimental models of osteoarthritis, in Moskowitz RW, Howell DS, Goldberg VM, et al (eds): *Osteoarthritis: Diagnosis and Management*. Philadelphia, WB Saunders, 1984, pp 109–128.

108. Adams ME, Billingham MEJ: Animal models of degenerative joint disease. *Curr Top Pathol* 1982;71:265–297.

109. Elmer RM, Moskowitz RW, Frankel VH: Meniscal regeneration and postmeniscectomy degenerative joint disease. *Clin Orthop* 1977;124:304–310.

110. Cox JS, Cordell LD: The degenerative effects of medial meniscus tears in dogs' knees. *Clin Orthop* 1977;125:236–242.

111. Lutfi AM: Morphological changes in the articular cartilage after meniscectomy: An experimental study in the monkey. *J Bone Joint Surg* 1975;57B:525–528.

112. Noble J, Hamblen DL: The pathology of the degenerate meniscus lesion. *J Bone Joint Surg* 1975;57B:180–186.

113. Ghosh P, Ingman AM, Taylor TKF: Variations in collagen, non-collagenous proteins, and hexosamine in menisci derived from osteoarthritic and rheumatoid arthritic knee joints. *J Rheumatol* 1975;2:100–107.

114. Herwig J, Egner E, Buddecke E: Chemical changes of human knee joint menisci in various stages of degeneration. *Ann Rheum Dis* 1984;43:635–640.

115. Howell DS: Diseases due to the deposition of calcium pyrophosphate and hydroxyapatite, in Kelley WN, Harris ED Jr, Ruddy S, et al (eds): *Textbook of Rheumatology*, ed 2. Philadelphia, WB Saunders, 1985, vol 2, pp 1398–1416.

116. Bjelle A: Cartilage matrix in hereditary pyrophosphate arthropathy. *J Rheumatol* 1981;8:959–964.

117. Meachim G: The state of knee meniscal fibrocartilage in Liverpool necropsies. *J Pathol* 1976;119:167–173.

118. Doyle JR, Eisenberg JH, Orth MW: Regeneration of knee menisci: A preliminary report. *J Trauma* 1966;6:50–55.

119. Evans DK: Repeated regeneration of a meniscus in the knee. *J Bone Joint Surg* 1963;45B:748–749.

120. King D: Regeneration of the semilunar cartilage. *Surg Gynecol Obstet* 1936;62:167–170.

121. Smillie IS: Observations on the regeneration of the semilunar cartilages in man. *Br J Surg* 1944;31:398–401.

122. DeYoung DJ, Flo GL, Tvedten H: Experimental medial meniscectomy in dogs undergoing cranial cruciate ligament repair. *J Am Anim Hosp Assoc* 1980;16:639–645.

123. Kim J-M, Moon MS: Effect of synovectomy upon regeneration of meniscus in rabbits. *Clin Orthop* 1979;141:287–294.

124. Espley AJ, Waugh W: Regeneration of menisci after total knee replacement: A report of five cases. *J Bone Joint Surg* 1981;63B:387–390.

125. Wigren A, Kolstad K, Brunk U: Formation of new menisci after polycentric knee arthroplasty: Report of four cases, one with a bucket handle tear. *Acta Orthop Scand* 1978;49:615–617.

126. Arnoczky SP, Warren RF, Kaplan N: Meniscal remodeling following partial meniscectomy: An experimental study in the dog. *Arthroscopy* 1985;1:247–252.

127. Schlehr FJ, Limbird TA, Swiontkowski MF, et al: The use of laser Doppler flowmetry to evaluate anterior cruciate blood flow. *J Orthop Res* 1987;5:150–153.

128. Nilsson GE, Tenland T, Öberg P: Evaluation of a laser Doppler flowmeter for measurement of tissue blood flow. *IEEE Trans Biomed Eng* 1980;BME-27:597–604.

129. Limbird T, Swiontkowski MF, Schlehr F, et al: Direct, in vivo measurement of meniscal blood flow. Presented at the interim meeting of the American Orthopaedic Society for Sports Medicine, San Francisco, California, Jan 21, 1987.

130. Arnoczky SP: Understanding the meniscus, editorial. *Cleve Clin Q* 1983;50:377–378.

131. Gershuni DH, Skyhar MJ, Danzig LA, et al: Healing of tears in the avascular segment of the canine lateral meniscus. *Trans Orthop Res Soc* 1985;10:294.

132. Henning CE, Lynch MA, Clark JR: Vascularity for healing of meniscus repairs. *Arthroscopy* 1987;3:13–18.

133. Webber RJ, Harris MG, Hough AJ Jr: Cell culture of rabbit meniscal fibrochondrocytes: Proliferative and synthetic response to growth factors and ascorbate. *J Orthop Res* 1985;3:36–42.

134. Arnoczky SP, McDevitt CA, Warren RF, et al: Meniscal repair using an exogenous fibrin clot: An experimental study in the dog. *Trans Orthop Res Soc* 1986;11:452.

135. Webber RJ, York L, Vander Schilden JL, et al: Fibrin clot invasion by rabbit meniscal fibrochondrocytes in organ culture. *Trans Orthop Res Soc* 1987;12:470.

Index